Electrophoresis —
Technical Applications
a Bibliography of Abstracts

Electrophoresis —
Technical
Applications

a Bibliography of Abstracts

By B. J. Haywood

**Research Chemist
Veterans Administration Hospital
Ann Arbor, Michigan**

ann arbor-humphrey science publishers inc.

ANN ARBOR  LONDON

Preface

Electrophoresis has proved to be a most valuable separation technique for clinician, biochemist and researcher. The vast number of publications in this field within the last decade attests to its usefulness and popularity.

This book includes two sections: abstracts of current literature and abstracts for background reference. The primary purpose of the book is to present a selective survey of the literature which describes electrophoretic techniques. An effort was made to present this survey in a form useful to persons actually performing electrophoretic separations. To this end, enough experimental detail is included to enable the reader to determine whether the technique described fits his purposes. Journal review articles are included in the "Books, Reviews, and Symposia" section. The journal abbreviations used throughout this book follow those of the American Chemical Society.

Grateful thanks go to Miss Nancy Lynn Morton of Gelman Instrument Company for her assistance with references and to Mrs. Doris Watkins for typing the manuscript.

B. J. HAYWOOD

ANN ARBOR, MICHIGAN
JUNE, 1969

Contents

Abstracts: 1965 - 1969

Books, Reviews, and Symposia

FIFTH INTERNATIONAL SYMPOSIUM ON CHROMATOGRAPHY AND ELECTROPHORESIS

Ann Arbor: Ann Arbor-Humphrey Science Publishers, Inc.; London: Ann Arbor-Humphrey Science Publishers, Ltd., 1969.

This book contains papers presented at the Symposium organized by the Belgian Pharmaceutical Society. Original lectures and plenary lectures in progress are included. Some of the speakers at the plenary session were Dr. M. Lederer, B. P. Lisboa, and György Pataki. Other papers in the book deal with the subjects of thin-layer chromatography of phenols, chamber saturation effect on separation, and carbonic anhydrase isoenzyme separation.

INTERNATIONAL SYMPOSIUM IV, CHROMATOGRAPHY, ELECTROPHO-RESE, 1966 PROCEEDINGS OF BRUSSELS SYMPOSIUM

Ann Arbor: Ann Arbor-Humphrey Science Publishers, Inc.; London: Ann Arbor-Humphrey Science Publishers, Ltd., 1968. (Originally published by Presses Académiques Européénnes, Brussels, 1968.)

Contained in this book are papers presented at the Symposium organized by the Belgian Pharmaceutical Society. Original papers as well as plenary lectures in progress are presented. Those lectures in progress were: Chroma-tography, by E. Lederer; Characterization and Isolation of Proteins by Elec-trophoresis, by N. Heimburger and H. G. Schwick; Ion Exchanges in Thin-Layer Chromatography, by K. Randerath; Juridicial Aspects of Chromatography, by G. B. Marini-Bettolo; and Systematic Analysis of Steroids by Thin-Layer Chromatography, by B. P. Lisboa.

CYTOLYTIC TOXINS OF BACTERIAL ORIGIN

Berheimer, A. W., Science **159,** 847–851 (1968).

Cytolytic toxins are proteins found extracellularly and which give rise to neutralizing antibodies. The α-toxin of *Staph. aureus* is emphasized in this review of current findings on cytolytic toxins. Chemical and physical characteristics, biological mechanisms of effects, and possible enzymatic activities of the toxins are considered.

ELECTROPHORESIS OF PROTEINS AND PEPTIDES IN STABILIZING MEDIA

Denucé, J. M., Phys. Tech. Biol. Res. **11A**, 151–216 (1968).

This is an excellent review article of principles of electrophoresis in stabilizing media as well as a discussion of the different kinds of stabilizing media. Filter paper, cellulose acetate, starch paste, agar gel, pectin gel, starch gel, polyacrylamide and density gradient columns are the support media generally used. The preparation of these media for electrophoretic use and designs of apparatus used were discussed. Miscellaneous media such as silica gel, gypsum, Sephadex®, polyvinyl chloride, and polyvinyl acetate may be used. The use of support media such as starch gel and cellulose acetate for immunoelectrophoresis and enzyme analysis are reviewed.

MULTIPLE MOLECULAR FORMS OF CARBONIC ANHYDRASE IN ERYTHROCYTES

Edsall, J. T., Ann. N.Y. Acad. Sci. **151**, 41–63 (1968).

This review article presents the separation techniques used to characterize carbonic anhydrase. Enzymatic activities and normal distribution of the isozymes are given.

THE P_i-SYSTEM: GENETIC VARIANTS OF SERUM α_1-ANTITRYPSIN

Fagerhol, M. K., and C. B. Laurell, Ser. Haematol. **1**, 153–161 (1968).

Acid hydrolyzed potato starch was used to electrophoretically separate P_i-system α_1-antitrypsin variants of serum. A discontinuous Tris-citrate–borate sodium hydroxide buffer system was used. Separation was achieved at 200 volts (8-10 volts/cm). The separated proteins (rare P_i phenotypes) were located by staining with Amido Black 10B in a mixture of methanol–water–acetic acid (5:5:1). The P_i-zones were distinctly revealed at the anode.

THE IDENTIFICATION AND CHARACTERIZATION OF AMINO ACIDS AND SMALL PEPTIDES

Leaf, G., Biochem. J. **107**, 11P–12P (1968).

This review article discusses the isolation-identification of amino acids and peptides by the use of electrophoretic methods, thin-layer and gas-liquid chromatography.

MOLECULAR BASIS FOR ISOENZYMES

Markert, C. L., Ann. N.Y. Acad. Sci. **151**, 14–40 (1968).

This review article presents data on the genetic interpretation of lactic dehydrogenase isozyme distribution in fish. The use of electrophoretic techniques to gain this kind of information is presented. The use of data obtained from *in vivo* and *in vitro* experiments using fish and their application to other vertebrates is discussed.

DISC ELECTROPHORESIS: THEORY AND APPLICATION OF DISCONTINUING POLYACRYLAMIDE GEL ELECTROPHORESIS

Maurer, H., Ann Arbor: Ann Arbor-Humphrey Science Publishers, Inc.; London: Ann Arbor-Humphrey Science Publishers, Ltd., available October, 1969 (originally published by Walter de Gruyter & Co., Berlin, 1968).

This is a well-written and well-organized book on the theory and practice of polyacrylamide-gel disc electrophoresis. The works of authors who have used the technique were reviewed in a fashion which will enable the reader to select the technique which suits his purposes. Chapter 2 is an excellent presentation of practical guide lines to the use of polyacrylamide as an analytical technique. A total of 600 references are included, making the book even more useful, especially as a research aid.

STABLE-FLOW BOUNDARY ELECTROPHORETIC STUDIES OF CELLS AND OTHER BIOLOGICAL PARTICLES

Mel, H. C., Abstracts of 155th ACS National Meeting, B-21 (1968).

The principles of stable-flow electrophoresis are reviewed, and its use in biological areas of investigation is presented. Preparative features of the technique include the collection of up to 48 fractions for further analytical analysis.

CHROMATOGRAPHIC AND ELECTROPHORETIC TECHNIQUES: II. ZONE ELECTROPHORESIS

Smith, I., New York: John Wiley and Sons, 1968.

The second of a two-volume series which presents a number of different zone electrophoresis methods, this volume presents familiar methods of zone electrophoresis, the application of these methods to routine separation, and other essential information.

THE EFFECT OF THE MEDIUM ON ELECTROPHORETIC SEPARATIONS

Bloemendal, H., Anal. Chim. Acta **38**, 169–177 (1967).

This is a brief review of electrophoretic methods useful in studying biological materials: moving-boundary, paper, continuous-flow, starch-cellulose and poly-acrylamide-gel electrophoresis as applied to the separation of eye lens protein. The advantages, disadvantages and uses of each medium are discussed.

RADIOACTIVE ISOTOPES IN INVESTIGATING TRANSPORT PHENOMENA BY PLASMA PROTEINS

Cohen, Y., J. Wepierre, and J. P. Rousselet, Ann. Biol. Clin. **25**, 81–106 (1967).

This is a review of methods for using radioisotopes to study transport in live animals. Electrophoresis, autoradiography, and spectrophotometry can be combined to give fine separation of the proteins involved.

ISOENZYMES OF HUMAN ALKALINE PHOSPHATASE

Fishman, W. H., and N. K. Ghosh, in Advan. Clin. Chem. **10**, 255–370 (1967). O. Bodansky and C. P. Stewart, editors. New York: Academic Press, 1967.

This comprehensive review of the subject gives methods of separation and detection of alkaline phosphatase. The distribution of the isoenzymes in lower animals, in disease states and in their association with blood group substances, is presented. The review includes references to original works on the subject.

EFFECT OF THE INTERACTION OF MACROMOLECULES IN GEL PERMEA-TION ELECTROPHORESIS AND ULTRACENTRIFUGATION

Gilbert, G. A., Anal. Chim. Acta **38**, 275–278 (1967).

One approach to studying interactions of macromolecules and gels is to build models to test in advance the feasibility of a given experiment. A model to determine the parameters of a reversibly reacting system was calculated and is explained in this article. This is an example to test pseudo-data and analyze it by the method of least squares.

IMMUNODIFFUSION TECHNIQUES IN CLINICAL MEDICINE

Lou, K., and E. Shanbron, J. Am. Med. Assoc. **200**, 171 (1967).

This is a brief review of the clinical applications of immunoelectrophoresis. No experimental or laboratory details are presented. Aberrant immunoelectro-phoretic patterns in serum are discussed.

CRITERIA FOR SUCCESSFUL SEPARATION BY CONTINUOUS ELECTRO-PHORESIS AND ELECTROCHROMATOGRAPHY IN BLOCKS AND COLUMNS

Ravoo, E., and P. J. Gellings, Anal. Chim. Acta **38,** 219–232 (1967).

The article is a discussion of heat dissipation and transverse spreading in continuous electrophoresis. A general theory correlating power input, time, and temperature rise in cylindrical and rectangular geometries is presented. Limitation of separation capacity can be predicted by application of the theory which is in agreement with experimental evidences.

HIGH RESOLUTION ACRYLAMIDE GEL ELECTROPHORESIS: SOME PRACTICAL ASPECTS OF ITS PROBLEMS

Ritchie, R. F., J. Maine Med. Assoc. **58,** 15–19 (1967).

This review article describes difficulties encountered when performing disc electrophoresis. Causes of suboptimal resolution are presented, and techniques for their avoidance suggested.

LES ISOZYMES DE LA LACTICO-DES HYDROGENASE DU GLOBULE ROUGE

Schapira, F., and S. Rosa. Nouv. Rev. Franc. Hematol. **7,** 109–114 (1967) **French.**

This review article discusses separation of RBC by density gradients, isoenzyme patterns in older and younger RBC, and genetic control of the subunits. Separation of young erythrocytes from old erythrocytes and the variation of lactate dehydrogenase isoenzyme pattern with the age of the cell are discussed. Eighteen references are given.

PAPER CHROMATOGRAPHY AND ELECTROPHORESIS: VOL. I. ELECTRO-PHORESIS IN STABILIZING MEDIA

Zweig, G., and J. R. Whitaker, New York: Academic Press, 1967.

This book presents the use of stabilizing media for electrophoresis. The first chapter discusses the theoretical aspects of electrophoresis, types of stabilizing media, apparatus, and other general information. The succeeding chapters deal with the separation of amines, amino acids and peptides, proteins, nucleic acids, derivatives and related compounds, carbohydrates, organic acids, organic compounds, and inorganic ions. The author has arranged this volume to make it valuable to the scientist engaged in these research areas. References regarding the different chapters are included.

QUANTITATIVE EVALUATION OF SERUM PROTEINS SEPARATED BY STARCH GEL ELECTROPHORESIS

Battistini, A., Lattante **37**, 97–135 (1966).

This is a review of quantitative techniques of serum electrophoresis. Agar, immuno-, continuous and cellulose-acetate electrophoresis are discussed.

ELECTROPHORESIS FOR THE PRACTICING PHYSICIAN, PART I

Koneman, E., Rocky M. Med. J. **63**, 47 (1966).

This review is a discussion of basic principles of electrophoresis and clinical applications for the physican. The pathology of selected abnormal sera is diagnosed electrophoretically and the physiologic principles are explained.

ELECTROPHORESIS FOR THE PRACTICING PHYSICIAN, PART II

Koneman, E., Rocky M. Med. J. **63**, 39 (1966).

Practical applications of routine serum electrophoresis in the diagnosis and management of disease are presented. The electrophoretic patterns of some general disease categories are discussed in detail. Use of the Analytrol in quantitation of serum proteins is described.

PAPER ELECTROPHORESIS

Lubran, M., J. Am. Med. Assoc. **197**, 360–361 (1966).

Electrophoresis and immunoelectrophoresis in the clinical laboratory can be used for diagnosis of many diseases. This topic is reviewed and examples are given.

SEPARATION OF VIRUSES

Markham, R., Brit. Med. Bull. **22**, 153 (1966).

This is a review of methods of purifying whole viruses from tisssue components, incomplete viruses, defective viruses, and viral component parts. Electrophoretic methods are useful for special separations which cannot be accomplished by centrifugation and chromatography, for example, the separation of the "satellite" tobacco necrosis virus from the "parent" virus or the "top" and "bottom" components of the wild cucumber virus.

RECENT ADVANCES IN ELECTROPHORETIC TECHNIQUES

McDougall, E. I., and R. L. Synge, Brit. Med. Bull. **22**, 115 (1966).

This review details preparative methods of electrophoresis, electrophoretic separation of organelles and cells, and electrophoresis in molecular scene gels. A 156-reference bibliography is included.

CHEMICAL STRUCTURE OF LIGHT CHAINS

Milstein, C., Proc. Roy. Soc. (London) Ser. B **166**, 139–149 (1966).

This is a discussion and review of chemical studies of the light chains of immunoglobulins. The monomer and dimer theories of light chain arrangement are discussed, and tables giving summaries of amino acid sequence studies are presented. Peptide maps using a peptic digest of the light chains are prepared by two-dimensional electrophoresis at pH 6.5. This technique is useful in studying the location of disulfide bridges.

CARBOHYDRATES AND MUCOID SUBSTANCES

Northcote, D. H., Brit. Med. Bull. **22**, 180–184 (1966).

This review article presents the various methods of fractionation of carbohydrates and mucoid substances depending upon the charge of the species. Monosaccharides and their separation by partition chromatography on paper, partition chromatography on ion exchange columns, thin-layer chromatography, gas-liquid chromatography, and electrophoresis are reviewed. The separation of oligasaccharides and mucopolysaccharides and separation methods such as electrophoresis, gel filtration and ion exchange chromatography are also reviewed.

GEL ELECTROPHORESIS IN BUFFERS CONTAINING UREA

Poulik, M.D., Methods Biochem. Analy. **14**, 455–495 (1966).

This is a review of the use of urea in starch and acrylamide-gel electrophoresis, both vertical and horizontal. Equipment and procedure received the greatest emphasis. Diagrams of apparatus are drawn. Electrophoretic patterns show the uses of various techniques. Fifty-three references are included.

POSSIBILITÁ E LIMITI DEL METHODO IMMUNOELETTROFORETICO PER LO STUDIO DELLE SIEROPROTEINE NELLAPRATICAL CLINICA

Ricci, P., Minerva Med. **57**, 2865–2870 (1966) **Italian.**

A review of the uses and limitations of immunoelectrophoresis for analyzing serum protein characteristics is presented. The use of immunoelectrophoresis to delineate pathological conditions as evidenced by serum protein characteristics is reviewed. Fifty-five references are cited .

GLYCOPROTEINS: METHODS OF STUDY AND CHANGES IN HEALTH AND DISEASE

Shetlar, M. R., Progr. Clin. Pathol. **1**, 419 (1966).

Isolation and analytical procedures for the quantitation of various serum glycoproteins are presented. The periodic acid Schiff reaction was adapted to detect oxidized protein-bound 1, 2-glycol groups following paper electrophoresis. This technique may be adapted to the study of tissue glycoproteins. Clinical alterations in normal glycoprotein levels are discussed.

ELECTROKINETIC PHENOMENA AS PURIFICATION TOOLS

Strickland, R. D., Ann. N. Y. Acad. Sci. **137**, 139 (1966).

This is a discussion of factors which affect the quality of electrophoresis, including inhomogeneities in the voltage gradient and the temperature, and variations in ionic strength, pH, and electro-osmotic flow. Bringing the bed into a steady state before use can minimize distortions caused by concentration tides.

ELECTROPHORESIS IN ACRYLAMIDE-AGAROSE GELS

Uriel, J., and J. Berges, Bull. Soc. Chim. Biol. **48**, 969–982 (1966).

This is a review of the properties of a mixed gel compared to the properties of acrylamide or agarose alone. Instructions for the preparation, standardization, and uses of the mixed gel are given. Examples showing the separation of proteins are presented.

CELL ELECTROPHORESIS

Ambrose, E. F., Boston: Little, Brown and Co., 1965.

These papers, presented at the British Biophysical Symposium in 1963, cover a wide range of cell electrophoresis topics. Apparatus and techniques are diagramed and briefly discussed. Methods of studying cell surfaces using electrophoresis after chemical alterations, and differences between normal and diseased or transformed cells are extensively covered. The influence of viruses or antibodies on electrophoresis of RBC is covered. Blood, liver, sperm, and bacterial cells are the principal cells used in the experiments reported.

THE DETERMINATION OF AMINO ACIDS BY HIGH-VOLTAGE PAPER ELECTROPHORESIS

Blackburn, S., Methods Biochem. Anal. **13**, 1–46 (1965).

A laboratory guide covering all phases of HVE of amino acids is presented. Apparatus and typical amino acid patterns are diagramed, detection and elution methods are reviewed, and specific recommendations are made.

IMMUNOELECTROPHORESIS: THEORY AND SOME CLINICAL APPLICATIONS

Cordes, R. A., Marquette Med. Rev. **31,** 173 (1965).

A review of the physics of electrophoresis is expressed in easily understood terms. A concise summary of abnormal serum protein levels and clinical problems is included.

HISTIOCHEMICAL AND ELECTROPHORETIC PROPERTIES OF CHOLINESTERASES AND NON-SPECIFIC ESTERASES IN THE RETINA OF SOME MAMMALS, INCLUDING MAN

Esilá, R., Acta Ophthalmol, Suppl. **77,** 111-113 (1965).

This is an extensive review of research of cholinesterases in the retina of the eye. The distribution of these esterases in sections of the eye is documented. Several starch-gel electrophoretic methods applicable to this topic are reviewed. There are 130 references.

MOBILITY DETERMINATION BY ZONE ELECTROPHORESIS AT CONSTANT CURRENT

Waldmann-Meyer, H., Methods Biochem. Analy. **13,** 47–49 (1965).

This is an extensive review of parameters of mobility important in zone electrophoresis. Mathematical expressions of mobility and factors influencing it are derived and explained. The procedure is detailed and mobility calculated for a sample protein.

Fundamental, Theoretical, and General

EFFECTS OF CATIONS ON ELECTROPHORETIC MOBILITY AND SUBSTRATE BINDING PROPERTIES OF PYRUVATE KINASE

Betts, G., and H. J. Evans, Biochim. Biophys. Acta **167,** 190–193 (1968).

The effect of cations on substrate binding and electrophoretic mobility of pyruvate kinase under different conditions was studied by free-flow electrophoresis using the Perkin-Elmer Model 38 electrophoresis unit at different concentrations of chloride. A 0.01M Tris HC1 buffer, pH 7.4, was used. To this buffer the cations 0.1M KCl, 0.1M LiCl, 0.1M KCl + 0.01M MgCl$_2$, and 0.1M LiCl + 0.01M MgCl$_2$ were added. Anodic migration was observed in the medium containing 0.1M LiCl or 0.1M KCl. A decrease in mobility was observed when 0.01M MgCl$_2$ was added to the LiCl- or KCl- containing medium. The influence of cations on mobility of the enzyme in the presence of substrate was investigated. A decrease in mobility was observed when cations were in the presence of the substrate phosphoenolpyruvate.

A THEORETICAL STUDY ON THE ZONE MOBILITY-pH CURVE IN PAPER ELECTROPHORESIS OF LOW MOLECULAR WEIGHT COMPOUNDS WITH A DISSOCIABLE PROTON AND ITS APPLICATION TO PHOSPHORUS COMPOUNDS

Kiso, Y., M. Kobayashi, Y. Kitaoki, K. Kawamoto, and J. Takoda, J. Chromatog. **36,** 215–228 (1968).

The zone mobility of a compound with a dissociable proton when plotted against pH value of the background solution was found to be a hyperbolic tangent function. It was possible to estimate the consecutive dissociation constants knowing this relationship. An equation was derived which was found useful for the estimation of the dissociation constant and/or molecular weight.

ELECTROPHORETIC MOBILITY OF SUSPENDED PARTICLES TREATED AS "LARGE IONS"

Guastalla, J., and A. Gouskov, Compt. Rend. Acad. Sci. (Paris) C **264,** 851–854 (1967).

An electrical layer, resulting from a layer of oriented water molecules, surrounds a silica powder in suspension. The charge of the particle is small so that the particle may be compared electrophoretically to a large ion. By applying Stokes Law, the relationship between particle mobility and applied electric field can be determined.

INFLUENCE OF ADSORBED WATER ON THE ELECTROPHORETIC MOBILITY OF TITANIUM DIOXIDE IN NON-AQUEOUS MEDIA

Griot, O., Trans. Faraday Soc. **62,** 2904–2908 (1966).

The influence of the interaction of water with activated TiO$_2$ and the influence of the degree of hydration on its mobility in organic solution was investigated using a Zeta meter. The instrument was equipped with a glass-Teflon® electrophoretic cell having a glass path of 4.64 mm x 10 cm and a vertical micrometer for use with nonaqueous solvents. Electrophoresis of samples prepared by heating a 25-30°C for 14 hours under 10^{-5} mm mercury was performed at an 80 volts/cm potential gradient. Results obtained indicated that the removal of adsorbed water was accompanied by increased mobility.

IONOPHORETIC STUDIES OF INORGANIC IONS ON PAPER

Mukheijee, S. K., Sci. Cult. **32,** 39 (1966).

When using glucose as an electro-osmatic indicator, two corrections should be made. The first correction is for electro-osmosis; the second is for "added migration path length," a correction which brings the mobility of the migrant up to that observed in free electrophoresis. Free mobilities of common cations are given.

A SUGGESTION FOR THE DEFINITION OF ZONE (OR BOUNDARY) RESO-LUTION IN PHYSIO-CHEMICAL SEPARATION TECHNIQUES

Svensson, H., Chromatog. **25,** 266–273 (1966).

Two mathematical methods of defining zones in separations are discussed: 1) Two adjacent zones of equal mass contents are said to be *just resolved* if the minimum between the peaks is $e^{-1/2}=0.61$ (peak heights). Further derivations to determine a shoulder on the main peak are presented. 2) *Just resolved* zones are defined as double zones with a horizontal inflection, a definition that depends upon mass ratios and is therefore inconvenient to use.

REMOVAL OF BACTERIA FROM RAW MATERIALS BY ELECTROPHORESIS: I. FACTORS AFFECTING THE ELECTROPHORETIC MOBILITY OF CERTAIN SPECIES OF BACTERIA

Tanikawa, E., T. Motohiro, and M. Akiba, J. Food Sci. **31,** 596–604 (1966).

The optimum conditions for electrophoresing bacteria away from food were evaluated for an electrophoresis cell of the Abramson type, a cell having agar bridge, and for a simple glass slide cell. Mobility of bacteria was found to be smaller at the top and bottom of cell and the velocity was proportional to the intensity of the electrical field. The mobility increased randomly with increased voltage. Mobility was increased by sodium tetradecyl sulfate. If increased, the ionic strength was found to decrease the mobility of bacteria.

THEORY OF MOVING BOUNDARY ELECTROPHORESIS OF REVERSIBILITY OF INTERACTING SYSTEMS: REACTION OF PROTEINS WITH A SMALL UNCHARGED MOLECULE SUCH AS UNDISSOCIATED BUFFER ACIDS

Cann, J. R., and W. B. Good, J. Biol. Chem. **240,** 148–155 (1965).

The theoretical evaluation of moving-boundary electrophoretic movement of proteins in a buffer atmosphere containing varying concentrations of macromolecular ions is presented. The author used a digital computer to show that two peaks may result as a consequence of gradients of uncomplexed macromolecules.

FUNDAMENTAL ASPECTS IN THE PRACTICE OF THIN-LAYER ELECTRO-PHORESIS

Criddle, W. J., G. J. Moody, and J. D. R. Thomas, Proc. Soc. Anal. Chem. Conf. **1965**, 135–140.

This is a discussion and review of the effects of heat generation and layer thickness on electrophoresis on Kieselguhr G thin layers. In 0.05M aqueous NaBH₄, a load of 0.21 μ/cm^2 is tolerable for 15 minutes. A lower load is recommended for longer term experiments. Lyophilization is a recommended means of eliminating random migration of zones due to drying at high temperatures.

CONTROLLING THE VARIABLES OF IMMUNOELECTROPHORESIS

Jordan, W. C., and W. White, Am. J. Med. Technol. **31**, 169–174 (1965).

This paper describes the variables encountered in immunoelectrophoresis and lists ways of controlling them. No significant changes were observed in patterns obtained if barbital-acetate buffer was used instead of barbital-sodium barbital. Ionic strength must be controlled at $0.025–0.03 \pm 0.2$. One percent agar at a thickness of 0.1 mm must be used.

MEASUREMENT OF IONIC MOBILITY FROM ELECTROMIGRATION ON PAPER

Lahiri, M. M., J. Indian Chem. Soc. **42**, 843–846 (1965).

Glucose was used as an electro-osmotic indicator for determining the mobilities of five inorganic cations, Cu^{+2}, Ni^{+2}, Co^{+2}, Zn^{+2} and Fe^{+2}. The electrolyte was 0.05M HC1, (pH 1.3, ionic strength 0.05); the voltage gradient was 200 volts for 30 minutes; paper was Whatman® #1. The ionophoresis equipment was modified for reproducible results. Corrections for electro-osmotic and tortuosity effects were made in calculating mobilities, which were of the same order as mobility in a free solution.

Techniques

DETECTION OF PERSULFATE IN ACRYLAMIDE GELS

Bennick, A., Anal. Biochem. **26**, 453–457 (1969).

The use of a 2% solution of benzidine in 10% acetic acid for the detection of persulfate in polyacrylamide gel is described. When newly prepared gels were stained with the solution, a deep blue color developed. Gels freed of persulfate by a preelectrophoresis step remained colorless when placed in contact with the benzidine solution. The use of this technique before protein separation should reduce the number of artifacts attributable to persulfate.

IMMUNOELECTROFOCUSING

Catsimpoolas, N., Clin. Chim. Acta **23**, 237-238 (1969).

The characterization of antigen-antibody by means of immunoelectrofocusing was performed using carrier ampholytes (Ampholines® LBK Instruments, Inc.). Electrofocusing was performed on 1% agarose gels prepared in 2% aqueous ampholytes and poured on microscope slides. The cathode vessel contained 5% EDTA and the anode vessel contained 5% phosphoric acid solution. A current of 2 mA per slide was applied for one hour. Antiserum was then added to the other well and allowed to diffuse for 24 hours. The first well was filled with a Tris-barbital buffer, pH 8.6, to insure pH atmosphere that was optimal for antigen-antibody reaction. A pH range of 5.0—8.0 was found to be most suitable for electrofocusing. The pH ranges were easily obtained by selecting the appropriate ampholyte.

LIQUID SCINTILLATION SPECTROMETRY OF TRITIUM-LABELED PROTEINS SEPARATED BY DISC ELECTROPHORESIS

Le Bouton, A. V., Anal. Biochem. **26,** 445–447 (1969).

A method which will allow measurement of tritium-label in protein fractions separated in polyacrylamide gels is described. The bands were dissected with a razor blade and the gel dissolved with 30% hydrogen peroxide. A solubilizing agent was added to permit complete contact with the scintillation fluid. The samples were then counted in a scintillation spectrometer. A counting efficiency of 24% is possible with this procedure.

POLYACRYLAMIDE GELS OF INCREASING CONCENTRATION GRADIENT FOR THE ELECTROPHORESIS OF LIPOPROTEINS

Pratt, J. J., and W. G. Dangerfield, Clin. Chim. Acta **23,** 189–201 (1969).

Plasma lipoproteins were separated into two classes on polyacrylamide gels of increasing concentration gradients. The gradient was made in a special Perspex® rectangular cell divided into two parts by a sloping partition. Electrophoresis was performed at 50 mA in a glycine-sodium hydroxide buffer (upper chamber) and Tris–HC1 (lower chamber). Resolution afforded by this method was found to be superior to conventional paper electrophoretic and ultracentrifugal procedures.

IMPROVED SOLUTION TECHNIC FOR SPECTROPHOTOMETRY OF PONCEAU S-STAINED CELLULOSE ACETATE ELECTROPHORETOGRAMS

Tech. Bull. Registry Med. Technologists **39,** 17–18 (1969).

Serum proteins separated on cellulose-acetate strips and stained with Ponceau S were found to be quantitatively eluted from the strips with a solution composed of one volume of formic acid and nine volumes of dimethyl sulfoxide. Values reported for percentages of protein fraction were higher for this technique than for the densitometric technique.

FACTS AND ARTIFACTS IN ACID PHOSPHATASE AND ESTERASE LOCALIZATION IN DISC ELECTROPHORESIS OF HUMAN SALIVA

Weinstein, E., and I. D. Mandel, Arch. Oral Biol. **14,** 1–6 (1969).

A discussion of pitfalls which lead to erroneous results when polyacrylamide gels are used for separation of esterases and phosphatases is presented. It was demonstrated that control of pH, temperature, and composition of gel used are necessary to obtain reproducibility.

POLYACRYLAMIDE GEL ELECTROPHORESIS IN A CONTINUOUS MOLECULAR SIEVE GRADIENT

Margolis, J., and K. G. Kendrick, Anal. Biochem. **25,** 347–362 (1968).

A method for the preparation of continuous linear gradients of polyacrylamide gel, by casting the gels in batches in a battery of flat glass cells, is described. Because of the increasing retardation of zones in the gradient, improved separation of materials was achieved using this procedure.

A NEW TECHNIQUE OF pH GRADIENT ELECTROPHORESIS AS APPLIED TO THE SEPARATION OF NUCLEIC ACID BASES

Narurkar, M. V., L. M. Narkurar, and M. B. Sahasrabudhe, Anal. Biochem. **26,** 174–205 (1968).

Purine and pyrimidine bases were separated using a new technique of gradient electrophoresis. Separation was achieved by passing the mixture of purines and pyrimidines through zones of preset pH gradient having pH values of 9.2, 9.45, 9.8, 9.85 and 10.1 at a voltage of 50 volts/cm. The apparatus used was a Perspex® box with a lid having two holes on the side for the electrodes. A solvent trough was placed at one end inside the box. Separation takes place on a sheet of Whatman® No. 1 paper (20 x 30 cm) cut into strips 5 cm long and 2 cm wide. Eight centimeters were left uncut at the starting end and tapered to dip into the solvent trough. The method was successfully applied to both RNA and DNA hydrolysates.

AN IMPROVED PROCEDURE FOR H³ AND C¹⁴ COUNTING IN ACRYLAMIDE GELS WITH A NONAQUEOUS SCINTILLATION SYSTEM

Spear, P. G., and B. Roizman, Anal. Biochem. **26**, 197–200 (1968).

A more efficient method for counting H^3- and C^{14}- labeled proteins separated on polyacrylamide gels is described. The separating polyacrylamide gels were prepared with ethylene diacrylate as the cross-linking agent. This diacrylate cross-linkage is unstable at high pH. Therefore the gels were hydrolyzed with concentrated ammonium hydroxide in scintillation vials. The insertion of glass fiber paper in the vial absorbed the sample. The papers were then dried, added to toluene scintillation fluid and counted. Efficiencies of 15-20% were obtained for H^3 and 65-70% for C^{14}.

ARTIFACT PRODUCED IN DISC ELECTROPHORESIS BY AMMONIUM PERSULFATE

Brewer, J. M., Science **156**, 256–257 (1967).

Ammonium persulfate inactivated yeast enolase to produce an extra electrophoretic protein band; the persulfate inactivated the enzyme only in the presence of 8M urea. Accessible oxidizable groups, the sulfhydryls, can be oxidized by persulfate in the absence of urea. Riboflavin and light are recommended for polymerizing the lower gels when doing enzyme studies.

COUNTERCURRENT ELECTROPHORESIS OF PROTEINS

Broome, J., Nature **214**, 849–850 (1967).

Preliminary experiments in countercurrent electrophoresis of a protein sample are described. An apparatus was designed for these experiments and diagrams are included. Migration coefficients which were derived are included.

SERUM PROTEINS IN THE GERM-FREE MOUSE AFTER ORAL CHALLENGE WITH CANDIDA ALBICANS

Phillips, A. W., and R. C. LaChapelle, Nature **213**, 709–710 (1967).

An analytical electrophoresis apparatus was used to separate and quantitate the proteins in pooled mouse serum. The mice were first challenged with live C. *albicans*. Values for serum protein were similar for germ-free and conventional mice. The germ-free mice had decreased β- and γ-globulin levels and increased albumin levels.

DOUBLE-DISC ELECTROPHORESIS OF PROTEINS

Rascusen, D., Nature **213**, 922 (1967).

A method whereby two discontinuous buffer systems were coupled and which allowed for the running and concentrating of both cationic and anionic proteins in the same sample is described. The cation running gel and anion running gels were joined horizontally with a 3-cm section of plastic. A small hole in the plastic permitted the introduction of the sample between the two gels.

FIXATION AFTER AGAR GEL ELECTROPHORESIS

Van Vreedendaal, M., Clin. Chim. Acta **15**, 359–360 (1967).

The solubility of proteins in various common mixtures was examined. In the usual 70% ethanol–5% acetic acid fixation mixture, many proteins are soluble, making this a poor fixative when quantitative work is done. A solution of 5% w/v trichloroacetic acid and 2½% (w/v) formaldehyde in water was a satisfactory fixative since the test serum proteins were insoluble.

QUANTITATIVE MICRODENSITOMETRIC MEASUREMENTS OF IMMUNO-LOGICAL PRECIPITATES IN AGAR GEL

Von Der Decken, A., Anal. Biochem. **18**, 444–452 (1967).

A simple method for the measuring of relative incorporation of amino acids into immunologically active proteins is described. A direct correlation was found between the density of the antigen-antibody precipitates and the protein content after agar-gel electrophoresis.

ACIDIC BUFFER SYSTEMS FOR UREA STARCH-GEL ELECTROPHORESIS

Azen, E. A., R. A. Nazhat, and O. Smithies, J. Lab. Clin. Med. **67**, 650–659 (1966).

The concentration and pH of the buffer were the most important factors in determining the character of the separation of erythrocyte stromal proteins. The cation concentration was usually in the range of 0.02M; pH was usually in the range of 3.2. The results were suggested as a possible guide to buffer selection for the electrophoresis of other materials.

TRANSVERSE GRADIENT ELECTROPHORESIS: PROTEIN HOMOGENEITY TEST AND SUBFRACTIONATION TECHNIQUE

Dubbs, C. A., Science **157**, 463–464 (1966).

A pH gradient was incorporated into starch gel transverse to the direction of protein migration. The technique can suggest pH conditions of further experiments for maximum purification.

CHANGES IN ELECTROPHORETIC MIGRATION DUE TO IMPREGNATION OF THE PAPER STRIPS WITH VARIOUS BUFFERS

Morales, J. A., Clin. Chim. Acta **14**, 654–660 (1966).

Before electrophoresis of serum proteins, the paper was impregnated with different buffers; however, other experimental conditions were constant. The buffers used were 1) Michaelis Buffer, pH 8.2, ionic strength (μ) 0.12 and 0.075; 2) Tris-succinate, pH 7.0, $\mu = 0.16$; 3) Tris–EDTA, pH 8.9, $\mu = 0.218$ Seventy-five, 50, and 25% dilutions of these buffers were tested. The protein patterns varied according to the buffer used, but were characteristic for varying concentrations of any one buffer. Therefore, protein migration in paper electrophoresis seems to be a special effect that depends on the migration media.

DISK ELECTROPHORESIS: FURTHER MODIFICATIONS IN THE TECHNIQUE OF SAMPLE INTRODUCTION

Narayan, K., S. Narayan, and F. Kummerow, Clin. Chim. Acta **14**, 227–232 (1966).

Greater precision in quantitative experiments was obtained by modifying how the sample was placed on the gel before disc electrophoresis. A solution of 0.03 ml of ovalbumin dissolved in a light solution β-sucrose (10 ml solution β, 40 ml 20% w/v sucrose, 30 ml water) was placed over the spacer gel. The large pore gel was layered over this and photopolymerized for 20 minutes.

REFINEMENTS OF ACRYLAMIDE ELECTROPHORESIS

Ritchie, R. F., J. G. Harter, and T. B. Baylis, J. Lab. Clin. Med. **68**, 842–850 (1966).

A modification of a polyacrylamide-gel technique which makes the technique suitable for preparative separations is presented. The procedure is based on the original technique of Raymond, Davis and Ornstein. The addition of Tween® 80 to all solutions except buffer insures smooth flow of water down the wet plastic. Polymerization was allowed to proceed overnight with water circulating. Improved resolution was obtained with the method described. An increase in the number of minor bands was noted.

CELLULOSE ACETATE ELECTROPHORESIS: I. A COMPARISON OF CELLULOSE ACETATE MEMBRANES

Yoichi, M., Nippon Eisekensa Gishikai Zasshi **15,** 73–76 (1966).

The author presents a comparison of results obtained with five commercial cellulose-acetate electrophoresis membranes. The membranes compared were Oxoid® (England), Millipore® (U.S.A.), Membranfolien® (Germany), Cellogel® (Italy) and Separex® (Japan). Separations were achieved in a 0.07M Veronal buffer, pH 8.6, for 60 minutes at a constant current of 0.6-0.8 amp/cm. The strips were stained with a solution of Ponceau 3R (0.8 g in 100 ml 6% aqueous solution of TCA) for one to four minutes. After treatment with 1% acetic acid and Decalin®, the separated proteins were assayed densitometrically at 500 mμ. After soaking in Decalin and liquid paraffin, Membranfolien, Millipore, Separex and Oxoid in that order were rated as having good transparency.

APPLICATION OF CHROMATOGRAPHIC AND ELECTROPHORETIC METHODS TO THE STUDY OF THE SZILARD-CHALMERS EFFECT

Adloff, J. P., Chromatog. Rev. **7,** 52–64 (1965).

The Szilard-Chalmers phenomenon is the chemical effects associated with nuclear transformations. The applications of paper and gas chromatography, paper electrophoresis, and adsorption and ion-exchange chromatography to separation of these compounds are revealed. There are 118 references.

EFFECT OF SERUM HANDLING ON ELECTROPHORETIC PATTERNS: PAPER STRIP AND MOVING BOUNDARY ANALYSIS

Blatt, W. F., J. Kerkay, and M. Mager, Am. J. Med. Technol. **31,** 349–354 (1965).

The effects of common laboratory methods of handling serum prior to moving-boundary and paper electrophoresis were investigated. Electrophoresis was performed for 18 hours on paper strips in the Spinco® Model R. A Veronal buffer, pH 8.6, ionic strength 0.075, and field strength of 12.3 volts/cm were employed. Moving-boundary electrophoresis was performed with a Spinco electrophoresis diffusion Model H apparatus. Differences in patterns of protein distribution were found. Larger albumin content was obtained by paper electrophoresis. Standards for laboratories can thus be obtained by minimal handling of fresh sera prior to storage.

RAPID REMOVAL OF BACKGROUND DYE FROM ZONE ELECTROPHORET-OGRAMS

Stanton, M. G., Anal. Biochem. **12**, 310 (1965).

The excess dye present after staining a starch of polyacrylamide gel was removed by a transversely applied electric field. Methanol–acetic acid–water was the electrolyte used over a dye-adsorbing bed of activated charcoal. A current of 1 amp/100 cm² surface cleared a starch gel in 20 minutes.

IMMUNOELECTROPHORESIS ON CELLULOSE ACETATE MEMBRANES: A METHOD WITH CONTROLLED APPLICATION OF ANTISERUM

Zec, J., Microchem. J. **9**, 510–521 (1965).

An apparatus for mounting a cellulose-acetate membrane for accurate spotting of the sample with a semiautomatic applicator has been developed. Diagrams for the construction of the apparatus and applicators are provided. Electrophoresis was done in a pH 8.6 barbital buffer B-2, ionic strength 0.038M, using a constant voltage of 150 volts for 60 to 90 minutes. Ponceau S was the fixative and stain. The method makes cellulose-acetate electrophoresis of a known quantity precise and reproducible.

Apparatus, Quantitation and Instrumentation

AN IMPROVED VERTICAL GEL APPARATUS FOR QUANTITATIVE ELECTROPHORESIS

Blattler, D. P., Anal. Biochem. **27**, 73–76 (1969).

An improved vertical gel apparatus which has several advantages over conventional equipment is described. The apparatus described is equipped with a thermistor and relay which regulate the temperature to $\pm$ 0.1°C when the operating temperature is 1-10°C below ambient temperature. The gel plates are glass, which prevents the acrylamide from sticking and also provides for more efficient transfer of heat.

AN APPARATUS FOR PREPARATIVE POLYACRYLAMIDE ELECTROPHORESIS: THE ISOLATION OF A RIBONUCLEASE INHIBITOR

Bont, W. S., J. Geels, and G. Rezelman, Anal. Biochem. **27,** 99–107 (1969).

A new apparatus for preparative gel electrophoresis is described. The apparatus consists of two concentric cylinders, the inner one connecting to the lower buffer vessel and the outer one connecting to the upper vessel. The running gel was obtained by pouring gel into a circular trough which fits between the cylinders. During electrophoresis the gels were cooled by a circulating water bath. The usefulness of this apparatus was demonstrated by the separation of a RNAase inhibitor isolated from a rat liver postmicrosomal fraction. The electrophoresis step resulted in further purification of the inhibitor.

APPARATUS FOR LARGE-SCALE PREPARATIVE POLYACRYLAMIDE GEL ELECTROPHORESIS

Hjerten, S., S. Jerstedt, and A. Tiselius, Anal. Biochem. **27,** 108–129 (1969).

A preparative gel electrophoresis apparatus having a maximum load of 1 gram is described. The substances separated migrate into a granular bed and are displaced by buffer flow to a fraction collector. Because cooling coils are placed inside the gel chamber, efficient cooling of the gels is possible. The resolution obtained with this apparatus is very close to the resolution obtained with analytical gels.

GEL ELECTROPHORESIS: A NEW CATALYST FOR ACID SYSTEMS

Jordan, E. M., and S. Raymond, Anal. Biochem. **27,** 205–211 (1969).

Ascorbic acid used with ferrous sulfate and hydrogen peroxide was found to be an efficient catalyst for promoting polyacrylamide gellation under acidic conditions. The catalyst described not only promotes polymerization but eliminates oxygen inhibition and acts as a trapping agent for residual oxidizing agents. The efficiency of this system was tested with highly basic proteins and found to yield reproducible results.

LIEBIG CONDENSER DISK ELECTROPHORESIS

Nerenberg, S. T., and G. Pogojeff, Am. J. Clin. Pathol. **51,** 229–237 (1969).

An apparatus for disc electrophoresis, constructed from jacketed Liebig condensers 250 or 800 mm in length, is described. The use of condensers provides a longer migration path, increased sample capacity and increased gel capacity. Efficient cooling of gels is possible with circulating water baths.

DEVICE FOR RAPID SAMPLE APPLICATION IN AGAR GEL PRECEDING ELECTROPHORESIS

Bradley, G. F., Tech. Bull. Registry Med. Technologists **38**, 64–66 (1968).

A simple apparatus was designed and built for semiautomatic cutting of the slit for sample application on microscope slides. The use of this apparatus increased accuracy of densitometric analysis over the conventional manual method of preparing the slit. The apparatus may be constructed with a minimum amount of equipment.

DISPOSABLE U-CELL FOR DETERMINING PARTICLE MOBILITY [IN PAINTS]

Brewer, G. E. F., and M. E. Horsch, Paint Varnish Prod. **58**, 54–56 (1968).

A disposable tube for the determination of the electrophoretic migration rate of colloidal particles in paints is described. The tube is U-shaped, 25 cm in length and constructed of clear plastic. The tube was placed on a U-shaped groove on a plastic holder which had a centimeter scale along the arms of the groove. The paint to be electrophoresed is inserted through the wall at the bottom of the U-tube.

A REFLECTIVE SCANNER FOR ELECTROPHORETIC STRIPS

Brian, P., S. African Med. J. **42**, 452–453 (1968).

An easily constructed reflectance scanner for cellulose acetate strips is described.

SLIDE RULE FOR THE RAPID QUANTITATION OF SERUM PROTEIN ELEC-TROPHORETIC PATTERNS

Burrows, S., Techn. Bull. Registry Med. Technologists **38**, 19–20 (1968).

A variable scale slide rule was modified to calculate the proportion and concentration of protein fractions of an electrophoretogram. The inert scale was replaced by a ½-inch wide measuring tape, with 1/16-inch markings along the lower edge and one millimeter markings along the top edge. The exact calibration of the upper scale depends on the electrophoretic system used in the laboratory.

A NEW DEVICE FOR SLICING POLYACRYLAMIDE GELS

Goldberger, R. F., Anal. Biochem. **25**, 46–54 (1968).

A new device for slicing polyacrylamide gels 6 mm x 7.0 cm into 50 slices each 1.5-mm thick is described. The slices are obtained by laying the gel in the trough-shaped device. Fifty thin circular blades, arranged parallel to each other, then rotate through the gel, slicing it to the desired thickness. The apparatus is made of lucite and stainless steel and is in one piece.

A HIGH-RESOLUTION METHOD FOR SLICING AND COUNTING RADIO-ACTIVITY IN POLYACRYLAMIDE GELS

Gray, R. H., and D. M. Steffensen, Anal. Biochem. **24**, 44–53 (1968).

A new method of slicing polyacrylamide gels and counting radioactivity in separated samples is presented. The polyacrylamide gels were dehydrated and sliced with a rotary microtone. The sliced gals were rendered soluble in hydrogen peroxide and counted in Bray's solution.

A "MACROTOME" FOR POLYACRYLAMIDE GEL AND TISSUE SLICING

Gressel, J., Anal. Biochem. **22**, 352–354 (1968).

This article discusses an instrument which was designed and constructed for ease in slicing low concentration (2.2–5%) acrylamide gel. A picture of the mechanism is given although there are no construction diagrams presented. A modified model of the instrument is now commercially available.

AN AUTOMATIC DEVICE FOR SECTIONING ANALYTICAL POLYACRYLA-MIDE GELS: RADIOACTIVE ESCHERICHIA COLI 50 S RIBOSOMAL PRO-TEINS.

Groves, W. E., F. C. Davis, and B. H. Sells, Anal. Biochem. **24**, 462–469 (1968).

An automatic electronic device for sectioning cylindrical analytical polyacryla-mide gels is described. Sections of gel 1-mm thick or less could be obtained.

ANALYTICAL AND PREPARATIVE DISC ELECTROPHORESIS

Hansl, R., Lab. Management **6**, #9, 18 (1968).

The use of disc electrophoresis, a separation technique of high application and resolution capacities, is presented. This technique combines the principles of gel electrophoresis with sieving and a pH discontinuity. The cost of appa-ratus for clinical laboratory and research laboratory operations is discussed. This is an excellent article for the novice.

PROGRAMMABLE POWER SUPPLY FOR POLYACRYLAMIDE GEL ELECTRO-PHORESIS

Hoagland, P. D., Anal. Biochem. **26,** 194–197 (1968).

Control of the voltage of the Kepco Model HB-2-AM power supply by placing a resistance of 100 Chm/V in the control circuit between terminals #11 and #14 on the back of the instrument gave a programmable power supply. Results obtained using this apparatus indicated that it is entirely reproducible and can be used for routine separations with vigorous control.

A MULTI-SAMPLE APPLICATOR FOR ZONE ELECTROPHORESIS

Kohn, J., Clin. Chim. Acta **18,** 65–68 (1968).

A multi-sample applicator consisting of a castellated thin metal plate with grooved projections and glass sample plate corresponding to castellation areas is described. The applicator is charged by bringing it in contact with serum samples placed in numbered areas on the sample plate. The applicator is then applied to the buffer-impregnated and blotted cellulose-acetate strip. This method of application was found to save considerable time, and it required little skill or practice.

SEPARATION TECHNIQUES IN THE BIOMEDICAL LABORATORY

Kopp, R. H., Lab. Management **6,** #9, 23 (1968).

A discussion of the cost, time and separation power of the major biochemical separation techniques is presented. Paper chromatography, acetate-strip electrophoresis, thin-layer chromatography, column chromatography, high-voltage electrophoresis, free-flowing electrophoresis, and counter-current distribution techniques are discussed and compared. Emphasis is placed on the biomedical research and clinical laboratories.

FLUOROMETRIC DETERMINATION OF THE ELECTROPHORETIC HETERO-GENITY OF γ-GLUTAMYL-TRANSPEPTIDASE

Laursen, T., and K. Jacyszyn, Clin. Chim. Acta **21,** 497–499 (1968).

A Vitraton fluorometer, containing a mercury lamp, scanner, and densitometer, was used to characterize γ-glutamyl transpeptidase. The enzyme was separated on agarose gels, and the activity was located by the sandwich technique. The fluorometric procedure seems to achieve more quantitative results than does colorimetry.

SERUM PROTEIN ELECTROPHORESIS: EVALUATION AND MODIFICATION OF THE MICROZONE SYSTEM: II. PONCEAU S

Luxton, G. C., Can. J. Med. Technol. **30**, 55–70 (1968).

The use of Ponceau S for staining cellulose-acetate strips used in the Beckman® MicroZone system is evaluated, and conditions for elution and quantitation are presented. It was concluded that the dye uptake by both human albumin and globulin is linear over the range up to 3 μg per sq. mm. Albumin was found to bind 1.46 times as much dye per unit concentration as gamma globulin. Staining did not seem to occur at protein concentrations lower than 0.025 μg per sq. mm.

SERUM PROTEIN ELECTROPHORESIS: EVALUATION AND MODIFICATION OF THE MICROZONE SYSTEM: III. POSSIBLE DYE SUBSTITUTES FOR PONCEAU S

Luxton, G., Can. J. Med. Technol. **30**, 83–101 (1968).

The MicroZone® system was evaluated in detail and judged to have the advantages of speed and ultramicro specimen sample. Coomassie Brilliant Blue R-250, Amido Black 10B, Light Green SF, Yellowish and Procion Brilliant Blue M-RS were studied as possible substitutes for Ponceau S as staining dyes. Procion Brilliant Blue M-RS appears to be the most promising substitute.

ELECTROPHORESIS ON CELLULOSE ACETATE

Meltzer, T. H., Lab. Management **6**, #9, 20 (1968).

The use of cellulose-acetate membranes for electrophoretic support medium is presented. Sharp, distinct separations are possible in a very short time. Methods of quantitation, biochemical applications, types of membranes and the use of Gelman Triacetate Metrical® Type GA-6 filter membranes for high voltage electrophoresis are discussed.

ELECTROPHORESIS

Moore, D. H., in Physical Techniques in Biological Research, **11A**, 121–148, Second Edition, D. H. Moore, ed. New York: Academic Press, 1968.

A review article on electrophoresis which discusses microscopic methods of observing electrophoresis patterns, early moving-boundary apparatus, optical systems, measurement of conductivity, and calculation of mobility is presented. A limited discussion of preparatory electrophoresis techniques and 51 references are included in the review.

AN APPARATUS FOR THE ELECTROPHORESIS OF PROTEINS ON POLY-ACRYLAMIDE GEL

Price, G. M., Lab. Pract. **17**, 467–470 (1968).

An inexpensive apparatus for vertical polyacrylamide-gel electrophoresis, constructed from Perspex®, is described. The use of this apparatus for separation of esterase, peptidase, alkaline phosphatase, acid phosphatase and phenol oxidase from the hemolymph of blowfly larvae is documented.

APPARATUS FOR PREPARATIVE ELECTROPHORESIS THROUGH GELS

Schenken, I., M. Levy, and P. Weis, Anal. Biochem. **25**, 387–395 (1968).

A preparative apparatus suitable for polyacrylamide starch or other gels is described. Separation takes place in a vertical cylinder of gel held in place by adhesion of closure with wet filter paper. The buffer solution is placed beneath this cylinder and is separated from the second electrolyte and its electrode by a semipermeable membrane.

A NEW APPARATUS FOR PREPARATIVE GEL ELECTROPHORESIS

Smith, J. K., and D. W. Moss, Anal. Biochem. **25**, 500–509 (1968).

A preparative gel electrophoresis apparatus which contains a new elution device is described. The apparatus was constructed from borosilicate glass and is essentially an outer tube surrounded by a 1 liter globe. The platinum electrode (cathode) was placed in a side-neck. A fixed condensor encloses the tube. Results obtained during the separation of alkaline phosphatases indicate good resolution, and quantitative recoveries were possible with the apparatus.

ÉLECTROPHORÈSE VERTICALE EN GEL D' ACRYLAMIDE DYSPOSITIF SIMPLIFIÉ

Szylit, M., Pathol. et Biol. (Paris) **16**, 247–256 (1968) **French.**

An apparatus for vertical acrylamide-gel electrophoresis was constructed. Diagrams are presented for duplication of the chamber; temperature and voltage relationships are tabulated. Sample separations done with lactate dehydrogenase, hemoglobin and serum proteins are pictured.

SEPARATION OF MYOINOSITOL PENTOPHOSPHATES BY MOVING-PAPER ELECTROPHORESIS (MPE)

Tate, M. E., Anal. Biochem. **23**, 141–149 (1968).

A ceramic apparatus was designed and constructed for improved separation in MPE. The apparatus permits separations over a meter at a voltage gradient of 35 volts/cm. The advantages of compactness and lower heat transfer are offered by this technique. Construction diagrams for the tank, cooling coils, glass guide, and paper roll are detailed. Sample separations are pictured.

A TECHNIQUE FOR MICROIMMUNOELECTROPHORESIS IN ACRYLAMIDE GELS

Van Orden, D. E., Immunochemistry **5**, 497–499 (1968).

A method for molding inexpensive templates for polyacrylamide gels to be used in immunoelectrophoresis is described. The patterns were cut in agar, and dental acrylic resin was poured in two layers to form the mold. Polyacrylamide gel was poured into this mold and the whole assembly heated for 20 minutes at 56°C. The template is reusable and requires a minimum of effort to assemble.

AUTOMATIC ANALYSIS OF ELECTROPHORESIS STRIPS BY MEANS OF A CYCLIC SCANNER

Vanzetti, G., A. Palatucci, and G. Cosci, Clin. Chim. Acta **20**, 215–225 (1968).

An electronic cyclic scanner for the automatic analysis of stained electrophoretic strips or cellulose acetate membranes is presented. The scanner employs a monochromatic light source and a cyclical scanning device for the rotating strip. Modulated pulses give rise to instant signals proportional to the area being analyzed. It is possible to display the electrophoretic pattern on the fluorescent screen of a cathode ray tube. Protein concentrations are computed and read on a digital dial.

APPARATUS FOR ZONE ELECTROPHORESIS IN VERTICAL COLUMN

Bergham, B., Sci. Tools **14**, 34–38 (1967).

A new apparatus which allows separation of up to 50 mg of protein by zone electrophoresis in a vertical column is described. The basic design was based on two symmetrical circulation couplings which make it possible to change columns easily when separation is completed.

UN APPAREILLAGE POUR L'ÉTUDE CINÉTIQUE DE L'ÉLECTROPHORÈSE DES SUBSTANCES COLORÉES EN GEL D'ACRYLAMIDE

Delcourt, P. R., Ann. Biol. Clin. (Paris) **25**, 1261–1267 (1967) **French.**

A vertical apparatus for electrophoresis on polyacrylamide gels is diagramed. The apparatus allows use of superimposed gels, two-dimensional electrophoresis and good temperature control. A photometric attachment permits scanning of colored substances during electrophoresis.

SPECTROPHOTOMETRIC DETERMINATION OF LACTIC DEHYDROGENASE ISOENZYMES AFTER ELECTROPHORETIC SEPARATION

Holeysovská, H., Clin. Chim. Acta **15**, 353–358 (1967).

Lactic dehydrogenase isoenzymes separated on agar gels and detected by staining with NBT were quantitated spectrophotometrically. The zones containing the formazan were cut out and solubulized at 100°C. The colored solutions were measured quantitatively at 550 mμ in the ERI-10-Carl Zeiss Jena 26910 DDR densitometer. The procedure appears to be more precise than densitometry, and reproducibility was good.

PREPARATIVE ELECTROPHORESIS IN LIQUID COLUMNS STABILIZED BY ELECTROMAGNETIC ROTATION: I. APPARATUS

Kolin, A., J. Chromatog **26**, 164–179 (1967).

Mixtures of substances which differ only in electrophoretic mobility were easily separated by injecting a fine stream of the mixture into a horizontal column of fluid having an angular shape. The column was stabilized against connection currents by electromagnetic rotation about the cylindrical axis. It is possible to induce each component of the mixture to follow a helical path if speed of axial flow, electric current and strength of magnetic flow are controlled.

REFLECTANCE DENSITOMETRY OF CELLULOSE ACETATE PROTEIN ELECTROPHORESIS

Kremers, B., R. O. Brierre, and J. G. Batsakis, Am. J. Med. Technol. **33**, 28–34 (1967).

Serum samples were diluted in a ½ volume of barbital buffer before cellulose-acetate electrophoresis, according to the method of Briere and Mull [Am. J. Clin. Pathol. **42**, 547 (1964)]. After staining and drying for one hour at 37°C, the strips were cleared in acid alcohol: 15% acetic acid in 95% ethanol. Densitometry readings were made on a reflectance densitometer. This method is reproducible and quantitative.

QUANTITATIVE ASSAY OF LACTATE DEHYDROGENASE ISOENZYMES BY REFLECTIVE DENSITOMETRY

Latner, A. L., and D. M. Turner, Clin. Chim. Acta **15**, 97–101 (1967).

Lactic dehydrogenase isoenzymes separated on starch gels were successfully quantitated using reflectance densitometry. The Joyce-Löebl Chromoscan was the instrument used. Reproducible results were obtainable if the total units of enzyme applied did not exceed 400 international units per liter.

ACRYLAMIDE GEL ELECTROPHORETOGRAMS BY MECHANICAL FRACTIONATION: RADIOACTIVE ADENO-VIRUS PROTEINS

Maizel, J. V., Science **151,** 988–990 (1967).

A device for automatically separating the zones of a gel column after electrophoresis has been developed and tested. Ten resolvable protein components of radioactive adenovirus were mechanically separated. Agreement between the automated and manual method was excellent in plating experiments and cpm. Diagrams for construction of the fractionator are detailed. The device could be constructed in most laboratories.

MAINS OPERATED APPARATUS WITH CONSTANT CURRENT AND VOLTAGE FOR DISC ELECTROPHORESIS

Neuhoff, B., B. Mühlberg, and J. Meier, Arzneimittel-Forsch. **17,** 649 (1967).

An apparatus which permits electrophoresis to be performed in 5 ul capillaries at currents up to 20 mA is described. These currents were obtained through the use of special diode switches. Voltages were measured with very high resistance value-voltmeters. An almost constant output up to 450 volts was obtained, with variations as low as $\pm$ 2%.

SIMPLE CELL-ELECTROPHORESIS APPARATUS FOR SMALL AMOUNTS OF MATERIAL

Sachtleben, P., D. Krämer, W. Schwarz, and G. V. F. Seaman, Klin. Wochschr. **45,** 425–428 (1967) **German.**

An apparatus designed especially for the separation of micro amounts of material is described. It has a specially designed flow-through cell, with a capacity of 0.3 ml, which can be used to separate blood, tissues, and serum antibodies. The cell is placed between two electrodes and a constant temperature is maintained by circulating water. Migration of the sample was observed microscopically.

SIMPLIFIED TECHNIQUE FOR PREPARATIVE DISC ELECTROPHORESIS: II. FURTHER IMPROVEMENTS IN APPARATUS AND SOME DETAILS OF PERFORMANCE

Sulitzeanu, D., M. Slavin, and E. Vecheskeli, Anal. Biochem. **21,** 57–67 (1967).

An apparatus was designed for large scale disc electrophoresis using the method of Ornstein and Davis [Ann N. Y. Acad. Sci. **121**, 404 (1964)]. It was possible to fractionate 200-400 mg of protein in three to four hours. Diagrams for construction of the apparatus are provided. Sample protein mixtures were separated with a high degree of purity.

AN AUTOMATED MICRODENSITOMETER FOR THE QUANTITATION OF PHEROGRAMS AND ZYMOGRAMS IN ACRYLAMIDE GELS

Allen, R. C., and G. R. Jamieson, Anal. Biochem. **16**, 450–456 (1966).

This paper describes an automated optical density analyzer for microdensitometric measurement of electrophoretic patterns on acrylamide gels. Traces of the electrophoretogram are automatically integrated and presented in digital form on paper tape. Bands as small as 35μ in width can be accurately quantitated over a range of 0.01 to 2.0 O.D. units.

A VERTICAL FLAT-BED DISCONTINUOUS ELECTROPHORESIS SYSTEM IN POLYACRYLAMIDE GEL

Allen, R.C., and D. J. Moore, Anal. Biochem. **16**, 457–465 (1966).

A technique for flat-bed vertical discontinuous separations in acrylamide is presented. This technique has two advantages: a) it gives a more precise comparison between various staining reactions on a single sample, and b) it provides a method for observing unstained protein zones for migration rates and zone location.

ELECTROPHORESIS ON POLYACRYLAMIDE

Gabler, F., and O. Rueker, Lab. Sci. (Milan) **14**, 29–34 (1966) **Italian.**

An apparatus that separates proteins, viruses, bacteria and other high molecular weight substances is described. After being mixed with grains of raw sugar, the sample was applied to the gel. Either a pH 8.5 Tris-glycine buffer or a pH 8.8 Tris-borate-EDTA buffer was used for routine separation of serum proteins.

A DENSITOMETRIC METHOD FOR THE QUANTITATIVE DETERMINATION OF PROTEIN FRACTIONS AFTER AGAR GEL ELECTROPHORESIS

Gijzen, A. H., E. M. M. A. van Nunen, and P. E. Verheesen, Clin. Chim. Acta **13**, 397–399 (1966).

A simple, accurate method of densitometric analysis of proteins after electrophoresis is described. The gel is made with 9 g Noble agar per 500 ml boiling water. The solid gel is cut into cubes of 1 cm and soaked in 50% ethanol for two washings of three hours each. The gel is then washed in 500 ml water and dissolved in 500 ml of buffer (pH 8.4, 17 g sodium diethyl barbiurate, 23 g 1N HCl per liter). After cooling, 2-mm slices are prepared. Electrophoresis is done at 150 volts per slide for 25 minutes at 20°C. The slides are fixed for one hour in acetic acid–ethanol–water (1:16:4). Proteins are stained for 30 minutes with Amido Black 10B solution. The slides were scanned with a densitometer.

IMPROVED METHODS OF MOLDING STARCH GEL FOR ZONE ELECTRO-PHORESIS

Hori, J., Anal. Biochem. **16,** 234–243 (1966).

Two methods of molding starch gel medium which practically remove zonal distortion during electrophoresis are discussed. Special frames for holding the gel and techniques for covering gels are described.

SEPARATION OF GUANIDINO COMPOUNDS BY COMBINATION OF ELECTROPHORESIS AND GEL FILTRATION: APPLICATION TO HUMAN URINE

Lauber, E., and S. Natelson, Microchem. J. **11,** 498–507 (1966).

An apparatus which allows continuous preparative electrophoresis in Sephadex® G-200 is described. The apparatus is 24 inches by 24 inches by 0.5 inches. The buffer flows either by gravity or by pumping. Electrophoresis proceeds laterally, and eluates may be collected from ports in the base. The instrument was used to separate guanidino compounds from urine with Sephadex G-200 and Biol gel P-10.

A SIMPLE METHOD FOR SLICING STARCH GELS

Lewis, G. P., J. Clin. Pathol. **19,** 521–522 (1966).

A dermatome blade uniformly sliced a starch block after electrophoresis. The blade is mounted on a Perspex® platform. Diagrams for the construction of the slicing apparatus are presented.

ELECTROPHORETIC STUDIES OF SALIVA

Mandel, I. D., J. Dental Res. **45,** Suppl. 3, 634–643 (1966).

This review article examines the use of moving-boundary electrophoresis, paper electrophoresis, agar, cellulose-acetate, starch-gel, and acrylamide-gel electrophoresis for the separation of saliva proteins in systemic diseases which affect the composition of saliva, plaque and calcus. The advantages and limitations of the various techniques and their application to the study of oral diseases and systemic diseases which affect oral diseases are reviewed.

DISC ELECTROPHORESIS: AN IMPROVED TECHNIQUE

Nelson, T. E., and A. Hale, J. Lab. Clin. Med. **68**, 838–839 (1966).

The development of a special apparatus for making disc for polyacrylamide-gel electrophoresis is described. A modification of Ornstein and Davis technique, this procedure eliminates the need for serum bottle stoppers. The apparatus seals the ends of tubes during polymerization.

ELECTROPHORESIS — A PRACTICAL LABORATORY MANUAL

Nerenberg, S. T., Philadelphia: F. A. Davis Company, 1966.

This book presents material suitable for teaching a course in zone electrophoresis. Basic theory and techniques are described with illustrative experiments in the advanced section. Hemoglobin and lactic dehydrogenase isoenzyme separations are covered extensively, and the use of polyacrylamide disc electrophoresis as a clinical tool is reviewed. This is a very good text in the practical application and performance of electrophoresis.

PREPARATIVE SCALE ELECTROPHORESIS ON ACRYLAMIDE GEL

Ng, W. C., and J. R. Brunner, J. Dairy Sci. **49**, 96–98 (1966).

A preparative electrophoretic cell and electrophoretic extractor for recovery of proteins from the gel have been designed. The recovery of 65–70% for a twice-run 3–5 ml sample of a 6% α_s-casein solution. A construction diagram of the apparatus is presented. The preparative electrophoresis was complete in 16 to 20 hours, in a potential gradient of 300–400 volts.

DISC ELECTROPHORESIS (DESTAINING, STAINING, PROTEIN RECOVERY)

Schwabe, C., Anal. Biochem. **17**, 201–209 (1966).

A device for destaining polyacrylamide gel disc after staining with Amido Black is described. Made of polyacrylate, the apparatus accommodates six gels and destains in six minutes.

A MODIFIED CELL FOR MOVING-BOUNDARY ELECTROPHORESIS

Smith, M. B., and S. J. Rose, Anal. Biochem. **17**, 236–243 (1966).

A modification of the Tiselius electrophoresis cell for moving-boundary separations is described. The cell was modified so that boundaries were formed by a flowing-junction method; solution and buffer were controlled by external valves. Gassing electrodes were used.

EFFECT OF TEMPERATURE UPON THE BARBITAL BUFFERS USED IN ELECTROPHORESIS

Strickland, R. D., and M. M. Anderson, Analyt. Chem. **38,** 980–982 (1966).

The effects of temperature upon the solubility and dissociation constant of 5, 5-diethylbarbituric acid and upon the pH and conductivity of barbital buffers of various ionic strengths was presented. These data were supplied with the aim of allowing one to predict the electrophoretic conditions needed to achieve separation in certain instances.

USE OF COOMASSIE BRILLIANT BLUE R250 FOR THE ELECTROPHORESIS OF MICROGRAM QUANTITIES OF PAROTID SALIVA PROTEINS ON ACRYLAMIDE STRIPS

Meyer, T. S., Biochim. Biophys. Acta **107,** 144–145 (1965).

A technique for the identifying of saliva proteins present in a concentration of approximately 0.1% is described. Twenty-five to fifty μl of sample were required. A total volume of 98 ml was obtained when 5.133 g acrylamide and methylenebisacrylamide (97.4:2.6 w/w) and 0.06 ml of $N_1N_1N'_1N'$-tetramethylethylenediamine were dissolved in 0.1M Tris–EDTA–boric buffer, pH 9.0. Two ml of fresh 5% (μ/v) ammonium persulfate were added for polymerization. Gel strips 260 x 60 x 3 mm were prepared. Electrophoresis was performed in a 0.05M Veronal buffer, pH 8.6, for three hours at 9 to 11 volts cm. The strips were then stained for 20 minutes with 0.25% Coomassie Brilliant Blue R250 in methanol–glacial acetic acid–water (5:1:5 v/v). Excess stain was removed by a modification of the technique of Ferris using 3.75% acetic acid [Am. J. Clin. Pathol. **38,** 383 (1962)].

ADAPTATION OF A SPECTROPHOTOMETER FOR SCANNING ELECTRO-PHORETIC PATTERNS ON CELLULOSE ACETATE

Mull, J. D., and J. D. Johnson, Am. J. Clin. Pathol. **43,** 77–80 (1965).

The Coleman® Junior spectrophotometer has been equipped with a hardwood dowel which fits into the cuvette well. One surface was flattened to allow the cellulose-acetate strip to pass before the photo cell. The strip is moved by means of a small synchronous motor fastened to the dowel. It is possible to focus light on the strip by means of lens cemented to the bottom of the dowel. The adaptation was found to be completely reproducible as evidenced by results of replicate scanning of a single cellulose-acetate strip.

A METHOD OF QUANTITATION OF LACTIC DEHYDROGENASE IN AGAR GEL ELECTROPHORESIS

Mull, J. D., and W. H. Starkweather, Tech. Bull. Registry Med. Technologists **35**, 231–233 (1965).

A scanning device for enzymes separated on agar gels is described. The scanning device consisted of a onocular microscope with a mechanical stage. A 4 rpm synchronous motor was connected to the horizontal drive of the stage. A green filter from the Spinco model RA Analytrol® was placed between the eyepiece and light source. The photocell of the Analytrol was connected to the eyepiece and the Analytrol. The motor moves the pattern across the light path.

ELECTROPHORETIC MOBILITY OF MILK FAT GLOBULES: I. CELL DESIGN

Tjepkema, R., and T. Richardson, J. Dairy Sci. **48**, 1391–1394 (1965).

An electrophoresis chamber with the large electrode wells and the acrylic plastic construction of the Reddick cell has been adapted. Because of the large electrode wells, electrodes having a large surface area can be used; therefore electrolyte solutions are not needed to avoid polymerization. Construction of the chamber is described. The validity of the cell is demonstrated by a mobility-depth curve. Red blood cells and milk fat globules were used in these experiments.

Agar-Gel Electrophoresis

A NEW RAPID ELECTROPHORETIC METHOD FOR DETERMINATION OF DEOXYRIBONUCLEIC ACID BASE COMPOSITION

Borokowski, T., J. Wojcierowski, and S. Kulesza, Anal. Biochem. **27**, 58–64 (1969).

Rapid analysis of DNA bases was possible by agar gel electrophoresis. DNA preparations were freed from RNA by alkaline hydrolysis; the DNA was then hydrolyzed with 72% $HClO_4$ for one hour. The DNA bases were rendered free of $HClO_4$ by evaporation under reduced pressure. The dry bases were dissolved in distilled water and subjected to electrophoresis on 0.6% agar gels on a 5 x 20 cm glass plate. A sodium acetate–acetic acid buffer (0.1M, pH 3,5, 3.7, or 3.9) was used both the prepare gels and as the electrolyte. Separation was complete in 45 minutes at 5 to 7 volts/cm. All four bases— guanine, adenine, cytidine and thymine—were well separated from each other at pH 3.7. The separated bases were detected by viewing under UV light. Quantitation was possible by cutting the spot from the gel, warming it in a water bath to dissolve agar, and reading the absorption at 260 mμ.

ISOLATION OF ALPHA, BETA AND GAMMA CONGLYCININS

Catsimpoolas, N., and C. Ekenstam, Arch. Biochem. Biophys. **129**, 490–497 (1969).

Alpha, beta and gamma conglycinins were isolated from soybean seeds by gel filtration on Sephadex® G-100, column chromatography on Bio-Gel® A-1.5M. Agar gel, disc, and disc immunoelectrophoresis were used to characterize the isolated proteins. Immunoelectrophoretic analyses employing anti-soybean water extract serum 123 and a nonspecific anti-γ-conglycinin serum 130, as well as the other electrophoretic studies, showed the homogeneity of the isolated protein.

ELECTROPHORESIS OF RIBOSOMES IN AGAR GEL

Dessen, G. N., C. D. Venkov, and R. G. Tsanev, European J. Biochem. **7**, 280–285 (1969).

Ribosomes and their subunits of rat liver and E. Coli were separated by electrophoresis on 1% agar using a 0.02M Tris-HC1 buffer. At a voltage gradient of 8 to 10 volts/cm the ribosomes moved as a single band. Two bands were seen if the ribosomes were treated with EDTA. It was possible to extract RNA from the agar zones which contained two subunits of rat liver by immersing the dried agar film in 1% acetic acid containing 0.01% sodium hydrogen phosphate. The faster migrating subunit contained only 18S RNA while the slower moving unit contained 18S and 28S RNA.

ISOLATION AND PROPERTIES OF γ-CHAIN FROM HUMAN FETAL HEMO-GLOBIN

Kajita, A., K. Tazutoshi, and R. Shukuya, Biochim. Biophys. Acta **175**, 411–448 (1969).

Fetal hemoglobin was isolated from erythrocyte homogenates by ion-exchange chromatography on Amberlite IRC 50 (or XE 64). This isolated hemoglobin was allowed to react with p-chloro-mercuribenzoate at pH 6.5, pH 6.0, pH 5.5, pH 5.0 and pH 4.7. The dissociation products formed by these treatments were separated by agar-gel electrophoresis according to the procedure described by Wieme [Clin. Chim. Acta **4**, 317 (1959)]. Electrophoresis was performed in a barbiturate buffer, pH $8.6/\mu = 0.05$). A potential gradient of 180 to 220 volts (2.5 mA/cm) was applied for 16 hours. In addition, the dissociation products were fractionated by column chromatography on CM-cellulose. Two components of hemoglobin F were found when hemoglobin was allowed to react with p-chloro-mercuribenzoate at pH 4.7. The second component was found to be a γ-chain.

ESTIMATION OF RELATIVE MOLECULAR LENGTH OF DNA BY ELECTRO-PHORESIS IN AGAROSE GEL

Takahashi, M., T. Ogino, and K. Baba, Biochim. Biophys. Acta **174**, 183–187 (1969).

Mechanically fragmented phage (T_2) DNA (labeled with C^{14} and H^3) of various lengths were isolated by density-gradient centrifugation. A sucrose gradient (5-20% in 0.1M sodium chloride, 0.05M phosphate buffer) was used to isolate DNA. The isolated material was then fractionated by agarose gel electrophoresis on a 1% gel. The gel was prepared in a 0.05M barbital buffer, pH 8.5. The same buffer was used as the electrolyte. The mobility of the fragmented DNA on agarose gel was inversely related to its sedimentation rate in a sucrose density gradient. The use of this finding for molecular weight estimation is discussed.

DIFFERENTIATION OF NATIVE AND RENATURED ALKALINE MILK PHOS-PHATASE BY MEANS OF SEPHADEX® GEL THIN LAYER CHROMATOG-RAPHY

Copius-Peereboom, J. W., and H. W. Beekes, J. Chromatog. **39**, 339–343 (1968).

The phosphatase activity of pasteurized cream and pasteurized cream to which magnesium chloride was added to reactivate the enzyme was investigated using agar-gel electrophoresis and thin-layer chromatography. Agar-gel electrophoresis was performed according to Wieme's technique [Agar Gel Electrophoresis, Amsterdam: Elsevier, 1965]. A sodium barbital buffer, pH 8.4, ionic strength 0.05, was used. Separation proceeded at 25 mA per glass slide. Phosphatase activity was detected by staining with a solution containing β-naphthyl phosphate as substrate. Alkaline phosphatases were separated by thin-layer chromatography on Sephadex® G-200 with phosphatase buffer pH 7.0 as the developer. In the serum phase of raw or under-pasteurized cream, α-phosphatase only was found. In the serum phase of reactivated cream, β_2 and β_3 alkaline phosphatases were detected.

APPLICATION DE L'ÉLECTROPHORÈSE ET DE L'IMMUNOELECTROPHORÈSE Á L'ETUDE DE QUELQUES ENZYMES Á USAGE THERAPEULIQUE

Dony, J., R. Bontinck, and H. Muyldermans, Int. Symp. IV Chromatographie and Electrophorèses Proc., 339–349 (1966) Presses Acad. Europ. Brussels (1968) **French.**

The use of agar-gel electrophoresis and immunoelectrophoresis for separation and identification of the biologically important substances α-chymotrypsin and hyaluronidase mucopolysaccharidase m_x was described. Generally, electrophoresis was performed in a pH 8.2 barbital buffer at a potential of 6 volts/cm on agar-gel-coated slides. Immunoelectrophoresis was performed using the micromethod of Scheidegger.

AGARELECTROPHORETISCHE STUDIEN ZUR FRAGE VON ISOENZYMEN DER AMYLASE

Götz, H., H. Wüst, and F. Raies, Clin. Chim. Acta **19**, 235–243 (1968) **German.**

Agar-gel electrophoresis characterized the amylase activity of the parotid gland, pancreas, human sera, human urine and bacterial amylase. A Michaelis buffer, pH 8.6, ionic strength 0.05, was used for electrophoresis. Two amylase fractions were generally found in human biological fluids. Serum amylase migrates with slow mobility between β- and α-globulins.

ELECTROPHORETIC SEPARATION OF GLUCOSE-6-PHOSPHATE DEHYDRO-GENASE FROM HUMAN ERYTHROCYTES WITH AGAR GELS

Haywood, B. J., W. H. Starkweather, H. H. Spencer, and C. J. Zarofonetis, J. Lab. Clin. Med. **71**, 324–327 (1968).

An inexpensive method employing agar gels for separation of erythrocyte glucose-6-phosphate dehydrogenase is described. Separation was achieved in a glycyl–glycine–NaOH–EDTA buffer, pH 8.6, ionic strength 0.04. A potential of 27 volts/cm for 40 minutes was used at 4°C. The separated enzymes were detected by staining with tetrazolium salts, with glucose-6-phosphate as substrate.

TOTAL LDH AND LDH ISOENZYME DISTRIBUTION IN THE SERUM OF NORMAL CHILDREN

Heiden, C. V. D., J. Desplanque, J. W. Stoop, and S. K. Wodman, Clin. Chim. Acta **22**, 409–415 (1968).

The total LDH activity and LDH isoenzyme distribution were determined in sera of 111 normal children, ages 4 to 13 years. Standard assay procedure which employed pyruvate as substrate was used to assess the total enzyme content. Electrophoretic separation was achieved on Noble® agar using Wieme's microprocedure. No significant differences were found when enzyme activity of normal children was compared to that of normal adults.

A METHOD FOR DETERMINATION OF THE HETEROGENEITY OF γ-GLUTAMYL-TRANSPEPTIDASE

Jacyszyn, L., and T. Laursen, Clin. Chim. Acta **19**, 345–352 (1968).

Agarose-gel electrophoresis performed in 0.04M barbiturate buffer, pH 8.4, separated the γ-glutamyl-transpeptidase of normal serum, urine and tissue. The heterogeneity of the separated enzyme was demonstrated by molding the agarose-containing substrate on the plate containing the enzyme. After a 15 to 20 minute incubation period, a third layer of agarose gel containing the diazonium salts was added. There was a definite migration of the enzymes toward the anode.

ELECTROPHORESIS OF PIG SERUM LIPOPROTEINS IN AGAROSE GEL

Kalab, M., and W. G. Martin, Anal. Biochem. **24**, 218–225 (1968).

The lipoproteins of pig serum were separated into three distinct zones by electrophoresis in agarose gel using a barbitone acetate buffer, pH 8.6, ionic strength 0.1. Separations were carried out in the cold with a current of 10 mA/slide. The separated lipoproteins were visualized by staining with a solution of Oil Red 0 and Fat Red 7B (1:4) saturated in 80% ethanol.

A RAPID SERUM SCREENING TEST FOR INCREASED OSTEOBLASTIC ACTIVITY

Kerkhoff, J. F., Clin. Chim. Acta **22**, 231–238 (1968).

Agar-gel (0.8%) electrophoresis in a Veronal buffer, $\mu = 0.1$, pH 8.4, distinguished bone alkaline phosphatase isoenzymes from liver, intestine and placenta alkaline phosphatase isoenzymes. Electrophoresis was performed for 90 minutes with a potential gradient of 150 volts (11 mA/cm agar). Heat inactivation as a simple distinguishing test compared well with results obtained by agar-gel electrophoresis.

THE ISOENZYMES OF SERUM ALKALINE PHOSPHATASE IN CATS

Kramer, J. W., and S. D. Sleight, Am. J. Vet. Clin. Pathol. **2**, 87–91 (1968).

Total alkaline phosphatase and alkaline phosphatase isoenzymes were evaluated in serum of normal mature cats, kittens, and kittens having an experimentally produced biliary obstruction. Electrophoretic separation was carried out an 1% agarose in a pH 8.6 Veronal buffer, $\mu=0.037$. Separation proceeded for 60 minutes at 150 volts. The enzyme was located using the standard histochemical technique with sodium naphthyl AS-MX phosphate as the substrate. Only one electrophoretically distinguishable enzyme was present in all sera. This enzyme migrated in the beta-1-globulin fraction. Total serum alkaline phosphate values were 1.6 ± 0.83 sigma units (mean values). Kittens having a biliary obstruction demonstrated no increase in alkaline phosphatase activity.

AN ABNORMAL HEMOGLOBIN IN RED CELLS OF DIABETICS

Rahbar, S., Clin. Chim. Acta **22**, 296–298 (1968).

Agar-gel electrophoresis in a citrate buffer, pH 6.2, according to the method of Robinson et al. [J. Lab. Clin. Med. **50**, 745 (1957)] separated an abnormal hemoglobin from the erythrocytes of diabetics. This fraction migrates in front of hemoglobin A to the cathode in the same position as hemoglobin F.

LIPOPROTEINELECTROPHORESE IN AGAROSEGEL

Rapp, W., and W. Kahlke, Clin. Chim. Acta **19**, 493–498 (1968).

Serum lipoproeins were successfully separated using 0.8% agarose gel and a sodium Veronal hydrochloric acid buffer, pH 8.2. A potential of 10 volts/cm was applied for 90 or 120 minutes. Upon staining the gel with Sudan Black, four bands of lipoprotein activity were observed in normal fasting serum. As many as nine bands were present in abnormal serum.

ISOLATION AND PROPERTIES OF A MOUSE SERUM PREALBUMIN EXCRETED IN URINE

Reuter, A. M., G. Hamoier, R. Marchand, and F. Kennes, Europ. J. Biochem. **5**, 233–238 (1968).

The fastest prealbumin fraction (PA) was isolated from serum and urine of Balb/c$^+$ male mice by first precipitating it with sodium chloride ($2°$ to $4°C$) and 5% phenol and then performing gel filtration on Sephadex® G-75. The material was concentrated by preparative starch-block electrophoresis. The discontinuous buffer was composed of 0.005M Tris–0.006 citric acid–0.0027M boric acid in the gel and 0.06M sodium hydroxide–0.9M boric acid in the electrode vessels. Agar-gel electrophoresis and starch-gel electrophoresis were employed to characterize the isolated material. Identical electrophoretic behavior was noted for both urinary and urine prealbumins. The molecular weight was estimated to be 20,200.

STUDIES ON TUBULAR PROTEINURIA

Walravens, P., E. C. Laterre, and J. F. Heremans, Clin. Chim. Acta **19,** 107–120 (1968).

The low molecular weight components isolated by gel filtration from the urine of patients with tubular proteinuria were characterized by electrophoresis (agar and disc) and immunoelectrophoresis. The electrophoretic technique used to identify the lysozyme was essentially that described by Wieme [Studies on Agar Electrophoresis, Brussels: Arscia (1959)]. Polyacrylamide-disc electrophoresis was performed in a Tris-glycine buffer, pH 8.5. Immunoelectrophoresis was performed in a 2% agar gel using a micromethod of Scheidegger [Intern. Arch. Allergy Appl. Immunol. **7,** 103 (1955)]. Normal urine was found to contain α_2, β_2 and post-γ proteins characteristic of tubular proteinuria. Several trace components present in normal serum were identified in CSF, normal urine, and tubular proteinuria.

STAPHYLOCOCCAL HYALURONATE LYSASE: MULTIPLE ELECTROPHORETIC AND CHROMATOGRAPHIC FORMS

Abramson, C., Arch. Biochem. Biophys. **121,** 103–106 (1967).

Semipreparative electrophoresis in agar gels separated four hyaluronidase active fractions from staphylococcal lysates. The lysates were subjected to ammonium sulfate fractionation followed by column chromatography on Sephadex® G-200. The partially purified preparation was then electrophoresed on microslides coated with 1% Noble Agar prepared in 0.025 ionic strength Veronal barbital buffer, pH 8.6. The same buffer at 0.1 ionic strength was used in the electrode chamber.

SEPARATION OF SWINE KIDNEY WORM (STEPHANURUS DENTATUS) ANTIGENS BY ZONE AND BY BARRIER ELECTROPHORESIS

Baisden, L. A., and F. G. Tromba, J. Parasitol. **53,** 100–104 (1967).

Zone electrophoresis on ion agar in a Veronal buffer (ionic strength 0.0375, pH 8.6 at 5.3 volts/cm) separated proteins of excretory gland extracts of swine kidney worms *(Stephanurus dentatus).* Immunoelectrophoresis characterized the specificity of these fractions.

LACTIC DEHYDROGENASE AND METABOLISM OF HUMAN LEUKOCYTES

Bloom, A. D., Science **156,** 979–981 (1967).

The agar-gel electrophoretic pattern of lactic dehydrogenase isozymes changed during culture of leukocytes from healthy donors. During transformation and division, the LDH isozymes migrated increasingly to the cathode. As the mitotic index declined, isozyme bands migrating to the anode became dominant, despite a tendency to anaerobic metabolism. Agar-gel electrophoresis was performed at 4°C with sodium lactate as the substrate, NAD as the coenzyme, and phenazine methosulfate as the electron carrier.

HEMAGGLUTININS (LECTINS) EXTRACTED FROM MACLURA POMIFERA

Jones, J. M., L. P. Cawley, and G. W. Teresa, Vox. Sanguinis **12,** 211–214 (1967).

Upon electrophoresis, saline extracts of *Maclura pomifera* seeds yielded protein zones, two of which had hemagglutinin activity. Electrophoretic separation was achieved on agar gels according to the procedure of Cawley *et al.* [Trans. 4, **441** (1964)]. The protein of the extract was also examined by gel filtration on Sephadex® G-50. Five protein peaks were found, two of which were active.

MULTIPLE ANALYSIS ON A SINGLE GEL ELECTROPHORESIS PREPARATION

Jordan, E. M., Nature **216,** 78 (1967).

This paper describes a modification of the Raymond vertical electrophoresis technique. When multiple samples were applied across the horizontal gel slab, several sera would be processed together for reliable comparisons. It was feasible to stain the gel slab with separate staining solutions without cutting it apart. An agar layer containing the stain was selectively placed on the slab, thus allowing different systems to be tested on one electrophoretic run.

A FAST ELECTROPHORETIC SERUM α_1-ANTITRYPSIN VARIANT

Laurell, C. B., and E. Gustavsson, Clin. Chim. Acta **15,** 361–362 (1967).

An anomalous fast α_1-antitrypsin variant was detected in serum using agar-gel electrophoresis. The anomaly is characterized by normal antitrypsin content but decreased amounts of α_1-antitrypsin which have normal mobility.

ELECTROPHORETIC BEHAVIOUR OF GLUTAMATE DEHYDROGENASE FROM RAT ORGANS

Pashev, I. G., Clin. Chim. Acta **16,** 127–130 (1967).

Glutamate dehydrogenase isoenzymes of rat liver, kidney and heart homogenates were fractionated by electrophoresis on agar gel. Electrophoresis was performed on microscope slides according to Wieme's technique. The separated isozymes were located histochemically by agar overlay technique and by the ultraviolet light test. When the histochemical detection procedure was used, the pattern was different for each tissue and was almost identical with lactic dehydrogenase patterns. The UV test revealed two fractions — a fast and a slow one — and both were in identical positions in all tissues.

CATABOLISM OF SERUM PROTEINS AFTER X-IRRADIATION

Reuter, A. M., G. B. Gerber, F. Kennes, and J. Remy-Defraigne, Radiation Res. **30**, 725–738 (1967).

Agar-gel electrophoresis was used to identify the breakdown products of catabolyzed proteins in X-irradiated rats. Serum was analyzed following *in vivo* experiments. H^3-phenylalamine and I^{131} were used to label the proteins. Electrophoresis of 5 μl of sera was done in pH 8.5 Veronal buffer, ionic strength 0.05, for 40 minutes at 175 volts. Densitometer measurements were made after staining with Amido Black. Protein areas on the gel were eluted with 0.5N NaOH in 50% ethanol. From another plate, the protein fractions were scraped into tubes to be solutionized for couintng. One ml of 1M methanol hyamine (37°C for two days) was used to solubilize the gel. Fluorosystem was added for counting Catabolism in the perfused liver was not altered by X-irradiation, but several protein fractions disappeared more rapidly from the blood.

CERULOPLASMIN ACTIVITY IN LIVER DISEASES: QUANTITATIVE DETERMINATION AND ELECTROPHORETIC CHARACTERIZATION

Secchi, G. C., A. Petrella, B. Lomanto, and N. Gervasine, Enzymol. Biol. Clin. **8**, 33–41 (1967).

Ceruloplasmin, which exerts an oxidase activity when appropriate substrate is used, was separated from serum using agar-gel electrophoresis. Electrophoresis on 1% agar plates (15 x 4 cm) in a sodium Veronal buffer (pH 8.2, ionic strength 0.05) for two hours at 100 volts was sufficient to separate the enzyme. Enzyme activity was demonstrated using p-phenylene-diamine in an acetic acid-sodium acetate buffer as a substrate and incubating plates for two hours at 37°C. Neoplastic tissue exhibited an increase in enzyme activity. No changes in mobility of the enzyme were noted in cirrohitic conditions when compared to normal patients.

HYPERSENSITIVITY TO ASCARIS ANTIGENS: I. SKIN-SENSITIZING ACTIVITY OF SERUM FRACTIONS FROM GUINEA PIGS SENSITIZED TO CRUDE EXTRACTS

Strejan, G., and D. H. Campbell, J. Immunol. **98**, 893–900 (1967).

Serum from guinea pigs immunized with a crude preparation of *Ascaris* serum extracts was separated by preparative electrophoresis in agar, and the isolated fractions were characterized by agar-gel immunoelectrophoresis, cellulose-acetate immunoelectrophoresis, and passive cutaneous anaphylaxis.

Preparative agar-gel electrophoresis was carried out on a 0.5% Noble agar gel in a pH sodium barbiturate-sodium acetate-hydrochloride buffer, ionic strength 0.025. A potential of 200 volts (22 mA) for 20 hours was applied. The gel was cut onto 10-mm transverse slices, frozen in dry ice for 30 minutes and thawed overnight at room temperature. The gel sections were then mashed and filtered, and the fractions were collected and subjected to immuno-electrophoresis at 120 volts (0.5 mA/cm) in a pH 8.9 Tris–EDTA buffer (cellulose-acetate strips 2.5 x 18 cm). Globulins 7S γ1- and γ2- were identified and had the power to elicit passive cutaneous anaphylaxis.

STUDIES ON HORSE ANTIBODIES: PART I. DEVELOPMENT OF ANTIBODY IN DIFFERENT FRACTIONS OF SERUM DURING IMMUNIZATION OF HORSE WITH DIPHTHERIA TOXOID

Acharya, U. S. V., K. P. Gunaga, and S. S. Rao, Indian J. Biochem. **3**, 33–37 (1966).

During immunization of horses there was a decrease in albumin with a concurrent increase in total serum proteins. Different fractions of the serum were separated by agar-gel electrophoresis in a pH 8.6 Veronal acetate buffer, ionic strength 0.05. Lipoproteins were separated the same way and stained with Sudan Black B. Antibody content of the fractions was tested *in vivo* using rabbits. Increases in the gamma globulin fraction were seen during the immunization process.

APPEARANCE OF THE PRE-α_2-GLOBULINS SOON AFTER THE VERY FIRST DOSE OF DIPHTHERIA TOXOID IN THE HORSE

Acharya, U. S. V., and S. S. Rao, Experientia **22**, 167–168 (1966).

Agar-gel electrophoresis of horse sera after immunization with diphtheria toxoid demonstrated pre-α_2-globulins. They were permanent during the course of immunization. A pH 8.6 Veronal acetate buffer, ionic strength 0.05, was used for electrophoresis.

A SIMPLIFIED METHOD TO PREPARE SOLUBLE BRAIN PROTEINS FOR ELECTROPHORESIS

Allegranza, A., P. Canevini, and P. Mocarelli, Clin. Chim. Acta **13**, 119–121 (1966).

Slices of human brain, 3 to 4 mm thick, were rapidly washed with saline, and 1 to 1.5 g of tissue were homogenized for five minutes at 2°C with a small amount of quartz powder. The homogenate was separated into five layers by high-speed centrifugation at 104,000 g for 60 minutes at 2°C. The second layer was protein, which was separated and recentrifuged at 10,000 g for ten minutes at 2°C. Agar-gel electrophoresis was performed on 0.01 ml of the supernate [Naturwiss. **44**, 112 (1957)]. The number of fractions varied according to the region of the brain sampled. This method avoids dilution and denaturation of the tissue proteins.

HUMAN ISOAMYLASES

Aw, S. E., and J. R. Hobbs, Biochem. J. **99**, 16 (1966).

Urinary amylase was separated by electrophoresis on agar and agarose in a Veronal buffer at 200 volts (20 mA per slide) at 8°C. Enzyme activity was detected by placing moistened cellulose-acetate strips over slides containing separated amylases, incubating the strips on starch plate, and then staining the strips with alcoholic iodine solution. Two bands of activity were found in urine, and one band was found in pancreatic juice obtained from fistula. The enzyme activity could not be detected in postmortem livers using this procedure.

ISOLATION OF IMMUNOGLOBULIN A (IgA) FROM HUMAN COLOSTRUM

Axelsson, H., B. G. Johansson, and L. Rymo, Acta Chem. Scand. **20**, 2339–2348 (1966).

The IgA in defatted, dialyzed human colostrum was isolated by column electrophoresis using Pevikon®, formaldehyde-treated cellulose, or Sephadex® G-25 "Fine" as the supporting medium. A pH 8.2 Tris-citrate buffer containing 0.05 or 0.2M Tris was the electrolyte. Small columns were run with 0.05M buffer for 24 hours at 600 volts; large columns in 0.2M buffer for 40 hours at 300 volts. Gel electrophoresis (1% w/v Noble agar, 0.05M sodium barbital buffer, pH 8.4, 8 volts/cm for 90 minutes) confirmed the purity of the isolated IgA. The IgA sedimentation coefficient was 11.6 S.

SOME PHYSICOCHEMICAL PROPERTIES OF ALBUMINS ISOLATED FROM HUMAN SERUM AND CEREBROSPINAL FLUID

Bocci, V., and D. B. Gammack, J. Neurochem. **13**, 875–878 (1966).

Agar-gel electrophoresis in a pH 8.6 barbital buffer, ionic strength 0.05, was used to distinguish electrophoretic mobility differences of albumin from CSF and serum. CSF and serum albumins were concentrated by ultrafiltration dialysis against barbital buffer, $\mu=0.075$. Approximately 200 mg of protein were used for the sample. Protein fractions were eluted from the gel by 0.15M NaCl for sedimentation analysis.

SEPARACION ELECTROFORETICA DE DOS PROTEINAS ENZYMATICAS EN PREPARALIONE PURIFICADAS DE ARGINASA DE BIGADO Y DE ERITROCITOS HUMANOS. MOVILIDAD DE LOS COMPONENTES RAPIDOS EN AMBAS PREPARACIONES

Cabello, J., V. Prajoux, and M. Plaza, Arch. Biol. Med. Exp. **3,** 7–13 (1966) **Spanish.**

Agar-gel electrophoresis performed at pH 8.0 and 8.6 in a 0.05M Tris–maleate buffer was used to characterize partially purified arginase enzymes of human liver and erythrocytes. Two enzyme proteins were observed. One remained near the starting well and migrated toward the cathode. The other moved quickly toward the cathode and constituted 95% of the total arginase activity.

ELECTROPHORESIS OF NORMAL SPINAL CORD FLUID ON AGAR GEL

Chankov, I., Lab. Delo **1966,** 296–298.

Spinal cord fluid from 58 persons was concentrated by dialysis versus 50% gum arabic at 4°C to a protein concentration of 165–400 mg/ml. Densitometry measurements were made after electrophoresis (1% agar, barbital buffer, pH 8.6) and staining with Amido Black 10B. The usual pattern showed ten zones, which were named and quantitated.

PURIFICATION OF FERRATIN-LABELLED IMMUNOGLOBULINS

Charles, A., Experientia **22,** 486–487 (1966).

Zone electrophoresis is a convenient means of separating conjugated ferratin before using it in electron microscope experiments. Ferratin was coupled to γ-globulin using 2,4 diisocyanate and dialyzed against 0.05M phosphate buffer, pH 7.5. Separation was achieved on agar-gel electrophoresis in a pH 8.2 Veronal buffer, ionic strength 0.05 (run overnight with 20-30 mA at 300 volts). The gel was stained with Ponceau Red. The conjugate stains brown and moves between the ferrates and the globulin, which remains at the origin. The conjugate was eluted from the gel with 0.05M phosphate, pH 7.5.

ELECTROPHORESIS OF LACTATE DEHYDROGENASE, ALKALINE PHOSPHATASE AND NON-SPECIFIC ESTERASE IN JEJUNAL BIOPSIES OF CONTROLS AND PATIENTS WITH MALABSORPTION SYNDROME

Fric, P., and Z. Lojda, Gastroenterologia **106,** 65–76 (1966).

Agar-gel electrophoresis was used to separate lactate dehydrogenase, alkaline phosphatase, and nonspecific esterases in 50 jejunal biopsies obtained with Crosby-Kugler capsules from controls and usbjects with malabsorption syndromes. LDH-5 was decreased and LDH-4 and LDH-3 were elevated in primary malabsorption syndrome, whereas alkaline phosphatase was decreased. Nonspecific esteroses were also decreased.

COMPARATIVE STUDY OF ELECTROPHORETIC AND RAPID CHROMATO-GRAPHIC METHODS FOR SEPARATION OF LACTIC DEHYDROGENASE ISOENZYMES

Gotts, R., and L. P. Skendzel, Clin. Chim. Acta **14**, 505–510 (1966).

Agarose-gel electrophoresis is a more accurate method than both Sephadex® DEAE-chromatography for quantitative determinations of LDH isoenzymes in clinical cases. Electrophoresis of the supernate fluid after Sephadex DEAE-chromatography demonstrated that some isoenzymes were not adsorbed by the resin, thus making the laboratory confirmation of clinical diagnosis difficult.

ELECTROPHORETIC HETEROGENEITY OF α_2-MACROGLOBULIN

Granrot, P. O., Clin. Chim. Acta **14**, 137–138 (1966).

Electrophoresis in agarose indicated that the α_2-macroglobulin in human sera may be heterogeneous. A 0.07M barbital buffer, pH 8.6 with 2mM calcium lactate was the electrolyte. Two-dimensional and immunoelectrophoresis gave comparable results of heterogeneity, which may be due to complexes formed between the α_2-macroglobulin and trypsin, somato-trypsin, plasmin, or insulin.

HAPTOGLOBIN ELECTROPHORESIS

Hyland Laboratories: Technical Discussion Number 5 (1966).

Agar gel, prepared in a 0.04M TEB solution (ph 8.75) and electrophoresed in a pH 8.6 barbital buffer, separated haptoglobin from haptoglobin-hemoglobin complexes. A potential of 175 volts for 15 minutes was sufficient to effect the separation.

ELECTROPHORETIC STUDY OF HUMAN ISOAMYLASES. A NEW SACCHA-ROGENIC STAINING METHOD AND PRELIMINARY RESULTS

Joseuph, R. R., E. Olivero, and N. Ressler, Gastroenterology **51**, 377–382 (1966).

A new saccharogenic method for the visualization of amylase isoenzymes separated on agar gels is described. Electrophoresis of serum, saliva and tissue homogenates was performed on glass plates (25 x 15 x 0.5cm) coated with 0.4% agar. A phosphate buffer, pH 7.25, and 75 volts at room temperature (10 to 14 hours) were used to effect separation. Enzyme activity was detected by incubating the agar-coated plates in a solution containing Glucostat® reagent, maltase and soluble starch for 30 minutes at 37°C. A blue precipitate formed at the site of enzyme activity. Liver, pancreatic and salivary amylases could be differentiated using this procedure.

AGAR ELECTROPHORESIS OF CEREBROSPINAL FUILD PROTEINS: SEMEIOLOGY OF THE GAMMA ZONE

Laterre, E. C., Acta Neurol. Psychiat. Belg. **66**, 289–304 (1966).

The γ-globulin fraction of CSF is composed of two proteins, αaT and β_2aT, and 7Sγ. The αaT protein can be identified by agar electrophoresis since it migrates to the anode more slowly than γ-globulin. In aged serum it migrates in the middle of the γ-globulin fraction. Tissue permeability changes can cause the αaT and B_2aT component to be found in blood protein.

LACTIC DEHYDROGENASE ISOENZYMES IN THE AORTIC WALL

Lojda, Z., and P. Fric, J. Atherosclerosis Res. **6**, 264–272 (1966).

The distribution of lactic dehydrogenase isoenzymes in aortae of various species was studied using the agar-gel electrophoresis technique of Wieme. Tissue homogenates prepared in 0.04M barbital buffer, pH 8,4, were electrophoresed forr 25 - 40 minutes at 140 volts. Enzyme activity was detected by staining, with INT or NBT used as electron acceptors. The enzyme pattern was found to be different from one species to another. In the cock and duck, only one cathodal migrating LDH was observed. Multiple bands of activity were observed in other species. Distinct LDH-5 was found only in human aorta. In the rabbit, guinea pig and pig, LDH-1 was pronounced. The distribution of LDH isoenzymes was different in individual layers of aorta.

DETERMINATION OF DEHYDROGENASE ISOENZYMES BY THE ENZYME-ELECTROPHORESIS METHOD

Markelovo, I. M., Ukr. Biokim. Zh. **38**, 334–336 (1966) **Russian.**

A microelectrophoresis chamber and a technique for the separation of water-soluble proteins as well as salt-soluble proteins are presented. Water-soluble proteins were separated on 1% agar in a pH 8.55 Veronal—acetate buffer, ionic strength 0.05. Salt-soluble proteins were separated using a glycine phosphate buffer, ionic strength 0.2. The dehydrogenase activity was detected by staining the agar slides with a tetrazolium solution containing lactate as substrate.

ENZYMATIC ACTIVITIES IN PRE-RIGOR AND POST-RIGOR SARCOPLASMIC EXTRACTS OF CHICKEN PECTORAL MUSCLE

Neelin, J. M., and D. J. Ecobichon, Can. J. Biochem. **44**, 735–741 (1966).

Myogen extracted in hypotonic salt solutions and sarcoplasmic from chicken pectoral muscle were fractionated by starch-gel electrophoresis. Sodium cacodylate-sodium chloride or sodium veronal–sodium chloride gradient buffers were used as were constant systems of sodium cacocylate–sodium citrate. Esterase, acid and alkaline phosphatase, lactate, malate and isocitric dehydrogenase activities were demonstrated by appropriate stains. Lactic dehydrogenase was detected in sarcoplasmic proteins. Eleven highly anionic zones predominated in nonspecific esterase profiles.

SERUM PROTEINS FROM MILK SERUM EXAMINED IN ELECTROPHORESIS IN AGAR GEL

Piccinini, D., Corriere Farmacista **21**, 72–77 (1966).

Three fractions from milk seurm were identified by agar gel electrophoresis (sodium Veronal–HC1 buffer, pH 8.2, for seven hours). Casein was precipitated by 10% ACOH and the milk serum concentrated by lyophilization.

ELECTROPHORETIC IDENTIFICATION OF IRON-BINDING PROTEINS OF THE MILK OF WOMEN

Ponzone, A., and G. Papa, Minerva Pediat. **18**, 842–846 (1966).

Milk serum was obtained enzymatically from defatted milk dialyzed and concentrated by ultrafiltration. Iron-59 (as citrate) was added (1.5–4.0 μc/ml), and the mixture was incubated for two hours at 37°C. After agar-gel electrophoresis, the plates were stained and the radioactivity measured. The principal iron carrier was α-casein. Non-ferriproteins and amino acids can interfere with the quantitation.

BIOACTIVATION ET ACTION ANTICHOLINESTERASIQUE DES INSECTICIDES ORGANO-PHOSPHORÉS AU NIVEAU DU SANG HUMAIN ET DU FOIE DE RAT "IN VITRO"

Salamé, M., Ann. Biol. Clin. **24**, 441–449 (1966) **French.**

Agar-gel electrophoresis (1%), run in a Veronal buffer (pH 8.4) for 30 to 35 minutes, was used to study the effect of organo-phosphorous insecticides on the esterases of tissue from rat serum and liver. The phosphates were more inhibitory toward esterases of liver than were the thiophosphates. The phosphates were active against the esterases of serum only after activation with bromine water.

ROENTGEN IRRADIATION EFFECTS ON MOUSE PROTEINS

Sassen, A., F. Kennes, and J. R. Maisin, Acta Radiol. Therapy Phys. Biol. **4**, 97–112 (1966).

Sera was obtained from mice before and after they received a single dose of roentgen rays. Agar-gel electrophoresis in pH 8.55 and pH 8.45 Veronal buffer separated the proteins in the sera. No obvious changes in the mobility of serum proteins, except in γ-globulin, were detected by agar-gel electrophoresis. Changes in the immunoelectrophoretic pattern were detected as early as six to eight hours postirradiation.

NEW METHODS OF IDENTIFYING SERUM SIDEROPHILIN USING RADIO-ACTIVE IRON

Slopek, S., J. Ladosz, K. Hryncewicz, and W. Brzuchowska, Nature **209**, 1036 (1966).

Agar-gel electrophoresis of sera mixed with iron-55 was used to identify siderophilin by combining the results of an autoradiogram and Amido Black protein stain. The siderophilin arch lies in the β_1-globulin region, a result which confirms those obtained with pure siderophilin. Isotope iron-59 has a higher energy of β-radiation than iron-55 and can be used when there is a need for shortened autoradiogram exposure.

THE LACTATE DEHYDROGENASE OF HEMAPOIETIC CELLS

Starkweather, W. H., H. H. Spencer, and H. K. Schoch, Blood **28**, 860–871 (1966).

The lactate dehydrogenase isoenzyme pattern of hemapoietic tissue was examined by the procedure of Wieme as modified by Starkweather. This technique has shown that the hemapoietic stem cell of man contains five isoenzymes. In addition it appears that cellular differentiation leads to the synthesis of new LDH or a shift in isoenzyme pattern.

THE ELECTROPHORETIC SEPARATION OF LACTATE DEHYDROGENASE ISOENZYMES AND THEIR EVALUATION IN CLINICAL MEDICINE

Starkweather, W. H., H. H. Spencer, E. L. Schwarz, and H. K. Schoch, J. Lab. Clin. Med. **67**, 329–343 (1966).

Improved separation of lactate dehydrogenase isoenzymes from blood and tissue homogenates was achieved on modified agar gels. The agar gels were prepared in a O.1M Tris, 0.028M diethyl barbituric acid buffer, pH 8.9 (ionic strength 0.1).

IMMUNOLOGIC IDENTIFICATION OF A WIDE RANGE OF PROTEINS IN HUMAN HEPATIC BILE

Stoica, G., Rev. Intern. Hepatol. **16**, 1237–1247 (1966).

High-voltage microelectrophoresis in agar gel (pH 8.4, 20–22 volts/cm), immunoelectrophoresis, and two-dimensional electrophoresis in agar gel were techniques used to study the proteins in human hepatic bile. Three proteins not normally found in human sera were found in the bile.

LACTATE DEHYDROGENASE ISOENZYMES IN DEVELOPING HUMAN MUSCLE

Takasu, T., and B. P. Hughes, Nature **212**, 696–610 (1966).

Agar-gel electrophoresis separated the LDH isoenzymes from infants at different stages of development. The isoenzymes were located by precipitation of a formazan following incubation with lactate and NAD in the presence of nitro-blue tetrazolium. Only a trace amount of isoenzyme 5 was detected in a five-month-old fetus but it was the most abundant isoenzyme in the full-term infant and adults. The amount of the LDH isoenzymes may be a function of the type of muscle.

EVALUATION OF NORMAL GASTRIC MUCOSAL ANTIGENS AND EN-ZYMES IN CANCEROUS HUMAN GASTRIC MUCOSA BY AGAR AND IMMUNOELECTROPHORESIS

Aronson, S. B., W. Rapp, I. Kushner, and P. Burtin. Int. Arch. Allergy Appl. Immunol. **26**, 327–332 (1965).

Gastric tumor extracts and nontumorous gastric extracts were examined by agar gel and immunoelectrophoresis. A 0.05M sodium Veronal buffer, pH 8.2, was used for electrophoresis separation at a potential of 6 volts/cm for two hours. The separated proteins were examined for carboxylic esterase activity and proteolytic activity. Immunoelectrophoresis evaluations were made by reacting against rabbit anti-mucosal extract (normal) serum two anti-tumor extract serum and two noninvolved antiserum. Results indicated that the proteolytic activities of four different pepsins were absent or diminished in tremor. Proteolytic activity was diminished but not absent in nontumorous mucosa. Carboxylic esterases were diminished in all extracts.

THE COMPLEXITY OF HOUSE DUST, WITH SPECIAL REFERENCE TO THE PRESENCE OF HUMAN DANDRUFF ALLERGEN

Berrens, L., J. H. Morris, and R. Versie, Intern. Arch. Allergy Appl. Immunol. **27**, 129–144 (1965).

Five different fractions were isolated from house dust allergens, purified and characterized chemically and electrophoretically. Electrophoresis was performed on agar gels using a pH 8.6 Veronal buffer, ionic strength 0.04; separation proceeded at 20 mA for 15 minutes. Chemically, house dust was found to be a complex mixture of proteins, glycoproteins, polysaccharides, poly-uronides, mucopolysaccharides and glycopeptides. Electrophoretically house dust revealed at least four different components when stained with a protein dye. Immunoelectrophoresis demonstrated extreme heterogeneity when tested against rabbit polyvalent antisera.

IDENTIFICATION DES GLYCOPROTÉINES DU SERUM SOLUBLES DANS L'ACIDE PHYTIQUE

Biserte, G., J. E. Courtois, R. Havez, J. Agneray, and A. Hayem-Levy, Bull. Soc. Chim. Biol. **47**, 1827–1833 (1965) **French.**

Serum glyproteins soluble in 0.01N phytic acid, pH 2.10, were separated by electrophoresis. Veronal-HCl buffer, pH 8.2, ionic strength 0.1, was used to prepare the agar gel. Electrophoresis was performed vertically using a discontinuous buffer. The buffer used was Tris–EDTA–borate, pH 8.9. Eleven different compounds werre separated, as seen by staining and immuno-electrophoresis.

FRACTIONATION AND CHARACTERIZATION OF PROTEINS AND LIPIDS IN BILE

Clausen, J., I. Kruse, and H. Dam, Scand. J. Clin. Lab. Invest. **17**, 325–335 (1965).

Human and chicken bile were characterized by their behavior on agar-gel microelectrophoresis, immunoelectrophoresis, quantitative precipitin reactions and gel filtration on Sephadex® G-25 and G-50. Human hepatic and bladder bile was shown to contain plasma albumin and γ-globulin on immunoelectro-phoretic slides. Quantitatively, equal amounts of albumin and γ-globulin were found in human and chicken bile. Human and chicken bile were shown to possess two components separable by gel filtration.

ELEKTROPHORETISCHE ISOENZYM-AUFTRENNUNGEN DER LACTATE-DEHYDROGENASE IM AGAR GEL

Frölich, C., Z. Klin. Chem. **3**, 137–141 (1965) **German.**

Starch-gel electrophoresis according to the method of Wieme in the LBK apparatus was used to separate the lactic dehydrogenase isoenzymes of human serum. A pH 8.8 Veronal–sodium barbiturate buffer, ionic strength 0.06, was used. Electrophoresis was carried out for 140 minutes at 250 volts and 28 mA. With the LBK immunoelectrophoresis apparatus, it was possible to carry out 18 determinations simultaneously. Patterns of serum LDH activity in various disease states are presented.

THE IMMUNOCHEMICAL HETEROGENEITY OF PROTEINS AND GLYCO-PROTEINS IN NORMAL HUMAN URINE

Grieble, H. G., J. Courcon, and P. Grabar, J. Lab. Clin. Med. **66**, 216–231 (1965).

Electrophoresis at pH 8.2 in agar gel, 0.025M Veronal buffer, demonstrated different mean mobilities of the albumin and gamma globulin from urine and serum. Selective excretion of glycoproteins in normal urine was suggested by DEAE-cellulose chromatography and agar-gel electrophoresis (0.025M acetic acid acetate buffer, pH 4.5). The limitations of electrophoresis of proteinuria in health and disease are discussed.

PROTEIN CONCENTRATION AND DYE UPTAKE AFTER AGAR GEL ELECTROPHORESIS

Kisutzer, H. J., Clin. Chim. Acta **12**, 575–578 (1965).

The nonlinearity of densitometer measurements of stained albumin following electrophoresis in agar gel was examined. Experiments without electrophoresis showed linearity at the higher concentrations of stained albumin. Previous difficulties in quantitative experiments are believed to be due to sulfur groups in the agar gel which can pick up the stain. Agar gel varied in the amount of stain it retained.

ISOLATION AND PARTIAL CHARACTERIZATION OF 'TRACE' PROTEINS AND IMMUNOGLOBULIN G FROM CEREBROSPINAL FLUID

Link, H., J. Neurol. Neurosurg. Psychiat. **28**, 552–559 (1965).

The finding of β-trace protein and γ-trace protein from human cerebrospinal fluid was confirmed. DEAE-cellulose and Sephadex® chromatography were used to further study these low molecular weight problems. An isolation procedure is described. Agar-gel electrophoresis was performed in a 0.9% agar solution in a continuous sodium pH 8.4 Veronal buffer, ionic strength 0.05. Electrophoresis was carried out for 25 minutes at 200 volts. Slides were stained with Amido Black 10B. Starch-gel electrophoresis was done in a 0.035M glycine buffer, pH 8.8, containing 8M urea. The electrode vessels contained 0.3M boric acid–0.06N NaOH, pH 8.2.

AGAR-GEL ELECTROPHORESIS OF CEREBROSPINAL FLUID OF LEPERS

Renders, J., Acta Neurol. Psychiat. Belg. **65**, 808–815 (1965).

Agar-gel electrophoresis was used to study the CSF of patients with tuberculoid or lepromatous forms of leprosy. The transferrin molecule in two-thirds of the patients seemed to be split in half. The amount of β- and γ-globulin was influenced by the form of the disease, presence of paralysis or amputations.

UNTERSUCHUNGEN AN MITTELS RIVANOL GEWONNEN SERUMFRAK-TIONEN

Schatz, H., Acta Med. Scand. **177**, 427–430 (1965) **German.**

Electrophoresis and immunoelectrophoresis on agar gels were employed to characterize the proteins fractionated from blood plasma with 0.4% Rivanol. A high γ_1A-globulin, γ_2-globulin and transferrin were detected in this fraction. The γ_1-macroglobulins were not detected in the Rivanol fraction. Rivanol was added to blood in a 1:3.5 proportion.

MULTIPLE FORMS OF SERUM AND LIVER ESTERASES IN THE NORMAL STATE AND IN CIRRHOSIS OF THE LIVER

Secchi, G. S., and N. Dioguardi, Biol. Clin. **5**, 29–36 (1965).

Electrophoresis on agar gels according to the technique of Grabar and Williams [Biochim. Biophys. Acta **10**, 193 (1955)] separated four esterases from serum and six or seven from liver. In cirrhosis a marked reduction of cholinesterase and aromatic esterases was observed.

ALTERATIONS OF LACTATE DEHYDROGENASE IN MAN

Starkweather, W. H., L. Cousineau, H. K. Schoch, and C. J. Zarafonetis, Blood **26**, 63–73 (1965).

Agar-gel electrophoresis performed in a pH 8.9 Tris–barbital buffer separated lactate dehydrogenase isoenzymes of erthocyte hemolysates. Five isoenzymes of LDH were found in erythrocyte of normal men and women. LDH-5 was found exclusively in young cells, whereas LDH-1 predominates in aged cells.

THE DEMONSTRATION OF THE PROTEIN FRACTIONS OF HUMAN TEARS BY MEANS OF MICROELECTROPHORESIS

Tapaszto, I., and Z. Vass, Acta Ophthalmol. **43**, 796–801 (1965).

Microelectrophoresis on 1.5–2% agar slides (77 x 25 mm) demonstrated the presence of five protein groups from human tears. Eight to twelve protein fractions were demonstrated in lysozyme of commercial origin. A sodium barbital-sodium acetate buffer, pH 8.2, ionic strength 0.05, was used. Separation was carried out at 110-400 volts for 1.5 to 2.0 hours.

SERUM PROTEINS OF THE GREY SQUIRREL (SCIURUS CAROLINENEIS)

Wild, A. E., Immunology **9**, 457–466 (1965).

Serum proteins of the grey squirrel were examined by fluid-agar and free-boundary electrophoresis. Fluid-agar electrophoresis was performed according to a modification of the method of Wild (Thesis, 1963, University of Wales). A pH 8.6 sodium barbitone–sodium acetate buffer, ionic strength 0.05, and 0.15% ionagar were used. Free-boundary electrophoresis was carried out in the Perkin-Elmer Model 38A Tiselius apparatus. Quantitative results as to the composition of serum proteins were obtained by these two procedures. Starch-gel electrophoresis and immunoelectrophoresis were performed in order to obtain qualitative information. A high proportion of prealbumin was found, thereby accounting for 4.3 to 9.5% of total protein. At least 17 antigenically different proteins were distinguished.

Cellulose-Acetate Electrophoresis

CREATINE KINASE ISOENZYMES IN SERUM OF CHILDREN WITH NEURO-LOGICAL DISORDERS

Cao, A., S. DeVirgiliis, C. Lippi, and N. Trabalza, Clin. Chim. Acta **23**, 475–478. (1969).

Creatine kinase isoenzymes of serum of children affected with acute encephalitis, tetanus, and acute febrile seizures cerebrospinal menigitis and of newborns affected with perinatal hypoxia were separated by electrophoresis on Cellogel® strips. Serum separations were carried out in a damp room on 4 x 17 cm Cellogel strips in a pH 8.6 Veronal buffer and constant voltage (8 volts/cm) for two hours. The isoenzymes were located by staining with the substrate solution of Rosalki [Nature **207**, 414 (1965)] as modified by Van der Veen and Willerbrands [Clin. Chim. Acta **13**, 312 (1966)]. The isoenzymes were increased in all of these disorders.

ELECTROPHORETIC STUDY OF THE SERUM PROTEINS OF NORMAL AND LYMPHOID TUMOUR-BEARING XENOPUS

Hadji-Azimi, I., Nature **221**, 264–265 (1969).

Normal serum proteins and serum proteins from tumor-bearing *Xenopus laevis* were separated on cellulose-acetate membranes using a barbital buffer solution, pH 8.6, ionic strength 0.075. Normal serum was found to contain four protein components. An increased amount of gamma globulin was found in the sera of tumour-bearing animals. This increase of gamma globulin usually paralleled a decrease in albumin. The gamma globulin peak showed a narrowing similar to that observed in human sera containing an "M" component.

DEMONSTRATION OF LACTATE DEHYDROGENASE ISOENZYME ON CELLULOSE ACETATE

Homer, G. M., B. Yott, and J. S. Lim, Tech. Bull. Registry Med. Technologists **39**, 11–15 (1969).

Electrophoretic separation and detection of lactate dehydrogenase isoenzymes was achieved on cellulose-acetate membranes in the Beckman MicroZone® equipment. Electrophoresis was performed at 200 volts for 75 minutes in Sorensen's buffer, pH 7.5. The isoenzymes were detected by staining the cellulose strips with tetrazolium.

DISTRIBUTION OF FRUCTOSE DIPHOSPHATE ALDOLASE VARIANTS IN BIOLOGICAL SYSTEMS

Lebherz, H. G., and W. J. Rutter, Biochemistry **8**, 109–121 (1969).

Cellulose-acetate electrophoresis was used to characterize the aldolases of thirteen vertebrate species, five invertebrate species, four plant species, two protoza, two fungi, and five microbial species. Three parental aldolases (F-1-6 DiP) were detected in all 13 vertebrate species. Single bands of class II enzymes were found in all bacteria and fungi tested. No multiplicity of tissue enzyme was absent in plants.

ELECTROPHORETIC STUDY OF PROTEIN CHANGES BETWEEN SEVERAL DEVELOPMENTAL STAGES OF THREE SPECIES OF HELIOTHIS (LEPIDOPTERA NOCTUIDAE)

Vinson, S. B., and W. J. Lewis, Comp. Biochem. Physiol. **28**, 215–220 (1969).

Proteins of *Heliothes zea H. virescens,* and *H. subflaxa* hemolymphs were examined by cellulose-acetate electrophoresis. A Tris–barbital–sodium barbitol Veronal buffer (0.027M, pH 8.8) was used. Electrophoresis continued for 45 minutes at 250 volts. Separated proteins were detected by staining with nigrosin. A decrease in the number of bands through the pupal stages was noted. Two dominant glycoprotein bands appeared in all three species in the late sixth instar phase and persisted through the pupal stage, disappearing in the adult stage. The adult exhibited an increase in the number of protein bands. Quantitative sex differences in the proteins were apparent.

LACTATE DEHYDROGENASE PATTERNS IN BLISTER FLUID AND IN EXFOLIATE DERMATITIS

Anderson, M., J. Invest. Dermatol. **51,** 283–285 (1968).

Cellulose-acetate electrophoretic separation of blister fluid and in serum of patients with exfoliative dermatitis in a pH 8.6 barbital buffer revealed no abnormalities of isoenzymes of lactate ·dehydrogenase. The blister fluids showed a greater amount of LDH-2 and 3.

PLASMA CELL LEUKEMIA WITH IgD PARAPROTEIN

Ben-Bassat, I., U. I. Frand, C. Isersky, and B. Ramot, Arch. Internal Med. **121,** 361–364 (1968).

Electrophoresis on paper, cellulose acetate, and starch gel was used to characterize proteins from urine and serum of patients with plasma cell leukemia. An abnormal protein which had the mobility of a fast γ-globulin was found in the serum. Precipitation arcs which would indicate monoclonal gammopathy were absent upon immunoelectrophoresis and diffusion against antiwhole serum anti-IgG, IgA, IgM, and light chains.

THE DIAGNOSTIC SIGNIFICANCE OF SERUM ELECTROPHORESIS IN ACUTE RENAL DISEASE

Berstein, S. H., J. I. Berkman, and J. Allerhand, Am. J. Med. Sci. **256,** 97–106 (1968).

Cellulose acetate separated serum proteins of patients with acute post-streptococcal glomerulonephritis and patients with the idiopathic nephrotic syndrome. Electrophoresis was performed in a pH 8.6 Veronal buffer, ionic strength 0.05. Separations proceeded for two hours at a constant current of 1.5 mA per strip. The electrophoretic patterns observed were different in the two diseases, so differentiation of the two diseases was possible.

SEPARATION AND QUANTITATIVE ANALYSIS OF SERUM LIPOPROTEINS BY MEANS OF ELECTROPHORESIS ON CELLULOSE ACETATE

Chin, H. P., and W. Blankenhorn, Clin. Chim. Acta **20,** 305–314 (1968).

A rapid method employing cellulose acetate (Gelman Sepraphore® III, or Schleicher and Schuell No. 2500, 1″ x 6¾″) is described for the separation of lipoproteins. A barbital buffer (pH 8.6, ionic strength 0.075) was used. Electrophoresis was carried out at 150 volts for 90 minutes. Complete resolution of chylomicrons, β-lipoproteins, pre-β-lipoproteins and α-lipoproteins into distinct bands was achieved. The lipoprotein fractions can be eluted and determined quantitatively.

CHONDROITIN SULFATE A AND SULFATED GLYCOPROTEINS IN DOG GASTRIC SECRETIONS FROM THE FUNDUS: I. ELECTROPHORETIC AND CHEMICAL CHARACTERIZATIONS

De-Graef, J., and G. B. J. Glass, Gasteroenterology **55**, 584–593 (1968).

Large amounts of sulfated glycosaminoglycuronoglycon were found in dog gastric juices. When this substance was subjected to electrophoresis on cellulose acetate in a pH 9.0 borate buffer, its electrophoretic mobility was found to be similar to chondroitin sulfate. Gelman Sepraphore III was used at a potential of 250 volts for two hours.

NEW SIMPLE TECHNIQUE FOR THE MULTIFRACTIONATION OF SERUM PROTEINS (18–21 FRACTIONS) ON CELLOGEL®-RS CELLULOSE ACETATE STRIPS

Del Campo, G. B., Clin. Chim. Acta **22**, 475–479 (1968).

A procedure employing Cellogel®-RS (a gelatinized cellulose acetate) strips for separation of serum (2.5 μl) is described. Eighteen to 21 fractions were separated in a pH 8.9 Tris–barbital buffer, 0.036M at 400 volts for two hours (5 x 23 cm strip). Multiple samples can be applied and run simultaneously. Resolution was comparable to that obtained on polyacrylamide gels.

NATURE OF THE HETEROZYGOTE BLOOD CATALASE IN A HYPOCATA-LASEMIC MOUSE MUTANT

Feinstein, R. N., J. T. Braun, and J. B. Howard, Biochem. Genetics **1**, 277–285 (1968).

Electrophoresis on both cellulose-polyacetate strips (Sepraphore® III, Gelman) and agar gels was used to investigate the blood catalase of a mutant mouse. Cellulose-acetate electrophoresis was performed in a cold Tris–barbital buffer (0.05M, pH 8.8). The separated enzymes were detected by exposing the strips to 3% H_2O_2 and then observing them for destruction of H_2O_2 in an area where catalase was present. Agar-gel electrophoresis was carried out for 3.5 hours at a constant voltage of 250 volts in a Veronal buffer, pH 8.6, ionic strength 0.025. The blood of both normal and mutant animals showed the same electrophoretic patterns on both media.

DEMYELINATING DISEASE IN A WOMAN FROM TROPICAL SOUTH AMERICA WITH FEATURES OF MULTIPLE SCLEROSIS AND NEUROMYELITIS OPTICA: CLINICAL, PROTEIN CHEMISTRY AND PATHOLOGICAL FINDINGS

Howell, D. A., E. H. Jellinek, and K. Gavrilescu, J. Neurol. Sci. **7,** 115–135 (1968).

The case history of a patient who developed a demyelinating disease in tropical South America is presented in detail. Clinical as well as morbid pathological findings are reviewed. A presentation of the findings of electrophoretic and immunoelectrophoretic separation of serum and cerebrospinal fluid are presented.

LACTATE DEHYDROGENASE ACTIVITY IN OOCYTES

Kessel, R. G., and E. G. Ellgard, Exp. Cell. Res. **45,** 243–246 (1968).

Electrophoretic separation of lactic dehydrogenase enzyme from the oocyte of *Necturus, Cambrus,* the lobster *Homarus,* and the ascidian *Ciona* was achieved. Separation was achieved on Sepraphore® III (Gelman) in the high resolution buffer at a potential of 300 volts for one hour at 25°C. Enzyme activity was localized on the strips by the decomposition of nitro blue tetrazolium formazan after incubating the strips, which had been soaked in a solution containing 10 cc sodium lactate, 30 cc NBT (1 mg/ml), 3 cc phenazine methosulfate (1 mg/ml), 10 cc distilled water, and 100 mg DPN for 30 minutes at 37°C. The LDH pattern was different from most other observed patterns in that only a single band was found.

DETERMINATION OF THE ISOELECTRIC POINT OF BOVINE PLASMA ALBUMIN BY CELLULOSE ACETATE PAPER ELECTROPHORESIS

Kubota, Y., and H. Ueki, J. Biochem. Tokyo **64,** 405–406 (1968).

The use of cellulose-acetate support medium for the determination of the isoelectric point of bovine plasma albumin is described. Electrophoresis was performed in three different buffers at 0.6 mA/cm for 30 minutes. The buffers used were: I. 0.01, 0.02, and 0.05M sodium hydroxide–formic acid, pH 4.2-4.6; II. 0.05, 0.10 and 0.20M sodium hydroxide–acetic acid, pH 4.2-5.0; III. 0.01, 0.02 and 0.5M sodium hydroxide–acetic acid, pH 4.2-5.0. Glucose was run as a standard to use in correcting migration due to electro-osmosis. Buffer III at ionic strrength .05 gave values which agree with published results.

IDENTIFICATION OF SPECIES IN RAW PROCESSED FISHERY PRODUCTS BY MEANS OF CELLULOSE POLYACETATE STRIP ELECTROPHORESIS

Lane, J. P., W. S. Hill, and R. J. Learson, Solutions **VII**, # 2 May 1968. Published by Gelman Instrument Company, Ann Arbor, Michigan.

Cellulose-acetate (Sepraphore® III) electrophoresis was used to identify species proteins in raw processed fishery products. Water extracts of fish protein were separated using this medium in a pH 8.6 Veronal buffer, ionic strength 0.05, at a potential gradient of 300 volts for 30 minutes. Ocean catfish, cod and cusk could be differentiated based upon the protein patterns obtained.

ELECTROPHORETIC STUDIES ON THE HAEMOGLOBINS OF AUSTRALIAN LAMPREYS

Patter, I. C., and P. I. Nicol, Austral. J. Biol. Med. Sci. **46**, 639–641 (1968).

Electrophoretic analysis of the hemoglobins of three species of Australian lampreys was reported. Electrophoresis was performed on Cellogel® in a Tris–EDTA–acetic acid–boric acid buffer, pH 9.5, according to the procedure of Nicol and O'Gower [Nature **216**, 684 (1967)]. Three major components were found in the ammocete and two in the adult of *Mordacia mordox*. The three bands found in the ammocete were seen at the beginning of metamorphosis and persisted until March. In April both adult and ammocete types of hemoglobins persisted. In May the adult type of hemoglobin predominated in the metamorphasing animals.

PROGRAMMED PROCESSING AND INTERPRETATION OF PROTEIN AND LDH ISOENZYME ELECTROPHORETIC PATTERNS FOR COMPUTER OR MANUAL USE

Pribor, H. C., W. R. Kirkman, and G. Fellows, Am. J. Clin. Pathol. **49**, 67–74 (1968).

Serum proteins and LDH isozymes separated on cellulose acetate (Sepraphore III®) were scanned with a densitometer. The analog signal was then submitted to a digital computer. Upon demand, the pattern could be displayed on a cathode ray tube and individual components marked. The computer compiled a list of pathological conditions associated with a pattern from data accumulated.

ANALYSIS OF THE PROTEINS IN HUMAN SEMINAL PLASMA

Quinlivan, W. L. G., Arch. Biochem. Biophys. **127**, 680–687 (1968).

Cellulose-acetate electrophoresis, immunoelectrophoresis, and disc-gel electrophoresis have been used to analyze the proteins of human seminal plasma. Cellulose-acetate electrophoresis was performed in a Beckman® MicroZone system using a barbital buffer (0.1M, pH 8.6). A potential gradient of 250 volts for 20 minutes at room temperature was used. Immunoelectrophoresis was performed in a Shandon® Universal Electrophoresis Apparatus on glass slides coated with agarose. An Owens modified barbitone buffer, pH 8.6, and a constant 50 mA current equivalent to 20 volts/cm for 4.5 hours were employed. Disc electrophoresis was performed according to the method of Davis [J. Ann. N. Y. Acad. Sci. **121**, 404 (1964)]. Cellulose-acetate electrophoresis demonstrated four protein bands while disc electrophoresis demonstrated 15. Immunoelectrophoresis identified six proteins, three of which were identical to those in human serum.

CHEMICAL STUDIES OF THE SPECIFIC FRAGMENT OF SHIGELLA SONNEI PHASE II

Romanowska, E., and M. Mulczyk, Eur. J. Biochem. **5**, 109–113 (1968).

The lipopolysaccharide fragments of *Shigalla sonnei* phase II were purified by fraction on Sephadex® G-25 and ECTEOLA-cellulose ET-30 after hydrolysis with sulfuric acid. The use of chemical tests as well as electrophoresis on Oxoid strips at pH 5.3 and pH 3.6 to characterize the active fragment isolated was recorded. Electrophoresis was performed at 20 to 25 volts/cm in either pyridine–acetic acid–water (20:30:472, pH 3.6), 0.05M citric acid–0.1M sodium hydrogen phosphate (410:90, pH 2.9) or pyridine–acetic acid–water (2:20:472, pH 3.6), 0.05M citric acid–0.1M sodium hydrogen phosphate (410:90, pH 2.9) or pyridine–acetic acid–water (2:20:228, pH 3.6).

DETECTION OF SERUM ALKALINE PHOSPHATASE ISOENZYMES WITH PHENOLPHTHALEIN MONOPHOSPHATE FOLLOWING CELLULOSE ACETATE ELECTROPHORESIS

Romel, W. C., S. J. La Mancusa, and J. K. DuFrene, Clin. Chem. **14**, 47–57 (1968).

Serum alkaline phosphatase isoenzymes were fractionated on cellulose-acetate membranes using a barbital buffer, pH 8.6, ionic strength 0.025. Phenolphthalein monophosphate, a chromogenic substrate, was used to detect the separated enzymes. Quantitation was achieved by elution and measurement of the absorbancy at 550 mμ.

EFFECTS OF SODIUM CHLORIDE CONCENTRATION ON SOLUBILITY AND ELECTROPHORETIC CHARACTERISTICS OF PROTEIN FROM THE EYE LENS NUCLEUS IN A YELLOWFIN TUNA (THUNNUS ALBACARE) AND IN DESERT WOOD RAT (NEOTOMA EPIDA)

Smith, A., Comp. Biochem. Physiol. **27**, 543–549 (1968).

The proteins of the yellowfin tuna and desert wood rat contained in water or

sodium chloride were fractionated by electrophoresis on Sepraphore III®
(Gelman) in a Beckman® B-1 buffer or Harleco® barbital buffer, pH 8.6, ionic
strength 0.050-0.52. Electrophoresis was performed in the Durrum® cell (Bechman
Model R) at 300 volts for 20 minutes. Thirty-three different salt concentrations
were used to prepare extracts. Results indicated that at 0.018 of 1% saline
no salt effect was produced by albumins and globulins. At higher salt concentrations, fusion of some proteins into single bands was observed.

A QUANTITATIVE METHOD FOR THE ESTIMATION OF TISSUE LACTATE DEHYDROGENASE ISOENZYMES ON CELLULOSE ACETATE

Stagg, B. H., and G. A. Whyley, Clin. Chim. Acta **19**, 139–145 (1968).

Isoenzymes of lactate dehydrogenase of *corpus uteri* and cervix homogenates
have been separated on cellulose acetate strips using 0.06M barbitone buffer,
pH 8.6. Separation was achieved at 4°C by applying a current of 0.4 mA/cm
for 2.5 hours. Losses of the heat-labile enzymes LDH_4 and LDH_5 during electrophoresis were prevented by the addition of human serum protein to the tissue
homogenates.

THE ELECTROPHORETIC INDEX: A PARAMETER OF THYROID FUNCTION

Taylor, K., Australasian Ann. Med. **17**, 49–55 (1968).

The electrophoretic index is the ratio of radioactivity in the α_1- to α_2- globulin
zone following cellulose-acetate electrophoresis of serum. I^{131} radiothyroxene
was diluted in saline containing Bromophenol Blue (1:10 v/v) before being
added to the serum. A final concentration of thyroxine (10 to 12 μg/100
ml of sera) was used. The commercial buffer Oxoid® barbitone acetate buffer,
pH 8.6, 0.05M was used for electrophoresis. The buffer was prepared fresh
every two weeks and stored at 4°C. A current of 2 mA/5 cm strip was run for
two to three hours. Before staining with 0.1% Lissamine Green in 3% sulphosalicylic acid for ten minutes, the strip was dried at 110°C for ten minutes.
The electrophoretic index was calculated as follows:

$$E.I. = \frac{number\ of\ counts\ in\ \alpha\ area - background \times 100}{total\ counts - 2 \times background}$$

The normal range of values was found to vary from 61 to 75%. Sera from
patients in different disease states were tested for their E.I.

THE ISOENZYMES OF LACTATE DEHYDROGENASE IN THE NEPHRON OF THE HEALTHY HUMAN KIDNEY

Thiele, K. G., and H. Mattenheimer, Z. Klin. Chem. **6**, 132–138
(1968).

The lactic dehydrogenase isoenzymes of micro-dissected anatomical and functional units of the nephron of the normal healthy kidney were studied by

cellulose-acetate electrophoresis. Electrophoresis was carried out on cellulose acetate strips at 150 volts (0.5–0.75 mA/2.5 cm) for 90 minutes. A pH 8.6 sodium acetate-sodium diethylbarbiturate-hydrochloride buffer was used. LDH-2 and LDH-5 were low in the glomeruli. In convoluted tubules and medullary rays, LDH-1 was the most active. The most active isoenzymes of the papilla were LDH-3 and LDH-4.

MULTIPLE FORMS OF CALF SERUM ADENOSINE DEAMINASE

Cory, J. G., G. Weinbaum, and R. J. Suhadolink, Arch. Biochem. Biophys. **118,** 428–433 (1967).

Four multiple forms of adenosine deaminase were detected by cellulose-acetate electrophoretic separation of ammonium sulfate fractionation and by DEAE-column chromatography of calf serum. Electrophoretic separation was achieved on cellulose-polyacetate strips in a barbital buffer, pH 8.6, ionic strength 0.075, for 16 hours at 7.2 volts/cm. Two major and two minor deaminase bands were detected.

EVIDENCE OF LACTATE DEHYDROGENASE IN ANAPLASMA MARGINALE

Darri, D. L., U. R. Wallace, and R. T. Dimopollos, J. Bacteriol. **93,** 806–810 (1967).

Extracts of *Anaplasma marginale* were separated on cellulose acetate strips, using Gelman® Sepraphore III as electrophoretic medium to separate the enzymes. A bimodal distribution of LDH activity as well as an altered LDH_1 were observed: one enzyme migrated toward the cathode and one migrated toward the anode. Both appeared to be nicotinamide-adenine dinucleotide dependent.

THE DETERMINATION OF LACTATE DEHYDROGENASE ISOENZYMES USING THE MICROZONE CELLULOSE ACETATE SYSTEM

Fisher, C. L., and J. C. Nixon, Clin. Biochem. **1,** 34–41 (1967).

The Beckman® MicroZone cellulose acetate system was used to separate serum lactate dehydrogenase. A barbital buffer, 0.075M, pH 8.6, was employed, and separation occurred at a constant current of 5 mA for a period of 20 minutes. The separated isozymes were detected by staining with a tetrazalium solution containing L− (+) sodium lactate.

ALTERATIONS IN SERUM ENZYMES AND ISOENZYMES IN VARIOUS SPECIES INDUCED BY EPINEPHRINE

Garbus, J., B. Highman, and P. D. Altmand, Comp. Biochem. Physiol. **22**, 507–516 (1967).

Cellulose-acetate electrophoresis performed in a barbital buffer, pH 8.6, ionic strength 0.052, separated lactic dehydrogenase from serum and tissue of the rat, guinea pig, rabbit and the dog. After injection of epinephrine, it was found that an elevation of lactic dehydrogenase was increased following the administration of epinephrine. The lactic dehydrogenase increase was found to be greatly due to a cardiac type that moved into the serum because of an increase in cellular permeability.

ELECTROPHORESIS OF HUMAN SERUM ISOAMYLASES ON CELLULOSE ACETATE

Solutions **VI**, 5–7 (1967) Publ. by Gelman Instrument Co., Ann Arbor, Michigan.

The separation of serum isoamylases on cellulose acetate (Sepraphore® III) was achieved using a high resolution buffer, pH 8.6, ionic strength 0.05. The samples were electrophoresed for 60 minutes at 360 volts at 5–10°C. The area of the strip corresponding to the albumin, α-region, β-region and γ-globulin were removed by cutting the membrane and placing in 0.85% saline. The saline eluates were then tested for amylase activity using the modified Somogyi procedure. Recovery rates of 95% of original activity were obtained.

FETUS-SPECIFIC SERUM PROTEINS IN SEVERAL MAMMALS AND THEIR RELATION TO HUMAN α-FETOPROTEIN

Gitlin, D., and M. Boseman, Comp. Biochem. Physiol. **21**, 327–336 (1967).

Cellulose-acetate strips were used to separate proteins of fetal and maternal sera from 12 species of mammals. Electrophoresis was performed in a pH 8.6 barbital buffer, ionic strength 0.075, using 35 volts/cm for 20 to 45 minutes at room temperature. In each species the fetal serum contained an α-globulin not present in maternal sera. The α-fetoproteins of several species possessed some of the antigenic determinants found in human α-fetoproteins.

NORMAL VALUES FOR CELLULOSE ACETATE ELECTROPHORESIS OF SPINAL FLUID PATIENTS

Izou, P. C., Am. J. Med. Technol. **33**, 501–503 (1967).

Spinal fluid was taken from 23 patients with nonneurological disorders and concentrated before electrophoresis. Proteins were separated using the Micro-

Zone® system. Fractions were quantitated using an analytical technique. Mean values obtained were: prealbumin, 4.1%; albumin, 62.4%; α-globulin, 5.3%; α_2-globulin, 8.2%; β-globulin, 12.8%; and γ-globulin, 7.2%.

ELECTROPHORESIS OF CEREBROSPINAL FLUID PROTEINS

Kaplan, A., Am. J. Med. Sci. **253**, 549–555 (1967).

The proteins in cerebrospinal fluid were separated by electrophoresis on cellulose acetate using the Beckman® MicroZone system. Proteins were concentrated by ultrafiltration through a collodion sac prior to electrophoresis. A pH 8.6 barbital buffer, 0.075 ionic strength, was used to achieve separation. The most common abnormality detected was an increase in the concentration of γ-globulin accompanied by a normal or elevated total protein.

POSTMORTEM CHEMISTRY OF HUMAN VITREOUS HUMOR

Leahy, M. S., and E. R. Farber, J. Forensic Sci. **12**, 214–222 (1967).

Because of its anatomically isolated position, ocular vitreous humor is the fluid-medium least subject to postmortem changes. Ten ml of fluid were collected postmortem using a 20-gauge needle. Routine chemistry was done on the autoanalyzer. Lactic dehydrogenase isoenzymes were separated by cellulose-acetate electrophoresis by the method of Preston [Am. J. Clin. Pathol. **43**, 256–260 (1965)].

GLYCERALDEHYDE-3-PHOSPHATE DEHYDROGENASE IN PHYLETICALLY DIVERSE ORGANISMS

Lebherz, H. G., and W. J. Rutter, Science **157**, 1198–1200 (1967).

Electrophoresis carried out on cellulose polyacetate Sepraphore III® (Gelman), separated distinct forms of glyceraldehyde-3-phosphate dehydrogenase from turtle, perch, trout, spinach and yeast. Electrophoresis was carried out at 4°C, 250 volts for 90 minutes. A 0.01M Tris, 0.001M EDTA, 0.07% β-mercapto-ethanol (pH 7.5), or a 0.05M Veronal buffer containing 0.07% β-mercapto-ethanol (pH 8.6) was used.

ELECTROPHORETIC SEPARATION OF SOLUBLE RAT LIVER CELL PROTEINS ON GELATINIZED CELLULOSE ACETATE

Le Bouton, A. V., Anal. Biochem. **20**, 550–552 (1967).

Superior separation of rat liver soluble proteins was achieved on a new type of gelatinized cellulose acetate (Cellogel®). Separations were obtained in a barbital buffer, pH 8.6, $\mu = 0.05$. Electrophoresis proceeded at 20 volts/cm for three hours. The cellulose acetate strips were fixed and stained with a 1%

solution of Fast Green FCF. With this technique, 17 bands were reproducibly obtained from soluble rat liver protein homogenates.

HEMOGLOBIN ELECTROPHORESIS

Levitt, E., Am. J. Med. Technol. **33**, 513–514 (1967).

It was possible to routinely determine hemoglobins A, S, and C by cellulose-acetate electrophoresis. Tris–EDTA–borate buffer was used. After spotting, the strips were quickly placed in the electrophoresis chamber to minimize movement of the hemoglobin. Three hundred volts were applied for 1.5 hours. Ponceau S was the stain used, and acid alcohol was used to decolorize after staining. Best interpretation of results was obtained when a control of the standard hemoglobin and the sample were run in combination.

THE ISOENZYMES OF TRYPSIN

Miller, J. M., Jr., A. W. Opher, and J. M. Miller, The Johns Hopkins Medical Journal **121**, 328–332 (1967).

Trypsin was separated into eight components on polyacetate Sepraphore® III. Separation was achieved using a barbital buffer, pH 9.5, at 250 volts for two hours. The major activity as demonstrated by hydrolysis of BAPA was found in the cathodic fraction.

SERUM PROTEIN ALTERATONS IN UREMIA

Morgan, J. M., R. E. Morgan, and G. E. Thomas, Alabama J. Med. Sci. **4**, 68–73 (1967).

Proteins of uremic sera were studied using cellulose-acetate electrophoresis and iodine titration. Electrophoretic separation was achieved on cellulose-acetate strips in a barbital buffer, pH 8.6. Iodine reaction was measured by incubating with iodine in a sodium acetate–acetic acid buffer, pH 5.7, for three hours. The remaining iodine was titrated with standard thiosulfate. An albumin-like fraction from uremic patients was found to bind an increased amount of iodine. An increase in the α_1-globulin was found and thought not to be related to the severity of the illness.

MULTIPLE FORMS OF YEAST PHOSPHOGLUCOSE ISOMERASE: I. RESOLUTION OF THE CRYSTALLINE ENZYME INTO THREE ISOENZYMES

Nakagawa, Y., and E. H. Noltman, J. Biol. Chem. **242**, 4782–4788 (1967).

Crystalline phosphoglucose isomerases were isolated from brewer's and from baker's yeast using the techniques of column chromatography on DEAE-cellulose. Elution of the enzyme was achieved with a gradient starting with 0.02M phosphate→0.07M sodium prosphate (reservoir), pH 6.0. The enzyme was examined by cellulose-acetate electrophoresis in a pH 6.5 phosphate buffer, μ=0.088 at 100 volts constant voltage. In addition, electrophoretic runs were performed at pH 4.7 and 5.2. Three isoenzymes were demonstrated. Multiple bands were demonstrated in the enzyme from three different sources.

HAEMOGLOBIN VARIATION IN ANDARA TRAPEZIA

Nicol, P. I., Nature **216**, 684 (1967).

Three buffer systems were used during Cellogel® strip electrophoresis to demonstrate hemoglobin variations in geographically separated populations of *Andara trapezia,* an arcid dam. The buffers used were: 1) Tris (hydroxyl-methyl) aminomethane—(Tris) ethylenediaminetetracetic acid (EDTA)—boric acid buffer, pH 9.5; 2) Tris—EDTA—boric acid buffer pH 9.2;3) Tris-hydrochloric acid buffer, pH 8.2. The first buffer required 400 volts for complete separation in 1 to 1½ hours; the second and third buffers required electrophoresis for four hours of separation. Amido Black 10B was the stain. The band patterns separated appear to be under the control of one pair of alleles.

IDENTIFICATION OF PROTEIN IN MOZZARELLA CHEESE BY ELECTROPHORESIS ON CELLULOSE ACETATE

Olbonico, F., and P. Resmini, Boll. Lab. Chim. Provinciali **18**, 143–152 (1967).

Cellulose-acetate electrophoresis in a Veronal-acetate continuous urea buffer, pH 8.2, ionic strength 0.035, was used to separate casein extracted from mozzarella cheese. This technique was used to identify the aldurants of the cheeses.

ELECTROPHORETIC STUDIES OF INTERACTIONS BETWEEN PROTEOLYTIC ENZYMES AND INHIBITORS

Osuga, D. T., and R. E. Feeney, Arch. Biochem. Biophys. **118**, 340–346 (1967).

Ovomucoids isolated from different species of fowl by TCA extraction and DEAE-cellulose chromatography were incubated with trypsin and α-chymotrypsin. Their reaction products were characterized by moving-boundary, paper, starch-gel, and polyacrylamide-gel electrophoresis. Horizontal starch-gel electrophoresis and polyacrylamide-gel electrophoresis were carried out at 4°C using the discontinuous buffer system of Poulik at 7 volts/cm and 15 mA or 36 volts/cm and 25/mA. The buffer used consisted of 0.076M Tris, 0.005M citric acid and 2.0M urea, pH 8.6, in the gel. The bridge buffer was 0.3M boric acid and 0.06M sodium hydroxide, pH 8.6. The ovomucoids were selectively stained by immersion of gel in a solution containing 5% α-naphthol dissolved in 95% ethanol and then in concentrated sulfuric acid until a red color developed. A technique for detecting enzymatic inhibitors is described.

IMMUNOLOGIC STUDIES IN NORMAL HUMAN SWEAT

Page, C. O., Jr., and J. S. Remington, J. Lab. Clin. Med. **69**, 634–650 (1967).

The major proteins of human sweat were separated by cellulose-acetate electrophoresis using a barbital buffer, pH 8.6, in a Beckman® MicroZone Model R-101 electrophoresis cell. An α-globulin, albumin and a protein migrating to cathodic γ-globulin were found. Immunoelectrophoretic studies revealed a pre-albumin in addition to the proteins separated above. IgA, IgG, IgD, transferrin, ceruloplasmin, or osmucoid were detected by immunodiffusion studies with specific antisera.

THE CHEMISTRY OF ANTIGENS: XIX. ON THE NUMBER OF ANTIGENS AND THE HOMOGENEITY OF THE ISOLATED ANTIGENS OF FRACTION CB-1A FROM CASTOR BEANS

Spies, J. R., Ann. Allergy **25**, 29–34 (1967).

Cellulose-acetate electrophoresis of castor bean extract CB-1A separated a number of fractions in which eight antigens were detected. Electrophoresis was performed according to conventional techniques [Spies, J. R., and E. J. Coulson, J. Biol. Chem. **239**, 1818 (1964)]. Antigens were detected by gel diffusion analysis of 7 mm disc cut from the cellulose acetate which contained the separated material. The major antigens I and II were highly purified and chemical determinations made on them.

SIMULTANEOUS ESTIMATION OF URINARY VANILMANDELIC ACID AND OF 3-METHOXY-4-HYDROPHENYLGLYCOL BY ELECTROPHORESIS ON BANDS OF CELLULOSE ACETATE

Cristol, P., Ann. Biol. Clin. **24**, 155–164 (1966).

Urine was hydrolyzed and extracted with ethyl acetate, and then vanilmandelic acid was extracted at pH 1.0 and β-methoxy-4-hydrophenylglycol was extracted at pH 9–11. The extracts were further purified by electrophoresis for three to four hours on cellulose acetate bands. Bands were stained with diazotized p-nitroaniline, then diluted with alkaline methanol and read at 520 mμ.

ÉTUDE D'UNE ULTRAMICROMETHOD D'ÉLECTROPHORÈSE SUR ACÉTATE DE CELLULOSE (MICROAONE ÉLECTROPHORÈSE)

Demaret, M., Ann. Biol. Clin. **24**, 369–382 (1966) **French.**

Experiments with a method of electrophoresis on cellulose acetate using the Beckman-Spinco® MicroZone system are presented. The technique allows the simultaneous separation of eight sera in 20 minutes. Quantities of serum as small as 25 μl were separated. A Veronal acetate buffer, pH 8.6, ionic strength 0.075, was used. Separation proceeded at 250 volts for 20 minutes. Quantitative elution of proteins stained with Ponceau S was possible.

USE OF THE GEL FILTRATION IN THE STUDY OF URINARY PROTEINS AND ENZYMES

Dioguardi, N., G. Fiorelli, G. Ideo, and G. Emanuelli, Farmaco Ed. Prat. **21**, 334–346 (1966).

Urine (200 ml) was concentrated 10 times by ultrafiltration and was purified on a G-25 or G-50 Sephadex® column with 0.01M Tris buffer, pH 7.5. Fractions adsorbing at 2537 A were electrophoresed on Cellogel® (Veronal buffer, ionic strength 0.1, pH 8.5, 6 volts/cm for 150 minutes). Three protein fractions, two alkaline phosphatases, and three lactic dehydrogenases were found in normal urine. Urine from renal diseases and β- and γ-myeloma were also analyzed.

LACTATE DEHYDROGENASE (LDH) GLUTAMIC OXALACETIC TRANSAMINASE (GOT) AND MALATE DEHYDROGENASE (MDH) ISOENZYMES IN LYMPHOCYTES FROM FOETAL AND ADULT THYMUS, SPLEEN AND FROM PERIPHERAL BLOOD

Dioguardi, N., G. Ideo, P. M. Mannucci, and G. Fiorelli, Enzym. Biol. Clin **6**, 324–338 (1966).

LDH, GOT, and MDH of lymphocytes from fetal and adult thymus, spleen, and peripheral blood were studied by cellulose-acetate electrophoresis and chromatography on DEAE-Sephadex®. Electrophoresis was carried out on isolated lymphocytes by the procedure of Barnett [Biochem. J. **84**, (1962)] on cellulose-acetate strips. Fetal spleen and thymus LDH exhibited an undifferentiated LDH pattern. GOT and MDH were characterized by cytoplasmic-type enzymes. Differences in isozyme pattern of MDH and GOT were found in the lymphocytes from adut thymus, spleen, and small circulating lymphocytes.

CELLULOSE ACETATE ELECTROPHORESIS OF THE SERUM PROTEINS OF SHEEP: STUDY OF DEVELOPMENTAL CHANGES IN YOUNG LAMBS

Dobson, C., Australian J. Esp. Biol. Med. Sci. **44**, 475–479 (1966).

The total protein was determined by the biuret method in serum taken from 28 nine-week-old Merino X lambs. The experiments continued for ten weeks. The proteins were separated by electrophoresis on cellulose-acetate strips in 0.7M barbiturate buffer, pH 8.6. In ten weeks there was a 17% increase in total serum protein, with the albumin fractions remaining constant. Gammaglobulins increased by 40% of the original value.

DISTINCTION BETWEEN THE VARIOUS SEXUAL PHENOTYPES AND GENOTYPES IN THE RAT BY IMMUNOELECTROPHORETIC ANALYSIS OF SPECIFIC URINARY PROTEINS

Dufour, D., Experientia **22**, 28–29 (1966) **French.**

Cellulose-acetate electrophoresis in phosphate buffer, pH 7.0, of rat urine (65 mg protein/ml) was used to study possible differences between the sexes and to study genetic purity within a sexual group. The method was useful in visually demonstrating differences in urine protein migration between groups.

RESOLUTION OF THE RABBIT IMMUNOGLOBULINS INTO MULTIPLE COMPONENTS BY ELECTROPHORESIS ON CELLULOSE ACETATE

Flechner, I., and A. L. Olitzki, Life Sci. **5,** 495–499 (1966).

The β_{-2} and γ-globulins of rabbit sera were resolved into a minimum of five components by electrophoresis on cellulose acetate. Cellulose-acetate strips 30 x 5 cm, run in a Tris–EDTA–boric acid buffer, pH 8.9, at 6 volts/cm, were used for separation. Strips were stained directly with 0.05 g% aqueous nigrosin solution containing 5% TCA and 5% sulfosalicylic acid. Immunoelectrophoresis on the separated proteins was carried out using the transfer technique. All five components were shown to possess some common antigenic determinants.

STAINING OF PROTEINS ON CELLULOSE ACETATE

Gadd, K. G., Nature **212,** 628 (1966).

Dimethyl sulphoxide was used as a solvent for quantitative elution of stained proteins from cellulose-acetate strips. After electrophoresis, the strips were stained for ten minutes in 0.3% Fast Green F.C.F. in 3% aqueous trichloroacetic acid. The strip was washed six to eight times in 2% aqueous acetic acid to remove the background. It was then dried over low heat and the stained bands and a blank were cut and put into test tubes. Five ml of dimethyl sulphoxide was added and the tubes were shaken for five minutes at room temperature. The protein-dye complex and cellulose acetate dissolved in the solvent. The tubes were read at 630 mm.

A RAPID AND DIRECT METHOD FOR THE MICRODETERMINATION OF URINARY AMV AND MHPG ON CELLULOSE ACETATE MEMBRANE

Ganon, J. P., Ann. Biol. Clin. Paris **24**, 1185–1189 (1966).

Urinary vanilmandelic acid (VMA) and methoxy-3-hydroxy-4-phenylglycol (MHPG) can be separated by electrophoresis on cellulose acetate. The VMA is detected with 0.1 ml of urine, and the MHPG is detected with 0.2 ml of hydrolyzed urine. Diazotized p-nitroaniline stained the separated compounds, and quantitative measurements were made at the 520 mμ level.

A FAMILIAR NEUROPATHY ASSOCIATED WITH A PARAPROTEIN IN THE SERUM, CEREBROSPINAL FLUID, AND URINE

Gibberd, F. B., Neurology **16**, 130–134 (1966).

An electrophoretically-homogenous, abnormal protein was found in a mother and son with chronic peripheral neuropathy. Cellulose-acetate electrophoresis [Clin. Chim. Acta **3**, 450 (1958)] was used to identify the abnormal proteins in the urine, CSF, and serum of the affected family members. The abnormal fraction migrated between the beta- and gammaglobulins; it constituted 3.7-11.6% of the total serum protein. The abnormal protein is related to the immunoglobulins but has a different antigenic specificity.

SERUM PROTEIN AND GLYCOPROTEIN–PATTERNS IN "SPONTANEOUS" MOUSE AMYLOIDOSIS

Gray, G. R., Arch. Pathol. **81**, 129–135 (1966).

The serum proteins and glycoproteins in a strain of mice showing a high incidence of amyloidosis with aging were examined. Cellulose-acetate electrophoresis was performed in a pH 8.6 barbital buffer, ionic strength 0.08, at a constant potential of 200 volts for two hours. Four μl of serum were used for total protein analysis. The stain used was 0.2% Ponceau-S in 3% trichloroacetic acid. Eight μl of serum were required for glycoprotein determination by the PAS technique. The development of amyloidosis was not associated with a significant increase in γ-globulin or serum protein-bound hexoses.

LOW MOLECULAR WEIGHT URINE PROTEIN INVESTIGATED BY GEL FILTRATION

Harrison, J. F., and B. E. Northam, Clin. Chim. Acta **14**, 679–688 (1966).

Urine was concentrated by ultrafiltration before quantitation of proteins lower than albumin in molecular weight. Cellulose-acetate and paper electrophoresis and gel filtration techniques were combined to characterize the molecular weight and mobility of these proteins. The excretion of low molecular weight proteins is increased in cases of renal tubular disease.

PARAPROTEINS IN FOLIC ACID DEFICIENCY

Hobbs, J. R., and A. V. Hoffenbraud, Lancet **1**, 714 (1966).

An examination of 62 patients with folic acid deficiency for paraprotein is reported. The serum of the 62 patients was separated by cellulose-acetate electrophoresis. The serum was electrophoresed within four hours of collection to minimize artifacts. In a patient with paraproteinemia and folate or vitamin B-12 deficiency, myeloma or macroglobulinemia was the usual diagnosis. A small narrow band with γ_2-mobility was found. This band was typed as having heavy chains and κ-light chains only by immunoelectrophoresis.

ELECTROPHORETIC STUDY OF PROTEIN IN THE ARTIFICIALLY-PROVOKED MAMMARY SECRETION IN VIRGIN BUFFALO HEIFERS

Intrieri, F., Acta Med. Vet. **12**, 413–425 (1966).

Cellulose-acetate electrophoresis with a sodium barbiturate buffer, pH 8.6, ionic strength 0.07, was used to examine the hormone-induced mammary secretions in virgin heifers. Amido Black stained a fast-moving component (casein and globulin) and a slow-moving component (α-lactalbumin and β-lactoglobulin). After induction, the secretion had the protein ratio of colostrum; after five months of continuous stimulation, the secretion was similar to normal cow's milk.

CONCENTRATION OF CEREBROSPINAL FLUID PROTEINS AND THEIR FRACTIONATION BY CELLULOSE ACETATE ELECTROPHORESIS

Kaplan, A., and M. Johnstone, Clin. Chem. **12**, 717–727 (1966).

Cellulose-acetate electrophoresis, using cellulose-acetate membranes and the MicroZone®system, separated cerebrospinal fluid. A barbital buffer, pH 8.6 and ionic strength 0.075, was used in the electrophoretic runs. Vacuum ultra-filtration through a collodion sac was found to be the most efficient means of pre-concentrating proteins.

THE MEASUREMENT OF HUMAN THYROXINE-BINDING PROTEINS BY ELECTROPHORESIS ON CELLULOSE ACETATE MEMBRANES: STUDIES ON A SPURIOUS FOURTH THYROXINE-BINDING COMPONENT

Launay, M. P., Canad. J. Biochem. **44**, 1657–1667 (1966).

Cellulose-acetate electrophoresis separated a poorly defined component be-tween the α_2 and α globulin when serum to which I^{131}-L-thyroxine was electro-phoresed. Electrophoresis was carried out at a constant voltage of 150 at room temperature for 4.5 hours. A Tris-EDTA-borate buffer, pH 9.0, was used. This component was found in all sera in which the affinity of thyroxine-binding prealbumin was less than 30%.

ELECTROPHORETIC DETERMINATION.OF HEMOGLOBIN AND MYOGLOBIN USING CELLULOSE ACETATE

Leyko, W., Abhandl. Deut. Akad. Wiss. Berlin, Kl. Med. **1966,** 191–195.

Hemoglobin, myoglobin and cytochrome C from cattle heart were separated by cellulose-acetate electrophoresis in a pH 8.6 Veronal buffer (ionic strength 0.05, 150 volts for two hours). Quantitation was performed by densitometric measurements of the Amido Black stained fractions. This method is useful for checking the purity of myoglobin and hemoglobin preparations.

THE OCCURRENCE OF ABNORMAL SERUM PROTEINS IN PATIENTS WITH EPITHELIAL NEOPLASMS

Lynch, W. J., J. Clin. Pathol. **19,** 461–463 (1966).

Serum proteins were separated by cellulose-acetate electrophoresis in a barbitone buffer [Brackenridge, Anal. Chem. **32,** 1357 (1960)]. Immunoelectrophoresis was performed in 1% agar using a barbitone buffer, pH 8.4, by the method of Scheidegger [Intern. Arch. Allergy Appl. Immunol. **7,** 103 (1955)]. Theoretical aspects of paraproteinemia in carcinoma patients are discussed.

UNSELECTIVE PROTEINURIA IN ACUTE ISCHAMIC RENAL FAILURE

MacLean, P. R., Clin. Sci. **30,** 91–102 (1966).

The urine of six patients with acute ischamic failure was studied by electrophoresis and immunoelectrophoresis. Before use, the urine was concentrated 10X to a concentration of 1-4 protein/100 ml by overnight dialysis against polyethylene glycol. Cellulose-acetate electrophoresis was performed in 0.033M barbitone buffer, pH 8.6 for 90 minutes at 2.5 mA/strip. The stain was 0.2% Ponceau S in 6% sulphosalicylic acid. Immunoelectrophoresis was carried out in a 0.10M barbitone buffer, pH 8.6, for 2.5 hours at 7.5 mA/slide. The slides were incubated for 48 hours in antihuman serum to develop precipitin lines. A high proportion of α_2- and β-globulins and high molecular weight proteins were found. All proteins visible by immunoelectrophoresis of serum could be identified in the urine.

THE USE OF CELLULOSE ACETATE FOR THE ELECTROPHORETIC SEPA-RATION AND QUANTITATION OF SERUM LACTIC DEHYDROGENASE ISOZYMES IN NORMAL AND PATHOLOGICAL STATES

Mager, M., W. F. Blatt, and W. H. Abelmann, Clin. Chim. Acta **14,** 689–697 (1966).

A technique for rapid separation and visualization of LDH isoenzymes is described. This method, which is useful for routine laboratory diagnosis, calls for the application of 6X serum to a cellulose-acetate strip equilibrated with Michaelis buffer, pH 8.6, ionic strength 0.05. This was also the electrolyte used in electrophoresis. A current of 3.0 mA per strip separated the isoenzymes in 75 minutes. Isoenzyme bands were identified by a substrate-staining mixture of 4 mg/ml p-iodonitro tetrazolium violet. The color was fixed in acetic acid, washed and dried; densitometry measurements were made at 485 mμ. Mean distribution patterns of disease states from serum, liver and heart are diagramed.

THE SUBUNIT POLYPEPTIDES OF HUMAN FIBRINOGEN

McKee, P. A., L. A. Rogers, E. Marler, and R. L. Hill, Arch. Biochem. Biophys. **116,** 271–279 (1966).

Three polypeptide chains isolated from S-sulfofibrinogen by chromatography on carboxymethylcellulose were examined by ultracentrifugation and cellu-lose-acetate electrophoresis. Electrophoresis was performed on gelatinized cellulose-acetate strips (Cellogel®) in a 0.2M sodium acetate buffer, pH 5.6, for three hours at 6 volts/cm. The molecular weight of the A-chain was found to be 47,000, the β-chain was 56,000, and the C-chain was 63,000. Each chain was found to correspond to one of the three electrophoretic components ob-served in the unfractionated material upon electrophoresis. Amino acid analysis of tryptic digests of the chains revealed that the amino acid content was unique.

ALTERATIONS IN SERUM PROTEINS FOLLOWING DELIPIDIZATION BY 2-OCTANOL

Miller, D. F., and L. L. Walter, Yale J. Biol. and Med. **38,** 410–416 (1966).

Serum proteins freed of lipids by treating serum with 2-octanol were electro-phoresed using the Beckman MicroZone system (Beckman Manuel RM-1-M-2 Aug., 1963). Results obtained indicated that the dye uptake in the alpha fraction was decreased and the beta globulin peak was observed to trail and smudge. Mucoproteins were not changed. The immunogenic reactivity of the lipid extracted proteins was not altered.

SOME ELECTROPHORETIC AND IMMUNOLOGICAL PROPERTIES OF HUMAN SERUM

Mischler, T. W., and E. P. Reineke, J. Reprod. Fertility **12**, 125–129 (1966).

Cellulose-acetate electrophoresis demonstrated the different proteins in human serum, semen, seminal plasma, and thrice-washed spermatozoa. Barbitone buffer with calcium lactate, pH 8.6, was the electrolyte; a current of 1 mA/strip for 1½ hours at room temperature was applied. The method of Ornstein and Davis was used for disc electrophoresis. In order to speed photopolymerization, N, N, N¹, N¹-tetramethylenediamine (1 μl/ml) was added to the large pore gel. Two hundred μg of sample protein was added to each gel. Using this technique it was difficult to distinguish differences in the patterns of human semen and washed human spermatozoa. At least ten fractions were produced. Using immunoelectrophoresis, ten antigenic components were identified in human semen, five of which were specific to serum.

SERUM PROTEINS IN URINE: AN EXAMINATION OF THE EFFECTS OF SOME METHODS USED TO CONCENTRATE THE URINE

Miyasato, F., and V. E. Pollak, J. Lab. Clin. Med. **67**, 1036–1043 (1966).

Osmotic concentration against polyvinyl pyrrolidine, preevaporation, and ultra-filtration plus lyophilization were the concentration methods compared for recovery of total serum proteins in urine. Ultrafiltration plus lyophilization gave the best reproducibility. Cellulose-acetate electrophoresis in barbital buffer, pH 8.6, ionic strength 0.05, separated the proteins in two hours using 1 mA per strip. The strips were stained with Ponceau S, eluted with chloroform–ethanol (90:10 v/v) and read at 520 mμ.

A RAPID MICROTECHNIC FOR HEMOGLOBIN ELECTROPHORESIS

Owens, J. B., A. P. Miller, W. G. Brown, and J. A. Stool, Am. J. Clin. Pathol. **46**, 144–147 (1966).

Cellulose-acetate electrophoresis separated hemoglobins A and S using a hemoglobin preparation from only 0.2 ml of blood. The RBC of 0.2 ml of fresh blood were washed three times in saline and then lyzed with 0.2 ml distilled water. The sample was 0.25 ml of this mixture. A pH 8.6 barbital buffer, ionic strength 0.0125, was the electrolyte. Electrophoresis was done for 30 minutes with 250 volts. Ponceau S was used as the stain.

CHEMICAL STUDIES ON INTERSTITIAL FLUID IN THE HARD TISSUES

Paunio, K., J. Dental Res. **45**, 630–633 (1966).

Interstitial fluid extracted from pig dentin and cartilage using centrifugal force were examined by electrophoresis on cellulose acetate in a barbiturate buffer, pH 8.65, ionic strength 0.125, at 110 volts. Mucropolysaccharides isolated after hydrolysis with papain for 24 hours at 65°C were separated by this procedure and detected by staining with Alcian Blue. Proteins were detected by staining with 0.005% nigrosine. Starch-gel electrophoretic comparison of interstitial fluid and serum was carried out in the discontinuous buffer system of Poulik. Interstitial fluid from dentin and cartilage was found to differ greatly from serum. The tissue fluid was found to contain a soluble collagen not found in serum.

MULTIPLE FORMS OF FRUCTOSE DIPHOSPHATE ALDOLASE IN MAMMALIAN TISSUES

Penhoet, E., T. Rajkumar, and W. J. Rutter, Proc. Nat. Acad. Sci. U.S. **56**, 1275–1282 (1966).

A new fructose diphosphate aldolase (aldolase C) was isolated from rabbit brain by the use of techniques developed by Rajkumar, Woodfin and Rutter (Methods of Enzymology V9). The isolated enzyme was characterized by catalytic studies, electrophoretic and chromatographic studies as well as immunochemical techniques. Zone electrophoresis on cellulose-acetate strips in a 0.06M barbital buffer, pH 8.6, 250 volts for 90 minutes, was used to evaluate the electrophoretic characteristics. Brain tissues had five isoenzymes; kidney and liver, four isozymes; spleen, heart, and muscle had one. Cross-reactivity between the species was not present.

SPECIFIC ACTIVITY OF TRYPSIN

Peterson, R. C., J. Pharm. Sci. **55**, 49–52 (1966).

Trypsin was fractionated into four bands by horizontal electrophoresis on cellulose-acetate strips using a formic acid–pyridine–water buffer (15:2.5:928.5), pH 2.65. The strips were prerun for 15 minutes with reversed poles. Electrophoretic runs were carried out at 7.8 volts/cm for three hours at room temperature. Specific activity of trypsin zone, which was lactated with protein stain, was assayed with BAPA chromogenic assay.This was found to be a reproducible method to assay commercial preparations of trypsin.

ELECTROPHORETIC AND IMMUNOELECTROPHORETIC CHARACTERIZATION OF NORMAL MINK SERUM PROTEINS

Porter, D. D., and F. J. Dixon, Am. J. Vet. Res. **27**, 335–338 (1966).

Serum electrophoretic proteins of mink (Mustela _vison) were characterized by electrophoresis on paper, cellulose acetate, and starch gel. Standard procedures for separation were employed. Four protein bands were identified by paper electrophoresis, 5 by cellulose-acetate electrophoresis, and 25 by immunoelectrophoresis on agar. Cross-reactivity between mink and human serum proteins was demonstrated.

HEMOGLOBIN ELECTROPHORESIS ON CELLULOSE ACETATE

Rosenbaum, D. L., Am. J. Med. Sci. **252**, 726–731 (1966).

Cellulose-acetate strips 1 x 6¾ inches were used to separate hemoglobin of erythrocyte hemolysates. A Tris–EDTA–borate buffer, pH 8.8 was used, and a voltage of 150 was applied for 2 to 2.5 hours. The separated proteins were stained with Ponceau S and quantitated with the Beckman® Analytrol. Hemoglobin A_2 was separated from S, but not clearly from C. Hemoglobin F was discernable though faintly stained.

DIE ISOENZYME DER LACTAT-DEHYDROGENASE IN THE RATTEN-NIERE

Thiele, K. G., and H. Mattenheimer, Z. Klin. Chem. **4**, 232–233 (1966) **German.**

Lactic dehydrogenase isozymes of cortex, inner and outer zone of the medulla, and papilla of the rat kidney were separated by cellulose-acetate electrophoresis. Separation was achieved in a pH 8.9 barbital buffer. Tissue homogenates were prepared in pH 8.6 barbital buffer. LDH-1 predominates in the cortex, LDH-5 in the papilla. Dissected glomeruli containing all five isozymes were lost in wet preparations of glomeruli.

STUDIES OF INTERSPECIFIC (RAT x MOUSE) SOMATIC HYBRIDS: II LACTATE DEHYDROGENASE AND β-GLUCURONIDASE

Weiss, M. C., N. Christoff, and S. Kochwa, Genetics **54**, 1111–1122 (1966).

Interspecific molecular hybridization could be demonstrated in vivo using cellulose-acetate electrophoresis. The strips were presoaked in the electrophoretic buffer (0.3M boric acid–NaOH, pH 8.6). Four to six μl of the extracts at potential of 200 volts were applied for one hour. The stain was prepared and filtered just prior to use. It consisted of: 0.025M Tris buffer, pH 7.4; O.1M d & L lactate Li salt; 0.005M KCN; 0.001M DPN; 0.05 mg/ml Nitro Blue Tetrazolium; and 20 μg/ml phenazine methosulfate. The strips were incubated for ten minutes in the dark and fixed (50% methanol, 10% acetic acid, and 40% water) for ten minutes and dried.

SERUM PROTEIN STUDIES IN MYOCARDIAL INFARCTION

Woodford-Williams, E., D. Webster, and B. Landless, Gerontol. Clin. **8**, 44–52 (1966).

In seven of ten cases of myocardial infarctions there was a decrease in total serum protein for one week following the infarction. Cellulose-acetate electrophoretic patterns showed increased globulin/albumin ratios of serum in all ten cases. A conversion of albumin into α-globulin following tissue damage was suggested on the basis of albumin decrease and α-globulin increase curves. Total serum protein returned to the original level rapidly.

CATIONIC PROTEINS OF POLYMORPHONUCLEAR LEUKOCYTE LYSOSOMES

Zeya, H. I., and J. K. Spitznagel, J. Bacteriol. **91**, 750–754 (1966).

Three bands of anti-bacterial cationic proteins were isolated following cellulose-acetate electrophoresis of a PMN lysosomal fraction. The cationic proteins were not lysozymes. Prior to electrophoresis, Oxoid® strips were soaked in 0.5% cetyltrimethyl ammonium bromide for three minutes and then washed for four minutes in acetate buffer, pH 4.0. The sample of the granule suspension was in 0.25M sucrose for application. Electrophoresis was done for one hour at room temperature using a 200 volt current. Acetate buffer, pH 4.0, ionic strength 0.05, was the electrolyte. A test strip was stained with Amido Black, and the remainder of the sample was eluted by dialysis with 0.5 ml of 0.01N HCl. Anti-bacterial activity was assayed by the method of Hirsch [J. Exp. Med. **108**, 924 (1958)].

CATONIC PROTEINS OF PMN LEUKOCYTE LYSOSOMES: II. COMPOSITION, PROPERTIES AND MECHANISM OF ANTIBACTERIAL ACTION

Zeya, H. I., and J. K. Spitznagel, J. Bacteriol. **91**, 755–762 (1966).

Cellulose-acetate paper electrophoresis was used to further characterize three anti-bacterial cationic proteins. Ethanol fractionation of the crude lyosome fraction demonstrated that the cationic proteins were free of contaminating lysozymes.

ELECTROPHORETIC PRECIPITIN TEST ON CELLULOSE ACETATE

Zydeck, F. A., E. E. Muirhead, and H. Schneider, Am. J. Clin. Pathol. **45**, 323–328 (1966).

Antiserum was separated electrophoretically on cellulose acetate followed by immediate application and migration of the appropriate antigen through the γ-globulin. Termed the electrophoretic precipitin test, this method is useful for detecting the homo- or heterogeneity of an antigen. A barbital buffer, ionic strength 0.05, pH 8.6, was used as the electrolyte in a current of 250 volts 5–7 mA for 30 minutes following applications of 5λ antiserum. The current was shut off and approximately 5λ of antigen was behind the point of antiserum application; the current was then run for 20 minutes. The strip was rinsed in water, then stained with Ponceau S.

ANALISI QUANTITATIVA RAPIDA DELLA FRAZIONE EMOGLOBINICA A₂ MEDIANTE ELETTROFORESI SU ACETATO DI CELLULOSA

Aicardi, G., Minerva Pediat. **17,** 1191–1195 (1965) **Italian.**

Cellulose-acetate (Oxoid) strips were used to separate hemoglobin A_2 of erythrocyte homogenates. A Veronal-sodium Veronal buffer, pH 8.6, ionic strength 0.05, was used. A potential of 160 volts was applied for 120 minutes. The A_2 band was easily visualized by staining with Amido Black. It migrated just behind $A_1 + A_3$ toward the cathode.

TISSUE LACTIC DEHYDROGENASE ISOZYMES: VARIATION IN RATS DURING PROLONGED COLD EXPOSURE

Blatt, W. F., J. Walker, and M. Mager, Am. J. Physiol. **209,** 785–789 (1965).

The lactic dehydrogenase activity of heart, liver, lungs and kidneys of rats exposed to $5°C \pm 2$ room temperature for various periods of time was examined by electrophoresis on cellulose acetate. A barbital buffer, pH 8.2, $\mu0.05$, was used. Separation was achieved at 2.5 mA per strip for 70 minutes. A depression LDH activity was noted in the heart and liver following one to two weeks exposure at 5°C.

RAPID QUALITATIVE AND QUANTITATIVE HEMOGLOBIN FRACTIONATION

Briere, R. O., T. Golias, and J. G. Batsakis, Am. J. Clin. Pathol. **44,** 695–701 (1965).

Cellulose-acetate electrophoresis gave distinct separation of hemoglobins A_3, A_1, F, A_2 and two nonhemoglobin fractions. A pH 8.8 Tris–EDTA–borate buffer (16.5 g Tris, 1.56 g Na_2 EDTA, 0.92 g boric acid to 1 liter with water) was used. Hemolysates were prepared by the method of Chernoff [New Eng. J. Med. **253**, 322, 365, 416 (1955)]. The sample was 0.75-1.0 λ of hemolysate. Electrophoresis was run for 75 minutes at 400 volts; 0.2% Ponceau S was used as a stain for five minutes, followed by decolorizing in 5% acetic acid. Quantitation was done by a reflectance densitometer with a 0.5-mm slit opening using a blue filter.

ACID MUCOPOLYSACCHARIDES IN THE RASK-NEILSEN TRANSPLANT-ABLE MOUSE MASTOCYTOMA

Brunish, R., Acta. Pathol. Microbiol. Scand. **65,** 185–191 (1965).

Mastocytoma portions, 103 mg in thickness, were frozen and then extracted by grinding with 0.12 ml phosphate buffer, pH 6.0. After centrifugation, five μl of the extract was placed on a cellulose-acetate strip for electrophoresis in 0.5M lithium acetate. Acid mucopolysaccharides were stained with Alcian Blue in 50% ethanol and buffered at pH 3.0.

IDENTIFICATION AND DETERMINATION OF MILK SERUM PROTEIN FRACTIONS FROM THE COW, SHEEP, AND GOAT

D'Amore, G., F. Licastro, and A. Gullotta, Atti. Soc. Petoritana Sci. Fis. Mat. Nat. **11,** 37–47 (1965).

Cellulose-acetate strips were spotted with milk serum for electrophoresis in a pH 8.6 Veronal buffer, ionic strength 0.08. Electrophoresis was run for two hours at 25°C, 7.3 volts/cm. The strips were stained with Amido Black 10B and densitometric quantitations were made. Differences in cow, sheep and goat milk were noted.

MULTIPLE MOLEKULARFORMEN DER LACTATEDEHYDROGENASE NORMALER UND LEUKÄMISCHER MENSCHLICHER LYMPHOZYTEN

Dioguardi, N., and A. Agostoni, Enzymol. Biol. Clin. **5,** 3–13 (1965) **German.**

Column chromatography on DEAE-Sephadex® and electrophoresis on Cellogel® strips were used to study lactic dehydrogenase isozymes of normal and leukemic leukocytes (lymphocytes). Electrophoretic separation was achieved on Cellogel® strips in a pH 8.6 Veronal buffer, 0.005M, at 7 volts/cm for 45 minutes. Chromatography on DEAE-Sephadex® was achieved by elution with a concentration gradient of Tris/HCl (1M, pH 7.5). Five isozymes were detected in both normal and leukemic lymphocytes. LDH-2 predominates in normal lymphocytes. Mature and small leukemic lymphocytes abound in LDH-3.

SERUM PROTEIN CHANGES ASSOCIATED WITH OESOPHAGOSTOMIUM COLUMBIANUM INFECTIONS IN SHEEP

Dobson, C., Nature **207,** 1304–1305 (1965).

Cellulose-acetate electrophoresis characterized protein changes occurring in sheep serum following infection with the helminth O. *columbianum.* Electrophoresis was achieved in a pH 8.6 barbiturate buffer, 0.07M at 5 mA/cm for 1.5 hours. After the first infection, total protein was found to decline, and the

β_2- and γ-globulins increased. The albumin was almost entirely lost at the end of the second infection.

DIFFERÉNCE DE SPÉCIFICITÉ IMMUNOLOGIQUE ENTRE LA PHOSPHATASE URINAIRE HUMAINE MALE ET FEMELLE

Dufour, D., A. Tremblay, and S. Lemieux, Rev. Franc. d'Ètudes Clin. Biol. **11,** 89–92 (1965) **French.**

Differences in the urinary phosphatase of men and women were detected by cellulose-acetate electrophoresis (phosphate buffer, pH 7.0). Horse antihuman male serum was the antibody used to form precipitin lines in immunoelectrophoresis [Grant, J. Clin. Pathol. **10,** 360 (1967)]. There were differences in the antigenicity of male and female urinary prosphatase. It was suggested that measurement of urinary phosphatase activity be a test for prostate metabolic activity.

LACTIC DEHYDROGENASE ISOENZYMES IN MYOCARDIAL INFARCTION

Freeman, I., and A. U. Opher, Am. J. Med. Sci. **250,** 131–136 (1965).

Electrophoresis on cellulose-acetate strips (Sepraphore® III, Gelman) separated isoenzymes of lactic dehydrogenase of serum obtained from patients with myocardial infarcts. Of patients who exhibited vague symptoms and no definite clinical manifestations of infarcts, 73% showed elevated isozyme patterns.

LYTIC ENZYMES OF SPORANGIUM SP: A COMPARISON OF SOME PHYSICAL PROPERTIES OF THE α- AND β-LYTIC PROTEASES

Jurásěk, L., and D. R. Whitaker, Can. J. Biochem. **43,** 1955–1960 (1965).

Cellulose-acetate and starch-gel electrophoretic studies were performed on lytic enzymes isolated from *Sporangium* culture filtrates. Electrophoresis at pH's 3–9 indicated that there was no heterogeneity. When the β enzyme was electrophoresed in acetate buffer after exposure to 7M urea, different results were obtained. Short exposure gave one component while exposure of 24 hours or longer showed two components.

MICROÉLECTROPHORÈSE DES PROTÉINS DU SÉRUM SUR ACÉTATE DE CELLULOSE GELATINEAUX

Paget, M., and P. Coustenoble, Ann. Biol. Clin. (Paris) **23,** 1209–1219 (1965) **French.**

Standardized conditions are presented for rapidly separating serum proteins on gelatinous cellulose. Best separation was achieved on 2.5 x 17 cm strips of gelatinous cellulose acetate in a pH 8.6 Veronal buffer, ionic strength 0.1. Quantitation after staining with Amido Black, Lissamine Green and Sulfo

Green was possible. It was possible to separate 2.5 to 4 μl serum at 10 amperes.

RAPID SEPARATION OF LACTATE DEHYDROGENASE

Preston, J. A., R. O. Briere, and J. G. Batsakis, Am. J. Clin. Pathol. **43**, 256–260 (1965).

Rapid separation of lactate dehydrogenase isozymes of human serum was achieved on cellulose acetate (Sepraphore® III, Gelman) using a 0.1M phosphate buffer (pH 7.5). Electrophoresis of a 5 μl sample was completed in 90 minutes when 200 volts were used.

ELECTROPHORESIS OF UNCONCENTRATED CEREBROSPINAL FLUID USING CELLULOSE ACETATE STRIPS AND THE DYE NIGROSIN

Rice, J. D., and B. Bleakney, Clin. Chim. Acta **12**, 343–348 (1965).

Cellulose-acetate strips (2.5 or 5.0 cm width) were streaked with 20 μl of unconcentrated CSF applied 3 mm from the cathode. Electrophoresis was done in barbital–sodium barbital buffer, pH 8.6, ionic strength 0.05, for 90 minutes with a current of 0.4 mA/cm of strip width. The strips were dried in hot air and stained overnight in a 0.0025% (v/v) solution of Nigrosin in 2% acetic acid (dilutions from a stock concentration of 2.5%). Four protein fractions were identified, with results comparable to older techniques requiring concentration of the fluid.

ELECTROPHORETIC STUDIES OF SOME PLANORBID EGG PROTEINS

Wright, C. A., and G. C. Ross, Bull. World Health Organ. **32**, 709–712 (1965).

Cellulose-acetate electrophoresis, carried out on strips 1 x 3.9 inches, characterized proteins of planorbid small eggs. Separations were carried out in Sorenson's glycine buffer (0.022M, pH 11.78). A potential of 120 volts (0.625 mA current strength) was applied for five hours. The analysis of eggs from 80 populations indicated that protein separations were indicative of variations in populations. Protein patterns varied widely from one species to another. The use of this kind of analysis for toxanomic purposes was discussed.

DIFFERENTIATION OF MYOLOGLOBINURIA FROM HEMOGLOBINURIA HEMOGLOBIN

Brodine, C. E., Precursors Metab. **1964**, 90–93.

Eighty per cent ammonium saturation of urine will precipitate hemoglobin while myoglobin will stay in solution. Electrophoresis of the supernate was done on cellulose-acetate strips. A benzidine-water solution was the stain. Myoglobin migrates to the cathode faster than does hemoglobin.

ELECTROPHORESIS OF SERUM PROTEINS ON CELLULOSE ACETATE

Ritts, R. E., and F. W. Ondrick, Am. J. Clin. Pathol. **41**, 321–331 (1964).

Cellulose-acetate electrophoresis of serum proteins using a pH 8.6 Veronal buffer, ionic strength 0.05, is described. The Shandon Universal Cell was used. The run was carried out at 8 mA per strip for two hours. Proteins were detected by staining with Ponceau S; glycoproteins werre stained with basic fuchsin; and liproteins were stained with ozone-Schiff reagent. It was possible to scan the proteins with the Analytrol® without clearing at 600 mμ.

Column Electrophoresis

ISOLATION OF IMMUNOGLOBULIN A (IgA) FROM HUMAN COLOSTRUM

Axelsson, H., B. G. Johansson, and L. Rymo, Acta Chem. Scand. **20**, 2399–2438 (1966).

The IgA in defatted dialyzed human colostrum was isolated by column electrophoresis using Pevikon®, formaldehyde-treated cellulose, or Sephadex® G-25 "fine" as the supporting medium. A pH 8.2 Tris-sulfate buffer, containing 0.05 or 0.2M Tris, was the electrolyte. Small columns were run in a 0.05M buffer for 24 hours at 600 volts, large columns in 0.2M buffer for 40 hours at 300 volts. Gel electrophoresis (1% w/v Noble agar, 0.05M sodium barbital buffer, pH 8.4, 8 volts/cm for 90 minutes) confirmed the purity of the isolated IgA. The IgA sedimentation coefficient was 11.6S.

STUDIES ON EXPERIMENTAL INSULIN IMMUNITY

Horino, M., S. Y. Yu, and H. T. Blumenthal, Diabetes **15**, 812–822 (1966).

The insulin half-life and the nature of insulin-binding antibodies were studied. Column electrophoresis was used to analyze the antigen-antibody complex and to detect changes in the nature of the immunoglobins produced with time. Electrophoresis was performed in a column packed with ethanolyzed cellulose powder using a pH 8.6 sodium barbital buffer, ionic strength 0.03. In the immunization process the insulin-binding antibody titer increases with the elevation of the 7S globulin level.

FRACTIONATION OF K-CASEIN BY COLUMN ELECTROPHORESIS

Schmidt, D. G., Protides Biol. Fluids, Proc. 14th Colloq. 671–675 (1966).

Whole K-casein A and B from cow's milk was separated by column chromatography on cellulose powder. Whatman® cellulose powder and a pH 6.5 barbiturate buffer containing 5M urea and 0.005M 2-mercaptoethanol, ionic strength 0.05, were used. Electrophoresis was carried out for 70 hours at a current of 30 to 40 mA. Elution of separated proteins was achieved at 30 ml/hour with barbital buffer. Subsequently, starch-gel electrophoresis was used to separate the K-caseins contained in the effluents.

PURIFICATION OF FOLLICLE STIMULATING HORMONE FROM HUMAN PITUITARY GLANDS

Saxena, B. B., and P. Rathman, J. Biol. Chem. **242**, 3769–3775 (1965).

A highly purified follicle-stimulating hormone from the pituitary gland was prepared by fractionation with ammonium sulfate and ethanol and subsequent chromatography on Sephadex® G-100, CM-cellulose, zone electrophoresis on cellulose columns, preparative polyacrylamide, and continuous-flow electrophoresis. Some of the physio-chemical properties of the purified hormone were determined. The $S_{20,\omega}$ was 2.04, and at pH 8.6 and 4.3 a single zone was observed. A total purification of 10,000-fold was obtained.

Continuous-Flow Electrophoresis

THE ELECTROPHORETIC ISOLATION OF PROTEIN ASSOCIATED WITH mRNA IN RAT LIVER NUCLEI

Schweiger, A., and K. Hannig, Z. Physiol. Chem. **349**, 943–944 (1968).

Carrier–free flow electrophoresis, using a 0.15M Tris-citrate buffer, separated RNA-protein complex of nuclei isolated from rat liver nuclei. The method of G. P. Georgiev et al. [Nature (London) **200**, 1291 (1966)] was used to extract isolated nuclei. Electrophoresis on polyacrylamide gel containing 8.0M urea indicated that the isolated protein had four bands (two main bands and two minor bands) of protein activity. All bands were located near the upper gel.

EXPERIMENTS WITH FACTOR VIII SEPARATED FROM FIBRINOGEN BY ELECTROPHORESIS IN FREE BUFFER FILM

Bidwell, E., G. W. R. Dike, and K. W. E. Denson, Brit. J. Haematol. **12**, 583–594 (1966).

Human and bovine Factor VIII were separated from fibrinogen by continuous-flow electrophoresis using an Elphor-VaP Mark I and the technique of Hannig [Z. Anal. Chem. **181**, 244 (1961)]. Tris-citrate buffer, pH 8.6, and borate glycine buffer, pH 8.47, were used. The mobility of human Factor VIII was near the albumin fraction. Gel filtration studies on Sephadex® G-200 indicate that Factor VIII probably has a molecular weight of 200,000.

PREPARATIVE ELECTROPHORETIC SEPARATION OF RABBIT SERUM PROTEINS AND ANTIBODIES

Freeman, M. J., Immunochemistry **3**, 257–266 (1966).

Large scale separation of rabbit serum proteins and antibody was accomplished by using medium-free, continuous flow electrophoresis. The buffer consisted of 125 ml triethylamine and 45 ml glacial acetic acid to 6 liters of water, pH adjusted to 8.6. The conductance of a 1:3 and 1:6 dilution of the buffer was $2.8 \times 10^{+3}$ and 1.4×10^3 μmho. These dilutions were used for the electrode and separation chambers. Separations were made with a constant voltage of 1800–3000 volts at 4° to 8°C. Buffer flow rate was 80 ml/1 with a 48-minute sample evacuation cycle.

SEPARATION OF GUANIDINO-COMPOUNDS BY COMBINATION OF ELECTROPHORESIS AND GEL FILTRATION: APPLICATION TO HUMAN URINE

Lauber, E. J., and S. Nateloon, Microchem. J. **11**, 498–507 (1966).

Continuous electrophoresis on Sephadex®-200 or Biogel P-10 in a 0.05 acetate buffer separated guanidino compounds of urine. The urine was injected continuously at a rate of 5 ml per hour, and eluates were collected at a rate of 15 ml per hour for four hours. Urine which yielded a single peak on chromatography was found to contain guanidino-succinic acid as well as guanidino-acetic acid.

CONTINUOUS FLOW ELECTROPHORESIS OF BRAIN PROTEINS

Cioffi, L. A., and C. Di Benedetta, Protides Biol. Fluids, Proc. 13th Colloq., 217–219 (1965).

Continuous-flow electrophoresis on Pevikon® separated protein components of the basal cortex, basal nuclei and cerebellum of bovine brain. A pH 8.6 Veronal buffer, .002M, was used. Continuous-flow electrophoresis was performed in the JKM apparatus (JKM Instrument Co.) at a flow rate of 200 ml/hour and an applied voltage of 300. Subsequently, starch-gel electrophoretic analyses were performed on the isolated eluates in order to further characterize the proteins. A number of constituents were revealed by this technique which could not be detected by other techniques.

Free-Flow Electrophoresis

FREE ELECTROPHORETIC ANALYSIS IN VARIOUS BUFFERS OF THE SOLUBLE PROTEINS FROM CRYSTALLINE LENS

Cobb, B. F., and V. L. Keening, Exp. Eye Res. **7**, 91–102 (1968).

Soluble proteins of crystalline lens were examined by free-flow electrophoresis using several Tris, several phosphate and two Veronal buffers. Chicken, fish, rabbit and bovine lens materials were separated adequately in Tris-phosphate, pH 9.5, Tris-Veronal, pH 8.7, and potassium phosphate, pH 7.7. Tris-Veronal, pH 8.57, gave the most satisfactory results.

THE SEPARATION OF THE α-CHAINS OF COLLAGEN BY FREE-FLOW ELECTROPHORESIS

Francois, C. J., and M. J. Glimcher, Biochem. J. **102**, 148–152 (1967).

Three individual fractions were isolated from α-components of collagen by combining free-flow electrophoresis and gel filtration. Free-flow electrophoresis was carried out using a Brinkmann apparatus. The buffer was prepared by adding 1320 ml of distilled water and 34 ml of trimethylacetic acid (melted at 50°C) to 4100 ml of 8M-urea. The pH was adjusted to 5.25 with sodium hydroxide, and electrophoresis was run at 4°C.

CARRIER-FREE ZONE ELECTROPHORESIS OF INFECTIOUS RIBONUCLEIC ACID DERIVED BY PHENOL AND SODIUM DODECYLSULFATE METHODS FROM PURIFIED FOOT-AND-MOUTH DISEASE VIRUS

Matheka, H. D., R. Trautman, and H. L. Bachrach, Arch. Biochem. Biophys. **121**, 325–330 (1967).

Ribonucleic acid prepared from foot-and-mouth virus by phenol and acid detergent procedures was subjected to carrier-free zone electrophoresis. The Beckman® model H electrophoresis-diffusion apparatus was used. A Veronal acetate buffer of pH 8.6, ionic strength 0.01, or pH 5.0, ionic strength 0.1, was employed. Negative mobilities were observed at both pH's.

A SEPARATION OF SPORES FROM DIPLOID CELLS OF YEAST BY STABLE FLOW FREE BOUNDARY ELECTROPHORESIS

Resnick, M. A., Science **158**, 803 (1967).

Staflo electrophoresis is a variation of zone electrophoresis in free solution that used a density gradient for stability. This is the first technique used that gives a pure (99.04%) aqueous suspension of *Saccharomyces cerevisiae* spores. A mixture of single spores and diploid cells in 2% sucrose containing 2.5 x 10^{-3} M Tris, 5 x 10^{-5}M EDTA, and 1.9 x 10^{-4}M H_3PO_4 was treated by the Staflo method. Purity of the product was tested metabolically and genetically.

THE SEPARATION OF NUCLEOTIDES AND THEIR DERIVATIVES BY CONTINUOUS FREE-FLOW ELECTROPHORESIS

Sulkowski, E., and M. Laskowski, Sr., Anal. Biochem. **20**, 94–101 (1967).

Mono-, di- and triphosphates of nucleosides, mononucleotides, pairs of some dinucleotides, and deoxyadenylic pentanucleotide digests were successfully separated by free-flow electrophoresis. The instrument used was the continuous-flow electrophoretic separator (Model FF, Brinkmann). Semimicro preparative samples were processed.

STABLE-FLOW FREE BOUNDARY ELECTROPHORESIS; THEORY AND PRACTICE AS APPLIED TO THE STUDY OF HUMAN SERUM LIPOPROTEINS

Tippetts, R. D., Dissertation Abstr. **B-27**, 789B (1967).

The existing methods of electrophoresis are reviewed. The inadequacies of these methods for the study of serum lipoproteins are pointed out and a novel technique, stable-flow free boundary (Staflo) theory of H. C. Mel, is reviewed. Standardized conditions for Staflo electrophoresis are presented. In addition, chemical techniques such as UV spectroscopy, lipid analysis and electron microscopy were employed as adducts to the Staflo technique.

ELECTROKINETIC CHANGES PRODUCED BY THERMAL TREATMENTS COMBINED WITH ULTRAVIOLET IRRADIATION IN COW MILK SERUM PROTEINS

Ambrosino, C., J. Liberatori, G. Papa, and C. Sarra, Minerva Pediat. **18**, 762–766 (1966).

Most mobile components were lost on free-phase electrophoresis in a Veronal-citrate buffer, pH 7.8, following treatment with UV light at 88°C for five to six seconds. Agar-gel electrophoresis demonstrated similar losses in milk serum proteins that were pasteurized, irradiated, and "uperized." Zone electrophoresis was useful in studying minor changes in the proteins caused by mild heat (76-80°C).

PREPARATIVE, CONTINUOUS-FLOW ELECTROPHORESIS OF ACIDIC MUCOPOLYSACCHARIDES

Mashburn, T. A., Jr., and P. Hoffman, Anal. Biochem. **16**, 267–276 (1966).

Free-flow electrophoresis using the Brinkmann® Continuous-Flow Electrophoretic Separator, Model FF, separated acid mucopolysaccharides. Separations were performed at 30 volts/cm with separating buffer solutions pumped at rates of 125–200 ml/hour. Artificial mixtures of hyaluronic acid and protein-polysaccharide of connective tissue, cartilage extract and several other mucopolysaccharides were separated. Buffer systems and pH conditions were described in detail.

PHYSIO-CHEMICAL PROPERTIES OF α-LACTALALBUMIN OF GOAT MILK

Chandhuri, S., Proc. Nucl. Phys. Solid State Phys. Symp. 271–279 (1965).

Physical constants of α-lactalalbumins were calculated. Data were collected using free electrophoresis (pH 4-6 range), free diffusion (pH 5.4), and sedimentation velocity. Constants for α-lactalalbumins from the goat, cow, and buffalo had similar values.

Gradient Electrophoresis

DOUBLY DISCONTINUOUS ELECTROPHORESIS ON SUCROSE GRADIENTS FOR THE ANALYSIS OF PLANT PEROXIDASES

Racusen, D., and M. Foote, Anal. Biochem. **25**, 164–171 (1968).

Horseradish root and bean leaf peroxidases were separated using the method of double disc electrophoresis in sucrose gradients. The electrophoretic run was performed in an apparatus consisting of a long vertically mounted glass double U-tube; each arm of the tube contained a different buffer gradient.

ZONE ELECTROPHORESIS AND ELECTRON MICROSCOPY OF ALLERTON VIRUS

Polson, A., and A. Kipps, Arch. Ges. Virusforsch. **20**, 198–199 (1967) **German.**

Allerton virus was purified by electrophoresis in sugar gradients. Electrophoresis was performed in an apparatus described by Polson and Deeks [J. Hyg. **60**, 217 (1962)]. A sugar gradient prepared in a pH 8.6 borate buffer was used. Bovine serum albumin was added to a concentration of 0.5%. A voltage gradient of 3.5 volts/cm for 5 to 18 hours was applied, and the temperature was maintained at 20°C. Electron micrographs of the purified virus showed it to have an inner body 90 mμ in diameter with a pentagonal outline suggesting that the ring-like structure was arranged icosahedrally.

DIE ELEKTROPHORETISCHE TRENNBARKEIT VON ADENOVIRUS-KOMPONENTEN IM DICHTEGRADIENTEN

Schmidt, W. A. K., Arch. Ges. Virusforsch. **20**, 11–19 (1967) **German.**

The behavior of Adenovirus 19 hemagglutinin, complement-fixing antigen and infectivity component was examined by gradient electrophoresis. Sucrose was used at a 40% level to form the gradient. A pH 8.4 sodium Veronal–sodium acetate buffer, ionic strengths 0.03, 0.06, 0.12 and field strength of 3.5 volts/cm, was employed. All components investigated had different mobilities.

STUDIES OF THE SERUM PROTEINS: VII. SUCROSE GRADIENT ELECTROPHORESIS

Sunderman, F. W., and M. W. Johnson, Am. J. Clin. Pathol. **45**, 381–397 (1966).

Sera from 2000 patients were fractionated electrophoretically using the Skeggs-Hochstrasser technique of sucrose gradient elution. Fractionations of sera from patients in 22 different disease categories were subjected to statistical analyses and compared to fractionations of normal sera. Alterations found in disease states were not identical by this technique. Replicate fractionations indicated that precision was good.

ANTIBACTERIAL ACTION OF PMN LYSOSOMAL CATIONIC PROTEINS RESOLVED BY DENSITY GRADIENT ELECTROPHORESIS

Zeya, H. I., J. K. Spitznagel, and J. H. Schwab, Proc. Soc. Exp. Biol. Med. **121,** 250–253 (1966).

Ascending electrophoresis was performed in a sucrose gradient column with a pH 4.0 acetate buffer, ionic strength 0.01, at 700 volts, 16 mA at 22°C. The lysosomal preparation was from inflammatory PMN of the rabbit peritoneum. The anode was filled with 50% sucrose, and the other vessel was filled with buffer. The gradient was formed by mixing and gravity flow. The fractions were assayed for anti-bacterial activity and identified using cellulose-acetate electrophoresis. The control enzymes were lysozyme, ribonuclease, deoxyribonuclease, acid phosphatase, and β-glucuronidase.

High-Voltage Electrophoresis

CHEMICAL TYPING OF IMMUNOGLOBULINS

Frangione, B., C. Milstein, and E. C. Franklin, Nature **221,** 149–151 (1969).

High-voltage electrophoresis combined with radioautography separated and characterized tryptic digests of partially reduced and C^{14}- iodoacetate alkylated myeloma proteins. Electrophoresis was performed on Whatman® 3MM paper for one hour at pH 3.5. A potential of 60 volts/cm was applied. Myeloma proteins containing each of the four types of γ-chains gave a distinctive autoradioautographic pattern for peptides of interchain disulphide bridges. Only a few bands were found in all 14 types of γ-globulin. It was possible to characterize the light chains with this technique during that same run.

ELECTROPHORETIC AND CHROMATOGRAPHIC SEPARATION AND FLUOR-OMETRIC ANALYSIS OF POLYNUCLEAR PHENOLS: APPLICATION TO AIR POLLUTION

Abstracts of 155 meeting of ACS, National Meeting U-33 (1968).

A number of polynuclear phenols were separated by thin-layer chromatography and low voltage and high voltage electrophoresis. The location of the separated compounds with OphthaLdehyde and 3-methyl-2-benzeothiozolinone hydrazone was described. Coupling this technique with fluorometric analysis of hydrolysis products, eight phenols were detected in coal tar.

ELECTROPHORESIS IN ZWITTERIONIC BUFFERS

Frigerio, N. A., and L. K. Kleiman, Abstracts of 156 meeting of ACS, National Meeting p-B-279 (1968).

The feasibility of high voltage gradients for electrophoretic separations by using Zwitterionic buffer solutions is discussed. The use of Zwitterionic buffers avoids the increased heat generation, diffusion, and thermal convection encountered with conventional buffers.

STARKE-GEL-HOCHSPANN UNGSELEKTROPHORESE BEI KLINISCHCHEMISCHEN SERUMUNTERSUCHUNGEN

Lange, V., Z. Klin. Chem. **5**, 168–175 (1968) **German.**

High voltage starch-gel electrophoresis was used to investigate pathological traits in serum of 140 persons. The gels were prepared in a 0.015M borate–0.006M sodium hydroxide buffer, pH 8.4. The gels were run at 18 volts/cm. The electrode buffer was 0.15M boric acid and 0.03M sodium hydroxide. The separated protein bands were revealed by staining with Amido Black. As many as 19 fractions were revealed in normal serum. Hepatocirrhosis serum was characterized by weak or absent haptoglobins and prealbumin combined with strong γ-globulins.

ELECTROPHORESIS ON CARRIERS AT HIGH ELECTRIC FIELD STRENGTH. SEPARATION OF AMINO ACIDS

Badzio, T., and T. Pompowki, Chem. Anal. **12**, 409–416 (1967).

Hydroxyproline, glutamic acid, methionine, leucine, aspartic acid, analine, glycine, lysine, and histidine were separated n an electrolyte composed of glacial acetic acid:formic acid (75 ml: 20.5 ml diluted to 1 liter, pH 1.9). A voltage of 410 volts/cm (paper strip 23 cm long) was applied for seven minutes. The construction of the apparatus, which will tolerate up to 450 volts, is described. It is possible to cool on both sides with pressure of 120 g/cm$_3$ applied to the top plate.

SELECTIVE PURIFICATION OF PHOSPHOSERINE PEPTIDES BY DIAGONAL ELECTROPHORESIS

Milstein, C. P., Nature **215**, 1190–1191 (1967).

A pepsin hydrolysate of ovalbumin was spotted on Whatman® No. 1 chromatography paper for electrophoresis at pH 6.5 with 53 volts/cm for one hour. The strip was then treated with alkaline phosphatase in 0.2 ammonium carbonate, pH 8.8. The reaction was incubated in a damp chamber at room temperature. The strip was dried and sewn to another piece of chromatography paper for electrophoresis at a right angle. Peptides digested by alkaline phosphatase lost the peptide group and became more basic. Their electrophoretic mobility was altered so that they did not appear on the diagonal line.

AMYLASES, PHOSPHORYLASE AND RELATED ENZYMES: DETECTION OF NANOGRAM AMOUNTS BY POLYACRYLAMIDE (PAA) ELECTROPHORESIS

Siepman, R., and H. Stegemann, Naturewissenschaften **54**, 116–117 (1967).

Soluble proteins of the potato tuber were separated by high voltage electrophoresis at 2°C and 150 to 300 volts (150 mA) in 5% PAA polymerized with either starch or glycogen. The gels were washed at 0°C with acid buffer and stained with 0.01N iodine in 0.014N potassium iodide. Amylases showed pale zones on a blue background; phosphorylases appeared on glycogen-PAA as blue zones on a colorless background. Down to 2 hg of α- and β-amylases and approximately 20 hg of phosphorylase in 0.2 μl of tuber juice could be identified.

AMINOACIDUREA SCREENING BY THIN-LAYER HIGH-VOLTAGE ELECTROPHORESIS AND CHROMATOGRAPHY ON MICROPLATES

Samuels, S., and S. S. Ward, J. Lab. Clin. Med. **67**, 669–677 (1966).

The amino acids in 6 to 10 μl of urine could be separated and identified by electrophoresis for five minutes on cellulose thin layers. The buffer of Kickhofen and Westphal was diluted 1/10 (pH 1.9) for electrophoresis. Further resolution was achieved by perpendicular chromatography of the dried plates for 15 minutes in n-propanol–methylethyl ketone–water–90% formic acid (10:6:3:1). The plates were dried and sprayed with a ninhydrin reagent (0.2% ninhydrin, 0.5% glacial acetic acid, 0.5% S-collidine in acetone). This reagent is prepared fresh daily. Color was developed at 70°C.

AMINO ACID ANALYSIS BY HIGH VOLTAGE ELECTROPHORESIS

Winter, E., Z. Anal. Chem. **223**, 98–107 (1966).

Paper electrophoresis of serium at 70 volts/cm in acetic acid–formic acid–water (3:1:16) separated 38 amino acids. Twenty-one of the compounds were separated into individual acids while 17 produced group-specific spots. These groups could be easily separated with propanol–2 N aqueous ammonia (7:3) or proponal–water (7:3). Quantitation of the separated compounds was possible photometrically after elution.

SEPARATION OF NUCLEOSIDES AND THEIR CORRESPONDING BASES BY HIGH VOLTAGE ELECTROPHORESIS IN BORATE BUFFER

Reinauer, H., J. Chromatog. **19**, 453–455 (1965).

Electrophoresis at 5000 volts/65 cm in 0.05M sodium borate buffer, pH 9.5–10, separated adenine guanine, hypoxanthine, cytosine, urasil, and the corresponding nucleosides. Whatman® No. 3 MM paper was the carrier. Separation was complete in 60 to 75 minutes.

HOSHSPANNUNGSELEKTROPHORESE VON ADENIN — UND INOSIN-NUCLEOTIDEN AUF CELLULOSE-SCHICHTEN

Schweiger, A., and H. Gunther, J. Chromatog. **19**, 201–203 (1965) **German.**

High-voltage electrophoresis was carried out on 0.5 mm layers of cellulose in a 0.05M citrate buffer at 30 volts/cm (40 mA) in order to separate adenine and inosine from enzymatic hydrolysates of di- and trinucleotides. Excellent separation was possible in 120 minutes.

Immunoelectrophoresis

THE SPECIFIC ANTIGENS OF HUMAN SEMINAL PLASMA

Quinlivan, W. L. G., Fertility Sterility **20**, 58–66 (1969).

Human seminal plasma, serum, and extracts of liver, spleen, heart, pancreas and brain were examined for cross-reactivity toward antiserums to semen and seminal plasma. Immunoelectrophoresis was performed on 1.5% Noble agarcoated microscope slides in a 0.1M barbitone acetate buffer, pH 8.6. A potential of 3 volts/cm for 4.5 hours was applied. Immunodiffusion studies were performed according to the standard Ouchterlony procedure. An average of six precipitin lines were formed by seminal plasma. Serum and tissue extracts produced only two precipitin lines. The use of adsorbed sera demonstrated also that only two antigens in seminal plasma were common to the tissue extracts examined.

IDENTIFICATION AND GENETIC CONTROL OF TWO RABBIT LOW-DENSITY LIPOPROTEIN ALLOTYPES

Albers, J. J., and S. Dray, Biochem. Genetics **2,** 25–35 (1968).

Immunoelectrophoresis on 1% agarose in a sodium barbital buffer, pH 8.4, revealed the presence of two allotypic specifications in the low-density lipoprotein fractions isolated from rabbit serum. Immunodiffusion studies on agarose distinguished the variants.

ISOLEMENT ET ÉTUDE DES LIPOPROTÉINES SÉRIQUES ANORMALES AU COURS DES ICTÈRES PAR RÉTENTION APRÈS FLOCULATION PAR LE POLYVINYLPYRROLIDONE

Burstein, M., and J. Caroli, Rev. Franc. d'Etudes. Clin. Biol. **13,** 387–391 (1968) **French.**

Zone electrophoresis in Veronal buffer, pH 8.6, was performed to study alterations in the mobility of serum lipoproteins after reaction with anti-β-lipoprotein antibody. Micro-immunoelectrophoresis was done in gels and paper electrophoresis was also done in further studies. Light Green was the protein stain, Sudan Black was the lipoprotein stain.

ANTIGENS OF THE LENS OF XENOPUS LAEVIS

Campbell, J. C., R. M. Clayton, and D. E. S. Truman, Exp. Eye Res. **7,** 4–10 (1968).

The antigens of the lens of Xenopus laevis were investigated by electrophoresis, immunoelectrophoresis and diffusion against heterologous antiserum prepared in rabbits. At least 11 soluble protein fractions were demonstrated by cellulose-acetate electrophoresis in a 0.05M Tris-borate buffer, pH 8.6, run for 1.75 hours at 20 volts/cm. Immunoelectrophoresis according to the technique of Scheidegger in the high resolution buffer of Aronsson and Gronwall, pH 8.9, showed at least 22 antigens present.

STUDIES ON THE CHARACTERIZATION OF SOYBEAN PROTEINS BY IMMUNOELECTROPHORESIS

Catsimpoolas, N., E. Leuthner, and E. W. Meyer, Arch. Biochem. Biophys. **127,** 338–345 (1968).

Soybean proteins were separated into at least 12 antigenetically distinct components by immunoelectrophoresis in agar gel. Immunoelectrophoresis was carried out according to the procedure of Grabar and Williams as modified by Scheidegger on microscope slides. The gel was made of 1% Ionagar No. 2 (oxoid) in a Tris-barbital buffer, pH 8.8, ionic strength 0.05. A potential gradient of 5 mA per microscope slide was used. The antigenic components were detected with antiserum prepared by injecting isolated soybean protein into New Zealand white rabbits. Monospecific antiserum detected two components, 11S and 7S.

RADIOIMMUNOELECTROPHORETIC ANALYSIS OF THYROXINE-BINDING PROTEINS

Miyai, K., K. F. Itoh, H. Abe, and Y. Kumahara, Clin. Chim. Acta **22**, 341–347 (1968).

A technique described as radioimmunoelectrophoresis was used to analyze serum-binding globulins. Normal serum was mixed with low concentrations of purified I^{131}-thyroxine and incubated at room temperature for one hour or at 4°C overnight. Immunoelectrophoresis was performed by using a modification of the technique of Grabar and Williams. A pH 7.4 phosphate buffer, ionic strength 0.05, and a 1% agar plate were used separately. Separation was achieved at 2.5 to 3.5 volts/cm for 2.0 to 2.5 hours. Antiserum was added to the proper well and allowed to diffuse for 24 to 48 hours. Radioautography was then performed on the washed plate. Using this technique, a line of identity was found in TBG-deficient serum which was not found in normal serum. Five distinct TBG components were identified in normal serum.

THE DISTRIBUTION OF SPECIFIC ANTIBODY AMONG THE IMMUNO-GLOBULINS IN WHEY FROM THE LOCALLY IMMUNIZED GLAND

Outteridge, P. M., D. D. S. Mackenzie, and A. K. Lascelles, Arch. Biochem. Biophys. **126**, 105–110 (1968).

Immunoelectrophoresis techniques were used to identify antibodies of serum samples and whey samples isolated by gel filtration on Sephadex® G-200 and DEZE-cellulose columns. Immunoelectrophoresis was performed using microscope slides coated with Noble Agar (Difco) and LBK® electrophoresis cell. A Veronal buffer, pH 8.6 was employed. The immunoelectrophoretic patterns were developed with antisera prepared in rabbits.

ANALYSIS OF THE PROTEINS IN HUMAN SEMINAL PLASMA

Quinlivan, W. L. G., Arch. Biochem. Biophys. **127**, 680–687 (1968).

Cellulose-acetate electrophoresis, immunoelectrophoresis, and disc-gel electrophoresis have been used to analyze the proteins of human seminal plasma. Cellulose-acetate electrophoresis was performed in a Beckman MicroZone® system using a barbital buffer (0.1M, pH 8.6). A potential gradient of 250 volts for 20 minutes at room temperature was used. Immunoelectrophoresis was performed in a Shandon Universal Electrophoresis Apparatus on glass slides coated with agarose. An Owens modified barbitone buffer, pH 8.6, and a constant 50 mA current equivalent to 20 volts/cm for 4.5 hours were employed. Disc electrophoresis was performed according to the method of Davis [J. Ann.

N. Y. Acad. Sci. **121**, 404 (1964)]. Cellulose acetate electrophoresis demonstrated four protein bands, while disc electrophoresis demonstrated fifteen. Immunoelectrophoresis identified six proteins, three of which were identical to those in human serum.

HIGH SPEED IMMUNOELECTROPHORESIS

Traill, M. A., Lab. Pract. **17**, 709 (1968).

A procedure whereby the electrophoretic phase of immunoelectrophoresis was carried out in five minutes is described. Separation was achieved on 2.5 x ⅞ inch coverslips on 1% buffered agar. The agar strip was floated on mercury which was cooled by iced water that moved through copper pipes. A current of 25–40 mA was applied, and pH 8.6 barbitone acetate buffer (ionic strength 0.1) was used.

A TECHNIQUE OF IMMUNOCHEMICAL DETECTION OF PROTEIN FRACTIONS AFTER ACRYLAMIDE GEL ELECTROPHORESIS

Bednařík, T., Clin. Chim. Acta **15**, 172–174 (1967).

Two techniques for the immunochemical detection and characterization of fractions separated on polyacrylamide gels were presented. The first technique allows for antibody reaction directly in the gel after horizontal gel electrophoresis. The second technique, which can be used with either vertical or horizontal gels, appears to be more advantageous and easier to manipulate. The acrylamide gel is removed and placed on an agar plate and allowed to diffuse into suitable antisera.

PRODUCTION OF AUTOLOGOUS AND HOMOLOGOUS ANTIBODIES AGAINST SOME WHEY PROTEINS IN THREE RUMINANT SPECIES

Caruolo, E. V., J. Dairy Sci. **50**, 1616–1620 (1967).

Experiments were designed to test the hypothesis that an autoallergic condition to milk proteins may be important in bovine mastitis. The results indicated that the bovine species does not produce antibodies to either homologous or autologous milk proteins and that this is not an agent in an autoallergic condition of mastitis. Rabbits and cows were immunized over a period of three months with whole milk suspended in Freund's adjuvant. Serum was collected and analyzed by immunoelectrophoresis. The rabbit sera were used to show that the milk was antigenic and to demonstrate a heterologous antibody. Electrophoresis was performed as previously described [J. Nutr. **77**, 349 (1962)] and [Res. Vet. Sci. **5**, 332 (1964)], except that Tris-glycine buffer, pH 8.50, ionic strength 0.05, was used and Amido Black was the stain.

POLYPEPTIDE CHAIN STRUCTURE OF RABBIT IMMUNOGLOBULINS: III. SECRETARY γA-IMMUNOGLOBULIN FROM COLOSTRUM

Cebra, J. J., and P. A. Small, Jr., Biochemistry **6**, 503–512 (1967).

Purified rabbit γA-immunoglobulin from colostrum was reduced and alkylated to study the characteristics of its component chains. A model structure of the molecule was proposed, suggesting that the molecule was assembled of four pairs of α (M.S. 64,000) light chains and one or two κ (M.W. 50,000) chains. A map of tryptic peptides of the heavy chains showed differences between γA-, γG-, and γM-immunoglobulins [Lamm, Biochemistry **5**, 267 (1966)].

ERFAHRUNGEN ÜBER DIE OPTIMALE DARSTELLUNG UND DIAGNOSTIK DER G_c-TYPEN

Hillgermann, R., Z. Klin. Chem. **5**, 317–321 (1967) **German.**

An immunoelectrophoretic technique which used agar (Difco Noble) and a barbital-calcium lactate buffer, pH 8.6, is described. The agar was prepared on 115 x 300 mm clear glass plates, and separation was achieved at 7 volts/cm for 2 to 2.5 hours. This technique revealed as many as 30 diffusion patterns simultaneously.

LIMITED HETEROGENEITY OF GAMMA GLOBULIN IN HYPOGAMMA-GLOBULINEMIA

Hong, R., and R. A. Good, Science **156**, 1102–1103 (1967).

Sera from 11 hypogammaglobulinemic patients were separated in 7.5% acrylamide gels, pH 9.5. The gels were sliced after electrophoresis and diffused against antiserum specific for the gamma chain of immunoglobulin G. The immunoglobulin G of agammaglobulinemics differed from that of normals. Results obtained imply an abnormality in protein synthesis by these patients.

FETAL α_1-GLOBULIN OF RAT SERUM

Kithier, K., and J. Prokes, Biochem. Biophys. Acta **127**, 390–399 (1967).

Eleven precipitin bands were detected when immunoelectrophoresis was performed on newborn rat serum. The newborn rat serum reacted against RAR_aF antiserum, the antibody of which was prepared by using unabsorbed serum of newborn rats. After absorption with serum of a healthy rat, three precipitin lines were detected.

MICROIMMUNOELECTROPHORESE AUF CELLULOSE ACETEGEL

Lomanto, B., and C. Vergani, Clin. Chim. Acta **15**, 169–171 (1967).

A method for performing immunoelectrophoresis on cellulose acetate was described. The optimum pH for electrophoresis was found to be 6.5–8.2. Veronal buffer, pH 8.2, ionic strength 0.05, was used routinely. A potential gradient of 160 volts, 0.5mA per cm, was employed to achieve separation.

QUANTITATIVE IMMUNOASSAY BY DISC ELECTROPHORESIS

Louis-Ferdinand, R., and W. F. Blatt, Clin. Chim. Acta **16**, 259–266 (1967).

By using polyacrylamide-gel electrophoresis, it was possible to quantitatively determine the extend of reaction of antigen to antibody. The antibody (anti-horse albumin) was mixed with a large-pore gel. Migration of albumin was retarded because of the combination of antigen and antibody. Quantitation was achieved by comparing the densitometric evaluation of stained protein bands of antibody-reacted material with a system in which the antibody was omitted. This technique was found to be valid only if the antibody was capable of fixing the antigen over a wide range of antigen concentration.

ELECTROIMMUNODIFFUSION (EID): A SIMPLE, RAPID METHOD FOR QUANTITATION OF IMMUNOGLOBULINS IN DILUTE BIOLOGICAL FLUIDS

Merrill, D., T. F. Hartley, and H. N. Claman, J. Lab. Clin. Med. **69**, 151–159 (1967).

A technique is described for diffusing antigen under an electric field into a layer of agar containing specific antiserum. The Ab-Ag precipitate length is proportional to the concentration of antigen and the duration of electrophoresis. Cerebrospinal fluid, a biological fluid with low γ-globulin concentration, was used as a test system. Microslides were prepared using 1% agar in a pH 8.6 Veronal buffer, ionic strength 0.05, which contained anti-γ-G. Four μL of CSF were applied to a well at the end of the slide. Electrophoresis was performed for 30 to 90 minutes at 150 volts in a pH 8.6 Veronal buffer, ionic strength 0.1. The lower limits of the technique are: γG 0.2 mg%; γA 0.5 mg%; γM 0.8 mg%. Normal CSF contained 1.0 to 3.1 mg/100 ml of γG.

EVALUATION OF LIGHT CHAIN ANTIGEN BINDING BY RADIOIMMUNO-ELECTROPHORESIS

Minden, P., H. Grey, and R. S. Farr, J. Immunol. **99**, 590–595 (1967).

Radioimmunoelectrophoresis demonstrated specific antigen binding by isolated light chain antibody. Anti-BSA was purified by the procedure of Uliky et al. [Immunochemistry **1**, 219 (1964)]. Electrophoresis of the purified preparation was carried out according to the procedure of Yagi. After slides were washed, I^{131} BSA was added to the wells, and allowed to diffuse for 24 hours, and radioautography was performed. Radioactive arcs and specific binding of I^{131} were demonstrated.

STUDIES IN SARCOIDOSIS: III. SERUM PROTEINS IN CASES WITH CONCOMITANT ERYTHEMA NODOSUM

Norberg, R., Acta Med. Scand. **181**, 101–114 (1967).

The proteins in 29 sera from patients with active erythema nodosum were quantitated by electrophoresis [Bottiger, Clin. Chim. Acta **5**, 664 (1960)]. Proteins, hexoses, hexosamines, and sialic acids in the electrophoretic fractions were also quantitated. Immunoelectrophoresis using commercial rabbit antihuman sera was performed to try to distinguish differences between patients and the controls. A pH 8.6 Veronal buffer, ionic strength 0.01, was used. Transferrin, the complement complex proteins, γA and γM immunoglobulins, and rapidly migrating γG migrated between the β- and the γ-globulin of erythema patients.

OPTIMAL CONDITIONS FOR IMMUNOELECTROPHORESIS OF SOME SOLUBLE BOVINE LENS ANTIGEN

Rathbun, W. B., M. A. Morrison, and R. M. Fusaro, Exp. Eye Res. **6**, 267–272 (1967).

The conventional techniques for the immunoelectrophoretic separation of bovine α-crystallin, β-crystallin and other lens proteins are evaluated. Conventional pH values, temperature and specific conductance values were found to be inadequate. Tris-phosphate buffer, pH 6.0–7.5 (specific conductance equivalent 0.025M KCl at $4°C$), gave improved results. Optimum conditions were different for each protein studied.

CORRELATION BETWEEN NET CHARGE OF ANTIGENS AND ELECTROPHORETIC MOBILITY OF IMMUNOGLOBULIN M ANTIBODIES

Robbins, J., Nature **213**, 1013–1014 (1967).

The net charge of the antigen influenced the electrophoretic mobility of IgM antibodies. A 0.05 sodium barbital buffer, pH 8.6, was used for electrophoresis and preparation of the 1.5% agar. A current of 220 volts, 31 mA, separated the proteins on 18 slides in 100 minutes. Rabbit sera was the sample. Goat antirabbit antisera precipitated the separated rabbit serum proteins. Acidic antigens tested were diphtheria toxoid and DNP-(T,G)-A-L; basic antigens were lysozyme, ribonuclease, uridine-A-L, and DNP-poly-L-lysine.

HETEROGENEITY AND SEQUENCE OF APPEARANCE OF RABBIT ANTI-BODIES TO HOG INTRINSIC FACTOR

Samloff, I. M., and E. V. Barnett, J. Immunol. **98,** 558–567 (1967).

The antibodies to hog intrinsic factor (HIF) were detected and separated by radioimmunoelectrophoresis. The medium used was 1% agar gel, and a pH 8.2 barbital buffer, 0.03M, was used for electrophoresis, which was run for two or three hours at 10 volts/cm of microscope slide. Following electrophoresis, the slides were flooded with a HIF concentrate labeled with Co^{60} vitamin B_{12}.

INVESTIGATIONS ON THE ISOLATION AND SPECIFICITY OF THE RHEUMATOID FACTOR

Swierczyńska, Z. and G. Wozneczko-Orlowska, Polish Med. J. **6,** 1090–1100 (1967).

Immunoelectrophoretic characterization of rheumatoid factor isolated from rheumatoid sera was achieved using the technique of Scheidegger. The separated factor was tested with immune sera prepared by injecting animals with mixed serum containing the rheumatoid factor, human γ-globulin and a serum having a high titer in the Waaler-Rose reaction. Reaction of the rheumatoid sera with anti-γ-globulin serum was observed.

IMMUNOLOGICAL STUDIES OF PROTEIN SYNTHESIS DURING SEA URCHIN DEVELOPMENT

Westin, M., H. Perlmann, and P. Perlmann, J. Exp. Zool. **166,** 331–346 (1967).

The incorporation of C^{14} amino acids into antigen at different stages of development was studied in the sea urchin. The tissues were extracted and assayed. Antigens were separated by the Ouchterlony technique and immunoelectrophoresis. 0.9% agarose and 0.2% agar were prepared in barbiturate buffer for immunoelectrophoresis. The protein concentration of the tissue extract was 2–3.5 mg/ml. Antisera was added undiluted.

ELECTROPHORETIC AND IMMUNOELECTROPHORETIC ANALYSES OF SERUMS FROM NORMAL COWS AND COWS EXPERIMENTALLY AND NATURALLY INFECTED WITH MYCOBACTERIA

Wright, G. L., Dissertation Abstr. **B-27,** 2240 (1967).

A number of electrophoretic techniques were employed to study the serum proteins of normal cattle and cattle infected with tuberculosis. A modified polyacrylamide-gel technique proved to be the most sensitive single technique. Thirty to thirty-eight different bands were detected in normal bovine serum. Cows infected with M. *bovis,* virulent group III, and cows naturally infected manifested a hypergammaglobulinemia and a lowered albumin fraction by all separation techniques.

ÉTUDE IMMUNOÉLECTROPHORÉTIQUE DE LA MOELLE ASSEUSE CHEZ LA RAT IRRADIÉ

Zaorska, B., and A. Czupryna, Rev. Franc. d'Etudes Clin. Biol. **12,** 169–175 (1967).

Bone marrow from control and irradiated rats was homogenized in Tyrode's solution and separated into five fractions by density-gradient centrifugation by the method of Goodman. Antisera were prepared using these five fractions. Electrophoresis of this antigen was performed according to the technique of Scheidegger using 200 volts and 35 mA. Following electrophoresis, the gel was flooded with the immune sera and incubated in the usual fashion. After irradiation, some antigens present in bone marrow disappeared.

IMMUNOELECTROPHORETIC STUDIES OF SERA FROM RABBITS EXPERIMENTALLY INFECTED WITH HISTOPLASMA CAPSULATUM

Adamson, D. M., and G. C. Cozad, J. Bacteriol. **92,** 887–891 (1966).

In five weeks precipitins developed in the sera of rabbits infected with *H. capsulatum.* Microimmunoelectrophoresis was a sensitive technique to demonstrate this antibody. A pH 8.2 barbital sodium buffer, ionic strength 0.1, was used as the electrolyte. A 2% agar solution was prepared in equal volumes of water and the buffer. A drop of 5 or 10X concentrated histoplasmin was electrophoresed at 40 volts (5.3 volts/cm) for 45 minutes. Immune sera formed precipitin lines with the histoplasmin after incubation in a moist chamber for 48 to 72 hours.

IMMUNOCHEMISCHES STUDIUM DER EICVEISSFLOTOFFE DES MAGNESAFES BEIM MENCHEM

Balkova, A., and E. Paluska, Z. Immunitaetsforsch. Allergie Klin. Immunol. **130,** 444–453 (1966) **German.**

The immunochemical differences between human gastric juice proteins and serum protein were investigated using immunoelectrophoresis and double-diffusion studies on agar gels. Agar-gel electrophoresis on 1% agar in a pH 8.6 Veronal buffer, ionic strength 0.05, and immunoelectrophoresis in the same buffer at 3.5 volts/cm were used to evaluate the differences. Eleven proteins were found in gastric juice, four of which were specific for gastric juice. Seven could be found in serum also.

ISOLATION AND CHARACTERIZATION OF TWO ANTIGENS OF CORYNE-BACTERIUM HOFMANNII

Banach, T. M., and R. Z. Hawirko, J. Bacteriol. **92,** 1304–1309 (1966).

Two seriologically-active substances extracted from sonic extracts of C. *hofmanni* were partially purified by column chromatography on DEAE-cellulose and by Sephadex® G-200 gel filtration. The active components isolated were characterized by immunodiffusion, immunoelectrophoresis and polyacrylamide-disc electrophoresis. Glutamic acid, aspartic acid, alanine, glycine, valine and leucine were the main amino acids present in antigen A.

SIMPLIFIED TECHNIQUE OF IMMUNOELECTROPHORETIC ASSAY OF HUMAN INTRINSIC FACTOR ON ACRYLAMIDE GEL

Bardhan, K. D., Gut **7,** 566– 568 (1966).

Electrophoresis was performed in 16-cm test tubes using 11 cm of the following gel mixture: 1 part water : 1 part solution A : 2 parts solution B : 4 parts solution C. The solutions were: (A) 48 ml 1N HCl; 36.6 g 2-amino-2-(hydroxymethyl)-1, 3-propanediol (Tris); 0.23 ml N.N.N^1, N^1-tetramethylethylenediamine; 100 ml water; (B) 28 g acrylamide; 0.74 g N,N^1-methylenebisacrylamide; 100 ml water; (C) 14 g ammonium persulfate; 100 ml water. Gastric juice was neutralized with 1.0N NaOH and filtered. One ml of the gastric juice was the sample placed on the gel. Tris-glycine stock buffer (6 g Tris–28.8 g glycine to 1 liter with water) was diluted 1:10 (pH 8.6) for the buffer. Electrophoresis was performed for 3½ to 4½ hours at 3.75–4 mA per tube.

ANTIGENIC STRUCTURE OF BRUCELLA SUIS SPHEROPLASTS

Baughn, R. E., and B. A. Freeman, J. Bacteriol. **92,** 1298–1303 (1966).

Immunoelectrophoresis was used to quantitate the antigens of normal cells of *Brucella suis* and of spheroplasts induced by penicillin and/or glycine. The supporting medium consisted of a 1% solution of Difco purified agar in 0.075M barbituric buffer, pH 8.6, pipetted in 2-ml quantities onto microscope slides. A 0.05M barbituric buffer, pH 8.6, was used in the electrode vessels. Electrophoresis was performed with a total of 150 volts for 100 minutes for eight slides. Glycine spheroplasts retained only six antigens, and penicillin spheroplasts eight, as contrasted with the normal bacteria which had at least thirteen antigens.

DENATURATION PROCESS IN ALBUMINS ISOLATED FROM THE SERUM OF HEALTHY WOMEN AND FROM UMBILICAL BLOOD SERUM OF THEIR NEWBORNS

Chojnowska, I., Ginekol. Polska **37**, 933–934 (1966).

Albumins were separated by TCA and ethanol precipitation. On starch gel and immunoelectrophoresis, maternal blood albumin showed two fractions while umbilical serum showed one component. Umbilical cord serum albumin was more easily denatured by heat and urea than was maternal albumin.

STRAIN VARIATIONS OF MOUSE 7Sγ_1 GLOBULINS

Coe, J. E., Immunochemistry **3**, 427–432 (1966).

Immunoelectrophoresis performed on glass slides coated with 2% ion agar in a pH 8.2 barbital buffer (ionic strength 0.04) demonstrated strain differences in the electrophoretic mobility of 7Sγ_1 globulin of mice. Samples were electrophoresed for 60 to 90 minutes at 6 volts/cm. Antiserum was added to the trough and allowed to diffuse for 24 to 48 hours at 4°C. Autoradiographs were obtained by adding HEA^{131}I to the trough after precipitin lines were developed with the antiserum. There seemed to be variation in mobility, which was probably hereditary and was not accompanied by antigenic dissimilarity.

LACK OF GAMMA-A IMMUNOGLOBULINS IN SERUM OF PATIENTS WITH STEATORRHEA

Crabbé, P. A., and J. F. Heremans, Gut **7**, 119–127 (1966).

The serum of three patients with steatorrhea was examined by immunoelectrophoresis according to the technique of Scheidegger. With the use of a multivalent horse antiserum against whole human serum and specific antiserum (rabbit) against the three human immunoglobulins, the authors demonstrate a deficiency of γA-immunoglobulins. Histochemical studies of the jejunal mucosal biopsy showed these patients to be almost completely devoid of plasma cells containing γA and γM-immunoglobulins in the lamina propria.

IMMUNOCHEMICAL STUDY OF PARAINFLUENZA VIRUS (TYPE 2) IN AMNION CELLS

De Vaux St. Cyr, C., and C. Howe, J. Bacteriol. **91**, 1911–1916 (1966).

Immunoelectrophoresis demonstrated that, in addition to hemagglutinin and enzyme, there are at least three infection-specific antigens that appear intracellularly during virus development. Electrophoresis of infected and control amnion cells was done in a 0.06M Veronal buffer, pH 8.2. Standard microscope slides coated with agar were used. Separation was carried out at

10 mA per slide at room temperature. High-titer rabbit anti-parainfluenza (Type 2) sera precipitated the antigens present in infected cells.

ÉTUDE IMMUNOLOGIQUE DU POLYMORPHISMI DES PHOSPHATASES URINAIRES HUMAINES

Dufour, D., A. Tremblay, and S. Lemieux, Rev. Franc. d'Etudes Clin. Biol. **11**, 536–538 (1966) **French.**

Various molecular forms of urinary phosphatases were demonstrated in human urine. Differences were seen on cellulose-acetate electrophoresis and immuno-electrophoresis. The techniques described earlier were used for these analyses.

ANTIGENIC COMPONENTS OF HUMAN SEMINAL PLASMA

Eyquem, A., Ann. Inst. Pasteur **Suppl.**, 89–94 (1966).

Thirteen antigens were detected in human seminal plasma after immunoelec-trophoresis using goat antiserum. One β_1 and α_2-globulins are antigens that were also found in spermatozoa.

IMMUNOFLUORESCENCE ELECTROPHORESIS

Ghetic, V., Studii Cercetari Biochim. **9**, 359–366 (1966).

Fluorescein isothiocyanate was conjugated to the antibody or antigen at pH 9.0 for 16 hours at 4°C. A 2% protein solution was mixed with the dye in a 1:25 ratio. A Sephadex® G-50 column separated the conjugate and free dye by elution with 0.145M NaCl. A 1% agarose gel solution was prepared for electrophoresis (Veronal buffer, pH 8.6, 8 to 10 volts/cm and 0.8 mA/cm for two hours). Ultraviolet light was used to detect the separated conjugates. Several applications and examples of this technique are presented.

IMMUNOELECTROPHORETIC ANALYSIS OF GUINEA PIG SERA

Godzinska, H., Arch. Immunol. Therap. Exp. **14**, 263–274 (1966).

Immunoelectrophoresis performed on 1.5% Japanese agar or Bacto-Difco agar according to the micromethod of Scheidegger [J. Int. Arch. Allergy **7**, 103 (1955)] characterized guinea pig serum proteins. Electrophoresis was per-formed in Laurell's buffer [Bull. Soc. Chim. Biol. Suppl. 1, **39**, 85 (1956)] at 6.5 volts/cm for 2.5 hours. The antiserum was then placed in a well and allowed to diffuse for 24 to 48 hours. Nineteen to twenty-two antigenically different proteins were detectable. A prealbumin not detected before in guinea pig sera was recorded.

A STUDY OF GAMMA GLOBULINS IN CHOLERA

Goldstein, H. B., B. N. Dale, and O. Felsenfeld, Proc. Soc. Exp. Biol. Med. **121,** 425–427 (1966).

Sera from cholera-infected patients and immunized control patients were studied to determine the relationship of anticholera globulins. A 1% agar gel solution was poured into microscope slides for the electrophoresis. A 0.1M Veronal buffer, pH 8.6, was the electrolyte; 2.25 volts/cm was applied for two hours. Horse antiserum was employed to detect separated proteins. Agglutination tests were done on the sera. All cholera patients sera had γ-G, A and M globulins. Amido Black was the stain used to detect the separated proteins.

QUANTITATION OF IMMUNOGLOBULINS IN CEREBROSPINAL FLUID

Hartley, T. F., Arch. Neurol. **15,** 472–479 (1966).

Immunoelectrophoresis was used to quantitate the proteins in 26 normal and 83 abnormal cerebrospinal fluids. Goat antisera specific for immunoglobulins γG, γA, and γM were obtained commercially. A 1% agar solution was spread onto microscope slides for electrophoresis. A pH 8.6 Veronal buffer was used for preparation of the agar and electrophoresis. A voltage of 150 volts for 60 minutes gave good separation. The lower limits of detection of this technique were γG–0.2 mg/100 ml; γA–0.7 mg/100 ml, and γM–1.0 mg/100 ml.

HEPATIC ANTIGEN IN SERA FROM PATIENTS WITH LIVER DISEASE

Hirayama, C., and T. Toda, Nature **212,** 1061–1062 (1966).

Human liver was homogenized in 0.25M sucrose and extracted with sodium deoxycholate (final concentration 0.5%). After centrifugation and dialysis against saline, the extract was lyphilized and stored frozen. This antigen was used to produce anti-liver rabbit antisera. The antisera was adsorbed with human sera and boiled guinea pig kidney. After immunoelectrophoresis [Scheidegger, Int. Arch. Allergy Appl. Immunol. **7,** 103 (1957)], an antigen could be identified in the hepatic extract that migrated in the β-globulin area.

SERIAL SERUM IMMUNOELECTROPHORETIC CHANGES AND ACTIVITY OF TUBERCULOSIS

Huhn, O. H., G. B. Elliott, and B. P. Boyd, Am. Rev. Respirat. Diseases **94,** 727–733 (1966).

The serum from 18 patients with bacteriologically confirmed active tuberculosis was separated by electrophoresis. Polyvalent goat antiserum reacted with the separated sera and the precipitin arcs examined. Electrophoresis was done in ion agar using a Veronal buffer, pH 8.4, ionic strength 0.5M, for 2½ hours in a 4 mA current. The beta 1 b arc (hemopexin) decreased in size as the patient improved. The IgM arc was shorter in these patients than in controls. If the patient became worse, the IgM arc could disappear.

IMMUNOELECTROPHORETIC STUDIES ON RELATIONSHIPS BETWEEN PROTEINS OF PORCINE COLOSTRUM, MILK AND BLOOD SERUM

Karlsson, B. W., Acta Pathol. Microbiol. Scand. **67**, 83–101 (1966).

Agar-gel electrophoresis was performed on whole milk, colostrum and blood serum to characterize and compare the proteins present. Antisera to these samples was prepared in hyperimmunized rabbits. Causes of the occurrence of serum proteins appearing in milk and colostrum are speculated.

IMMUNOELECTROPHORETIC STUDY OF BIOLOGICAL FALSE-POSITIVE SERUM REACTIONS FOR SYPHILIS

Kiraly, K., Acta Dermato-Venereol. **46**, 506–510 (1966).

Paper electrophoresis in Michaelis buffer, pH 8.6, was done on sera from patients with a biological false positive to the *Treponema pallidum* immobilization test. The micro method of Scheidegger [Intern. Arch. Allergy Appl. Immunol. **7**, 103 (1955)] was used for immunoelectrophoresis at a potential gradient of 3.5 volts/cm. Horse or rabbit anti-human sera were used to form the precipitin lines. A hyperactive state of the immune system, characterized by increase level of γM and γG and decrease in albumin, was frequently found in biologically false-positive sera.

DERMATITIS HERPETIFORMIS AND HERPES GESTATIONSIS: ANALYSIS OF A γ_A AND γ_M SERUM PROTEIN BY IMMUNOELECTROPHORESIS

Kjartannson, S., R. M. Fusaro, and W. C. Peterson, Jr., J. Invest. Dermatol. **46**, 480–483 (1966).

When serum from patients with *dermatitis herpetiformis, herpes gestationsis,* and *pemphigus vulgaris* was examined immunoelectrophoretically, an increase in the γ_A globulin fraction of serum was observed. Antihuman serum-horse serum, antihuman-rabbit serum, antiherpes dermatitis-rabbit serum, and anti-dermatitis herpetiformis rabbit serum were used to demonstrate the increases. The immunoelectrophoretic method of Lim and Fusaro [J. Invest. Dermatol. **39**, 303 (1962)] was used.

IMMUNOELECTROPHORETIC ANALYSIS IN ACRODERMATITIS CHRONICA ATROPHICANS

Kraus, Z., and F. Mateja, Acta Dermato-Venereol. **46**, 217–223 (1966).

A pH 8.6 sodium barbital Veronal buffer, ionic strength 0.05, was used for immunoelectrophoresis of serum from nine treated and ten untreated acrodermatitis chronica atrophicans patients. The 1.5% agar solution was spread on 5 x 5 cm disposable slides and a voltage of 30 volts was applied. Horse anti-human sera formed precipitin lines with the separated sera. Amido Black 10B was the stain. An increase in all immunoglobulins was found in 12 patients; an increase in 2M globulins was found in seven.

THE OVERTAKING OR REACTION ELECTROPHORESIS AS A QUANTITATIVE IMMUNOLOGICAL METHOD

Lang, N., Protides Biol. Fluids, Proc. 14th Colloq. 517–526 (1966).

The technique of overtaking or reaction electrophoresis performed in a barbitone acetate buffer, pH 8.6, ionic strength $\mu 0.1$, is described. The faster migrating component was applied to the electrophoresis strip behind the slower substance. During electrophoresis the fast migrating substance can catch up with the slower substance and react. If one or both of the substances were labeled, it was possible to determine quantitatively the amount of reacting substances. Only a few μl or μg of reacting components were needed to perform the technique.

COMPARATIVE BIOCHEMICAL STUDIES OF MILKS: III. IMMUNOELECTROPHORETIC COMPARISONS OF MILK PROTEINS OF THE ARTIODACTYLA

Lyster, R. I., R. Jenness, N. I. Phillips, and R. E. Sloan, Comp. Biochem. Physiol. **17**, 967–971 (1966).

The milk proteins of 19 species of artiodactyls were studied by immunoelectrophoresis using rabbit antisera from *Bos taurus*. Electrophoresis was performed on lantern slide cover glasses coated with a 1% agar solution. A 0.05M Veronal buffer, pH 8.6, with 10 mM EDTA to improve resolution of the casein was used for electrophoresis. The milk proteins of both ruminant and non-ruminant species reacted with anti-cow serum albumin and anti-cow casein sera.

IMMUNOELECTROPHORETIC EVALUATIONS OF PRESERVED BLOOD PROTEINS

Mackiewicz, U., and J. Pech, Farm. Polska, **22**, 271–273 (1966).

Blood was stored at 4°C. Plasma was prepared by centrifuging, and then compared to fresh citrated plasma. Tests were done on the day of collection and on the third through twenty-eighth day of storage. Electrophoresis was in 2% agar gel. No decrease in the protein fractions was seen in the first 14 days of storage, after which there was a gradual decrease in the β-2-macro-globulins and β-1-lipoprotein. The albumin 7S γ-globulin, transferrin and fibrinogen fractions remained constant. Fractions were detected by precipitation of multi-valent immune antiserum.

GROUP-SPECIFIC COMPONENTS AND TRANSFERRINS IN HUMAN FETAL SERA

Melartin, Lüsa, Scand. J. Hematol. **3**, 117–122 (1966).

Studies of 24 pairs of mother-fetal sera indicate that the human fetus can synthesize its own group-specific components beginning with the tenth week of gestation. Transferrin in the fetus sera is less concentrated than in the maternal sera. Group specific components were identified using horse anti-Gc sera in immunoelectrophoresis [method of Hirschfeld, Progr. Allergy **6**, 155–186 (1962)]. Starch-gel electrophoresis by the method of Beckman [Acta Genet. (Basel) **11**, 106–110 (1961)] was used for the identification of transferrin.

A QUANTITATIVE IMMUNOELECTROPHORESIS METHOD (LAURELL ELECTROPHORESIS)

Minchin-Clarke, H. G., and T. Freeman, Protides Biol. Fluids, 14th Colloq. 503–509 (1966).

After initially separating proteins by simple electrophoresis, the separated proteins were transferred to another plate on the strip of gel. Anti-whole human serum was poured onto the second plate. Electrophoresis at right angles to the initial separation forced the serum antigen into the antibody gel. Simple zone electrophoresis was performed on 2% agarose gel in a sodium barbitone-calcium lactate buffer, pH 8.6, at 15 volts/cm for one hour. During this time the gel was cooled with water. Electrophoresis in the second dimension was performed on agarose in a Veronal buffer at 6 volts/cm for 22 hours. The fractions migrated until they were precipitated by antiserum. A number of overlapping but discrete peaks were formed. The areas under each peak were calculated by projecting the plate onto a white card, drawing, cutting out, and weighing each area. The method gave quantitative data about all proteins with mobility greater than that of γ-globulin.

ANTIGENIC CONSTITUENTS OF BASIC PROTEINS FROM HUMAN BRAIN

Rajam, P. C., S. Bogoch, M. A. Rushworth, and P. C. Forrester, Immunology **11**, 217–221 (1966).

Immunoelectrophoresis confirmed the further fractionation of a neutral extract from human brain. Rabbit anti-A absorbed with human serum formed precipitin arcs with the separate antigens. A pH 8.6 Veronal buffer, ionic strength 0.1, was used for electrophoresis at 25°C (20 mA, 250 volts for 45 minutes). Amido Black 10B stained the precipitin arcs.

BRAIN ANTIGENS: COMPONENTS OF SUBFRACTIONS FROM HUMAN GREY MATTER

Rajam, P. C., and S. Bogoch, Immunology **11**, 211–215 (1966).

DEAE-cellulose chromatography was used to fractionate a neutral, low ionic strength extract of human grey matter. Immunoelectrophoresis confirmed other findings of five basic group antigens and seven to eight acidic antigens. A pH 8.6 Veronal buffer, ionic strength 0.1, was used for the agar electrophoresis (20 mA, 250 volts for 54 minutes at 75°C).

THE SEPARATION AND CHARACTERIZATION OF PROTEINS INTRINSIC TO NASAL SECRETION

Rossen, R. D., R. H. Alford, W. T. Butler, and W. E. Vannier, J. Immunol. **97**, 369–378 (1966).

Immunoelectrophoresis and double diffusion were used to characterize proteins isolated from nasal secretions. The nasal proteins were isolated by gel filtration on Sephadex® G-200. Elution was achieved with 0.1M sodium chloride buffered at pH 8.0 with 10% borate. In addition, fractionation was possible on DEAE-cellulose with stepwise elution with 0.01M sodium phosphate, pH 6.2, and 0.3M sodium phosphate, pH 4.8. Specific antiserum (rabbit) identified six intrinsic components in the secretions. Gel filtration studies indicated that a low molecular weight substance was associated with the isozyme component.

IDENTIFICATION OF RAT URINARY PROTEINS BY ZONE AND IMMUNO-ELECTROPHORESIS

Roy, A. K., and O. W. Nevhas, Proc. Soc. Exp. Biol. Med. **121**, 894–899 (1966).

Proteinuria is a normal condition in rats. Paper, agar, starch-gel and immuno-electrophoresis were combined to study the origin and nature of these urinary proteins. Urine was pooled, concentrated and dialyzed against distilled water. Whatman® 3MM paper was electrophoresed in a pH 8.6 Veronal buffer, ionic strength 0.075. A 0.1% methanolic solution of Bromophenol Blue and a periodic acid Schiff reagent were used as stains. Starch-gel electrophoresis was done in a discontinuous buffer of glycine-NaOH and borate, pH 9.0, or Tris-citrate and borate, pH 8.4. Amido Black was the stain. Agar electrophoresis was performed in a pH 8.6 Veronal buffer, ionic strength 0.05 (250 volts for 40 to 50 minutes, Bromophenol Blue stain). Fifteen individual proteins were identified: seven were derived from serum, the others were unique to urine.

DIFFERENTIAL STAINING OF CERULOPLASMIN AS AN AID TO INTERPRETATION IN IMMUNOELECTROPHORESIS

Schen, R. J., and M. Rabinovitz, Clin. Chim. Acta **13**, 537–538 (1966).

Prussian Blue was used as a selective stain for the arc of ceruloplasmin which separates in the α_2-protein region in electrophoresis of serum. After electrophoresis, the excess proteins were eluted and the slide dried at low temperature. The slide was immersed for ten minutes at $37°C$ in a 0.08% solution of ferrous sulfate in 0.1M acetate buffer, pH 5.7. The slide was washed in water and immersed in a 1:1 solution of 2% potassium ferrocyanide and 0.2N HC1. The slide was again washed in water and dried. The usual protein stains may be used as a counterstain.

THE METHOD OF SIMULTANEOUS ELECTROPHORESIS OF ANTISERUM AND ANTIGEN (IMMUNOOSMOPHORESIS) APPLIED TO LENS ANTIGENS

Swanborn, P. L., Exp. Eye Res. **5**, 302–308 (1966).

Vertebrate α-crystallins were investigated using the technique of immunoosomophoresis. Antigen and antibody were electrophoresed simultaneously and met to form precipitates in the zone of contact. The technique was carried out on microscope slides coated with 1.5% Noble Agar in Veronal buffer (ionic strength 0.05, pH 8.4). Electrophoresis was performed at 20 volts/cm at $8°C$ for 15 minutes. A pre-α-crystallin group was found only in mammalian species.

ALLOTYPIC SPECIFICITIES IN PIGS

Trávnícek, J., Folia Microbiol. **11**, 11–13 (1966).

The formation of isoantibodies in the pig was investigated by the injection of gamma globulin isolated from pig serum in 50 mg doses. A total of 500 mg gamma globulin was given each animal. Antibodies formed were characterized by immunoelectrophoresis and classified as gamma M globulins. At least two allotypic specificities of gamma M globulin in pigs was demonstrated.

IMMUNE REACTIONS IN ELDERLY HUMANS

Virag, S., and L. Kochar, Zh. Microbiol. Epidermol. Immunobiol. **43**, 99–103 (1966).

Elderly humans and rabbits revealed a decreased response to tetanus toxoid as compared to younger subjects. The γ-globulin level was measured by radioimmunoelectrophoresis using I^{131}-labeled antibody.

EMBRYO-SPECIFIC SERUM PROTEINS IN THE CHICK

Weller, E. M., Texas Rep. Biol. Med. **24**, 164–172 (1966).

Immunoelectrophoresis on 1% lonagar-coated lantern slides using a discontinuous Veronal acetate buffer was used to characterize the proteins of chickens. The lonagar was prepared in 0.06M sodium barbital and 0.007 barbituric acid, pH 8.2, ionic strength 0.05. The electrode compartment consisted of 0.04M sodium barbital and 0.06M sodium acetate, pH 7.8, ionic strength 0.1. Separation was achieved at 20 mA for two hours at room temperature. Antiserum prepared in rabbits was used to detect specific proteins. Results obtained with heterologous and homogeneous antisera revealed existence of adult-like as well as embyro-specific proteins.

IMMUNOELECTROPHORETIC ANALYSIS OF CYTOPLASMIC PROTEINS OF NEUROSPORA CRASSA

Williams, C. A., and E. L. Tatum, J. Gen. Microbiol. **44**, 59–68 (1966).

Immunoelectrophoretic separation of cytoplasmic proteins of *N. crassa* was achieved on microscope slides by the procedure as described by Graber and Williams [Biochem. Biophys. Acta **17**, 67 (1955)]. Electrophoresis was for 90 minutes at 5 volts/cm in a Veronal buffer (ionic strength 0.038, pH 8.2). Catalase and other nonspecific proteins were identified according to the method of Uriel (Methods in ImmunoChemistry Immunol **2**, Ed. by M. S. Chase and C. A. Williams, Ch. 14, New York: Academic Press). Antisera prepared by immunization of rabbits with *N. crassa* strain 37401-116-3 and strain 17-3A (wild) were used to detect species-specific proteins. The technique revealed more than 30 antigenic components in cytoplasmic proteins. In addition a phyletic relationship within the class ascomycetes was detectable from results obtained with antisera prepared against *N. crassa*.

PROTEIN-IRON LEGANDS AND THE TRANSPORT OF IRON IN HUMAN MILK. IN VITRO RESEARCH USING RADIOACTIVE IRON AND AUTO-RADIOGRAPHS

Ambrosino, C., A. Ponzone, and G. Papa, Minerva Med. **56**, 4165–4167 (1965).

The iron-binding capacity of defatted human milk was analyzed by electrophoresis. Two iron-binding components were found in the β-lactoglobulin fraction, one in the α-globulin fraction, and another in an immunoglobulin.

IMMUNOELECTROPHORETIC PATTERNS OF C-REACTIVE PROTEIN IN SERUM AND IN ASCITIC FLUID FROM PATIENTS WITH CANCER

Anzai, T., K. Sato, M. Fukuda, and C. M. Carpenter, Proc. Soc. Exp. Biol. Med. **120**, 94–98 (1965).

Immunoelectrophoresis performed according to Scheidegger's microdiffusion technique [Intern. Arch. Allergy Appl. Immunol. **7**, 103 (1955)] in a pH 8.2 Veronal buffer revealed three types of C-proteins — β, γ, and 2H — in serum of cancer patients. These proteins were also found in non-cancer patients. There was no dissimilarity exhibited upon gel diffusion precipitin analysis.

IMMUNOCHEMICAL STUDIES ON BONE MARROW OF THE RAT

Beernink, K. D., J. Courcon, and P. Grabar, Immunology **9**, 377–389 (1965).

This article describes a comparison made between a water-soluble extract of rat bone marrow and extracts of heart, spleen and liver. The comparison was made by using immunoelectrophoresis, agar-gel electrophoresis and selective staining to determine enzymatic content. Antiserum prepared against the extract exhibited gross cross reactivity. Immunoelectrophoretic results showed the commonness of antigens to bone marrow, spleen peritoneal exudates, thymus, heart and liver. Transferrin was the only serum protein detectable in quantity in washed bone marrow extracts.

GLI ASPETTI DELLA MATURAZIONE PROTEICA SERICA NEL NEONATO A TERMINE E NEL PREMATURO NELLE PRIME SETTIMANE

Borrone, C., Minerva Pediat. **17**, 1633–1642 (1965) **Italian.**

Immunoelectrophoresis demonstrated an incomplete spectrum of serum proteins in 8 full-term and 55 premature infants in the first two months following birth. The technique of Scheidegger [Int. Arch. Allergy Appl. Immunol. **7**, 103 (1955)] was used. Plasma protein development varied with the individual, suggesting an independent selective synthesis of certain proteins regardless of the developmental state.

IMMUNOCHEMICAL COMPARISON OF PROTEINS IN HUMAN GINGIVAL POCKET FLUID, SERUM AND SALIVA

Brandtzaeg, P., Arch. Oral Biol. **10**, 795–803 (1965).

Immunoelectrophoretic studies carried out according to the method of Scheidegger but adapted to the LBK-6800A equipment was used to characterize the proteins of gingival fluid, serum and saliva. Albumin, γG-, γA- and γM-globulins, and fibrinogen were identified in gingival pocket fluid. Only γG-globulin, γA-globulin and albumin were detected in unconcentrated saliva.

SIMPLIFIED GEL ELECTROPHORESIS: II. APPLICATION OF IMMUNO-ELECTROPHORESIS

Cawley, L. P., L. Eberhardt, and .D. Schneider, J. Lab. Clin. Med. **65,** 342–354 (1965).

A rapid method for performing immunoelectrophoresis in an agar-gel matrix is described. A 0.6% agar gel prepared in Veronal buffer, pH 8.6, ionic strength 0.050, was supported on 35 mm Cranar polyester photographic film leader. Electrophoresis was performed in a Durrum® Cell for 30 to 45 minutes at 200 volts. Antiserum can be added to the performed troughs to detect the antigen. The reliability of the technique was tested by immunoelectrophoretic characterization of three sera with gammaglobulinopathies.

ISOLATION AND PARTIAL CHARACTERIZATION OF TWO LOW MOLECULAR SIZED PROTEINS AND OF IMMUNOGLOBULIN G FROM CEREBROSPINAL FLUID

Fink, H., Protides Biol. Fluids, Proc. 13th Colloq. 221–225 (1965).

Low molecular weight proteins (trace proteins) of cerebrospinal fluid isolated by gel filtration with Sephadex® G-200 were characterized by immunoelectrophoresis, agar-gel, and paper electrophoresis. Gel filtration succeeded in separating protein components which could be identified by diffusing against specific antiserum.

IMMUNOELECTROPHORESIS OF SOLUBLE PROTEINS ISOLATED FROM CELLULAR FRACTIONS OF LIVER FROM RATS POISONED WITH CARBON TETRACHLORIDE

Gazzaniga, P. P., J. Pathol. Bacteriol. **90,** 682–685 (1965).

Fatty changes induced by carbon tetrachloride alter the mitochondrial and cytoplasmic proteins in the liver. No qualitative changes in the antigens of the liver were observed although there was an increase in two antigens of the α-globulin type. Fatty and control livers were extracted with 0.09M NaCl, 0.12M phosphate buffer (pH 7.4), 0.011M EDTA and 0.046M glucose at 4°C. Following differential centrifugation, the cytoplasmic and mitochondrial fractions were used to prepare antisera in rabbits. The γ-globulin of the antisera was isolated by precipitation with ammonium sulfate. Immunoelectrophoresis of the antisera was performed in a 0.065M sodium diethylbarbiturate–0.017M HC1 buffer, pH 8.2, ionic strength 0.05M. Three volts/cm were applied for three hours, then 6 volts/cm for five hours.

THE QUANTITATION OF HUMAN SERUM ALBUMIN BY IMMUNOELEC-TROPHORESIS

Griffiths, B. W., B. G. Sparkes, and L. Greenberg, Can. J. Biochem. Physiol. **43**, 1915–1917 (1965).

A method for analysis of HSA which depends upon detecting areas of antigen excess, equivalence, and antibody excess is described. A pH 8.6 Veronal buffer, ionic strength 0.05, was the electrolyte in agar-gel electrophoresis; 250 volts was applied for one hour. Goat antisera was applied and the bands allowed to develop for 20 hours. The antisera was titrated for estimation of the amount of HSA.

DIFFERENTIATION OF LEUKOCYTIC FIBRINOLYTIC ENZYMES FROM PLAS-MIN BY THE USE OF PLASMATIC PROTEOLYTIC INHIBITORS

Hermann, G., and P. A. Miescher, Intern. Arch. Allergy Appl. Immunol. **27**, 346–354 (1965).

Immunoelectrophoresis performed on agar plates into which fibrinogen had been incorporated was used to distinguish plasmin from fibrinogen. A pH 8.2 barbital buffer, ionic strength 0.05, contained 0.0082N calcium chloride. Extracts of normal human leukocytes and lymphocytes from patients with lymphocytic leukemia were shown to contain fibrinolytic enzymes. The mobility of these enzymes was in the alpha-1 to beta-2 region.

IMMUNOCHEMISCHE BEFUNDE BEI HEPATITIS EPIDEMICA: II. DIE IM-MUNOLOGIE DER β-LIPOPROTEINE

Kellen, V. J., Acta Hepato-Splenol. **12**, 228–231 (1965) **German**.

Qualitative and quantitative differences in the β-lipoproteins were found by performing immunoelectrophoresis on the serum of patients with hepatitis and normal controls. Agar electrophoresis was performed in sodium barbital buffer, pH 8.6, ionic strength 0.05, with a voltage of 5 volts/cm for 100 minutes. Commercial antihuman sera was used to form the precipitin lines.

MERCURIC BROMOPHENOL BLUE STAINING OF PRECIPITIN LINES IN AGAR

LaVelle, A., Stain Technol. **40**, 347–349 (1965).

Mercuric Bromophenol Blue is an excellent stain for antibody-antigen precipitin lines in agar. The precipitin-carrying slides were incubated overnight in saline at 4°C and then washed in saline for two hours at 25°C. Soaking in saline stopped the precipitin reaction. The slides were rinsed with water for ten minutes and then dried overnight at 37°C. Filter paper should not be used for drying. The precipitin lines were fixed in 95% ethanol and hydrated for five minutes in water. The slides were stained for ten minutes in the MBB mixture which consisted of 10 mg HgCl$_2$, 0.1 mg Bromophenol Blue; 100 ml 95% ethanol. The slides were rinsed through 95% and absolute alcohol and xylene before mounting with resin.

ISOLATION AND PARTIAL CHARACTERIZATION OF TRACE PROTEINS AND IMMUNOGLOBULIN G FROM CEREBROSPINAL FLUID

Link, H., J. Neurol. Neurosurg. Psychiat. **28**, 552–559 (1965).

The finding of β-trace protein and γ-trace proteins from human cerebrospinal fluid was confirmed. DEAE cellulose and Sephadex® chromatography were used to further study these low molecular weight proteins. An isolation procedure is described. Agar-gel electrophoresis was performed in a 0.9% agar solution in a continuous sodium Veronal buffer, pH 8.4, ionic strength 0.05. Electrophoresis was done for 25 minutes at 200 volts. Slides were stained with Amido Black DB. Starch-gel electrophoresis was done in 0.035M glycine buffer, pH 8.8, containing 8M urea. The electrode vessels contained 0.3M boric acid — 0.06N NaOH, pH 8.2.

IMMUNOELEKTROPHORETISCHE DETEKTION UND EIGENSCHAFTEN DER OSCENDOGENEN SERUM-LIPOPROTEINLIPASE

Losticky, C., Hoppe-Seylers Z. Physiol. Chem. **342**, 13–19 (1965) **German.**

Agar-gel electrophoresis and immunoelectrophoresis on agar gel, using a pH 8.3 Veronal-citrate buffer, μ .05, were used to demonstrate plasma lipase activity to Tween-60 in normal plasma and serum. Lipoprotein lipase was shown to be the cause of this activity through the use of the inhibitors protamine sulfate, heparin and sodium desoxycholate.

AN ALBUMIN AS γ-GLOBULIN CHARACTERISTIC OF BOVINE CEREBRO-SPINAL FLUID

MacPherson, C. F. C., and M. Saffran, J. Immunol. **95**, 629–634 (1965).

Agar-gel immunoelectrophoresis by the method of Scheidegger [Intern. Arch. Allergy Appl. Immunol. **7**, 103 (1955)] was used to identify an albumin and γ-globulin in bovine cerebrospinal fluid. High titer rabbit antibovine-CSF sera, adsorbed with guinea pig anti-CSF sera, was used to form the precipitin lines.

UBER DIE RADIOIMMUNOLOGISCHE BESTIMMUNG VON INSULIN IM BLUT

Melani, F., H. Ditschuneit, K. M. Bartlett, H. Friedrich, and E. F. Pfeiffer, Klin. Wochschr. **43**, 1000–1007 (1965) **German.**

The authors describe various methods for isolating antibody-bound I^{131} insulin.

Excellent separation of free and antibody-bound insulin was achieved using agar-gel electrophoresis, anion exchange resin Dowex® 1 and Amberlite® CG 4001I.

DETECTION OF CARBONIC ANHYDRASE ACTIVITY AFTER ELECTROPHORESIS AND IMMUNOELECTROPHORETIC ANALYSIS

Micheli, A., Enzymol. Biol. Clin. **5,** 175–178 (1965).

A histochemical technique for the detection of carbonic anhydrase isozymes of erythrocyte hemolysates is described. After paper, agar or immunoelectrophoresis, the following incubation solution was used:

A: 1.0 ml $CoSO_4$ — 0.1M,
6.0 ml H_2SO_4 — 0.05M
B: Dissolve 1.0 g $NaHCO_3$ in 50 ml Na_2SO_4 — 0.1M

A and B were mixed just before incubation. The solution was brought in contact with the electrophoresis plate or by first impregnating filter paper and laying over electrophoresis plate. At least two isozymes were detected by this procedure.

IDENTIFICATION OF INTRINSIC FACTOR AUTOANTIBODY AND INTRINSIC FACTOR IN MAN BY RADIOIMMUNODIFFUSION AND RADIOIMMUNOELECTROPHORESIS

Samloff, I. M., and E. V. Barnett, J. Immunol. **95,** 536–541 (1965).

Radioautography following immunoelectrophoresis was used to detect intrinsic factor and an autoantibody to it. Anti-whole human serum was prepared in horses to use in immunoelectrophoretic analyses. Rabbit antisera for specific globulin fractions was also used. Immunoelectrophoresis was performed in agar gel at pH 8.6 [Intern. Arch. Allergy Appl. Immunol. **7,** 103 (1955)]. The vitamin B_{12} was labeled with Co^{60}. Autoantibody was found in the gastric juices of 12 patients with pernicious anemia.

ZUR FRAGE DER GEFRIEDENATURIERUNG VON SERUMPROTEINEN: PAPIER- UND IMMUNOELEKTROPHORETISCHE UNTERSUCHUNGEN

Scheiffarth, F., H. Götz, and R. Creny, Z. Klin. Chem. **3,** 81–84 (1965)
German.

Paper or agar-gel supports for immunoelectrophoresis were employed to study the effects of freezing on serum proteins. Detectable changes in the electrophoretic pattern were evident in the proteins after they were frozen and thawed 12 times. Quantitative changes were evident after this freeze-thaw cycle was performed 20 times. These changes were only detectable by immunoelectrophoresis. Relative decreases were detectable in the α-precipitate, β-siderophilin, γ-globulin, and in some cases in loss of γ_1-A and γ_1-M components.

IMMUNOCHEMICAL COMPOSITION OF PEPTIC FRAGMENTS AND FAB AND FC FRAGMENTS OF NORMAL HUMAN GAMMA GLOBULIN

Skvaril, F., and V. J. Brummelova, J. Hyg. Epidemiol. Microbiol. Immunol. **9**, 265–269 (1965).

Immunoelectrophoresis was used to study the fragments remaining after spontaneous decomposition of human gamma globulin and fragments formed by pepsin digestion of IgG. Scheidegger's technique was used on a micro-scale [Intern. Arch. Allergy Appl. Immunol. **7**, 103 (1955)]. The outer line of the fragment FAB and the outer line from the peptic digest seem to be identical; the inner line of the peptic digest appears identical with fragment FC. Possible structures are discussed.

THE INFLUENCE OF DIFFERENT METHODS OF DISSOCIATING THE RHEUMATOID FACTOR, STUDIED BY ULTRACENTRIFUGATION, SHEEP-CELL TESTS AND IMMUNOELECTROPHORESIS

Svartz, N., H. Schatz, and S. Hedman, Acta Med. Scand. **177**, 213–217 (1965).

The changes which occur in the structure of the rheumatoid factor of serum on precipitation in cold, fractionation with ammonium sulphate, and treatment with cysteine or mercapethanol were examined. Immunoelectrophoresis and ultracentrifugation sheep cell tests were the techniques used. Cold fractionation did not alter sedimentation pattern, the hemagglutinating ability of the component for sheep cells, or the immunoelectrophoretic pattern. Fractionation with ammonium sulfate resulted in changes in the sedimentation pattern, although hemagglutinin tests were unaltered. Treatment of the isolated material with cysteine or mercapethanol completely abolished its characteristic behavior.

Isoelectric Focusing

MICRO ISOELECTRIC FOCUSING IN POLYACRYLAMIDE GEL COLUMNS

Castimpoolas, N., Anal. Biochem. **26**, 480–482 (1969).

Microisoelectric focusing on polyacrylamide gels prepared with 40% ampholyte (Ampholyne®) is described. The isoelectric focusing was performed on gels prepared in the Canalco® disc electrophoresis apparatus. The upper bath contained 5% phosphoric acid and the lower one contained 5% ethylene-diamine solution. Electrofocusing was performed at a starting current of 5 mA per gel for one hour at 150 volts. The current drops during the run to 0.5 mA per gel. Most of the separated protein bands could be visualized by placing the gel in 12% trichloroacetic acid solution.

ISOLATION OF THE PI 4.5 SOYBEAN TRYPSIN INHIBITOR BY ISOELECTRIC FOCUSING

Catsimpoolas, N., C. Ekenstam, and E. W. Meyer, Biochim. Biophys. Acta **175**, 76–81 (1969).

Soybean trypsin inhibitor was isolated in a pure state using isoelectric focusing in the region pH 3.0–pH 10.0. An LBK 8102® electrofocusing column with a capacity of 440 ml column was used. Carrier ampholyte Ampholyne® was used to prepare density gradients. Electrofocusing for 16 hours at a potential of 500 volts (10°C) was sufficient to separate the inhibitor. Polyacrylamide-disc electrophoresis was used to demonstrate the purity of the isolated material.

DEMONSTRATION OF THE SUBUNIT STRUCTURE OF ALPHA-CRYSTALLIN BY ISOELECTRIC FOCUSING

Bloemendal, H., and J. G. G. Shoenmakers, Sci. Tools **15**, 6–7 (1968).

Bovine α-crystallin was separated by electrofocusing in a sucrose gradient containing 1% ampholytes (pH 5–8) and 6M urea. Electrophoresis was performed for two hours at 500 volts and subsequently at 800 volts for 48 hours. Fractions were collected and analyzed by polyacrylamide-gel electrophoresis. Considerable heterogeneity of the fractions separated was evident.

ISOELECTRIC FOCUSING IN POLYACRYLAMIDE GELS

Dale, G., Lancet **1**, 847–848 (1968).

Isoelectric focusing takes place when a potential is applied in an electrolyte system in which the pH steadily increases from the anode to the cathode. Thus, proteins separate as each accumulates in the region of its isoelectric point. A method is described for stabilizing the pH gradient for polyacrylamide-disc electrophoresis. The results demonstrated separation of 40 bands in human serum and more LDH isoenzymes that can be seen with conventional techniques.

ISOELECTRIC FOCUSING OF PLANT PIGMENTS

Jonsson, M., Sci. Tools **15**, 2–6 (1968).

The pigments of raw saps of red beet, bilberry, black currant and strawberry were separated by isoelectric focusing in a pH gradient in a region of 3–10.

The gradients were created with Carrier ampholytes (Ampholyne®). An electrolysis column of 110 ml volume was used. Electrolysis was performed at 700 volts for the Carrier ampholyte gradients, 200–500 volts for the acid gradients for four to six hours at 4°C. Separation was excellent.

THE ISOELECTRIC FRACTIONATION OF HEN'S EGG OVOTRANSFERRIN

Wenn, R. V., and J. Williams, Biochem. J. **108**, 69–74 (1968).

Six protein fractions were separated from ovotransferrin, half saturated with iron, by isoelectric focusing. Ampholyte solutions and the apparatus of LBK® company were used. The ampholyte solutions had a pH range of 5 to 8 and a current strength of 500 volts for 24 hours. Starch-gel electrophoresis was performed on the fractions. Fractions 1, 3, and 5 showed mobility similar to that of the slower, major component. Fractions 2, 4, and 6 possessed fast mobility. Amino acid analyses of peak 5 (pH_I 5.78) and peak 6 (pH_I 5.62) showed all the amino acids as the unfractionated material except tryptophan and an amide. Labeling of ovotransferrin with 59_{Fe} and then subjecting it to isoelectric fractionation revealed that two major peaks of 59_{Fe} activity were present, one at pH_I 5.2 and pH_I 5.5.

ISOELECTRIC FRACTIONATION, ANALYSIS AND CHARACTERIZATION OF AMPHOLYTES IN NATURAL PH GRADIENTS

Vesterberg, O., and H. Svenson, Acta Chem. Scand. **20**, 820–834 (1966).

Myoglobulins with isoelectric point (pl) differences of only 0.06 units were separated by electrophoresis using a mixture of low-molecular aliphatic poly-amino–polycarboxylic acids as an ampholyte carrier (LBK–Produkter AB, Stockholm–Bromma). The ampholyte was easily dialyzable, colorless, a buffer, and non-light-absorbing at 280 mm. Zone breadth of myoglobulin separation satisfied a theoretically derived equation. A mathematical definition of resolving power was also derived.

Moving-Boundary Electrophoresis

MOVING BOUNDARY ELECTROPHORESIS OF FOOD STABILIZERS

Hidalgo, J., and P. M. T. Hansen, J. Food Sci. **33**, 7–11 (1968).

Moving-boundary electrophoresis using the Perkin–Elmer Model 38 A apparatus of Tiselius design, was used to characterize locust bean gum, guar gum, arobic gum carboxymethylcellulose, K and λ-carrageenan, alginate and non-dialyzable fraction of corn syrup solids. A sodium phosphate buffer, pH 7.0, ionic strength 0.02, was used. Both descending and ascending migration were used. Locust bean gum and guar gum did not migrate; thus separation'was poor. The ascending boundary of the remaining substances was sharp and mobilities were calculated.

LACTATE DEHYDROGENASE STUDIES IN PUERTO RICAN BATS

Valdivieso, D., E. Conde, and J. R. Tamsitt, Comp. Biochem. Physiol. **27**, 133–138 (1968).

The lactic dehydrogenase isoenzymes of tissues of Jamaican fruit bats *(Artibeus jamaicensis),* brown flower bat *(Erophylla bombifrons)*, and large mastiff bats *(Molassus fortis)* were investigated using moving-boundary electrophoresis. Separation was achieved using the Durrum-type chamber (Beckman Instruments and Sepraphore® III 2.5 x 9 cm cellulose-polyacetate strips). A pH 8.6 Veronal buffer, ionic strength 0.075, was used, and a current of 1.5 mA per strip was applied for two hours. Heart and liver isozyme patterns exhibited similar distribution. Liver of fetal and adult *E. bombifrons* have identical LDH profiles, whereas fetal heart exhibited a predominance of cathodal (M) forms. H forms (anodal) predominate in adult heart. LDH_x was absent in testicular tissue of male *A. jamaicensis.*

PHYSICAL AND CHEMICAL PROPERTIES OF BOVINE MILK AND COLOS-TRUM WHEY M-1 GLYCOPROTEINS

Bezkorovainy, A., and C. M. Gerbeck, J. Dairy Sci. **50**, 1368–1375 (1967).

An acidic glycoprotein from bovine milk and colostrum, termed M-1, was studied to determine its physical and chemical characteristics. Moving-boundary electrophoresis in barbital buffer, pH 8.6, or acetate buffer, pH 4.5, both at ionic strength 0.1, indicated that the M-1 acidic glycoprotein fractions were homogenous. This electrophoresis was performed at 4°C at 5.1 mA in a 15 ml cell. Ultracentrifugation data also indicated that the preparation was homogeneous. Polyacrylamide-gel electrophoresis was done using a 5% gel prepared in 0.38 M Tris–0.06M HCl buffer at pH 8.6 Glass slides, 8.2 x 10.2 cm, were covered with a 3 mm layer of the gel; 0.0248M Tris–0.192M glycine, pH 8.6, was the buffer used in the chamber. A 50 μl sample of M-1, protein 30 mg/ml, was separated in 140 minutes using a gradient of 36 volts/cm at OC. The preparation was stained with periodate-fuchsin. Several bands were detected using this method, which suggests that the M-1 preparation is a mixture of closely related molecular species.

ELECTROPHORETIC AND BUOYANT DENSITY VARIANTS OF SOUTHERN BEAN MOSIAC VIRUS

Magdoff-Fairchild, B. S., Virology **31**, 142–153 (1967).

Moving-boundary electrophoresis performed in a Perkin–Elmer apparatus 38A equipped wiith Schierlen and fringe optics was used to characterize five variants of the Pittsburgh straing of bean mosiac virus. The buffers 0.067M phosphate, pH 7.25, and 0.05M sodium acetate, pH 3.63, were used. In the 0.10 ionic strength 0.10 buffers the isoelectric paints were found to range between 3.95 and 6.03. When sodium chloride was incorporated into the buffer, the isoelectric points were lowered. Buoyant density centrifugation in cesium chloride showed the viruses to be homogenous.

ELECTROPHORETIC BEHAVIOR OF MAMMALIAN-TYPE CYTOCHROME C

Barlow, G. H., and E. Margoliash, J. Biol. Chem. **241**, 1473–1477 (1966).

Human heart cytochrome c and cytochrome c from other species were examined by moving-boundary electrophoresis using a 0.407mM Tris-HCl buffer, pH 6.9, $\mu=0.1$, and a 0.163mM Tris-HCl buffer, pH 6.9, $\mu=0.2$. Isionic points were determined in a cell comparable to the one described by Katz and Ellinger [Biochemistry **2**, 406 (1963)] Anomalous boundaries were obtained and were attributed to the binding of a buffer constituent by cytochrome c. The isionic point of all cytochrome c tested was found to be pH 10.4 ± 0.04.

MULTIPLE ELECTROPHORETIC ZONES ARISING FROM PROTEIN-BUFFER INTERACTION

Cann, J. R., Biochemistry **5**, 1108 (1966).

In moving-boundary electrophoresis, bovine serum albumin interacted reversibly with phosphate–borate and borate buffers in a pH range 6.2–9.2. Two or more electrophoretic peaks were seen. These multiple zones on cellulose acetate are not indicative of heterogeneity but are due solely to protein-buffer interaction.

EFFECT OF FREEZING RED BLOOD CELLS AND HEMOLYSATES AT DIFFERENT TEMPERATURES

Chanutin, A., and R. R. Curnish, Arch. Biochem. Biophys. **113**, 114–121 (1966).

The change in the electrophoretic pattern following freezing of whole RBC at −13°C, −20°C, and −79°C was studied by electrophoresis using a Tiselius apparatus. A protective effect was found when glucose, sucrose, glycerol, or dimethylsulfoxide was mixed with the RBC prior to freezing.

ÉTUDE COMPARATIVE DES MOBILITÉS ELECTROPHORETIQUES DES PRO-TÉINES MYELOMATEUSES SÉUQUES ET URINAIRES

Creyssel, R., J. Site, and G. B. Richard, Rev. Franc. d'Études Clin. Biol. **11**, 290–293 (1966).

Moving-boundary electrophoresis was used to measure the mobility of serum-M components and urinary Bence-Jones proteins. A pH 8.6 Veronal buffer, ionic strength 0.1, was the electrolyte. The mobility of the Bence-Jones proteins is unrelated to that of the serum-M component. The results were further analyzed by patient group, and theoretical aspects discussed.

SOME CHEMICAL, IMMUNOCHEMICAL AND ELECTROPHORETIC PROP-ERTIES OF BOVINE FOLLICULAR FLUID

DesJardins, C., K. T. Kirton, and H. D. Hafs, J. Reprod. Fertility **11**, 237–244 (1966).

Follicular fluids and blood sera from nonpregnant cows were studied by agar-gel electrophoresis, moving-boundary electrophoresis and immunoelectropho-resis on agar gel. Agar-gel electrophoresis was performed on 3 μl of follicular fluid and serum in 0.85% agar using a barbital buffer, pH 8.6, ionic strength 0.0375, and 200 volts (60 mA) for 30 minutes. Moving-boundary electro-phoresis was performed in the Tiselius apparatus at 5°C. Five different buffer systems were used. A pH 8.6 barbital buffer gave best results. Eight different components were detected by moving-boundary electrophoresis. Rabbit anti-sera against antifollicular fluid, blood serum or plasma detected at least seven or eight lines of identity.

ON THE HETEROGENEITY OF BEEF HEART CYTOCHROME C

Flatmark, T., Acta Chem. Scand. **20**, 1476–1486 (1966).

The heterogeneity of cytochrome c was confirmed by moving-boundary electrophoresis in the pH range 8.0–11.2. The isoelectric points (pI 0°C) in borate buffer were: Cy I:10.78 (10.8), Cy II:10.58 (10.6), and Cy III:10.36 (10.22) in the ferric (ferrous) form. An estimate of the number and nature of ionizable groups was calculated from the function of mobility vs. pH.

INFLUENCE OF DIET AND HEREDITY ON THE SERUM PROTEIN COMPO-NENTS OF THE RAT

Lakshmanan, F. L., and M. W. Marshall, Proc. Soc. Exp. Biol. Med. **122**, 535–539 (1966).

Blood serum of rats was diluted 1:5 with a pH 8.6 Veronal buffer, ionic strength 0.1, and dialyzed against this buffer overnight in the cold. Moving-boundary electrophoresis was performed on this preparation. A significant difference between rat stain and diet was found for both the pre-albumin and beta globulin fractions. The concentrations and mobilities were calculated from the Schlieren patterns.

ISOLATION AND CHARACTERIZATION OF MILK PROTEIN COMPONENT "8"

Kolar, C., and J. R. Brunner, J. Dairy Sci. **48,** 772 (1965).

In free-boundary electrophoresis, component "8" was the leading peak of heat-treated acid whey. In pH 8.6 Veronal buffer, the electrophoretic mobility was -8. Other characteristics were S_{20} κ 1.0 and 40,000 in phosphate buffer pH 7.0; nitrogen 14%, phosphorus 1.2%, and carbohydrate positive.

ELECTROPHORETIC STUDIES ON β-LACTOGLOBULIN OF BUFFALO MILK: COMPARISON OF β-LACTOGLOBULINS OF COW AND BUFFALO MILK

Sen, A., Bull. Nat. Inst. Sci. India **29,** 289–300 (1965).

The electrophoretic mobilities of cow and buffalo crystalline β-lactoglobulin B were compared. Moving-boundary electrophoresis was done in 0.1N acetate and phosphate buffer (pH range 4 to 8) at 1°C with a potential gradient of 5 volts/cm. Buffalo β-lactoglobulin B had a lower electrophoretic mobility than did the cow protein, and the isoelectric points differed (5.28 vs. 5.23).

Paper Electrophoresis

ALLERGIC ENCEPHALOMYELITIS: ISOLATION AND CHARACTERIZATION OF ENCEPHALITOGENIC PEPTIDES FROM THE BASIC PROTEIN OF BOVINE SPINAL CORD

Hashim, G. A., and E. H. Eylar, Arch. Biochem. Biophys. **129,** 645–654 (1969).

Bovine spinal cord encephalitogen was digested with pepsin for two hours and subjected to gel filtration on a Sephadex® G-25 + G-75 column. After further fractionation of the peptides could be demonstrated by high-voltage electrophoresis on Whatman® 3MM paper in a buffer system of pyridine–

acetic acid−n-butanol−water (1:1:2:36), separation was carried out at 2500 volts for one hour. Eleven peptidic substances were recovered after a desalting step on Sephadex® G-25 + G-75.

SEPARATION OF LACTATE DEHYDROGENASE ISOZYMES BY PAPER ELECTROPHORESIS

Morales-Malva, J. A., A. Vallega-Magasich, J. M. Uribe-Echewarria, and M. Sapag-Hager, Proc. Soc. Exp. Biol. Med. **130**, 224−226 (1969).

Improved separation of serum and tissue lactate dehydrogenase isozyme was obtained by incorporating 5 μl of human serum into homogenates. The homogenates were prepared in 0.9% saline and electrophoresed on Whatman® 3MM paper strips (30 x 4 cm) using Michaelis Veronal buffer, pH 8.2, ionic strength 0.075. The run was carried out at 300 volts for five hours. In addition, bovine serum albumin and human serum albumin were found to improve the resolution obtained on paper.

FILTER PAPER ELECTROPHORESIS OF PURINES AND PYRIMIDINES: MOBILITY DATA

Adams, W. S., and M. Nakatani, J. Chromatog. **37**, 343−347 (1968).

Paper electrophoresis on paper strips 3 x 30.4 cm (Schleicher and Schüll 2043-A®) separated and identified purines, pyrimidines and nucleosides. A pH 9.2 sodium borate buffer, ionic strength 0.1, was used as the electrolyte. Mobility data for 94 authentic compounds are recorded.

PAPER ELECTROPHORESIS OF MONOHYDRIC ALCOHOLS AND HYDROXY ACIDS: SEPARATION AS ZANTHATES

Frahn, J. L., J. Chromatog. **37**, 279−291 (1968).

Paper electrophoresis performed on Whatman® No. 4 (13.5 x 61 cm) paper strips separated potassium salts of zanthates derived from monohydric alcohols. Several electrolyte solutions were used; however, sodium hydrogen carbonate (0.1M), pH 8.4, was used most often. The electrophoretic run proceeded at 21 volts/cm at 4°C for one hour. The relative mobility rates for 21 compounds were recorded.

SEPARATION PAR ÉLECTROPHORÈSE SUR PAPIER DE Ge^{IV}, As^{+++}, As^{+5} ET DE Ge^{++++}, $77As^{+3}$, As^{V}. ETUDE CINÉTIQUE DE L'OXYDATION DU RADIOARSENIC PAR LE PEROXYDE D' HYDROGÈNE

Genet, M., and C. Ferrandini, J. Chromatog. **37**, 527−532 (1968) **French.**

Macro amounts of As^{o}, Ge^{IV}, As $^{+3}$ and As^{+4} were separated by paper electrophoresis. Separation was achieved on Whatman® 3MM paper (39 x 3 cm) impregnated with 0.05N solution of potassium salt; 0.05N potassium salt was also used as the electrolyte. A potential gradient of 600 volts was applied for one hour. When tracer amounts of trivalent arsenic were used according to the following equation:

$$^{77}As^{+3} + H_2O_2 \xrightarrow{Ki} {}^{77}As^{V} + 2OH^{-}$$

the rate constant (Ki at pH 14) was found to be $2.3.10^2$ M^{-1} Sec^{-1}.

N-TERMINUS OF α-CRYSTALLIN

Hoenders, H. J., J. G. G. Schoenmakers, J. J. T. Gerding, G. I. Tesser, and H. Bloemendal, Exp. Eye Res. **7**, 291–300 (1968).

An N-terminus peptide was isolated from bovine lens α-crystallin (after digestion with pronase) by cation exchange on Dowex® 50 W-X2 and high-voltage paper electrophoresis. The first electrophoretic purification step was carried out in pyridine–glacial acetic acid–water (100:10:890 v/v/v), pH 6.5. The second step was carried out at pH 3.5 in pyridine + glacial acetic acid + water (10:100:1890). A potential of 25 to 30 volts/cm for two hours was used; MN 214 paper 30 x 40 cm was used. The structure of the compound was elucidated by TLC, gas chromatography and NMR studies, and it is probably N-acetyl-Met-Asp. One acetyl group per 21,000 g of α-crystallin was found.

IMMUNOGLOBULIN DISORDER ASSOCIATED WITH FELINE EOSINOPHILIC GRANULOMATOSIS

Howard, E. B., and C. C. Jannke, Am. J. Vet. Clin. Pathol. **2**, 21–26 (1968).

Paper and polyacrylamide-disc electrophoresis characterized the serum proteins of cats with multiple eosinophilic granuloma. The immunoglobulin fraction was greatly elevated on paper electrophoretograms. Resolution obtained on a 7% polyacrylamide gel using a Tris-glycine buffer shows a very heterogeneous population of proteins in the gamma globulin area.

BEHAVIOUR OF PHENYLPHOSPHINIC AND PHENYLPHOSPHONIC ACIDS IN PAPER ELECTROPHORESIS

Kiso, Y., M. Kobayaski, Y. Kitaoki, K. Kawamoto, and J. Takada, J. Chromatog. **33**, 561–563 (1968).

The paper electrophoretic behavior of phenylphosphinic and phenylphosphonic acids was investigated using ammonium acetate buffer, ionic strength 0.1 over the pH range 1.0–11. A potential gradient of 500 volts/34 cm was applied for 30 minutes. The relative mobility of the acids was plotted against pH. Phenylphosphonous acids were characterized by one inflection point, phenylphosphonic acids by two.

DETERMINATION OF THE MAJOR PROTEINS AND MUCOPROTEINS IN THE DUODENAL FLUIDS OF CYSTIC FIBROSIS AND CONTROL SUBJECTS

Knauff, R. E., and J. A. Adams, Clin. Chem. Acta **19**, 19–24 (1968).

Paper electrophoresis using the Beckman-Spinco® Model R system was used to separate proteins of duodenal fluids. The electrophoretic strip containing 60 μl of duodenal fluid was subjected to 2.5 mA per cell for 20 to 30 hours at 4°C in order to effect separations. After completion of the electrophoretic run, the proteins were stained with Bromophenol Blue, and the mucoprotein fractions were stained with periodic acid Schiff reagent. Quantitation was achieved by scanning with the Analytrol. Fifteen different proteins and muco-proteins were found in duodenal fluid. Marked differences between cystic fibrosis and control subjects were found.

PROTEINS AND MUCOPROTEINS IN THE DUODENAL FLUIDS OF CYSTIC FIBROSIS AND CONTROL SUBJECTS

Kanuff, R. E., and J. A. Adams, Clin. Chim. Acta **19**, 245–248 (1968).

Nine major proteins and mucoproteins were identified in duodenal fluid from control subjects and cystic fibrotic subjects. These proteins were observed in duodenal fluids separated by paper electrophoresis using the Beckman-Spinco® Model R.

NEW METHOD OF FIBRINOGEN DETERMINATION BY THE PAPER ELEC-TROPHORESIS OF PLASMA PROTEINS

Lozsa, A., Kiserl. Orvostud. **20**, 209–215 (1968).

Paper electrophoresis performed in a pH 8.6 Veronal–acetate–HCl buffer with 0.8 g Tween 80 separated plasma into seven fractions in 17 hours. Fibrinogen was found to migrate between the point of application and γ-globulin. The separated fibrinogen and other proteins were stained with acid fuchsin and could be quantitated by elution or spectrophotometry. A correction factor of 1.33 must be applied for the fibrinogen band because it absorbs more dye than other separated fractions. Separation was achieved at 130 volts for the first two hours; thereafter the voltage was 180–200 volts.

ELECTROPHORETISCHE TRENNUNGEN VON HEXOSAMIN- UND HEXU-RONSÄUREDERIVATEN ALS MOLYBDATKOMPLEXE

Mayer, H., and O. Westphal, J. Chromatog. **33**, 514–525 (1968) **German.**

Natural hexosamines and hexuronic acid derivatives which occur in the lipo-polysaccharides of E. coli (O_7:K_7:H_4) were separated by paper electrophoresis in a molybdate buffer, pH 5.0. Separation was achieved at 45 to 47 volts/cm.

The principle of separation is the formation of negatively charged molybdate complexes at pH 5.0.

THE PAPER ELECTROPHORETIC STUDY OF ION PAIR FORMATION: III. ION PAIR FORMATION WITH QUARTERNARY AMMONIUM IONS

Mazzei, M., and M. Lederer, J. Chromatog. **37,** 292–296 (1968).

High-voltage electrophoresis using a Camag apparatus was used to study the mobility of ion pairs formed between quarternary ammonium ions and various anions. Whatman® No. 1 paper was used. A potential of 1500 volts for 30 minutes at 15°C was employed. Several compounds of pharmaceutical interest were studied: decamethonium, succinylcholine, sincurarine and d-tubocurarine. Typical electrolyte systems were sodium dichloracetate (1 N, pH 4.5,) and dichloro-acetate (0.1 N, pH 4.5). Little ion pair formation was evident in 0.1 N solutions but was greatly increased in 1 N solutions.

CREATINE KINASE ISOENZYMES IN NORMAL AND PATHOLOGICAL SERA

Menache, R., I. Rubenstein, L. Gaist, and I. Marziur, Clin. Chim. Acta **19,** 33–35 (1968).

Paper electrophoresis performed in a barbital buffer (0.1M, pH 8.6) at 70 to 80 volts for 18 hours separated the creatine kinase isoenzymes of muscle and cardiac tissue, of normal and pathological sera, and of muscular dystrophy. Isoenzyme patterns characteristic of each tissue were observed. Normal sera yielded three fractions, whereas skeletal and cardiac muscles were characterized by four isoenzymes.

THE DETECTION OF IMIDAZOLES INCLUDING HISTIDINE AND SOME OF ITS DERIVATIVES IN BIOLOGICAL FLUIDS

Pasieka, A. E., and M. E. Thomas, Can. J. Biochem. **47,** 7–10 (1968).

High-voltage paper electrophoresis separated histidine and histidine-containing compounds from urine. Whatman® 3MM filter paper and an acid buffer, pH 1.9, 3.4, 5.8 and 9.2, were used to effect separation at 3000 volts/cm for 1.5 hours (20 mA). The separated compounds were detected by dipping the dried paper in n-butanol (dry) then into a solution of mercuric chloride in alcohol. The electrophoretogram was then dipped into a 0.2% solution of Bromophenol Blue in 95% ethyl alcohol. The reaction was found to be specific for histidine and imidazoles. It was possible to detect the compounds in 10 microliters of a 1 mg/100 ml solution.

PROTEIN, ENZYME, AND ISOENZYME DISTRIBUTION IN HUMAN SERUM FRACTIONS OBTAINED BY ULTRACENTRIFUGATION

Pruden, E. L., P. L. Creason, M. F. Creason, and W. D. Block, Clin. Chim. Acta **20**, 173–180 (1968).

Three gross fractions of both normal and abnormal human sera were separated by paper electrophoresis in an albumin–EDTA–barbital buffer. Separation was achieved at a potential of 100 volts for 18 hours. The separated proteins were stained with Oil Red O.

ZUR PAPIERELECTROPHORETISCHEN TRENNUNG DES PSYCHOPHARMAKONS TOFANIL UND EINIGER SEINER METABOLITEN

Schneider, G., and G. Schneider, J. Chromatog. **37**, 348–349 (1968) **German.**

Paper electrophoresis performed in three buffer systems was employed to separate the metabolites of tofranil, DMI, DDMI, and IDB. The metabolites of DDMI and IDB were Imipramin, DMI, DDMI and IDB when the electrolyte was 1% acetic acid.

BASIC PROTEINS OF COHN FRACTION III OF HUMAN PLASMA PROTEINS

Suzuki, K., and K. Schmid, Arch. Biochem. Biophys. **123**, 421–422 (1968).

Basic proteins of Cohn fraction III were purified by column chromatography on a CM-cellulose column equilibrated against pH 5.5 (ionic strength 0.005) sodium acetate buffer and eluted with an ionic strength 0.005 → 1.0; pH 5.5 → 8.6 sodium acetate buffer. Paper electrophoresis was used to characterize the proteins eluted. Diethyl-barbiturate-citrate buffer, pH 8.6, ionic strength 0.1, was used to electrophorese the separated fractions. Two different basic proteins with different cationic mobility were observed.

PROCEDURE FOR THE PHENOTYPING OF β-CASEIN AND β-LACTOGLOBULIN VARIANTS IN COW'S MILK

Arave, C., J. Daily Sci. **50**, 1320–1322 (1967).

This paper describes a modification of Aschaffenburg's [J. Dairy Sci. **49**, 1284 (1966)] technique which makes possible simultaneous phenotyping of β-casein and β-lactoglobulin. The same buffer is used, except 30 ml of 90.7% formic acid replaced 30 ml of 98.0% formic acid. Skimmed milk was mixed with an equal volume of a solution of 10 g urea, 100 ml water, 5 ml 60% ethanol and 0.1 g methyl red. Since β-lactoglobulins deteriorate rapidly, the milk should be either fresh or thawed from freshly frozen milk. Electrophoresis was performed at 4°C at a potential of 180 volts after the sample, on a dried piece of filter paper, had been embedded into the gel. After 15 minutes, the sample inserts were removed and electrophoresis continued at 200 volts for about 20 hours. A methyl red indicator could be incorporated into the gel to serve as a marker. Electrophoresis was stopped when the methyl red migrated 12 cm. Proteins were detected by staining with Amido Black 10B.

INTERACTION OF HISTONES WITH PHOSPHOPROTEINS AND CERTAIN OTHER PHOSPHORUS-CONTAINING COMPOUNDS

Ashmarin, I. P., and A. I. Komkova, Biokhimiya **32**, 640–645 (1967), cited in Biochemistry **32**, 526–531 (1967).

Paper electrophoresis using a medinal acetate buffer (pH 8.6, μ 0.1) or a phosphate buffer (pH 4.5, μ 0.1 at a potential gradient of 7 volts/cm) was used to examine the binding of phosphoproteins to histones. Of the proteins investigated, casein was found to coprecipitate with thymus histones. Phosphoprotein phosphatase was found to form a soluble complex with histones. This complex was identified by electrophoresis. Pepsin and albumin did not form soluble complexes with histones.

PEPTIDES RELEASED DURING THE ACTIVATION OF PRORENNIN

Bundy, H. F., L. D. Albizati, and D. M. Hogancamp, Arch. Biochem. Biophys. **118**, 536–541 (1967).

Horizontal-zone electrophoresis on Whatman® 3MM paper was used to follow the release of peptides from prorennin upon activation with acetic acid. A barbital buffer 0.05M, pH 7.6, was used. Electrophoresis continued for three hours with a potential of 90 volts (21 cm length). After completion of the run, the strips were dried and stained with Bromophenol Blue. Two basic peptides were released. One peptide had alanine as the NH_2-terminal amino acid. After gel filtration of the activation mixture, four peptides were formed.

SOME EFFECTS OF HYDROGEN PEROXIDE ON CASEIN AND ITS IMPLICATIONS IN CHEESE MAKING

Fox, P. F., and F. V. Kosckowski, J. Dairy Sci. **50**, 1183–1188 (1967).

H_2O_2 can be used in the United States as a sterilizing agent of milk prior to making cheddar and Swiss cheese. The effect of this treatment upon casein and the proteolytic digestion of it were investigated. It was demonstrated that increased softness in cheddar cheese made from H_2O_2-treated milk may be due to higher susceptibility of the proteins to proteolytic enzymes. Paper electrophoresis was one technique used to analyze the proteins in the cheese and milk. Whatman® 3MM paper, the citric acid–phosphate–urea buffer system of Aschaffenburg [Nature **192**, 431 (1961)] and a potential gradient of 10 volts/cm were the conditions of electrophoresis. The time of electrophoresis varied from 20 to 36 hours. One percent Amido Black in 10% acetic acid in methanol stained the strip in ten minutes. Ten percent acetic acid in methanol removed the excess dye.

LIPID-PROTEIN INTERRELATIONSHIP IN PREGNANCY

Gupta, A. N., Am. J. Med. Sci. **253,** 469–472 (1967).

During pregnancy there is an increase in serum lipids attributed to hormonal changes in the body. An alteration in the protein level of sera is attributed to hypervolemia. This study was performed to demonstrate possible changes in the lipid–protein balance during pregnancy. The micro-Kjeldhal method was used to measure total protein. Electrophoresis [Kumar, J. Sci. Indust. Res. **20C,** 236 (1961)] was used to fractionate and quantitate proteins and lipoproteins.

THE LABORATORY DIAGNOSIS OF MYOGLOBINURIA

Kal'nova, L. I., Klin. Med. **45,** 116–121 (1967).

Myoglobin and hemoglobin in urine and serum were separated by paper electrophoresis on Whatman® No. 3 paper for 20 hours at 250 volts (barbiturate buffer, pH 8.6, ionic strength 0.05). The papers were dried at 90°C for 30 minutes and stained with Amido Black 10B or 2% benzidine in 20% acetic acid. Myoglobin was not bound to other proteins and moved one-half to one-third as far as hemoglobin.

SEPARATION OF URINARY SUGARS BY PAPER ELECTROPHORESIS

Kennedy, D. A., Anal. Biochem. **18,** 180–181 (1967).

D-glucose, D-galactose and D-fructose in untreated urine were separated by paper electrophoresis. The electrolyte was made of equal volumes of sodium molybdate dihydrate A. R. (20.8 g/liter) and ammonium molybdate tetrahydrate A. R. 20.6 g/liter), the pH adjusted to 5.0 with sulfuric acid. The paper was Whatman® 3MM and was spotted with 50 μl and electrophoresed for two hours at 300 volts. Brown sugar bands were visualized by heating the strip at 105°C for 15 minutes.

ISOLATION OF AN α_1-GLYCOPROTEIN RESEMBLING "OROSOMUCOID": A SIMPLE METHOD FOR ITS ESTIMATION IN BLOOD AND TISSUE

Kernel, L., R. E. Schrohenlober, and R. C. Caldwell, Arch. Biochem. Biophys. **122,** 280–288 (1967).

A glycoprotein isolated from plasma and tissue by chloroform extraction was characterized by paper electrophoresis, ultracentrifugation, chemical analysis and immunological methods. Paper electrophoresis was performed in the Beckman® Model R apparatus in a 0.5M Veronal buffer, pH 8.6, and in pyridine–acetic acid buffer, pH 6.4, at 250 volts and 1.5 mA for six hours. High-voltage paper electrophoresis was carried out in the Ensco high voltage apparatus at 0.9 mA/cm paper at ten different pH values. The mobility of the substance was found to be similar to that of an α_1-globulin. Disc electrophoresis showed three to four faintly staining minor components. Immunoelectrophoretic analysis demonstrated a precipitin line in the area of the α_1-globulin.

CONTRIBUTION A' l'ÉTUDE DES PROTÉINES DE LA BILE PAR ÉLECTROPHORÈSE ET CHROMATOGRAPHIE SUR GEL DE SEPHADEX®

Lambilliotte, J. P., and M. C. Waussen-Cole, Acta Gastro-Enterol. Belg. **30**, 56–68 (1967) **French.**

The proteins of normal and pathological bile samples were separated by column chromatography on Sephadex® G-25, G-75, G-100 and G-200. Elution of the bile substances was accomplished with 0.002M sodium chloride. Both paper electrophoresis and agar-gel electrophoresis were used to further examine the column eluates. Agar-gel electrophoretic separation was achieved on 0.1% Noble Agar in a Veronal buffer, pH 8.4. A current potential of 30 mA was applied for 20 minutes. Paper electrophoresis was carried out for eight hours in a Veronal buffer, pH 8.6 ($\mu=0.05$), at a potential of 120 volts. Proteins were detected by staining with Amido Black; lipids with Sudan Black; mucopolysaccharides with periodate Schiff reagent; and acid mucopolysaccharides with Alcyan Blue.

SERUM PROTEINS IN THE HEDGEHOG

Larsen, B., and O. Tönder, Acta Physiol. Scand. **69**, 262–269 (1967).

Hedgehog serum proteins were separated by paper electrophoresis, electrophoresis on Pevikon® and immunoelectrophoresis. Paper electrophoresis in a 0.1M Veronal buffer, pH 8.6, revealed two beta-bands. Beta-2 was more pronounced. The rapid lipoprotein-line found upon immunoelectrophoresis varied with temperature and storage.

ACETOACETYLATION OF RIBONUCLEASE A

Marzotto, A., Biochem. Biophys. Res. Commun. **26**, 517–521 (1967).

Acetylated ribonuclease was used as a blocking agent to elucidate the primary structure of the enzyme. Native, acetylated and deacetylated ribonuclease A were subjected to paper electrophoresis on Whatman® No. 1 paper. Electrophoresis was performed in a 25% acetic acid buffer, pH 1.9, for 2.5 hours at

18 volts/cm. There were no differences in the mobility of native and deacety-lated ribonuclease A.

ELECTROPHORETIC SEPARATION OF THE REDUCTION PRODUCTS OF S-SULFONATED INSULIN CHAINS

Massaglia, A., F. Pennisi, and U. Rosa, J. Chromatog. **28,** 495–498 (1967).

The reduction products of I^{125}-labeled and sulphonated glycyl and phenyla-lanyl chains of insulin were separated by paper electrophoresis. A 10M urea–acetic acid buffer, pH 3.2, was used. The separations were carried out at 4°C at 7 volts/cm (1.5 mA). Radioautography detected the separated components.

PAPER ELECTROPHORESIS OF METAL IONS IN ACETATE BUFFERS

Mazzei, M., and M. Lederer, Anal. Letters **1,** 67–70 (1967).

The paper electrophoretic separation of 26 ions in 0.1 N, 0.5 N, and 1.0 N acetic acid-sodium acetate buffer is reported. Whatman® sheets 40 x 10 cm were used in a Camag Cell at 150 volts for 30 minutes. The ions were detected by spraying with ammoniacial oxine, exposing to H_2S, or spraying with rhodizonic acid.

STUDIES ON PHOSPHATE BINDING SITES OF INOSINIC ACID DEHY-DROGENASE AND ADENYLOSUCCINATE SYNTHETASE

Nichol, A. W., Biochemistry **6,** 1008–1015 (1967).

Paper chromatography and electrophoresis (Whatman® No. 1, 30 volts/cm, 0.04M phosphate buffer, pH 7.4) were used to confirm the purity of inosine 5'-methylphosphonate and inosine 5'-phosphofluoridate, which were used as inhibitors in binding-site studies. The reagents did not react with the phosphate binding sites of inosine 5'-phosphate dehydrogenase.

USE OF PAPER ELECTROPHORESIS FOR SEPARATING SOME ACIDS OF THE CITRIC ACID CYCLE IN BIOLOGICAL FLUIDS

Petrun'kina, A. M., Lab. Delo **1967,** 23–25.

High-voltage paper electrophoresis was used to detect intermediates of the citric acid cycle in whole urine and ethanol–ether extracts of urine. Six hundred volts were applied for three hours to separate intermediates in 50 μl of sample. Bromophenol Blue and p-dimethylaminobenzaldehyde reagents were used as stains. Hippuric, succinic, citric, and cis-aconitic acids could be identified.

PROTEIN FRACTIONS IN HUMAN MILK: I. EFFECT OF DURATION OF LACTATION AND DIETARY SUPPLEMENTATION ON THE PROTEIN FRACTION

Rao, P. Z., Indian J. Med. Res. **55,** 174–178 (1967).

Paper electrophoresis was used to fractionate 175 samples of human milk. The serum albumin fraction was in highest concentration in mature milk after one month lactation. A basic protein, which eventually reached a concentration of 2.7% was found in the seventh month of lactation. The total protein content decreased from first to tenth month. Dietary supplements or a high calorie diet did not alter the protein fractions.

PLASMA PROTEINS AND LIPIDS DURING NORMAL GESTATION AND IN THE POSTPARTUM PERIOD

Reboud, P., P. Groslambert, C. Oliver, and J. Groulade, Ann. Biol. Clin. **25,** 383–403 (1967).

Paper electrophoresis assays were made for proteins, glyco- and lipoproteins, fibrinogen, and lipids in the plasma from women between the ninth week of gestation and two months postpartum. Albumin and γ-globulin decreased in gestation while fibrinogen, β-lipoproteins, and lipids increased. There was a general increase in all fractions postpartum with a maximum on the eighth day.

COMPOSITION AND PROPERTIES OF THE TAMM-HORSFALL GLYCOPROTEIN

Rozenfel'd, E. L., and N. A. Yusipova, Biokhimiya **32,** 111–118 (1967), cited in Biochemistry **32,** 91–98 (1967).

The glycoprotein of Tamm-Horsfall was isolated from the urine of normal subjects and of subjects with nephronephritis, predominant nephritis, cirrhosis of the liver, and pancreopathy-sucrosuria. This urine was then characterized electrophoretically, chemically and physically. Electrophoretic studies were carried out on paper in a Veronal buffer, pH 8.6, and an acetate buffer, pH 4.5. A potential gradient of four volts/cm and a current strength of three to six mA for 24 hours was applied. The glycoprotein was found in concentrations of 10 to 12 mg/liter in normal urines, but rose to as much as 27 g/liter in disease states. There was a rise in the content of hexoses in nephrosis and sucrosia.

ELECTROPHORETIC MOBILITIES OF SERUM γ-GLUTAMYL TRANSPEPTIDASE AND ITS CLINICAL APPLICATION TO HEPATOBILIARY DISEASE

Rutenberg, A. M., E. E. Smith, and J. W. Fishbein, J. Lab. Clin. Med. **69,** 504–517 (1967).

Whatman® 3MM paper strip electrophoresis in a pH 8.6 Veronal buffer

(μ=0.075) at 20–25°C separated the γ-glutamyl transpeptidases into two components. One component migrated in the albumin α-1 globulin region and the other migrated in the α-2 globulin region. Patients with hepatic involvement showed a large increase in α-1 globulin band. A dominant G-T-3 peak was also evident in patients with biliary cirrhosis.

PAPER AND THIN-LAYER ELECTROPHORETIC SEPARATIONS OF POLY-NUCLEAR AZA HETEROCYCLIC COMPOUNDS

Sawicki, E., M. Guyer, and C. R. Engel, J. Chromatog. **30**, 522–527 (1967).

A large number of aza heterocyclic compounds were separated by paper and thin-layer electrophoresis. Paper electrophoretic separation was achieved on Whatman® No. 1 paper at 500 volts for 75 minutes. Separated compounds were detected by examination of the electrophoretogram under ultraviolet light. Thin-layer electrophoretic separation was achieved on 8 x 8-inch cellulose thin-layer plates at 500 volts. Formic acid buffer (31.2 ml formic acid + 59.2 ml acetic acid), pH 2.0, was employed in both instances. Quantitation of the separated compounds was possible with an Aminco-Bowman spectrophotofluorometer.

HIGH-MOLECULAR-WEIGHT γG-PARAPROTEIN IN THE CEREBROSPINAL FLUID IN PLASMOCYTOMA

Scheurlen, P. G., Klin. Wochschr. **45**, 419–422 (1967).

A high molecular weight protein from a patient with myeloma was identified by paper electrophoresis, gel filtration, and ultracentrifugation. The γG-globulin had a molecular weight between 155,000–160,000 and a sedimentation rate of 8.9S.

DETERMINATION OF ELECTROPHORETIC DISTRIBUTION OF CHOLESTEROL-BEARING PROTEINS IN SERUM

Searcy, R. L., Clin. Chim. Acta **15**, 73–76 (1967).

Electrophoresis performed on paper in the inverted-V Durrum type cell using a pH 8.6 Veronal buffer, ionic strength 0.075, was used to evaluate cholesterol-bearing lipoproteins of serum. The cholesterol-bearing proteins of hyperlipemic serum were found in the β-lipoprotein fraction.

THE QUANTITATIVE ESTIMATION AND QUALITATIVE IDENTIFICATION OF MICRO AMOUNTS OF D-ALLOSE

Short, W. A., J. W. Woods, and B. R. Baker, Alabama J. Med. Sci. **4,** 279–281 (1967).

Electrophoretic separation of D-allose on Whatman® 3MM paper impregnated with borax (0.05M, pH 9.2) was possible at 250 volts for five to six hours. This technique provides a useful adjunct to paper chromatographic separation techniques.

RELATION BETWEEN GAMMAGLOBULINS AND PERIPHERAL BLOOD INDICES IN PATIENTS WITH EXACERBATED CHRONIC LEAD POISONING

Veliev, B. A., Gigiena Truda i Prof. Zabolevaniya **11,** 57–59 (1967).

When serum proteins from control patients and patients with chronic lead poisoning were compared by quantitative paper electrophoresis, differences were revealed. The per cent values were (normals in parentheses): total serum protein 7.60 (7.93); albumin 4.34 (5.10); α_1-0.32 (0.30); α_2-0.65 (0.49); β-0.86 (0.75); and γ-globulin 1.43 (1.29).

THE ELECTROPHORETIC PATTERN OF HEMOGLOBIN IN NEWBORN BABIES, AND ABNORMALITIES OF HEMOGLOBIN F SYNTHESIS IN ADULTS

Vella, F., and T. A. Cunningham, Can. Med. Assoc. J. **96,** 398–401 (1967).

A search was made for variants of hemoglobin F. Paper electrophoresis in a discontinuous buffer system was performed at 400 volts for three to five hours. A pH 8.8 Veronal buffer, ionic strength 0.06, was used in the vessels. A pH 8.9 Tris–EDTA–borate buffer, 0.084M with respect to Tris, was used on the paper. Starch-gel electrophoresis was performed in a 0.001M phosphate buffer, pH 7.0, with 300 volts for four hours. Benzidine/H_2O_2 reagent was used as a stain for the gel. No abnormal varients of hemoglobin F were detected in 600 sera.

A SINGLE AMINO ACID SUBSTITUTION (ASPARAGINE TO ASPARTIC ACID) BETWEEN NORMAL (B+) AND THE COMMON NEGRO VARIANT (A+) OF HUMAN GLUCOSE-6-PHOSPHATE DEHYDROGENASE

Yoshida, A., Proc. Nat. Acad. Sci. (U.S.) **57,** 835–840 (1967).

A tryptic digest of a normal and a variant form of glucose-6-phosphate showed two peptides that differed in mobility on a two-dimensional chromatogram. The first dimension was chromatography in butanol–acetic acid–water (4:1:5 upper layer). The second dimension was electrophoresis in pyridine–acetic acid–water (1:10:289, pH 3.6–3.7) at 2000 volts for 60 minutes. The two peptides were then eluted from the paper and further digested with leucine amino peptidase, carboxypeptidase B and carboxypeptidase A. A two-dimensional amino acid fingerprint of these peptides was performed after digestion. Electrophoresis was performed in the first dimension on Whatman® 3MM paper in formic acid–pyridine–water (4:03:96 v/v pH 2.2) at 2000 volts for 60 minutes. Chromatography in the second dimension used ethanol–water (77:23 v/v) for 16 hours. It was possible to distinguish asparagine, aspartic acid, glutamine, and glutamic acid in this system.

DOSAGE ET DIFFÉRENCIATION DES ACIDS AMINÉS URINAIRES CHEZ LE SUBJECT NORMAL

Ambert, J. P., Ann. Biol. Clin. (Paris) **24,** 17–40 (1966) **French.**

Electrophoresis at pH 5.4, two-dimensional chromatography, and electrochromatography were used to separate and identify the amino acids in normal urine. The pH 5.4 buffer was composed of 20 ml pyridine, 8 ml acetic acid, and 2 liters water. Whatman® No. 3 paper was spotted and electrophoresed at 350 volts for two hours. The amino acids were stained with ninhydrin. For electrochromatography, the paper was first electrophoresed at 400 volts for four hours in pH 1.9 buffer (60 ml formic acid, 360 ml acetic acid, 6 liters water). Chromatography was done in the second dimension using butanol–acetic acid–water buffer (120:30:50). The dried paper was stained with 0.2% ninhydrin in acetone.

ELECTROPHORETIC INVESTIGATIONS OF THE INFLUENCE OF VARIOUS TREATMENTS UNDERGONE BY COW MILK ON WHEY PROTEINS

Aparico, M., Intern. Dairy Congr., 17th Proc. **2,** 349–355 (1966).

The wheys of raw, pasteurized, and sterilized milks and reconstituted milk powders were compared by electrophoretic analysis. Sterilization altered the protein pattern because of coagulation at high temperature. Pasteurized milk was the same as raw milk. Whey protein in powdered milk varied with the preparation of the powder.

CHOLESTEROLO MIGRANTE CON ALCUNE FRAZIONI PROTEICHE NEL SIERO DEI RATTI WISTAR A DIETA NORMALE E A DIETA EPERLIPIDICA IPERPROTIDICA E SOTTOPOSTI AD ESH E AD INIEZIONE DI ACTH

Biondi, A., and F. Badolato, La Ric. Scient. Rend. Sez. B. **6,** 259–266 (1966), **Italian.**

Paper electrophoresis performed in a pH 8.6 barbital buffer, ionic strength 0.075, for 18 hours at 220 volts (1 mA/strip) separated serum proteins of rat serum from rats given a diet rich in cholesterol as well as from rats given normal diets. High lipid diets caused an increase in cholesterol bound with β- and γ-globulins. A decrease in cholesterol bound with albumin was noted.

LOCATION OF DISULFIDE BRIDGES BY DIAGONAL PAPER ELECTROPHO-RESIS

Brown, J. R., and B. S. Bartley, Biochem. J. **101**, 214–228 (1966).

A new method for fingerprinting cysteic acid peptides obtained from disulphide bridges Chymotrypsinogen by enzymatic digestion is described. Electrophoresis was performed at right angles to the first direction using pyridine–acetic acid–water (25:1:225, pH 6.5). Asymmetrical distribution of the peptides was observed.

AN ELECTROPHORETIC COMPARISON OF THE SERUM PROTEINS OF FETAL AND ADULT NUTRIA (MYOCASTOR COYPUS).

Brown, L., Comp. Biochem. Physiol. **19**, 479–481 (1966).

The sera of six adult and two fetal nutria (Myocastor coypus) were separated by paper electrophoresis in a barbiturate buffer (ionic strength 0.05, pH 8.6). A constant current of 15 mA was maintained for six hours. The separated proteins were dried at 120°C for 30 minutes and stained with Bromophenol Blue — zinc sulfate. One protein band was found in adults, which was not present in fetal sera. The concentrations of globulins and albumins were both greater in adult sera.

PAPER ELECTROPHORESIS OF SERUM IN RATS FED VARIOUS CARBOHY-DRATE DIETS

Chang, Y. O., and T. R. Varnell, Proc. Soc. Exp. Biol. Med. **121**, 524–526 (1966).

Paper electrophoresis separated five serum protein fractions from serum of rats maintained on carbohydrate diets with and without lysine. Separations were achieved on paper strips in a Durrum® cell using a pH 8.6 Veronal buffer, ionic strength 0.75. Lysine was found to greatly affect the protein pattern (albumin). Without lysine, albumin was decreased by feeding cooked potato starch, raw potato starch, or sucrose.

ELECTROPHORETIC SEPARATION OF MONONUCLEOTIDES FROM ALKA-LINE RNA HYDROLYSATES

Click, R. E., Biochim. Biophys. Acta **129**, 424–426 (1966).

A buffer system which permits the separation of mononucleotides from alkaline hydrolysates of RNA without prior neutralizations is discussed. The electrophoretic separation was achieved on Whatman® 3MM paper at 2500 volts for 2½ to 3 hours in a 0.4M sodium acetate buffer, pH 3.8. Good resolution and migration from the origin were possible at this pH.

STUDIES IN MULTIPLE SCLEROSIS: II. COMPARISON OF THE BETA-GAMMA GLOBULIN RATIO, GAMMA GLOBULIN ELEVATION, AND FIRST-ZONE COLLOIDAL GOLD CURVE IN THE CEREBROSPINAL FLUID

Cosgrove, J. B. R., Neurology **16**, 197–204 (1966).

A paper electrophoresis study of the cerebrospinal fluid proteins from multiple sclerosis and normal patients was done. Increased gamma globulin and decreased beta globulin levels were found in multiple sclerosis patients. The beta-gamma globulin ratio can be used as an index of abnormality. Following electrophoresis, proteins were stained with Amido Black.

THE GENETIC CONTROL OF ALCOHOL DEHYDROGENASE AND OCTANOL DEHYDROGENASE ENZYMES IN DROSOPHILA

Courtright, J. B., Genetics **54**, 1251–1260 (1966).

Extracts of flies (400 mg/ml) were prepared and directly electrophoresed for detection of dehydrogenases. It was possible to assay the enzymes of one fly by crushing it on 3-mm wide strips of Whatman® 3MM paper and inserting the paper directly into the 0.9% agar gel. Gels were stained by a mixture of 50 mg NAD, 6.25 mg Nitro Blue Tetrazolium, 0.5 mg phenazine methosulfate, 21 ml of 0.2M Tris-HCl pH 9.0, and 5 ml n-octanol. Final pH 9.0. Staining was complete in 45 to 60 minutes. Ethanol–glacial acetic acid–water (15:1:5) was used to fix the gel.

ÉTUDES SUR LA STRUCTURE DES γ_G-GLOBULINES NORMALES ET MYÉLO-MATEUSES: III. COMPOSITION EN AMINO-ACIDES EN HYDROLYSE TRYP-SIQUE DES FRACTIONS LIBÉRÉES AU COURS DE LA RÉDUCTION DES GLOBULINES NORMALES ET MYÉLOMATEUSES

Dautrevoux, M., Pathol. Biol. (Paris) **14**, 816–820 (1966) **French.**

The amino acid composition of normal heavy and light chains of γ_G-globulins differ in serine and threomine content. Peptide maps of trypsin hydrolyzed chains were made by combining chromatography and electrophoresis. Whatman® No. 3 paper was spotted and electrophoresized for 16 hours at 40 volts in pyridine: acetic acid: water (30:110:4.870), pH 3.9. Chromatography was done in the second dimension using butanol: acetic acid: water (4:1:5) for 18 hours. Ninhydrin was used to stain the dried chromatogram.

FRACTIONS OF PLASMA PROTEINS OF THE SEMINAL FLUID OF HEALTHY MEN AND DURING PATHOLOGIC COURSE OF THE CLIMACTERIC PERIOD

Demchenko, A. N., Urol. i Nefrol. **31**, 40–46 (1966).

Protein fractions of the seminal fluid of three groups were examined by paper electrophoresis (Veronal–medinal buffer). The groups were: A) healthy men 45-60 years old; B) healthy men 45-60 years old; C) patients 45-60 years old with a pathologic course in the climacteric period. Four fractions were separated in all samples — albumin, and the α, β, and γ-globulins. The change in quantity of the protein fractions with age was less in Group C. The changes could be related to the degree of hormonal disorder.

SEPARATION OF AMINO ACIDS

Eastoe, J. E., Brit. Med. Bull. **22**, 174 (1966).

High-voltage electrophoresis quantitatively separated amino acids on filter paper strips. Acetic and basic amino acids were separated in pyridine:acetic acid:water buffer at pH 5.2, and most neutral amino acids were separated in acetic acid:formic buffer, pH 1.85.

SEPARATION OF URINARY METABOLITES BY LOW VOLTAGE "TWO SOLUTIONS PAPER ELECTROPHORESIS"

Eichhorn, F., and A. Rutenberg, Israel J. Med. Sci. **2**, 640–648 (1968).

A method is described for the separation and identification of urinary compounds, including phenolic acids, amines and indolic and imidazolic compounds. The urine was hydrolyzed to dissociate conjugated compounds, and 1/20,000 of 24 hour diaresis was applied to the paper. The anode electrolyte was 0.75% acetic acid, the cathode electrolyte 0.25% acetic acid. Electrophoresis was at 350 volts, the time variable. The paper was dried at 90°C for ten minutes and examined under UV light. Quantitative results can be obtained by eluting fluorescent compounds with tridistilled water and reading in a spectrophotometer. Vanillic acid, metanephrines, dopa, dopamine, 3-hydroxykynurenine, anthranilic acid, 3-hydroxyanthranilic acid, xanthurenic acid, kynurenic acid, indican, and amphetamine can be identified by this method. Standards were run concurrently for identification.

INHERITANCE OF MULTIPLE HEMOGLOBINS IN PEROMYSCUS

Foreman, C. W., Genetics **54**, 10007–1012 (1966).

The hemoglobin of erythrocytes was characterized by paper electrophoresis. A pH 8.6 barbital buffer, ionic strength 0.05, was used in a potential of 5 volts/cm. Relative concentrations of multiple hemoglobins were estimated using cellulose-acetate electrophoresis. Ten grams of Tris, 1.4 g EDTA, 8.0 g boric acid, and 1 liter of water composed the pH 8.1 buffers used for presoaking the cellulose-acetate strips. The electrode buffer was made by titrating 0.33M boric acid to pH 7.8 with 7N sodium hydroxide. Separation was achieved in 100 minutes using an initial potential of 10 volts/cm. The strips were stained with Ponceau S. Bands were eluted in 0.1N sodium hydroxide. Colorimetric readings were made at 560 mμ.

AN ESTIMATION OF THE NONELUTED DYE-PROTEIN COMPLEX FROM PAPER ELECTROPHEROGRAMS THROUGH THE APPLICATION OF LANGMUIR'S ISOTHERM EQUATION

Gaunce, A. P., Anal. Biochem **17**, 357–364 (1966).

A part of the dye-protein complex (DPC) of blood sera stained after separation by paper electrophoresis was not eluted by 1 N hydrochloric acid at 100°C. It was assumed that the complex was absorbed by the paper and the noneluted DPC could be estimated by applying the Langmuir isotherm equation,

$$X = M + \frac{\alpha C}{1 + \beta C}$$

where X is the amount of dye adsorbed by Mgm of filter paper, C is the concentration of DPC in solution, and α and β are constants.

ELECTROPHORETIC STUDY OF SERUM AND THE CEREBROSPINAL FLUID IN TUBERCULOSIS MENINGITIS OF CHILDREN

Grepaey, A., A. Kovacs, S. Kiss, and M. Kerkes., Rev. Med. (Targu-Mores) **12**, 18–21 (1966).

Paper electrophoresis of serum of children with tuberculous meningitis showed decreased albumin and increased α_1- and α_2-globulins. In the CSF, the albumin was decreased and the total proteins and α_2-globulins were increased. Values returned to normal with the cessation of the disease.

THE ELECTROPHORESIS OF PROTEINS OF AMNIOTIC FLUID

Heron, H. J., Obstet. Gynaecol. Brit. Commonwealth **73**, 91–98 (1966).

One hundred seventeen amniotic fluids were analyzed by paper electrophoresis on Whatman® No. 1 paper in a sodium diethyl–barbiton–barbituric acid, pH 8.6. The bands were defined with a variable recorder. The albumin–globulin ratio of the fluid was shifted as maturation of the fetus occurred. Lipoproteins were totally absent. β_1-globulin (siderophillin) was present in large amounts in amniotic fluids. The γ-globulin present may be different from that found in serum.

QUANTITATIVE MEASUREMENT OF INDIVIDUAL FREE AMINO ACIDS IN URINE BY MEANS OF A HIGH VOLTAGE PAPER ELECTROPHORESIS

Juul, P., Scand. J. Clin. Lab. Invest. **18**, 629–637 (1966).

Massive screening of mentally retarded patients' urine for detection of abnormal amino acid content is reported. Fifty μl of urine was high-voltage electrophoresed (2500 volts, or 65 volts/cm) for 60 minutes in formic acid–acetic acid–water buffer (15:83:902), pH 1.9. Ninhydrin positive spots were quantitated on a densitometer. Taurine, glutamine, serine, alanine, glycine, histidine, and lysine were the amino acids examined. No conclusions could be drawn, and the problem must be investigated further.

MONOCLONAL γ-GLOBULINS IN FERRETS WITH LYMPHOPROLIFERATIVE LESIONS

Kenyon, A. J., Proc. Soc. Exp. Biol. Med. **123**, 510–513 (1966).

The sera of 92 ferrets from three populations were examined to quantitate the level of gamma globulins and to determine the incidence of myeloma-like immunoglobulins. Gamma globulins were quantitated by paper electrophoresis in a pH 8.6 Veronal buffer, ionic strength 0.075, at 75 volts for 16 hours. The strips were stained with Bromophenol Blue and densitometry measurements made. Cellulose-acetate electrophoresis was performed at 250 volts for 20 minutes in the above buffer to identify the myeloma-like globulins. Ponceau S was the stain. Immunoelectrophoresis was done by the method of Scheidegger, [Intern. Arch. Allergy Appl. Immunol. **7**, 103 (1955)] using rabbit antisera to detect antigenic activity.

ON THE PRESENCE OF TWO DISTINCT PROTEOLYTIC COMPONENTS IN PANCREATIC CRYSTALLINE ELASTASE

Ling, V., and R. A. Anwar, Biochem. Biophys. Res. Commun. **24**, 593–597 (1966).

Both paper and borate starch-gel and formate-urea starch-gel electrophoresis were employed to separate a second component from crystalline elastase. Paper electrophoresis was performed on Whatman® No. 3 HR paper in a pyridine–acetic acid–water buffer, pH 3.6, at 38 volts/cm for two hours. Starch-gel electrophoresis was performed according to the procedure of Smithies in a borate buffer, pH 8.T, or a formate buffer, pH 3.1, at 4 volts/cm for 20 hours. Starch-gel electrophoresis at pH 8.8 revealed two major bands. Crystalline elastase was separated into three bands on urea-starch gels.

PAPER ELECTROPHORESIS

Lubran, M., J. Am. Med. Assoc. **197,** 138 (1966).

The use of paper electrophoresis, performed in pH 8.6 diethylbarbituric acid to separate serum proteins, is reviewed as to its clinical significance. The changes in the electrophoretic pattern of serum as seen in different disease states are presented. Basic abnormalities in serum protein patterns are very well detailed.

STUDIES ON THE ELECTROPHORETIC PROPERTIES OF HAEMOGLOBIN AND PLASMA PROTEINS OF MICROTIDAE (CL. GLAREOLUS, M. ARVALIS, M. AGRESTIS)

Marchuwska-Koj, A., Folia Biol. Warsaw **14,** 176–181 (1966).

Paper electrophoresis was used to examine hemoglobin and serum proteins of three species of the family *Microtidae*. Electrophoresis was performed in pH 8.6 Veronal buffer for 12 hours at a current of 4 mA and 280 volts. Whatman® No. 1 paper was used. After electrophoresis, the dried strips were stained with Bromophenol Blue in order to visualize the separated proteins. Hemoglobin migrated as one band and exhibited the same mobility in all three species. Five protein fractions were found when serum was electrophoresed. Albumin moved farthest away in the anode direction followed by α_1, α_2, β and γ.

CRYOGLOBULINEMIA – A STUDY OF TWENTY-NINE PATIENTS: I. IgG AND IgM CRYOGLOBULINS AND FACTORS AFFECTING CRYOPRECIPITABILITY

Meltzer, M., and E. C. Franklin, Am. J. Med. **40,** 828–836 (1966).

The proteins that reversibly precipitate from blood on cooling are known to be associated with multiple myeloma, macroglobulinemia or malignant lymphoma. The nature of these cryoglobulins was studied. Extremes of pH, ionic strength and 2M urea inhibited their precipitation. Paper electrophoresis in a pH 8.6 barbital buffer, ionic strength 0.05, was the technique used to identify and differentiate proteins from diagnostic groups of patients.

CHANGES IN ELECTROPHORESIS MIGRATION DUE TO IMPREGNATION OF THE PAPER STRIPS WITH NON-BUFFER SOLUTIONS

Morales-Malva, J. A., M. Sapag-Hagar, and S. Israel-Budnick, Clin. Chim. Acta **14,** 661–666 (1966).

Paper strips impregnated with non-buffered saline showed five typical bands after electrophoresis (110 volts for 16 hours) of serum. Salt solutions of equal ionic strength and the same inorganic anion but different cations gave better resolution with increasing ionic radii of the cation. Two bands were separated with distilled water or a nonpolar solvent impregnating the paper. Michaelis buffer, pH 8.2, ionic strength 0.075, was the electrolyte used in all experiments.

STUDIES ON THE SERUM γA-GLOBULIN LEVEL: III. THE FREQUENCY OF A-γ A-GLOBULINEMIA

Bachmann, R., Scand. J. Clin. Lab. Invest. **17,** 316–320 (1965).

A total of 7000 sera from a population of normal Swedish adults were screened for subnormal γA-globulin levels. Paper electrophoresis was used to quantitate γ-globulin. Immunoelectrophoresis was used to estimate the γM-globulin concentration.

STUDIES ON THE SERUM γ_{1A}-GLOBULIN LEVEL

Bachmann, R., Scand. J. Clin. Lab. Invest. **17,** 39–45 (1965).

Paper electrophoresis was used to quantitatively measure the γ_{1A}-globulin level in 99 patients with hypergammaglobulinemia. Barbital buffer 0.05M, pH 8.6, and calcium lactate 1.7mM were buffer systems used. In no instances was an M component demonstrable. Quantitation of γ_{1A}-globulin was also done by double diffusion and a modification of Cridin's technique. No definitive conclusions relating hypergammaglobulinemia and γ_{1A}-globulin levels were made.

THE EFFECT OF PASTEURIZATION AND STERILIZATION ON THE ELECTRO-PHORETIC PROTEIN PATTERN OF HUMAN MILK

Nicola, P., and A. Ponzone, Minerva Pediat. **17,** 785–791 (1965).

Human milk from lactating women was divided into three aliquots, each being treated separately. The first sample was stored at 4°C, the second pasteurized at 80°C for 15 minutes, and the third sterilized at 120°C for 30 minutes. Milk serums were prepared. Electrophoresis (sodium–Veronal–citrate–NaCl buffer, pH 7.8, 170 volts and 18 mA at 2°C) showed that heat treatment did not alter the electrophoretic pattern but did change the absolute mobility of serum albumins and lactoglobulins. The immunoglobulins and α-lactalalbumins were unchanged. Results indicated a thermal resistance in human milk proteins.

PROTEIN ELECTROPHORESIS IN ACUTE LEUKEMIA, MYELOID LEUKEMIA, AND CHRONIC LEUKEMIA

Perolini, G., L. Castano, and F. Anchisi, Minerva Med. **57,** 1942–1946 (1966).

Serum proteins from 73 patients were electrophoresed in a pH 8.6 Veronal buffer, ionic strength 0.1μ. When compared with serum from 50 normal patients, serum from acute leukemia patients showed increased globulins and decreased total protein and albumin. Chronic lymphatic leukemia patients showed decreased γ-globulin.

PROTEINURIA IN CHRONIC CADMIUM POISONING

Piscator, M., Arch. Environ. Health **12**, 335–344 (1966).

Paper electrophoresis was used to examine the urine of cadmium workers. Barbital buffer, pH 8.6, was used for both paper and immunoelectrophoresis. The microtechnique of Scheidegger was used for immunoelectrophoresis [Intern. Arch. Allergy Appl. Immunol. **7**, 103 (1955)]. Starch-gel electrophoresis was done by the method of Ferguson [Nature **190**, 629–630 (1961)]. Immunoelectrophoresis disclosed some 30 proteins in the urinary colloids of cadmium workers.

NEUCLEOTIDE SYNTHESIS UNDER POSSIBLE PRIMITIVE EARTH CONDITIONS

Ponnamperuma, C., Science **148**, 1221–1223 (1965).

C^{14}-Nucleosides were heated with inorganic phosphate, and nucleoside monophosphates were formed in appreciable yield. The products were identified by autoradiographs following paper electrophoresis in a pH 3.5 buffer at 1500 volts for one hour. For further confirmation, paper chromatography in the second dimension followed electrophoresis. Theoretically, it is of interest in chemical evolution that water was not required for the formation of the nucleoside monophosphates.

PREPARATION AND CHARACTERIZATION OF A SEX-DEPENDENT RAT URINARY PROTEIN

Roy, A. K., O. W. Neuhaus, and O. W. Harmison, Biochim. Biophys. Acta **127**, 72–81 (1966).

A urinary protein from the male rat was purified by ammonium sulfate fractionation and by chromatography on DEAE-cellulose. Paper electrophoresis (Whatman® 3MM paper, barbital buffer, pH 8.6, ionic strength 0.075) demonstrated the efficiency of the fractionation procedures. Electrophoresis in starch gel (discontinuous buffer system of glycine-NaOH-borate, pH 9.0) showed that the urinary protein migrated similarly to an α_{24}-protein from liver. This similarity was also seen from immunoelectrophoresis data in 1% ion-agar No. 2 in 0.05M barbital buffer at pH 8.2 (2.5 hours at 4 volts/cm).

ABSENCE OF SERUM ALPHA-2 GLOBULIN IN NIEMANN-PICK DISEASE: A MEASURE OF HEPATOCELLULAR INVOLVEMENT

Scheneck, L., A. Saifer, H. B. Warshall, and B. W. Volk, Proc. Soc. Exp. Biol. Med. **122,** 1295–1298 (1966).

The alpha globulins of patients with Niemann-Pick disease were searched to find an available index of hepatic malfunction. Since it can separate the alpha globulins into three fractions on paper electrophoresis, a Tris-borate buffer was used for the electrophoresis. This buffer was achieved by diluting 750 ml of the stock solution (484 g Tris, 48 g EDTA, 36 g boric acid diluted to 2 liters with distilled water) to 2 liters for use. Ten lambda of fasting serum was tested at 185 volts for 20 hours; 0.3% Lissamine Green was the stain. Strips were then decolorized in 2% acetic acid. The alpha-2 globulin fraction was missing in ten sera tested.

ELECTROPHORETIC BEHAVIOR OF SERUM AMYLASE IN VARIOUS MAMMALIAN SPECIES

Searcy, R. L., S. Hayashi, J. E. Berk, and H. Stern., Proc. Soc. Exp. Biol. Med. **122,** 1291–1295 (1966).

The electrophoretic distribution of amylase from 11 mammalian species was examined by paper electrophoresis. Whatman® 3MM filter paper strips (6 x 35 cm) and a pH 8.6 Veronal buffer, ionic strength 0.075, was used. Electrophoresis was performed in a horizontal direction at 80 volts for a period of 16 hours. Monomorphic patterns were found in human, sheep, rabbit, guinea pig, cat, and horse serum. Polymorphic patterns were exhibited by rat, pig, dog, and goat sera.

SERUM PROTEIN ELECTROPHORESIS AND SERUM TRANSAMINASE ACTIVITY OF DOGS WITH CANINE DISTEMPER

Snow, L. M., M. J. Burns, and C. H. Clark, Am. J. Vet. Res. **27,** 70–73 (1966).

Paper-strip electrophoresis, using the Beckman-Spinco® Model B apparatus and buffer system, was used to study the serum proteins of dogs with canine distemper. A decrease in albumin fraction, with a concomitant increase in the alpha-2 globulin and a delayed increase in gamma globulin was noted. Total transaminase values were normal up to the fifth day after infection; an increase on day five was observed.

SERUM PROTEIN ALTERATIONS INDUCED BY LISTERIA MONOCYTOGENES INFECTIONS

Sword, C. P., J. Immunol. **96**, 790–796 (1966).

Paper and immunoelectrophoresis were used to study possible changes in serum proteins during listerosis infection in mice. Paper electrophoresis was carried out on a Beckman® Model R system according to the manufacturer's instruction manual. Bromophenol Blue and periodic acid Schiff were the stains used. Immunoelectrophoresis was performed in 1% agar gel in a pH 8.2 Veronal buffer, ionic strength 0.05, at 6 volts/cm for 45 minutes. High titer antisera was prepared in rabbits. Paper electrophoresis showed an increase in the α- and β-glycoproteins with infection. Increases in albumin mobility, α_2-macroglobulin, haptoglobulin, and transferrin were seen in immunoelectrophoresis. Conversion of β_{1C}-globulin to β_{1P}-globulin was noted. There was little change in the γ-globulin.

INSULIN ANTIBODIES IN PREGNANT GUINEA PIGS AND IN THEIR OFF-SPRING

Thorell, J. I., Acta Endocrinol. **52**, 255–267 (1966).

Paper electrophoresis, performed with an LBK-paper electrophoresis apparatus on Schleicher and Schull No. 2043 in 0.1M Veronal buffer, pH 8.6, was used to study insulin antibodies in guinea pigs. Electrophoresis of guinea pig plasma, which had been preincubated with insulin I^{131} was performed at 100 volts (2.5 volts/cm) for two to five hours in order to separate ^{131}iodide from proteins and for 6 to 24 hours in order to separate the proteins. Activity was measured with a Geiger-Mueller counter. Radioimmunoelectrophoresis was performed according to the procedure of Scheidegger. Antiserum was allowed to react with separated proteins for 16 to 24 hours. Paper electrophoresis detected an insulin-binding globulin in the mother and offspring. Three types of insulin-binding antibodies were found in maternal plasma, whereas fetal plasma showed activity in only the γ_1 and γ_2 globulins. The insulin-binding capacities were identical in maternal and fetal plasma.

A CHEMICAL DIFFERENCE BETWEEN HUMAN TRANSFERRIN B₂ AND C

Wang, A. C., Am. J. Human Genet. **18**, 454–458 (1966).

Peptide maps, prepared by electrophoresis and chromatography, were made to study the tryptic digest of human transferrins. Pyridine–acetic acid–water buffer, pH 6.4 (25:1:224), was used for electrophoresis of the Whatman® 3MM paper. Descending chromatography in the second dimension was done in pyridine–isoamyl alcohol–water (14:14:11). The peptides were identified with 0.2% ninhydrin in acetone; diazotized sulfanilic acids located the histidine-containing peptides; Sakazuchi reagent located the arginine-containing pep-

tides; p-dimethylaminobenzaldehyde located the tryptophan-containing peptides, and α-nitroso-β-naphthol located the tyrosine-containing peptides.

FREE FRUCTOSE IN HUMAN CEREBROSPINAL FLUID

Wray, H. L., and A. I. Winegrad, Diabetologia **2**, 82–85 (1966).

High–voltage paper electrophoresis confirmed an enzymatic assay that
detected free fructose in the cerebrospinal fluid of 42 subjects. There was a
linear relationship between the glucose and fructose concentration. The
source and role of these free sugars in the cerebrospinal fluids is discussed.

SEPARATION OF HISTAMINE FROM AMINO ACIDS BY IONOPHORESIS

Wronski, A., Chemia. Anal. (Warsaw) **11**, 957–960 (1966).

Histamine was successfully separated from a mixture of amino acids by ionophoresis on Whatman® No. 1 paper. A pyridine buffer (pyridine–acetic acid–
water–glycol, 5:3:190:2) and a potential of 12 volts/cm (1.1 mA) were used.
and separation proceeded for 50 minutes. A 90 minute time period was sufficient to separate the amino acids into acidic, neutral and basic groups. The
application of the method to pharmaceutical and biological preparations is
discussed.

DETERMINATION OF PROTEIN AND PROTEIN FRACTIONS IN MILK

Akhundov, K. R., Tr. Azerb. Nauchn Issled. Vet. Inst. **19**, 177–181
(1965).

Milk proteins were separated by paer electrophoresis in 0.01M Veronal
buger, pH 8.6, at 0.12–0.13 mA/cm for 16 hours. Bromophenol Blue
stained the proteins, which were then eluted for colorimetric analysis. The total
and nonprotein nitrogen were determined using the Nessler reagent.

THREE HEMOGLOBINS K: WOOLWICH, AN ABNORMAL, CAMEROON AND IBADAN, TWO UNUSUAL VARIANTS OF HUMAN HEMOGLOBIN A

Allan, N., D. Beale, D. Irvine, and H. Lehmann, Nature **208**, 658–661
(1965).

The second incidence of a hemoglobin K in combination with hemoglobin S
is described. Paper electrophoresis at alkaline pH 8.6 separated and identified
three hemoglobins by their mobility. Peptide maps and amino acid analysis
further defined differences in the three hemoglobins.

DISTRIBUTION OF SERUM AMYLASE IN MAN AND ANIMALS

Berk, J. E., R. L. Searcy, H. Shinichiro, and I. Ujihira, J. Am. Med. Assoc. **192**, 389–393 (1965).

The distribution of amylase isoenzymes of human and canine serum was studied using horizontal electrophoresis on 3 x 35 cm filter paper strips. Electrophoretic separation was achieved in a barbital buffer, pH 8.6, ionic strength 0.075, at a constant potential of 80 volts. Following electrophoresis for 16 hours, the separated amylases were detected using the technique of Baker [Scand. J. Clin. Lab. Invest. **6**, 94–101 (1954)] and by the sacchrogenic technique of Summer [J. Biol Chem. **47**, 5–9 (1921)]. Most of the amylase activity of normal man, rabbit and dog was located in the α-globulin fraction.

CHANGES IN WHEY PROTEINS BETWEEN DRYING AND COLOSTRUM FORMATION

Carroll, E. J., F. A. Murphy, and O. Aalund, J. Dairy Sci. **48**, 1246–1249 (1965).

A protein migrating in an electric field between α-lactalbumin and the immune globulin fraction was demonstrated in the whey of a cow 45 days postpartum. The further studies described in this article suggest a fundamental difference in this protein when found in dry-cow secretions and colostrum. Wheys were fractionated on DEAE-cellulose using the step-wise elution scheme with NaCl in a phosphate buffer [J. Dairy Sci. **44**, 589 (1961)]. Further fractionations were done in DEAE-cellulose (0.02M phosphate buffer, pH 8.0, gradient to 0.03M) DEAE-Sephadex®, and Sephadex® G-75 eluting with 0.1M acetate buffer, pH 4.6. Casein was precipitated in the acid from the wheys before paper electrophoresis (Veronal buffer, pH 8.6, ionic strength 0.075). The nature of the unknown protein was not confirmed by these studies, although it is different in dry-cow secretions and colostrum.

SIMPLE PHYSIO-CHEMICAL ESTIMATION OF THIAMIN, PYRIDOXAL, NICOTINAMIDE AND RIBOFLAVIN IN ORAL DRUGS

Chakraborty, M., and S. K. Dutta, J. Proc. Inst. Chemists **37**, 114–124 (1965).

Gel filtration on Sephadex® G-50 was used to separate B vitamins in pharmaceuticals from flavoring and thickening agents, coloring additives, and preservatives. Distilled water saturated with butanol was the eluant used. Paper electrophoresis in a pyridine–formic acid–water buffer, pH 3.35, at 110 volts (8–10 mA for six hours) was used to separate subject compounds from the concentrated eluants. The separated compounds were located by viewing under UV light. Thiamin was retained on the gel column.

ELECTROPHORESIS OF ANTIGENS OF ATYPICAL MYCOBACTERIA

Chapman, J. S., J. Clark, and M. Speight, Am. Rev. Respirat. Diseases **92,** 73–84 (1965).

Paper-strip electrophoresis, performed in a pH 8.6 Veronal buffer for 17 hours using the Spinco® Model R Electrophoresis cell, was used to separate proteins of nine strains of Runyon's atypical mycobacteria. In some experiments, a potassium acid phthalate buffer, pH 5.9, and formic acetic acid buffer, pH 2.0, were used. Results obtained allowed the authors to group the organisms on bases of similarities of patterns obtained. Runyon's PI, P-6 and P-2 strains were similar. Two kinds of Group III (P-2) were observed.

STUDIES IN MULTIPLE SCLEROSIS: I. NORMAL VALUES FOR PAPER ELECTROPHORESIS OF SERUM AND CEREBROSPINAL FLUID PROTEINS; DESCRIPTION OF THE METHOD WITH SPECIAL REFERENCE TO A PRACTICAL CONCENTRATION TECHNIQUE FOR CSF

Cosgrove, J. B. R., and P. Agius, Neurology **15,** 1155–60 (1965).

Two to four ml of CSF were concentrated almost to dryness by using negative pressure through a colloidin bag. The concentrate was diluted to 0.01 ml with Veronal buffer, pH 8.8. The bag was suspended in a tube containing Elliott's solution during concentration. Paper electrophoresis was carried out in a pH 8.6 Veronal buffer, ionic strength 0.075, at a constant current of 0.104 mA/cm paper width for 16 hours. The paper strips were dried at 120°C for 30 minutes and stained with Amido Black 10B (method of Pucar, 1954). Normal electrophoretic values for serum and CSF are presented.

A MODIFICATION SYSTEM OF FINGERPRINTING HAEMOGLOBIN

Efremov, G., Nature **208,** 298–299 (1965).

New buffer and solvent systems for electrophoresis and chromatography for fingerprinting of hemoglobin were developed. They are an improvement over older systems since pyridine is not required. The electrophoresis buffer is equal parts of 1M ammonia and 0.5N formic acid, final pH 6.5. (For electrophoresis alone, equal parts 0.5M ammonia and 1N formic acid, final pH 3.6, was the preferred buffer.) The chromatography solvent was 2,6 Lutidine-wtter (122–60 v/v). Whatman® 3MM paper was used. Thirty volts/cm for 2½ hours gave good electrophoretic separation. The paper was dried at 60°C before the descending chromatography was performed. The staining agent was 0.2% ninhydrin in acetone.

ELECTROPHORESIS OF CHYLE IN BRONCHOGENIC CARCINOMA

Falor, W. H., J. J. Dettling, A. Kerkian, W. Batchel, E. Krismann, and M. A. Thomas, Arch. Surg., **91**, 671–677 (1965).

The protein pattern of serum, chyle and lymph of 80 patients who had bilateral scalene-lymph node excision and bilateral duct excision was studied. Paper electrophoresis using the Spinco® Model R system and Spinco® buffer, B-2 (pH 8.6, ionic strength 0.050), was used. Both groups of test patients' serum pattern was characterized by a hypoalbuminemia, an elevation of the α_2-globulin and the γ-globulin. The chyle showed a decrease in the trailing fraction. Glycoprotein fraction exhibited a decrease in the albumin fraction, an increase in the α_1 fraction; an increase in the α_1 fraction and α_2 fraction; a decrease in the β- and γ- fractions. Glycoproteins in chyle showed an increase in α_1-globulins.

IDENTIFICATION OF AROMATIC SUBSTANCES BY ELECTROPHORETIC SPECTRA USING PAPER CHROMATOGRAPHY

Franc, J., and V. Kovar, J. Chromatog. **18**, 100–115 (1965).

The use of an electrophoresis chamber with anode and cathode compartments having ten different chambers filled with an electrolyte of different pH is reported. This kind of chamber was used for paper separation of 75 aromatic compounds. The separation was achieved at 250 volts for three hours. The relationship among type, number and position of functional group on the compound was investigated for 75 strongly carcinogenic and biologically important aromatics.

ELECTROPHORETIC BEHAVIOR OF HUMAN URINARY AMYLASE

Franzini, C., J. Clin. Pathol. **18**, 664–665 (1965).

Amylase activity in human urine is associated with the gamma globulin electro-phoretic zone when separated on paper (Veronal buffer, pH 8.6). The enzyme was eluted from the paper in 3.0 ml of 0.02M phosphate buffer, pH 7.0. Incubation of 0.2 ml of the eluate and 0.2 ml of 2% soluble starch solution in 0.15M NaCl in 0.02M phosphate buffer, pH 7.0, for 30 minutes at 37°C, was the assay used to determine the amylase activity. After 0.3 ml phenylhydra-zine reagent was added, the tubes were centrifuged. The supernate was heated in a 100°C water bath for 30 minutes. Osazone formation was mea-sured at 395 mμ after cooling.

EXISTENCE OF TWO AMYLOLYTIC ENZYMES IN HUMAN URINE: A STUDY OF ELECTROPHORESIS AND GEL FILTRATION

Franzini, C., Biochim. Appl. **12,** 101–111 (1965).

Paper electrophoresis and gel filtration on Sephadex® G-100 showed the presence of two amylolytic enzymes in urine. One of the two enzymes acts like an α-amylase. The second enzyme appears to have a molecular weight greater than 200,000.

HUMAN URINARY AMYLOLYTIC ENZYMES IN ACUTE HEPATITIS

Franzini, C., J. Clin. Pathol. **18,** 775–776 (1965).

Paper electrophoresis revealed two amylolytic enzymes in human urine. The main peak was in the gamma globulin region and the minor peak in the beta globulin region. Possible origins of the second peak are discussed.

PROTEINURIA

Gabe, F., Protides Biol. Fluids, Proc. 12th Colloq. 322–336 (1965).

Physical exercise increased the amount of protein in the urine of 5% of the subjects tested. The increase was probably due to increased renal blood flow from which renal tubules are unable to absorb the excessive protein. On paper electrophoresis, the urine developed a "serum-like" protein pattern after strenuous exercise.

PROTEOLYTIC ACTIVITY OF HUMAN GASTRIC JUICES AT VARIOUS pH's AND ITS ESTIMATION BY HEMOGLOBIN AND ELECTROPHORETIC METHODS

Glass, G. B., Am. J. Digest. Diseases **10,** 766–789 (1965).

Gastric juices were analyzed for proteolytic activity at pH 1.5-1.8 and at pH 3.5. There was no correlation between proteolytic activity and the pH of the gastric juice. Pepsin before and after proteolysis was quantitated by paper electrophoresis in a pH 9.0 borate buffer, $\Gamma/0.12$. Amido Black 10B was the stain. Density readings were done at 575 mμ to quantitate separate fractions.

C-REACTIVE PROTEIN: A MOLECULE COMPOSED OF SUBUNITS

Gotschlick, E. C., Proc. Nat. Acad. Sci. U.S. **54,** 558–566 (1965).

C-Reactive protein (CRP) appears in human sera in the course of certain diseases. This paper presents evidence that CRP is an aggregate of subunits of molecular weight 21,500 held together by noncovalent interactions. Starch-gel electrophoresis in 8M urea sodium formate buffer demonstrated that CRP is different from reduced and alkylated Bence-Jones protein. At pH 9.5, CRP was dissociated. In the absence of urea, the CRP separated into six bands during starch-gel electrophoresis in a Tris-citrate buffer. This subunit showed no positive ninhydrin reaction when high-voltage paper electrophoresis was done at pH 4.7 in pyridine acetate buffer. A model of six identical subunits, with a total molecular weight of 129,000, is proposed.

PROTEINS IN BILE

Hanton, J. C., Biliary System Symp. NATO Advan. Study Inst., 1963, 443–448 (1965).

Paper and agar-gel electrophoresis were used to study bile from man and some animals. Agar-gel electrophoresis was performed in a $(NH_4)_2$ CO_3 buffer, pH 8.4, ionic strength 0.1, at 40 mA for six hours. The gall bladder was punctured and the bile removed. It was centrifuged, dialyzed for two hours vs. phosphate buffer (pH 7.5, ionic strength 0.01), and concentrated 5 to 10 fold at $4°C$ before electrophoresis.

STUDIES ON β-PAROTIN: IV. EXTRACTION OF β-PAROTIN FROM AUTO-LYZED BOVINE PAROTID GLAND (STUDIES ON THE PHYSIOLOGICAL CHEMISTRY OF THE SALIVARY GLAND LXIV)

Ito, Y., S. Okabe, and S. Namba, Endocrinol. Japon. **12**, 63–68 (1965).

Fresh bovine parotid gland was homogenized and two to three volumes of water added to make a paste. Toluene was added and the mixture incubated for autolysis at $15°$ or $37°C$ at pH 5.0, 6.5 or 8.0. The incubation times ranged from 5 to 30 hours. The autolyzed gland was extracted by stirring for three hours in five volumes of water, final pH 8.0. After centrifugation, the pH of the supernate was adjusted to pH 5.0 with 5% HCl. A precipitate formed during overnight-incubation in a cold room. This was removed and the supernate adjusted to pH 2.5 with 10% HCl. Acetone was added at $0°C$ to make an 83% concentration. The resulting precipitate was separated, washed with acetone and ether, and dried in a vacuum. Paper electrophoresis of the acetone-extracted parotid gland showed five spots when stained with Bromophenol Blue. A pH 8.6 Veronal buffer, ionic strength 0.1, was used as the electrolyte. Conditions were 1 mA/cm for six hours.

ELECTROPHORETIC ANALYSIS OF SERUM PROTEINS OF SOME SPECIES OF CHUB OF THE GENUS NEOGOBIUS FROM THE SEA OF AZOV

Kulikova, N. I., Dokl.-Biochem. Sect. (English Transl.) **163**, 229–232 (1965).

The serum proteins of U. melanostomus Pall, N. sysman Nordm and N. fluviatilis Pall were separated by paper electrophoresis using a buffer system described by Gurvich [Paper electrophoresis of serum proteins (Moscow), 1959]. Results obtained indicated that a component similar in mobility to serum albumin was absent from the serum of N. *melanostomus* and was present in very small amounts in N. *sysman* and N. *fluviatilis*. The α- and β-globulins constituted the main fractions of the chub sera. The amount of β-globulin exceeded the amount of α-globulin in all three species. Each species was characteristic in its protein pattern, and one species could be easily distinguished from another.

ELECTROPHORETIC SEPARATION OF UREA AND THIOUREA DERIVATIVES BY COMPLEX FORMATION

Lucci, G., Ann. Univ. Ferrara **2**, 117–123 (1965) .

Paper electrophoresis (Whatman® 3 MM) in a 0.01N lactic acid buffer was used to separate urea, thiourea, semicarbazide and thiosemicarbazide. A voltage of 200–600 was necessary to cause semicarbazide to migrate. Thiosemicarbazide migrated at 800 volts. Urea and thiourea were not moved by even higher voltage. If $PdCl_2$ was incorporated into the system and 300 volts applied, urea did not migrate but the other two components moved toward the cathode.

BLOOD CHEMISTRY OF THE BOTTLENOSE DOLPHIN L TURSIOPS TRUNCATUS

Medway, W., and J. R. Gerachi, Am. J. Physiol. **209**, 169–172 (1965).

Whole blood, plasma and serum constituents of blood of six bottlenose dolphins was examined. Proteins were separated by paper electrophoresis in a Spinco model R Durrum® type cell with a Veronal buffer, pH 8.6, ionic strength 0.075. Electrophoresis was carried out for 16 hours at room temperature and stained with Bromophenol Blue. Other chemical measurements were performed using routine procedures. The sodium and chloride of plasma and serum albumin levels appear to be lower than those found in dogs and horses. The α_1, α_2 and β-globulins were lower than those found in dogs and horses.

FRACTIONATION OF HUMAN SKIN COMPONENTS BY PAPER ELECTROPHORESIS IN DICHLOROACETIC ACID

Milch, R. A., M. A. Naughton, and R. A. Murray, Gerontologia **11**, 153–168 (1965).

Untreated human skin was dissolved in 100% dichloroacetic acid and then electrophoresed on glass or filter paper strips with 75% dichloroacetic acid. Five or six distinct bands separated; two stained positive with Bromophenol Blue.

ÜBER STEROID-KONJUGATE IN PLASMA XIV: ZUR BINDUNG VON SULFOKONJUGIERTEN STEROIDEN AN PLASMAPROTEIN

Oertel, G. W., K. Groot, and P. Brühl, Hoppe-Seylers Z. Physiol. Chem. **341**, 10–21 (1965) **German.**

Pregnancy plasma and normal plasma were analyzed for steroid conjugates after paper electrophoresis flotation and column chromatography fractionation were performed. Electrophoresis was performed on Schleicher and Schull paper No. 2043a or cellulose acetate in a 0.1M Veronal buffer, H 8.6. Preparative electrophoretic separation was achieved on Polyvinyl chloride using a Veronal buffer, pH 8.6. Steroid binding was demonstrated by incubation of plasma or plasma fractions with ammonium $[7\alpha\text{-}^3H]$ -dehydroepiandrosteron sulfate. Most of the radioactivity was found to reside in the albumin fraction. Flotation of low and high density lipoproteins from plasma left most of activity in the plasma. DEAE-Sephadex® column chromatography of the residual plasma separated a substance which migrated as albumin.

IATROGENIC URINARY AMINO ACID DERIVED FROM PENICILLIN

Perry, T. L., G. H. Dixon, and S. Hansen, Nature **206**, 895–995 (1965).

Two dimensional paper electrophoresis of urine from a patient receiving penicillin demonstrated at least three ninhydren-negative precursors of penicillamine sulphonic acid. Whatman® 3 HR paper was electrophoresed at pH 3.6 for 45 minutes at 60 volts/cm. The paper was dried and reelectrophoresed at right angles in the same buffer. The authors emphasize cautions to be taken when doing amino acid analysis of urine from patients taking antibiotics.

ELECTROPHORETIC INVESTIGATION OF SERUM PROTEINS OF GERBILLUS PYRAMIDUM GEOFFROY 1825

Rösler, B., Acta Biol. Med. Ger. **15**, 537–547 (1965).

Gerbillus serum was separated five fractions by paper electrophoresis in a Veronal buffer. These were albumin, γ-globulin, two β-globulins and a γ-globulin fraction. Albumin was separated into two fractions and α-globulin was further split into three fractions when electrophoresis was performed in a Tris–EDTA–borate buffer. Seventeen serum fractions were detected on starch gels.

IDENTIFICATION OF RAT URINARY PROTEINS BY ZONE AND IMMUNO-ELECTROPHORESIS

Roy, A. K., and O. W. Neuhaus, Proc. Soc. Exp. Biol. Med. **121**, 894–899 (1965).

Paper electrophoresis was performed on Whatman® 3 MM paper in a Veronal buffer, pH 8.6, μ 0.075. Starch-gel electrophoresis performed in a discontinuous buffer of glycine–NaOH and borate pH 9.0, as well as immunoelectrophoresis, detected 15 protein bands. Seven of the bands detected were derived from serum, while the remainder appeared unique to urine. Two prealbumins, α_1-globulin, 2 α_2-globulins, one β-globulin, one γ-globulin and one protein with albumin-like mobility were unique to rat urine.

PAPER ELECTROPHORESIS AND SOME OTHER TECHNIQUES FOR SEPARATING ACYL HYDRAZIDES

Russel, D., J. Chromatog. **19**, 199–201 (1965).

Acyl hydrazides were separated by paper electrophoresis at 80 volts/cm for one hour using 20% acetic acid (pH 2.0) as electrolyte. The compounds were detected by converting the hydrazides to their picryl derivatives. This is easily accomplished by spraying the dried chromatogram with 0.5% picric acid.

COMPARISON OF THE RECOVERY OF INORGANIC: I. BY PAPER ELECTROPHORESIS OR CHROMATOGRAPHY

Schimoda, S., Endocrinology **77**, 401–405 (1965).

Electrophoresis and paper chromatography were compared as to efficiency in recovery of carrier-free I^{131}. Three systems, n-butanal–acetic acid–water (4:1:5), n-butanol–ab. ethanol–0.5M ammonium hydroxide (5:1:2), and n-butanol–dioxane–0.5M ammonium hydroxide (4:1:4), separated the compound with only 5% loss of activity. Electrophoresis on Whatman® No. 3 filter paper using a Tris-maleate buffer, pH 8.6 (200 volts for 70-80 minutes), yielded recovery rates of 98%.

ELECTROPHORETIC ASSAY FOR URINARY CATECHOLAMINES AS A SCREENING TEST FOR PHENCHROMOCYTOMA

Sheridan, A. J., Mendel Bull. **1965**, 23–27.

Paper electrophoresis in a barbital buffer, pH 8.6, ionic strength 0.1, of 24-hour urine containing 0.2 mg creatinine was carried out for three hours at 400 volts. The paper (S&S 2043A filter paper) was shaped into a "V" and the sample applied to the center. Aqueous diazotized p-nitroaniline stained the catechol amines blue-purple. In normal urine, the bands were too light to be detected.

HOMOSERINE IN URINE OF PATIENTS WITH NEUROBLASTOMA

von Studnitz, W., Scand. J. Clin. Lab. Invest. **17,** 558–564 (1965).

An unknown amino acid was found by chromatography of urine from neuroblastoma patients. Staining with ninhydrin and isatin and analyzing the electrophoretic behavior demonstrated that the unknown was homoserine. Whatman® No. 3 MM paper was spotted with a urine volume having 60 μg of creatinine. Formic acid–water–acetic acid (150:750:100, pH 1.2) was the electrophoresis buffer. Conditions were 50 volts/cm for 2½ hours. Chromatography in the second dimension was done in phenol-water (4:1). Preliminary observations suggest that either homoserine or cystathionine was present in the urine of patients with widespread metastases.

PAPIERELECTROPHORETISCHE UNTERSUCHUNGEN ÜBER DIE EIWEISSZU-SAMMENSETZUNG DER MENSCHLICHEN TRÄNEN

Tapesto, I., Z. Vass, and K. Szlamka, Arch. Klin. Exp. Ophthalmol. **168,** 468–473 (1965) **German.**

Paper electrophoresis of human tears separated eight protein fractions within four protein groups (2 albumins, 1 α_y- and 3 β-globulins, and two lysozymes). Electrophoretic separation was achieved on Machary-Nagel 214 filter paper in a sodium barbital–sodium acetate buffer, pH 8.6, ionic strength 0.1. The temperature during the electrophoretic run was maintained at 4°C. The α- and γ-fractions of serum were absent in tear samples.

VERGLEICHENDE ELEKTROPHORETISCHE UNTERSUCHUNGEN VON GAL-LÉNSAFT UND SERUM

Wilde, J., and I. Thümmel, Acta Hepato-Splenol. **12,** 299–313 (1965) **German.**

Quantitative electrophoresis was used to determine the total protein, albumin, individual globulin, and glycoprotein concentrations of bile and serum from 103 cholecystectomy patients. The results suggest a common source for the different protein fractions. Bile and serum were similar qualitatively in the protein bands found. Liver and bladder bile differed from each other in alpha$_2$ and gamma globulin content.

CHARACTERISTICS OF POTATO PROTEINS IN RELATION TO POTATO VARIETIES

Zwartz, J. A., Bibl. Nutr. Dieta **7**, 221–232 (1965).

The proteins of potato tuber juice of 57 varieties of potatoes were separated by paper electrophoresis in a 0.04M barbital buffer, pH 8.3, ionic strength 0.04. A voltage of 220 (7 volts/cm, 2 mA per strip) for five hours was used. A specific protein pattern was found for each variety tested, and at least six soluble proteins were found for each variety. A statistical analysis of each variety was performed.

POSSIBILITIES OF THE USE OF PAPER ELECTROPHORESIS IN THE ASSAY OF HEMOGLOBIN FRACTIONS

Dettori, M., Riv. Clin. Pediat. **73**, 129–136 (1965).

The usefulness of a method to distinguish microcytemic and normal patients on the basis of the A_2 hemoglobin fraction is discussed. Paper electrophoresis was run with pH 8.6 diethylbarbituric acid–sodium–diethylbarbiturate buffer, ionic strength 0.04 (12 hours 220–225 volts) as a developer. The strips were stained with 0.5% Amido Black in MeOH–HOAC and decolorized with 15% phenol with 10% acetic acid added. Quantitation was made by densitometry or elution of the spots with a mixture of 2 g Na_2CO_3, 30 ml methanol and 70 ml water. The hemolyzates were prepared for electrophoresis by the Singer method.

Polyacrylamide Electrophoresis

HUMAN CHORIONIC GONADOTROPIN: I. PURIFICATION AND PHYSIO-CHEMICAL PROPERTIES

Bahl, O. P., J. Biol. Chem. **244**, 567–573 (1969).

Analytical polyacrylamide-disc electrophoresis and immunoelectrophoresis were used to characterize a preparation of human chorionic gonadotropin. The hormone was purified by column chromatography on DEAE-Sephadex® and Sephadex® G-100. A continuous gradient of 0.04M Tris-phosphate buffer, pH 8.7, started the elution of the hormone from DEAE-Sephadex. A change to 0.1M NaCl and 0.2M NaCl–0.04M Tris-phosphate buffer, pH 8.7 to 0.2M NaCl 0.04M Tris. Phosphate, pH 9.0, completed the gradient. Disc electrophoresis of the material eluted indicated that a high degree of purity was obtained.

FACTORS STIMULATING TRANSCRIPTION BY RNA POLYMERASE

Burgess, R. R., A. A. Travers, J. J. Dunn, and K. F. Bautz, Nature **221**, 43–46 (1969).

Polyacrylamide-gel electrophoresis was used to characterise a protein component associated with RNA polymerase of E. coli K-12 cells. The substances separated from RNA polymerase using phosphocellulose column chromatography was applied to a 7.5% polyacrylamide gel with and without urea and run according to the method of Shapiro [Biochem. Biophys. Acta **28**, 815 (1967)]. The isolated factor was found to greatly stimulate the synthesis of RNA. The greatest stimulation of synthesis was observed in the presence of T_4 template.

FRACTIONATION OF OLIGODEOXYNUCLEOTIDES BY POLYACRYLAMIDE GEL ELECTROPHORESIS

Doe, J., and M. Jones, Anal. Biochem. **27**, 193–204 (1969).

Analytical and preparative polyacrylamide gel electrophoresis was used to fractionate mixtures of oligomers produced by enzymic digestion of dAT (alternating copolymer of deoxythymidylate and deoxygadenylate). The gels were prepared according to Jovins procedure [Anal. Biochem. **9**, 351 (1964)]. High resolution of fractions with 6 to 40 degrees of polymerization was possible.

INTERACTION OF TRYPSIN AND CHYMOTRYPSIN WITH A SOYBEAN PROTEINASE INHIBITOR

Frattali, V., and R. F. Steiner, Biochim. Biophys. Acta **174**, 480–487 (1969).

Two modified forms of the Bowman-Birk inhibitor were detected by polyacrylamide-gel electrophoresis of a reaction mixture containing inhibitor and trypsin or chymotrypsin. (The components were reacted at acid pH). Electrophoretic separation was carried out at pH 8.35 in a buffer system of 0.089M Tris-HCl, 0.089M boric acid and 2.7mM EDTA. Separation was continued for two hours. The modified products did not retain inhibitory power.

COMPARISON OF THE EFFECTS OF PARASYMPATHETIC AND SYMPATHETIC NERVOUS STIMULATION ON CAT SUBMAXILLARY GLAND SALIVA

Kahn, N., I. Mandel, J. Licking, A. Wasserman, and D. Morea, Proc. Soc. Exp. Biol. Med. **130**, 314–318 (1969).

The proteins of cat parasympathetic and sympathetic saliva were examined using vertical disc electrophoresis and flat-bed electrophoresis. Disc electrophoresis was performed on a 7.5% gel in a 0.1M Tris-glycine buffer, pH 8.3, at a current of 5mA per tube for one hour. Flat-bed electrophoresis was performed on a 5% acrylamide gel in 0.05M barbital buffer, pH 8.6. Proteins were detected by staining with Coomassie Brilliant Blue. Parasympathetic saliva samples showed heavy staining, cathodic migrating components upon flat bed electrophoresis. These substances were absent from sympathetic saliva. Disc electrophoretic results revealed quantitative and qualitative differences between sympathetic and parasympathetic samples. In sympathetic nerve samples one to three rapidly migrating components were observed. These substances stained positively with Periodic-Schiff reagent, thereby indicating glycoproteins.

CHARACTERIZATION OF SARCOTUBULAR MEMBRANE PROTEIN

Masoro, E. J., and B. P. Yu, Biochem. Biophys. Res. Commun. **34**, 686–693 (1969).

Highly purified sarcotubular protein was characterized by gel filtration on Sepharose 4B®, analytical ultracentrifugation and by vertical polyacrylamide-agarose gel electrophoresis. Two fractions were obtained from the Sepharose 4B column. Fraction two was submitted to electrophoretic analysis on gel composed of 3% Cyanogum 41®–0.5% agarose gel in a buffer containing 85mM Tris, 80mM boric acid and 2.5mM EDTA (pH 9.4). Separation proceeded for two hours at 300 volts. Molecular weight estimates from ultra-centrifuge data yielded a value of 17,000. The electrophoretic technique yielded two bands. Electrophoresis of the fraction II materials subjected to treatment with dissociating agents indicated that similar peptide units were formed.

MICROSLIDE ACRYLAMIDE GEL ELECTROPHORESIS TECHNIQUE FOR TISSUE LACTIC DEHYDROGENASE

Allred, R. J., and H. J. Kentel, J. Lab. Clin. Med. **71**, 179 (1968).

Lactic dehydrogenase was extracted from tissue homogenates with Tris-buffer, pH 8.6, 0.05M, for electrophoretic separation of its isoenzymes. Microslides were prepared of a gel composed of: 2.72 ml riboflavin (0.005%), 0.22 ml $Na_2S_2O_3$ (0.64%), 10.0 ml acrylamide (8%), 7.06 ml Tris buffer (pH 8.6, 0.05M). The gel was photopolymerized in a carbon dioxide atmosphere.

Electrophoresis in a pH 8.6 barbitone acetate buffer at 150 volts for one hour gave clear separation. The isozymes were stained by a modification of Markert and Ursprung's technique.

CHEMICAL DETERIORATION OF FROZEN BOVINE MUSCLE AT −4°C

Awad, A., W. D. Pourie, and O. Feenema, J. Food Sci. **33**, 227–235 (1968).

Polyacrylamide-disc electrophoresis of protein, lipid analysis by TLC, total nitrogen, and drip volume measurements were performed on bovine muscle stored for eight-week periods at 4°C. The electrophoretic analyses were performed according to the procedure of Ornstein and Davis on actomyosin, sarcoplasmic and myofibrillar proteins extracted from bovine muscle. An increase in the free fatty acid content over an eight-week period was observed. Total protein declined from 91% to 51% over this eight-week period. The electrophoretic analysis revealed that freezing altered the mobility of sarcoplasmic protein within two weeks. Actomyosin stored two weeks separated into four bands, and after being frozen eight weeks separated into three bands.

ISOLATION OF TRIOSE PHOSPHATE ISOMERASE FROM THE LENS

Burton, P. M., and S. G. Waley, Exp. Eye Res. **7**, 189–195 (1968).

Triose phosphate isomerase was purified and crystallized from bovine lens. The purified enzyme was examined by disc and starch electrophoresis, and its mobility was compared to that of rabbit muscle triose isomerase. Disc electrophoresis in a Tris-EDTA-borate buffer showed one major band. Starch-gel electrophoresis revealed mobility differences between the two enzymes.

ISOLATION AND CHARACTERIZATION OF AN ACTIVE FRAGMENT FROM ENZYMATIC DEGRADATION ON ENCEPHALITOGENIC PROTEIN

Chao, L. P., and R. Einstein, J. Biol. Chem. **243**, 6050–6055 (1968).

Encephalitogenic protein of spinal cord and its proteinase digestion products were isolated and characterized by polyacrylamide-gel electrophoresis (disc), gel filtration, amino acid content, minimum molecular weight, peptide mapping and amino and carboxy-terminal group analysis. Electrophoretic separation of the active fragments was achieved at pH 4.5 in the manner described by Heisfeld, Lewis and Williams [Nature **195**, 281 (1962)] at 6 mA per gel for one hour. The isolated active fragment was found to be homogenous.

A HIGH-RESOLUTION ACRYLAMIDE-GEL ELECTROPHORESIS TECHNIQUE FOR THE ANALYSIS OF COLLAGEN POLYMER COMPONENTS

Clark, C. C., and A. Veis, Biochim. Biophys. Acta **154**, 175–182 (1968).

Acrylamide-gel electrophoresis, using a discontinuous gel system, resolved the α, β and γ components of collagen. A "pre-α" region and a "higher-polymer" can also be observed in some samples. A pH 4.8 sodium acetate buffer, ionic strength 0.05, was employed.

ISOLATION AND CHARACTERIZATION OF INSOLUBLE PROTEINS OF THE SYNAPTIC PLASMA MEMBRANE

Cotman, C. W., H. R. Mahler, and T. E. Hugli, Arch. Biochem. Biophys. **126**, 821–837 (1968).

Polyacrylamide-disc electrophoresis was employed to characterize the insoluble proteins of the synaptosome plasma membrane isolated by the procedure of Green et al. [Biochem. Biophys. Res. Commun. **5**, 109 (1961)]. The insoluble protein was resolved into five major bands and an equal number of minor bands. The insoluble protein of brain mitochondria and myelin exhibited patterns significantly more complex than the insoluble protein of synaptic plasma membrane.

A STUDY OF THE INDIGOGENIC DETECTION OF SERUM ACID PHOS-PHATASE IN POLYACRYLAMIDE GEL ELECTROPHORESIS

Epstein, E., P. L. Wolf, and B. Zak, Enzymologia **35**, 258–262 (1968).

Serum acid phosphatase enzymes were separated in polyacrylamide-gel disc) electrophoresis using 2-amino-2-methyl 1,3 propane-diol and boric acid at pH 9.0 and Tris (hydroxymethyl) methyl-2-amino-ethane sulfonic acid, pH 7.5. The separated enzymes were detected using a single step, self-indicating substrate, p-toluidinium-5-bromo-4-chloro-3-indolyl phosphate. Separations were achieved at 2.5 mA per tube at 5–6°C. With the substrate used, artifacts were absent.

STUDIES ON THE LOW MOLECULAR WEIGHT PROTEIN COMPONENTS IN RABBIT SKELETAL MUSCLE

Gaetiens, E., K. Bárány, G. Balin, H. Oppenheimer, and M. Bárnáy, Arch. Biochem. Biophys. **123**, 82–96 (1968).

Myosin and heavy meromyosin was purified to the extent of being free of 5'-AMP deaminase, adenylate kinase and nucleic acid by chromatography on DEAE-cellulose, and cellulose-phosphate. Vertical electrophoresis on poly-acrylamide (7%) gel separated the small protein of purified myosin and meromysin into three main components. A Tris–EDTA–borate buffer, pH 9.2, and a voltage of 12 volts/cm (two to four hours) were used to obtain separation.

NATURE OF THE HIGH MOLECULAR WEIGHT FRACTION OF FIBRINOLYTIC DIGESTS OF HUMAN FIBRINOGEN

Jamieson, G. A., and J. Gaffney, Jr., Biochim. Biophys. Acta **154,** 96–109 (1968).

Human fibrinogen isolated from pooled human plasma by the method of Pert and Johnson was characterized by both vertical-gel electrophoresis and acrylamide-slab or disc electrophoresis. Vertical-gel electrophoresis was performed in a Tris–EDTA–borate or Tris–glycine buffer system on either polyacrylamide discs or slabs. A potential of 5 mA/tube for three hours was employed. Results obtained before and after digestion indicated striking differences in the clotting times of the digested material as opposed to the undigested material. A complex pattern of bands were observed on both starch and acrylamide gels.

A NEW TECHNIQUE FOR THE ELECTROPHORESIS OF PROTEINS

Leabeck, D. H., and A. C. Rutter, Biochem. J. **108,** 19P (1968).

A new technique employing a 3% polyacrylamide-gel slab and carrier ampholytes such as Ampholines (LKB Instruments, Ltd.) is described. The technique has successfully separated the common hemoglobins after three to six hours. Convection currents and endosmosis are minimal in this technique.

MITOCHRONDRIAL STRUCTURAL PROTEINS: I. METHODS OF PREPARATION AND PURIFICATION: CHARACTERIZATION BY GEL ELECTROPHORESIS

Lenaz, G., N. F. Haad, A. Lauwers, D. W. Allman, and D. E. Green, Arch. Biochem. Biophys. **126,** 746–752 (1968).

Structural protein prepared from heavy beef heart mitochrondria by six different methods was characterized by electrophoresis on polyacrylamide gels. Electrophoresis was carried out in 7% acetic acid according to the procedure of Takayama et al. [Arch. Biochem. Biophys. **114,** 223 (1966)]. All preparations were shown to be heterogeneous.

PURIFICATION OF BOVINE GROWTH HORMONE AND PROLACTIN BY PREPARATIVE ELECTROPHORESIS

Lewis, U. J., E. V. Cheever, and B. K. Seavey, Anal. Biochem. **24,** 162–175 (1968).

Multiple components of bovine prolactin and growth hormone were isolated by preparative electrophoresis on polyacrylamide gel. Preparative electrophoresis was performed in the Buchler Poly-Prep® instrument, using the Tris–glycine buffer system of Ornstein [Ann. N. Y. Acad. Sci. **121**, 321 (1964)]. Immunodiffusion, bioassay and NH_2-terminal amino acid analyses were performed on the purified material.

PROTEIN CONSTITUENTS OF PLANT RIBOSOMES

Lyttleton, J. W., Biochim. Biophys. Acta **154**, 145–149 (1968).

Basic proteins obtained from plant ribosomes by urea dissociation have been studied by disc electrophoresis. A 12% acrylamide gel and a buffer of citrate phosphate, pH 4.5 was employed. Unrelated plant species gave recognizably different protein patterns. Cytoplasmic ribosomes from different organs of the pea plant gave similar patterns. Chloroplast cytoplasmic fraction of ribosomes gave quite different results.

ACRYLAMIDE GEL ELECTROPHORESIS OF SERUM PROTEINS IN INFANCY AND PREGNANCY

Man, E. B., and R. J. Whitehead, Jr., Clin. Chem. **14**, 1002–1009 (1968).

Polyacrylamide-gel vertical electrophoresis was modified so that 7.5 g Cyanogum®, 0.3 ml dimethylaminopropionitrile, 0.3 ammonium sulfate in 5% buffer was used to study protein patterns of neonates and mothers. A 2-amino–2-methyl–1,3-propanediol amino acetic buffer, pH 9.36, was used.

ELECTROPHORETIC MOBILITY OF EHRLICH ASCITES CARCINOMA CELLS GROWN IN VITRO OR IN VIVO

Mayhew, E., Cancer Res. **28**, 1590–1595 (1968).

Changes in the electrophoretic mobility of Ehrlich ascites tumor cells of mice during *in vivo* growth or in cell culture was investigated using the cylindrical tube apparatus described by Bangham *et al.* [Nature **182**, 642–644 (1958)]. The mobility of cells grown *in vivo* was found to be extremely variable, reaching maximum one to two days after tumor inoculation and declining thereafter until the death of the animal. Mobility of cells grown *in vivo* was found to decrease after treatment with neuraminidase and ribonuclease.

PHENOL OXIDASES OF A LOZENGE MUTANT OF DROSOPHILA

Peeples, E. E., D. R. Barnett, and C. P. Oliver, Science **159**, 548–552 (1968).

Polyacrylamide-gel electrophoresis was performed on single Drosophila pupae homogenates in order to separate phenol oxidases. The gel was prepared in a 0.1M Tris-glycine buffer, pH 8.9. Electrophoresis was performed in the same buffer at 400 volts for 90 minutes, with the temperature kept at 17°C. The proenzyme was located in the gel by incubating in an activator solution for four hours, after which incubation with copa or tyrosine yielded dark melanin bands at the site of enzyme activity. Monophenol oxidase activity appears to be absent from lozenge-glossy females and males of Drosophila.

ANALYSIS OF THE PROTEINS IN HUMAN SEMINAL PLASMA

Quinlivan, L. L. G., Arch. Biochem. Biophys. **127**, 680–687 (1968).

Cellulose-acetate electrophoresis, immunoelectrophoresis, and disc-gel electrophoresis have been used to analyze the proteins of human seminal plasma. Cellulose-acetate electrophoresis was performed in a Beckman MicroZone® system using a barbital buffer (0.1M, pH 8.6). A potential gradient of 250 volts for 20 minutes at room temperature was used. Immunoelectrophoresis was performed in a Shandon Universal Electrophoresis Apparatus on glass slides coated with agarose. An Owens modified barbitone buffer, pH 8.6, and a constant 50 mA current equivalent to 20 volts/cm for 4.5 hours were employed. Disc electrophoresis was performed according to the method of Davis [J. Ann. N. Y. Acad. Sci. **121**, 404 (1964)]. Cellulose acetate electrophoresis demonstrated four protein bands while disc electrophoresis demonstrated fifteen. Immunoelectrophoresis identified six proteins, three of which were identical to those in human serum.

INTERNAL COMPONENTS OF ADENOVIRUSES

Russell, W. C., W. G. Laver, and P. J. Sanderson, Nature **219**, 1127–1130 (1968).

Five percent polyacrylamide gels, run in a 0.01M Tris-glycine buffer, pH 9.0, containing 0.6M urea, separated protein components of purified adenovirus and "cores." A potential gradient of 150 volts/cm applied for 90 minutes was sufficient to provide separation. Five major bands and a number of minor protein bands were visible upon staining with Amido Black. When electrophoresis was performed at pH 2.4 in 8M urea on selective acid extracts of purified virus, two components rich in arginine were revealed.

SPECIFIC DIMERIZATION OF THE LIGHT CHAINS OF HUMAN IMMUNO-GLOBULIN

Stevenson, G. T., and D. Straus, Biochem. J. **108**, 275–282 (1968).

Human immunoglobulin light chains were allowed to dimerize in vitro in the presence of acetic acid and urea. The dimerized globulins were then separated by electrophoresis on polyacrylamide gel, with and without urea. A pH 8.8 buffer system of 0.05M glycine–sodium hydroxide was used to prepare the gels. The electrode buffer was 0.3M boric acid–0.6M sodium hydroxide, pH

9.5. A potential of 4 volts/cm for 16 hours was applied for gels prepared with urea. Five volts/cm were applied to separations in the absence of urea. Up to nine regularly spaced bands were found.

RIBOSOMAL RNA OF ARBACIA PUNCTULATA

Sy, J., and K. S. McCarty, Biochem. Biophys. Acta **166**, 571–574 (1968).

Electrophoresis on 2.4% polyacrylamide gels, using a 0.04M Tris-acetic acid (pH 7.2)–0.02M sodium acetate, 1 mm EDTA buffer, separated RNA of the sea urchin *(Arbacia punctulata)*. Rat 29S ribosomes, pea seedling 25S ribosomes, and *E. coli* 23S ribosomes were electrophoresed as control samples, and the distance migrated, in centimeters, was plotted against sedimentation constants. Sea urchin rRNA had a sedimentation coefficient of 26S–18S.

GEL FILTRATION OF CHICK LENS PROTEINS

Truman, D. E. S., Exp. Eye Res. **7**, 358–368 (1968).

Chick lens proteins were separated using gel filtration on Bio-Gel-P 300, Bio-Gel-P 150, and Bio-Gel-A 1.5M. The proteins thus obtained were characterized by immunoelectrophoresis and polyacrylamide-disc electrophoresis. Disc electrophoresis was performed according to the procedure of Ornstein with the modification that the gel was made in 0.38M Tris buffer, pH 8.9, and the tank buffer used was 0.038M glycine, pH 8.3. A potential of 50 volts was applied for the first 15 minutes, then increased to 300 volts (5 mA/tube) for 90 minutes. Immunoelectrophoresis was performed according to the procedure of Campbell, Clayton and Truman [Exp. Eye Res. **7**, 4 (1968)]. Gel filtration achieved partial separation of α, β and γ-crystallins. Disc electrophoresis showed that each fraction was heterogeneous, the least heterogeneous being the α-crystallin.

ELECTROPHORETIC STUDIES ON THE HETEROGENEITY OF THE CHICKEN LENS CRYSTALLINS

Zwaan, J., Exp. Eye Res. **7**, 461–472 (1968).

Two-dimensional polyacrylamide-gel electrophoresis and immunoelectrophoresis were employed to characterize α-crystallin, γ-crystallin, and β-crystallin of chicken lens. Polyacrylamide electrophoresis was performed on gels made in 0.18M Tris-borate buffer, pH 9.2 or 8.3, containing 0.0025M disodium ethylenediamine tetracetate. Molecular weights of the different fractions were estimated by the way they migrated two-dimensionally; individual proteins were identified by immunodiffusion after they were punched out of gels. Seventeen fractions were identified in chicken lens, and nine were β-crystallins.

The majority had molecular weights of around 55,000, but mobilities were different.

A PROCEDURE FOR RAPID AND SENSITIVE STAINING OF PROTEIN FRACTIONATED BY POLYACRYLAMIDE GEL ELECTROPHORESIS

Chrambach, A., R. A. Reisfeld, M. Wyckoff, and J. Zaccari, Anal. Biochem. **20**, 150–154 (1967).

Increased sensitivity of stained proteins separated on polyacrylamide gel was achieved with Coomassie Blue. The stain compares with Amido Black without destaining; however, far superior results were obtained if gels were destained.

RESOLUTION OF INSOLUBLE PROTEINS IN RAT BRAIN SUBCELLAR FRACTIONS

Cotman, C. W., and R. H. Mahler, Arch. Biochem. Biophys. **120**, 384–396 (1967).

The proteins of rat brain were isolated by density gradient ultra-centrifugation and examined by disc electrophoresis. The electrophoresis procedure of Takayama was used [Arch. Biochem. Biophys. **114**, 223 (1966)]. Acrylamide gels (7½%, 55 mm long) were prepared and allowed to equilibrate in a solvent of phenol–acetic acid–water (2:1:1) and 5M urea for two days prior to use. The protein of myelin was separated into four bands; synaptic muscle protein was separated into sixteen bands. The microsomal factions (microsomal membranes, microsomal vesicles and microsome) exhibit patterns similar to synoptic vesicles; however, these were different as exhibited by differences in bands. Synaptosome membranes were seventeen bands.

QUANTITATION OF AMYLASES IN DROSOPHILA SEPARATED BY ACRYLAMIDE GEL ELECTROPHORESIS

Doane, W. W., J. Exp. Zool. **164**, 363–378 (1967).

Following electrophoresis of an amylase enzyme extract, the acrylamide gel was placed against a starch-acrylamide film attached to a glass plate. This starch-acrylamide film was later stained with iodine, covered with a glass plate, and scanned with a densitometer. Electrophoresis was performed according to the method of Ornstein and Davis with the following modifications: 1) $K_4Fe(CN)_6$ was omitted from the small-pore gel; 2) one-half the amount of TEMED was added to the small-pore gels; 3) 0.47 Tris-phosphate buffer was substituted for Tris-HCl buffer in the large-pore gels, and 4) 1 μl of 10% 3-dimethylaminopropionitrite was added to the gel to insure polymerization. The sample was mixed with 0.1 ml of large-pore gel and placed directly on the spacer gel. Electrophoresis was performed at 4°C at a constant current of 3.0–3.5 mA/tube for 50 to 60 minutes. Bromophenol Blue was used as a tracer dye.

GLUCOSYLTRANSFERASE ISOZYMES IN ALGAE

Frederick, J. F., Phytochemistry **6,** 1041–1046 (1967).

Purified phosphorylase preparations from blue-green, green, and red algae were subjected to one-dimensional electrophoretic separation on 7% polyacrylamide. Five closely associated enzymatic proteins were separated at pH 9.3. Two-dimensional separation revealed that the five proteins were mixtures of three isozymes. The first four proteins contained two enzymes, and the fifth protein contained two or more similar enzymes.

ACRYLAMIDE-GEL ELECTROPHORESIS OF SERUM PROTEINS

Groulade, J., Ann. Biol. Clin. **25,** 371–381 (1967).

Serum proteins were easily separated by electrophoresis of acrylamide-gel plates. The usefulness of this technique is discussed. Alterations in the pH of the buffer can give altered mobility patterns of serum components. The effect of pH is reviewed.

SEX DIFFERENCE IN PITUITARY LACTIC DEHYDROGENASE CONCENTRATION IN RATS

Hoch-Ligeti, C., T. J. Brown, Jr., and H. H. Grantham, Jr., Endocrinology **80,** 483–489 (1967).

Pituitary LDH concentration was found to be higher in the sexually mature female rat than in the male, a difference coincidental with the opening of the vagina. Five isoenzymes, which do not appear the same on polyacrylamide-gel electrophoresis, are present in both sexes [Ornstein, L., Ann. N. Y. Acad. Sci. **121,** 321 (1964); Nachlas, M., *ibid.,* 404].

DISCONTINUOUS ACRYLAMIDE-GEL PLATE ELECTROPHORESIS

Holmes, R., Biochim. Biophys. Acta **133,** 174–177 (1967).

Human serum was separated into 28 protein components using thin-layer electrophoresis on polyacrylamide matrices mounted on glass projection slides. A 7.5% running gel was used. Electrophoresis was performed in a 0.1675M sodium borate buffer, pH 8.6. Variations in separation were obtained when gel concentration was increased or decreased.

PROPERTIES OF POLYPHENOL OXIDASES PRODUCED IN SWEET PO-TATO TISSUE AFTER WOUNDING

Hyodo, A., and I. Uritani, Arch. Biochem. Biophys. **122**, 299–309 (1967).

Two polyphenol oxidases produced in cut tissue of sweet potato roots have been isolated from acetone powder by ammonium sulfate extraction, DEAE-cellulose column chromatography and Sephadex® G-100. Polyacryla-mide-gel electrophoresis characterized the isolated enzymes. A 5% gel prepared in 0.02M Tris HCl buffer, pH 9.3, was the matrix. The electrode buffer was 0.02M Tris HCl buffer, pH 9.3. Electrophoresis was carried out at 16 volts/cm and 0.4 mA for 12 hours. Three kinds of polyphenol oxidases (IIa, IIb, and IIc) were present in healthy tissue. In addition two more oxidases were produced in injured tissue.

MOBILITY-MOLECULAR WEIGHT RELATIONSHIPS OF SMALL PROTEINS AND PEPTIDES IN ACRYLAMIDE-GEL ELECTROPHORESIS

Ingram, L., M. P. Tombs, and A. Hurst, Anal. Biochem. **20**, 24–29 (1967).

Polyacrylamide-gel electrophoresis separated small molecular weight proteins and peptides. The mobility of these proteins and peptides in different gel concentrations is used to estimate molecular weights of unknown peptides.

MAMMALIAN RIBOSOMAL PROTEIN: ANALYSIS BY ELECTROPHORESIS ON POLYACRYLAMIDE GEL

Low, R. B., Science **155**, 1180–1182 (1967).

Ribosomal protein, analyzed by discontinuous electrophoresis in 10% poly-acrylamide gel at pH 4.5 (150 minutes, 3°C, 3 mA), yielded 24 bands. Twelve to fourteen protein bands could be further resolved by electrophoresis in a 7.5% gel. E. coli ribosomal protein at pH 4.5 gave 29 bands in a different pattern than those from mammalian tissue. At pH 8.3, mammalian proteins did not separate, while those of E. coli formed eight to nine bands. The structure of genes responsible for synthesis of ribosomal protein is thought to be the same in mammals.

ELECTROPHORETIC PATTERNS OF LENS PROTEINS FROM GENETICALLY CAUSED CATARACT IN THE MOUSE

Moser, G. C., and S. Gluecksohn-Waelsch, Exp. Eye Res. **6**, 297–298 (1967).

The electrophoretic pattern of soluble lens proteins of cataractal lens of the inbred strain F 72 Kinky, phenotypic normal F 72 Kinky (an unrelated normal and normal inbred isogenic strain of mice) was studied using the technique of Konyokhov and Wachtel [Exp. Eye Res. **2**, 325 (1963)]. The electrophoretic pattern of normal lens protein inbred mice was identical to that pattern obtained for cataractal protein in the inbred strain. The inbred stain exhibited electrophoretic differences from that of the unrelated strain.

STUDIES ON HUMAN CASEIN: I. FRACTIONATION OF HUMAN CASEIN BY DIETHYLAMINOETHYL CELLULOSE COLUMN CHROMATOGRAPHY

Nagasawa, T., T. Ryoki, I. Kiyosawa and K. Kuwahara, Arch. Biochem. Biophys. **121**, 502–507 (1967).

Polyacrylamide-gel electrophoresis characterized human caseins separated by DEAE-cellulose column chromatography. Acrylamide-gel electrophoresis was performed horizontally in an 8% gel with a discontinuous buffer containing 7M urea as described by Boyer and Fainer [Science **140**, 1228 (1963)]. Nine heterogeneous fractions were observed.

PURIFICATION AND PROPERTIES OF YEAST INVERTASE

Neumann, N. P., Biochem. **6**, 468–475 (1967).

Polyacrylamide-gel electrophoresis at pH 6.5 demonstrated the homogeneity of a preparation of yeast invertase. Electrophoresis on cellulose acetate in the pH range 2–12 also showed one protein peak which corresponded to enzyme activity. Sedimentation equilibrium measurements gave a molecular weight of 270,000.

RESOLUTION OF MULTIPLE RIBONUCLEIC ACID SPECIES BY POLYACRYLAMIDE GEL ELECTROPHORESIS

Peacock, A. C., and C. W. Dingman, Biochemistry **6**, 818–827 (1967).

A number of RNA bands were resolved from extracts of rat liver, kidney and brain. Initial fractionation was achieved by preparative zone sedimentation. Electrophoresis was performed in 3.5% and 10% polyacrylamide gels at pH 8.3.

ADENOSINE DEAMINASE FROM CALF SPLEEN: I. PURIFICATION

Pfrogner, N., Arch. Biochem. Biophys. **119**, 141–146 (1967).

The adenosine deaminase from calf spleen was purified 500-fold by isoelectric precipitation, ammonium sulfate fractionation and chromatography on Bio-Rex-70, DEAE-Sephadex® A-50 and SE-Sephadex C-50. It was then examined for homogeneity by polyacrylamide-disc electrophoresis. Electrophoresis was performed at pH's 3.8, 8.3 and 7.2 as described by Duesberg and Reuckert

[Anal. Biochem. **11**, 342 (1965)]. Results obtained indicated that the enzyme was homogenous.

HOCHMOLEKULARES γG-PARAPROTEIN IM LIQUOR CEREBROSPINALIS BEI PLASMACYTOM

Scheurlen, P. G., H. Felgenhaver, and A. Pappas, Klin. Wochschr. **45,** 419–422 (1967) **German.**

The use of polyacrylamide-gel electrophoresis, starch-gel electrophoresis and immunoelectrophoresis to characterize a paraprotein found in the cerebrospinal fluid of a myeloma patient is recorded. Polyacrylamide-gel electrophoresis was performed in a pH 8.6 Veronal buffer according to the technique of Ornstein and Davis. Starch-gel electrophoresis was performed using the procedure of Smithies, and immunoelectrophoresis was carried out using the procedure of Scheidegger. The cerebrospinal fluid was obtained from a myeloma patient with extra and intradural localization. The abnormal protein was found in the cerebrospinal fluid and not in the serum. It behaved immunologically as a γ-G globulin and possessed a high sedimentation rate.

ACRYLAMIDE GEL–SUCROSE GRADIENT ELECTROPHORESIS: A USEFUL PREPARATIVE METHOD

Schuster, L., and B. K. Schier, Anal. Biochem. **19,** 280 (1967).

A preparative method combining the good resolution of zone electrophoresis and the good recovery of sucrose gradient electrophoresis is described. No special equipment is required. Electrophoresis was done in a U-tube filled with a linear gradient (60–20% w/w) of sucrose in 0.37M Tris HCl buffer, pH 8.8, containing 10^{-3}M EDTA. The fractions were eluted from the gel and precipitated with acid. The precipitate was collected on a Millipore® filter for a Lowry protein determination.

ESTIMATION OF MOLECULAR WEIGHTS OF PROTEINS BY POLYACRYLAMIDE GEL ELECTROPHORESIS

Zwann, J., Anal. Biochem. **21,** 155 (1967).

By using the Raymond technique, a linear relationship was found between gel concentration (5–10%) and migration rates in several proteins. The ratio of absolute mobilities in two different gel concentrations was linearly related to log molecular weight. Tris–EDTA–boric acid buffer, pH 9.2, 6–8°C, was the electrolyte used. Voltage and time of electrophoresis were dependent on the gel concentration. The results are probably based on the use of "ideal" proteins.

MULTIPLE FORM OF HYDROGENASES

Ackrell, A. C., R. N. Asato, and H. F. Mower, J. Bacteriol. **92,** 828–837 (1966).

Electrophoresis of bacterial extracts on polyacrylamide gels, 7.5% in a pH 8.3 Tris-glycine buffer, at 2.5 mA per column (room temperature) separated multiple hydrogenases. Thirty-three bacterial species were examined by this technique and found to contain hydrogenases of different mobilities and band patterns unique to the species studied.

THE EFFECTS OF CRYPTORCHISM IN THE GUINEA PIG ON THE ISO-ENZYMES OF TESTICULAR LACTATE DEHYDROGENASE

Blackshaw, A. W., and J. I. Samisoni, Australian J. Biol. Sci. **19,** 841–848 (1966).

Electrophoresis performed in 7.5% acrylamide gels was used to examine scrotal and testicular lactate dehydrogenase isoenzymes of guinea pigs rendered cryptorchisic by returning one testes to the abdomen. The Tris-glycine buffer of Ornstein and Davis was used. Electrophoresis was performed at 40°C at 2.5 mA per gel. Scrotal testis contained two enzymes not found in heart and muscle. Cryptorchism caused a rapid decrease in the amounts of these enzymes.

CYANATE FORMATION AND ELECTROPHORETIC BEHAVIOUR OF PROTEINS IN GELS CONTAINING UREA

Cole, E. G., and D. K. Mecham, Anal. Biochem. **14,** 215–222 (1966).

Horizontal electrophoresis, performed in 5% polyacrylamide gels which incorporated 7.5M urea and cyanate, separated water-soluble wheat flour proteins and crude casein. Exposure to cyanate in urea solutions altered the electrophoretic patterns of the proteins. Carbamylation of the amino groups causes this change in mobility.

DETERMINATION OF SERUM HEMOGLOBIN BINDING CAPACITY AND HAPTOGLOBIN TYPE BY ACRYLAMIDE GEL ELECTROPHORESIS (SLAB) HAS BEEN ACHIEVED

Ferris, T. G., R. E. Easterling, K. J. Nelson, and R. E. Budd, Am. J. Clin. Pathol. **46,** 385–388 (1966).

Electrophoresis was performed in an Arden model 500 cell, using a Tris–EDTA–boric acid buffer, pH 8.3–8.4. A 7% gel was used, and separation was achieved by applying a potential of 200 volts for three hours. Detection of bound hemoglobin (Hb-haptoglobin complexes) was achieved by staining the electrophoresed gel with a solution prepared by dissolving 400 mg O-dianisidine in 360 ml of water (heated to just below boiling). To this solution was added 40 ml of 1.5M acetate buffer, pH 4.7. Five millileters of 3% hydrogen peroxide was added just prior to use. The complexes appeared as red-brown bands.

A MOLECULAR APPROACH TO THE STUDY OF GENETIC HETEROZYGOSITY IN NATURAL POPULATIONS: I. THE NUMBER OF ALLELES AT DIFFERENT LOCI IN DROSOPHILA PSEUDOOBSCULARA

Hubby, J. L., and R. C. Lewontin, Genetics **54**, 577–594 (1966).

Acrylamide-gel electrophoresis was used to demonstrate the heterogeneity of esterases, glucose-6-phosphate dehydrogenase and alkaline phosphatase. The electrophoretic separations were achieved on polyacrylamide gels at pH 6.5 in a borate buffer. Esterase activity was demonstrated by staining with a solution of α-naphthyl acetate and Fast Red. Malate dehydrogenase and glucose-6-phosphate dehydrogenase were detected with tetrazolium procedures with the appropriate substrate. The usefulness of electrophoresis for genetic studies is discussed thoroughly.

PROTEIN SEPARATION OF THE CELLULAR LEVEL BY MICRO DISC ELECTROPHORESIS

Hydin, H., K. Bjurstam, and B. McEwen, Anal. Biochem. **17**, 1–15 (1966).

A procedure is described for the electrophoretic separation of 10^{-7} to 10^{-9} g of proteins on polyacrylamide gel in $200\,\mu$ diameter glass capillaries. Tris–glycine buffer, pH 8.4, was the electrode buffer used, and separations were achieved at a potiential of 250 volts. The gels were stained with Amido Black and scanned with a specially constructed microdensitometer.

CHANGES OF THE PROTEINS IN CHEDDAR CHEESE MADE FROM MILK HEATED AT DIFFERENT TEMPERATURES

Melachouris, N. P., and S. L. Tuckey, J. Dairy Sci. **49**, 800–805 (1966).

Cheddar cheese was made from four lots of milk heated to 61.7°C for 30 minutes, and 93.3°C, 110.0°C, 126.7°C for 2.08 seconds. The protein changes in these cheeses were studied using polyacrylamide electrophoresis. Seven grams of acrylamide and 0.35 g of bisacrylamide were dissolved in 100 ml Veronal buffer, pH 8.6, in 7M urea. Electrophoresis was done for 17 hours at 55°C with a constant voltage of 100 volts per gel. The buffer was the same as that used in preparing the gel plates. Heat treatment of the milk did not produce any qualitative difference in the electrophoretic patterns of the cheese proteins.

ELECTROPHORESIS USING POLYACRYLAMIDE GEL AS A SUPPORTING MEDIUM: IV. ANALYSIS OF PROTEIN FRACTIONS OF SEROUS OTITIS MEDIA EFFUSIONS BY MEANS OF DISK ELECTROPHORESIS

Mogi, G., Seibutsu Butsuri Kaoaku **11**, 245–248 (1966).

Polyacrylamide-gel disc electrophoresis separated proteins in serous effusions from otitis media (Tris–glycine buffer, pH 8.3). Fewer protein fractions were observed in the serous effusion than in serum from the same patient.

LIPOPROTEIN PATTERNS IN ACRYLAMIDE GEL ELECTROPHORESIS

Raymond, S., J. L. Miles, and J. C. J. Lee, Science **151**, 346–347 (1966).

This is a report of an improved technique for detection of serum lipoproteins by electrophoresis. Samples were stained prior to electrophoresis with 1% lipid crimson in diethylene glycol. Excess dye was removed by centrifugation before the sample was placed on a Raymond vertical cell for electrophoresis. The buffer contained 45 g EDTA and 460 g Tris in 45 liters of water, pH 9.0. Fifty ml of an 8% solution was made in this buffer (including 1% each TEMED and ammonium persulfate) and served as a support for the 3% electrophoresis gel. Both were polymerized in the tube. The lipoproteins were separated In 2.5 hours at 150 volts, when one or more bands in the β-lipoprotein region and two diffuse bands in the region of α-lipoprotein and albumin were visible.

SOME PROPERTIES OF THE PROTEIN FORMING THE OUTER FIBERS OF CILIA

Renaud, F. L., and A. J. Rowe, J. Cell Biol. **31**, 92A–93A (1966).

Outer fibers of cilia obtained from *Tetrahymena pyriformis* were examined by ultracentrifugation, electron microscopy and polyacrylamide-gel electrophoresis after reduction and alkylation in 8M urea. At pH 8.6, 90–95% of the protein migrated as a single band on polyacrylamide. When dissolved in 0.6M KCl, pH 8.3, or in 0.05M borate buffer, the protein sediments in the ultracentrifuge at 4S (single component) and heterogeneous aggregates at 6S to 30S. Molecular weight was found to be 60,000±5000.

ELECTROPHORESIS OF HEMOGLOBINS FOLLOWING THE APPLICATION OF WHOLE BLOOD SAMPLES TO ACRYLAMIDE GEL

Smith, E. W., and B. L. Evatt, J. Lab. Clin. Med. **69**, 1018–1024 (1966).

A method which separates hemoglobin after the application of a sample of whole blood is described. Gels were made from Cyanogum® at a concentration of 10%. Blood was placed on paper disc containing a hemolyzing agent (Photoflo). This was then applied to the polyacrylamide disc. Electrophoresis was performed on tubes which had been pre-run at 2 mA/tube for 30 to 60 minutes. Gels could be pre-made and stored for a period up to 14 days, a distinct advantage for clinical applications.

ACRYLAMIDE-GEL ELECTROPHORESIS OF β-LACTOGLOBULINS STORED IN SOLUTIONS AT pH 8.7

Akroyd, P., Nature **208**, 488 (1965).

Progress formation of at least eleven new species of β-lactoglobulin were found on acrylamide gel disc electrophoresis when -A or -B or -A/B genetic variants were stored in alkaline solution. An 11% gel suspension, Cyanogum 41®, was prepared in 0.076M Tris, pH 8.6, with citric acid. The electrolyte was 0.3M boric acid, pH 8.5, with sodium hydroxide. Ten mg/1 ml of crystalline enzyme was stored on buffer (0.076M Tris, pH 8.6, with HCl with 0.1M NaCl) for several days at 2°C. The electrophoretic pattern showed a progressive formation of slower migrating proteins as the storage period increased.

POLYMORPHISM IN THE RED PROTEIN ISOLATED FROM MILK OF IN-DIVIDUAL COWS

Groves, M. L., Nature **207**, 1007–1008 (1965).

Disc-gel electrophoresis revealed that red protein, or lactotransferrin, was polymorphous. The mobility of the red protein differs from that of serum transferrin. A 5% gel concentration was used, and zone electrophoresis was performed at pH 9.0 according to the methof of Raymond [Anal. Biochem. **2**, 23 (1962)].

ELECTROPHORESIS OF HEMOGLOBIN IN SINGLE ERYTHROCYTES

Matiole, G. T., Science **150**, 1824–1826 (1965).

A technique for electrophoretic analysis of the hemoglobulin mixtures from single erythrocytes is described. Polyacrylamide gel was used as the supporting medium for separation of hemoglobin fractions in picogram (10^{-12}) quantities with results comparable to large scale polyacrylamide electrophoresis. A special chamber was built with glass slides and tubes. A 1% agarose solution in 0.03M glycine, HCl–Tris buffer, pH 8.5, filled the chamber. An acrylamide erythrocyte mixture was prepared for dispersing the sample. A discontinuous buffer system (glycine–HCl–Tris and glycine–Tris), with added polyethylenimine to retard diffusion, gave best results. Five minutes at 1000 volts/cm gave good separation of hemoglobin A and A_2.

CYANOGUM ELECTROPHORETIC STUDIES OF SERUM AND SYNOVIAL FLUID

Hermans, P. E., W. P. Beetham, W. F. McGuckin and B. F. McKenzie, Proc. Staff Meetings Mayo Clinic **37**, 311 (1962).

Polyacrylamide gel, used in the technique of Raymond and Wang, has separated synovial fluid proteins. The haptoglobin components increased in rheumatoid arthritis. A description of the serum proteins encountered is presented.

Polyacrylamide (Disc) Electrophoresis

MULTIPLE FORMS OF TYROSINASE FROM HUMAN MELANOMA

Burnett, J. K., and H. J. Seiler, Invest. Dermatol. **52**, 199–203 (1969).

Two soluble and a macromolecular form of tyrosinace of human metastatic melanoma homogenates were examined by disc electrophoresis. The technique of Davis [Ann. N. Y. Acad. Sci. **121**, 104 (1964)] was used. The enzymes were detected by staining the gels with a solution containing 0.15% L-DOPA in 0.1M phosphate buffer (pH 6.8) for one to four hours. Further character-ization of the enzyme was performed histochemically on frozen sections. The electrophoretic mobility of the tyrosinase of the human melanoma exam-ined corresponded greatly to that of murine melanoma tyrosinase.

ALLERGIC ENCEPHALOMYELITIS: ENZYMIC DEGRADATION OF THE EN-CEPHALITOGENIC BASIC PROTEIN FROM BOVINE SPINAL CORD

Eylar, E. H., and M. Thompson, Biophys. **129**, 635–644 (1969).

Trypsin, pronase, chymotrypsin and pepsin digests of encephalitogen were characterized by polyacrylamide (disc) electrophoresis and high-voltage paper electrophoresis after chromatography on Sephadex® G-25 and G-75. Poly-acrylamide-disc electrophoresis was carried out at pH 4.2 according to technique of Reisfeld [Nature **195**, 281 (1963)]. Separation was achieved at 1 mA/tube for 15 minutes. High-voltage paper electrophoresis of the digests was performed on Whatman® 3MM filter paper for one hour at 2500 volts in pyridine–acetic acid–n-butanol–water (1:1:2:36), pH 4.7. The number of peptides released by these enzymes were 28, 54, 20, 14 (trypsin, pronase, chymotrypsin and pepsin respectively). Only the pepsin digests had encephalitogenic activity.

ALLERGIC ENCEPHALOMYELITIS: THE PHYSIO-CHEMICAL PROPERTIES OF THE BASIC PROTEIN ENCEPHALITOGEN FROM BOVINE SPINAL CORD

Eylar, E. H., and M. Thompson, Arch. Biochem. Biophys. **129**, 468–479 (1969).

Encephalitogen was isolated from bovine spinal cord by a procedure employing gel filtration of Sephadex® G-25 + G-75 and Cellex-P. The isolated protein was characterized by viscosity measurements, amino acid analyses, and electrophoresis on polyacrylamide discs at pH 4.7 and pH 10. Electrophoresis was performed on a 15% gel in a buffer composed of pyridine–acetic acid–n-butanol–water (1:1:2:36). The protein was found to be very basic and contained 38 moles of basic amino acids per mole of protein. The intrinsic viscosity was 9.27 ml/g. Polyacrylamide-disc electrophoresis revealed no change upon treatment with 8M urea at 100°C for one hour.

THE ENZYMIC DEACYLATION OF ESTERIFIED MONO- AND DI-SACCHARIDES: I. THE ISOLATION AND PURIFICATION OF AN ESTERASE FROM WHEAT GERM LIPASE

Fink, A. L., and G. W. Hay, Can. J. Biochem. **47**, 135–142 (1969).

The isolation of α-glucosidase from wheat germ lipase by preparative gel electrophoresis is reported. The electrophoretic run was made in an apparatus with a buffer capacity of 90 liters. The polyacrylamide gels were prepared according to the method of Williams and Reisfeld [Ann. N. Y. Acad. Sci. **121**, 373 (1964)] and Altschul et al. [Life Sci. Space Res. **3**, 611 (1964)]. Electrophoresis continued for two to four days at 400 volts and 130 mA. The enzyme was detected by a standard assay procedure using α- and β-PNP hexopyranosides as substrates. Using this technique, extensive fractionation of the wheat germ lipase was possible. The active esterase could be recovered by assaying material extruded from gel through a membrane. A Tris–barbital discontinuous buffer was used. The running gel had a pH of 7.5.

LACTATE DEHYDROGENASE ISOENZYMES IN HUMAN AND RAT FETAL LIVER AND LUNG

Francesconi, R. P., and C. A. Villee, Life Sci. **8**, 33–37 (1969).

Polyacrylamide-gel disc electrophoresis was used to study the distribution of lactate dehydrogenase in human and rat fetal liver and lung. Purified LDH-1, 2, and 3 in human and rat fetal tissues examined appeared similar. LDH-4 was absent in lung tissue of 13 to 17-week fetuses. LDH-4, in contrast, was present in the rat tissues. Hypothesis for these kinds of synthesis were advanced.

LIMITED MOLECULAR HETEROGENEITY OF PLANT HISTONES

Fambrough, D. M., and J. Bonner, Biochim. Biophys. Acta **175**, 113–122 (1969).

Pea-bud chromatin histone isolated and purified by ion-exchange column chromatography, gel-filtration and by preparative disc electrophoresis was characterized by analytical polyacrylamide-gel electrophoresis. Preparative gel electrophoresis separation was achieved in a Canalco Prep-Disc® apparatus. The upper gel (7.5%) contained 6M urea. A β-alaime buffer, pH 4.3, was used for separation. The histone prep was applied in 8M urea. Separation was carried out at 35–40 mA for three to four hours. Analytical polyacrylamide-gel separations were performed using a modification of the technique of Reisfeld, Lewis and Williams [Nature **195**, 421 (1962)]. Limited heterogeneity was found. There were eight molecular species detected by electrophoretic analysis. Other physical and chemical data also indicated limited heterogeneity.

PROTEINS OF ALTERED ELECTROPHORETIC MOBILITY FOUND IN HUMAN SERA AFTER POTASSIUM BROMIDE TREATMENT

Holmes, R., Biochim. Biophys. Acta **175**, 209–211 (1969).

Discontinuous acrylamide-gel electrophoresis was used to examine the alterations of proteins which occurred when human sera were allowed to react with potassium bromide. Electrophoresis was carried out on gels which were pre-electrophoresed in order to eliminate artifacts caused by interaction of the persulfate, used to polymerize the gel, with the proteins. Pre-electrophoresis was carried out in Aronsson's buffer, adjusted to pH 6.0. The actual run was made in sodium borate buffer, pH 8.6. Two new proteins were detected by this procedure. They do not appear to be lipoproteins.

LACTATE DEHYDROGENASE ISOENZYME PATTERNS IN THE TUMOR-BEARING COLON

Langvad, E., Int. J. Cancer **3**, 17–29 (1969).

Separation of lactate dehydrogenase isoenzymes of the human tumor-bearing colon was achieved on polyacrylamide discs using a 7.5% separation gel at pH 8.8 and a 3% spacer gel at pH 6.5. Enzyme activity was detected using the standard tetrazolium procedures. A preponderance of M-LDH in malignant tissue was confirmed. The distribution of the isoenzyme ratio was not normal in the nontumor-bearing tissue or in the tumor-positive tissue.

PEPTIDE-BOND HYDROLYSIS EQUILIBRA IN NATIVE PROTEINS. CONVERSION OF VIRGIN INTO MODIFIED SOYBEAN TRYPSIN INHIBITOR

Niekamp, C. W., H. F. Hitson, Jr., and M. Laskowski, J. Biochem. **8,** 16–22 (1969).

Preparative and analytical polyacrylamide-disc electrophoresis was used to examine virgin soybean trypsin inhibitor and pure modified inhibitor. Electrophoresis was carried out on small pore gels prepared according to the techniques of Davis for analytical results. The samples were placed on the gel between two 1 cm layers of Sephadex® G-200 in pH. Preparate separations were carried out on polyacrylamide slab according to Raymond's technique. These two techniques revealed the presence of the same products of reaction when trypsin was allowed to react at pH 4.0 with modified inhibitor or virgin inhibitor.

COMPARATIVE DISC ELECTROPHORESIS OF HAIR KERATEINES

Shechter, Y., J. W. Landeau, and V.D. Newcomer, J. Invest. Dermatol. **52,** 57–61 (1969).

Polyacrylamide (disc) electrophoresis was used to separate the S-carboxymethyl SCMKB and SCMKA keratines of the hair. A 7.5% polyacrylamide gel containing 6M urea and Tris–glycine buffer, pH 8.3, ionic strength 0.01, was used. Separation was achieved at a constant current of 5 mA per gel. The SCMKA extract of human hair yielded nine protein bands, monkey hair yielded 14, dog 13, and guinea pig 9.

PURIFICATION OF ESTRADIOL RECEPTOR FROM RAT UTERUS AND BLOCKADE OF ITS ESTROGEN-BINDING FUNCTION BY SPECIFIC ANTIBODY

Soloff, M. S., and C. M. Szego, Biochem. Biophys. Res. Commun. **34,** 141–147 (1969).

Preparative Sephadex® G-200 gel filtration and repeated ion exchange chromatography were used to purify rat uterus protein fractions which selectively bind ^{3}H-estradiol-17β. Disc electrophoresis at pH 8.1 or 4.5 and immunoelectrophoresis indicate that a single component is present in the preparation.

IDENTIFICATION OF L-FORMS BY POLYACRYLAMIDE-GEL ELECTROPHORESIS

Theodore, T. S., J. R. King, and R. Cole, J. Bacteriol. **97,** 495–499 (1969).

Proteins isolated from distilled water lysates of cell membranes of L-forms of

Proteous, Streptobacillus, Staphlococcus, and Streptococcus were studied using polyacrylamide (disc) electrophoresis. Electrophoretic separation of the proteins was achieved on 75% polyacrylamide gel containing 8M urea. A stacking gel of 2.5% acrylamide was used. A β-alanine buffer, pH 4.5 was used as electrolyte and for preparation of the gel. Distinct intergeneric differences were found among L-forms examined. Differences within the genus could not be detected.

α_{s_1}-CASEIN A (BOS TAURUS): A PROBABLE SEQUENTIAL DELETION OF EIGHT AMINO ACID RESIDUES AND ITS EFFECT ON PHYSICAL PROPERTIES

Thompson, M. P., H. M. Farrell, and R. Greenberg, Comp. Biochem. Physiol. **28,** 471–475 (1969).

The physical-chemical properties of the α_{s_1} casein A were compared to those of casein β using polyacrylamide-gel electrophoresis (slab), amino acid analysis, and ultracentrifugal studies. Polyacrylamide-gel electrophoresis was carried out at pH 9.1 in a Tris–citrate–borate buffer containing 4.5M urea. Amino acid and peptide analyses were carried out on chymotryptic digests made at pH 8.0. α_{s_1}-Casein A was found to be devoid of eight amino acid residues. At pH 3.3 the A variant was found to migrate faster than B or C. A deletion of peptide 26 in the A mutant was also felt to be of importance.

ANALYTICAL AND PREPARATIVE SEPARATION OF ACIDIC GLYCOSAMINOGLYCANS BY ELECTROPHORESIS IN BARIUM ACETATE

Wessler, E., Anal. Biochem. **26,** 439–444 (1969).

Analytical as well as preparative electrophoresis was used to separate acid glycosaminoglycans. Analytical electrophoresis was performed on cellulose-acetate strips (40 x 300 mm) using a sodium barbital buffer, pH 8.6, containing 0.1M barium acetate. Separation was achieved with a potential of 5 volts/cm for three hours. The substances were detected by staining with Alcian Blue. Preparative electrophoresis was carried out on a Pevikon® block in 0.1M barium acetate as the electrolyte. A potential of 3 volts/cm for 90 minutes was used to achieve separation. Uronic acid was detected by eluting 25 mm segments of the block with 0.1M hydrochloric acid and reacting with carbazole reagent containing borate. These techniques fractionated polysaccharides into three groups: heparin sulfate–heparin, dermatan sulfate–hyaluronic acid, and chondroitin sulfates.

THE CONJUGATED PLASMA PROTEINS IN ADULT FEMALES OF PERI-PLANETA AMERICANA (L) UNDER STARVATION AND OTHER STRESS

Adiyodi, K. G., and K. K. Nayar, Comp. Biochem. Physiol. **27**, 95–104 (1968).

The cell free-plasma of the female cockroach was studied for protein composition by polyacrylamide-disc electrophoresis. The simple and conjugated proteins were assayed after electrophoresis by the method of Adiyodi and Nayar [Current Sci. (India) **35**, 587–588 (1967)]. Bound lipids were lost during starvation; glycoprotein levels were irregular after food withdrawal but rose after ten days. Hypothermia resulted in hyperglycoproteinemia.

ANWENDUNG DER DISK-ELEKTROPHORESE ZUR TRENNUNG DER ISO-ENZYME DER ALKALISCHEN PHOSPHATASE

Akhtar, A., A. Hansen, and K. H. Kärcher, Z. Klin. Chem. **6**, 334–337 (1968) **German.**

Disc electrophoresis on polyacrylamide gel separated three main types of alkaline phosphatase isoenzymes from rat serum. Separation was achieved in a Tris–glycine buffer (14.4 g glycine and 3.0 Tris/1000 ml water) at 2 mA for 90 to 95 minutes. The separated isoenzymes were classified as liver, kidney and bone types. Irradiation of these tissues produced no significant changes in the serum distribution of the phosphatases.

PENGUIN BLOOD SERUM PROTEINS

Allison, R. G., and R. E. Feeney, Arch. Biochem. Biophys. **124**, 548–555 (1968).

The blood serum of two Antarctic penguins, the *Adelie* and *Emperor* were studied by starch-gel electrophoresis, disc electrophoresis, immunoelectrophoresis, immunodiffusion, autoradiography, gel filtrations and ion-exchange chromatography. Horizontal starch-gel electrophoresis was performed in the discontinuous buffer of Poulik [Nature **180**, 1477 (1967)]. Disc electrophoresis was performed in a continuous buffer system (0.1M glycine-NaOH, pH 8.65) without a stacking gel. Gel filtration on Sephadex®G-200 separated the sera into three main components. Strong cross-reactivity between species was demonstrated by immunoelectrophoresis.

ENZYMES OF THERMOPHILIC AEROBIC SPOREFORMING BACTERIA

Baille, A., and P. D. Walker, J. Appl. Bacteriol. **31**, 114–119 (1968).

Ultrasonic disintegrates of thermophilic aerobic sporeforming bacteria were examined by low voltage electrophoresis on polyacrylamide gel and starch gels. Preparative high voltage electrophoresis was performed in the Elphor

Vap 2 instrument. Separation in this instrument was achieved using an Oxoid® barbitone acetate buffer (pH 8.6, μ-0.1) which was fed into the chamber at a rate of 1.5 ml/hour with a buffer flow rate of 230 ml/hour. A 2Kv and current of 200 mA were maintained. Esterase patterns of 217 cultures were examined. Classification of the organism by patterns obtained on starch and polyacrylamide gels corresponded very closely to those based on biochemical tests.

ELECTROPHORETIC INVESTIGATIONS OF HUMAN MYOGLOBIN

Blessing, M. H., and H. Goette, Z. Ges. Exp. Med. **148,** 90–98 (1968).

Human myoglobin, isolated by a standard procedure, was transformed into cyanometmyoglobin by the addition of 5% potassium ferricyanide and 5% potassium cyanide. The cyanometmyoglobin was fractioned by electrophoresis on polyacrylamide (Cyanogum® 41). Five anodic fractions were seen in the normal. No electrophoretic differences were found between the myoglobins of adult, newborn or human fetuses.

IMMUNOCHEMICAL STUDY OF CHANGES IN RESERVE PROTEINS OF GERMINATING SOYBEAN PROTEINS

Catsimpoolas, N., T. G. Campbell, and E. W. Meyer, Plant Physiol. **43,** 799–805 (1968).

The changes occurring in the reserve proteins of soybean seeds *(Glycine max)* were investigated using disc electrophoresis as well as disc immunoelectrophoresis. Disc electrophoresis was performed according to the procedures of Ornstein and Davis. A pH 8.3 Tris–glycine buffer, ionic strength 0.01, was used with a current of 5 mA per gel disc. Immunoelectrophoresis was carried out after the electrophoretic separation as described above. Then the unstained gel was embedded in a gel medium consisting of 1% Ionagar No. 2 in a pH 8.8 Tris–barbital buffer. Trenches were cut after the agar solidified and antiserum was added to the trenches. Diffusion was allowed to proceed for three to five days. At least six antigenically distinct components were found in proteins of isolated soybean bodies. (Antisera used were anti-whole soybean extract, an anti-11S soybean protein and an anti-7S soybean protein monospecific. Rates of metabolism of these species were different for each species during the germination period.

T-2 DNA-DEPENDENT SYNTHESIS OF BACTERIOPHAGE-RELATED PROTEINS

Celis, J. E., and T. Conway, Proc. Nat. Acad. Sci. U.S. **59,** 923–929 (1968).

The proteins synthesized by a T-2 DNA-dependent amino acid incorporating bacteriophage were examined by polyacrylamide-disc electrophoresis, finger-print techniques and pulse labeling. Disc electrophoresis was performed using 7% gels in 0.6 x 12 cm tubes according to the technique of Davis [Ann. N. Y. Acad. Sci. **121**, 414 (1964)]. A pH of 8.3 and a current of 5 mA/tube were applied. Fingerprint studies, performed on tryptic digests of labeled samples, were made according to the method of Katz [J. Biol. Chem. **234**, 2897 (1959)]. Electrophoresis was performed at 2000 volts in pyridine–acetic acid–water (1:10:289) ph 3.7. Chromatography was then performed in the second direction (descendingly) in butanol–acetic acid–water (4:1:5) for six hours. A low molecular weight protein represented 80 to 90% of the *in vitro* product, and its tryptic peptides resembled peptides derived from low-molecular weight, phage-induced protein.

EMPLOI DE DIVERS SUPPORTS SOLIDES POUR L'ETUDE ÉLECTROPHORÉ-TIQUE DES HISTONES TOTALES ET DE LEURS FRACTIONS

Coirault, Y., K. B. Tan, and R. Vendrely, Int. Symp. IV Chromatographie and Electrophoreses Proc., 262–267 (1966) Presses. Acad. Europ. Brussels (1968) **French.**

Polyacrylamide (disc) and starch-gel electrophoresis as well as electrophoresis on starch + Sephadex® G-100 were used to examine the histones of erythrocytes and thymus of various species. Starch-gel electrophoresis was performed using a 0.01N solution of chlorohydrin, pH 2.2. A potential of 4.7 volts/cm was applied for eight hours. Polyacrylamide-disc electrophoresis was performed using a 0.04M potassium acetate buffer, pH 2.9. A potential of 16 volts/cm was applied.

EFFECT OF CULTIVATION TEMPERATURE ON PEROXIDASE ISOZYMES OF PLANT CELLS GROWN IN SUSPENSION

De Jong, D. W., A. C. Olsen, K. M. Hawker, and E. F. Jansen, Plant Physiol. **43**, 841–844 (1968).

The peroxidase isozymes elaborated by stable tobacco cells (WR-132 cultures) when grown at 13°, 25°, and 35°C were precipitated from growth media with 90% ammonium sulfate and analyzed by polyacrylamide-disc electrophoresis. A 7% gel and a Tris-buffer, pH 9.5, were used in the standard Canalco® system. Peroxidase isozymes were located by staining tubes with 1% H_2O_2 and 0.1M benzidine–HCl for 15 to 30 minutes. Different isoenzyme patterns were found for cells grown at different temperatures. Specific peroxidase activities were highest in cold-grown cells and lowest in warm-grown cells.

A DISC ELECTROPHORETIC STUDY OF PROTEINS OF BLUE-GREEN ALGAE

Derbyshire, E., and B. A. Whitton, Phytochemistry **7**, 1355–1360 (1968).

Proteins extracted from cultures of fourteen strains of blue-green algae were separated by disc electrophoresis using the technique of Ornstein and Davis. All extracts exhibited similar electrophoretic patterns with minor differences evident in Microcystis aeruginosa, Chlorogloea fritschii, and Anaceptis nidulans. Cultures grown on nitro-gen-free media were found to lack a set of three bands.

ISOLATION OF NUCLEI AND HISTONES FROM ROOTS OF VICIA FABA

Dick, C., Arch. Biochem. Biophys. **124**, 431–435 (1968).

Polyacrylamide-gel electrophoresis at pH 4.5 was used to study the mobility and heterogeneity of isolated histones from roots of Vicia faba. Electrophoretic analysis of extracts made above pH 2.8 revealed no appreciable histone. In addition, a number of basic proteins were detected by the electrophoretic technique and different patterns were obtained with different pH extracts.

PROTEIN COMPOSITION OF RIBOSOMES AND RIBOSOMAL SUBUNITS FROM ANIMAL TISSUES

DiGirolamo, M., and P. Commarano, Biochim. Biophys. Acta **168**, 181–194 (1968).

Polyacrylamide (disc) electrophoresis, using standard techniques, was used to characterize ribosomal proteins extracted with lithium chloride-urea from tissues. Within a given tissue, ribosomal proteins isolated from monomes and from polysomes of different aggregate size appeared electrophoretically similar. Ribosomal proteins isolated from the same tissue of animals of various orders and genera were electrophoretically different.

USE OF PORE SIZE CONCENTRATION GRADIENT IN ELECTROPHORESIS

Epstein, E., Y. Houvras, and B. Zak, Clin. Chim. Acta **20**, 335–339 (1968).

A technique is described for preparing a concentration gradient at the sample end of the gel in the disc electrophoresis process. The gradient was achieved by injecting buffer or water into the poured gel prior to gelation. The sample was applied by layering it between the buffer and gel. Thus a simple gel was obtained whereby the sample is concentrated in the buffer layer quickly. The globulins of serum were resolved at the slow end of the gel.

POLYACRYLAMIDE GEL ELECTROPHORESIS OF HIGHLY PURIFIED CHICK INTERFERON

Fantes, K. H., and I. G. Furminger, Nature **216,** 71 (1967).

Chick interferon in equal volume of 40% sucrose was put on a 7% poly-acrylamide gel column for separation. After electrophoresis, the column was cut lengthwise. Half was cut into 2 mm slices placed in a 1 ml of 0.5 KPO_4 buffer (pH 7.5) containing BSA (500 $\mu g/ml$) and Tween® 80 (20 $\mu g/ml$) to elute the interferon. Staining the gel column showed two protein bands which coincided with antiviral activity detected in the eluted slices.

SYNTHESIS OF EMBROYONIC HEMOGLOBINS DURING ERYTHROID CELL DEVELOPMENT IN FETAL MICE

Fantoni, A., A. Delachopelle and P. A. Marks, J. Biol. Chem. **244,** 675–681 (1968).

The nature of hemoglobins formed in erythroid cells derived from the yolk sac blood island in fetal mice of the C 57 BL/6J strain of mice were studied by using polyacrylamide-disc electrophoresis. The buffer systems described by Davis were used at a pH of 7.5 for the separating gel. Three embroyonic hemoglobins were formed as indicated by electrophoretic studies. The relative rates of synthesis were found to change as the cells differentiated. Synthesis of the hemoglobins in yolk sacs was not inhibited by actinomycin.

STREPTOCOCCAL GROUP SPECIFIC ANTIBODIES: OCCURRENCE OF A RESTRICTED POPULATION FOLLOWING SECONDARY IMMUNIZATION

Fleischman, J. B., D. G. Braun, and R. M. Krause, Proc. Nat. Sci. U. S. **60,** 134–139 (1968).

Polyacrylamide-disc electrophoresis was used to examine the relative homo-geneity of antibodies and their light chains formed by rabbits upon secondary immunization with streptococcal vaccines. Disc electrophoresis was performed at pH 8.3 in acrylamide gels, according to the method of Davis. Primary intravenous injections with group A variants did not produce group-specific antibodies with restricted heterogeneity. Immunoelectrophoresis revealed that antibodies in primary response serum were more heterogeneous than those in the secondary response.

DISC AND CELLULOSE ACETATE ELECTROPHORESIS OF HUMAN PLACENTAL PROTEINS

Gershbein, L. L., and J. Al-Wattar, J. Chromatog. **34,** 485–497 (1968).

Extracts of human term placenta were subjected to cellulose-acetate electro-phoresis as well as disc electrophoresis in order to analyze the protein composition. Cellulose-acetate electrophoresis was performed in an Owens buffer, pH 8.6 (composition: 5.0 g sodium barbital, 1.9 g sodium acetate, 0.38 g calcium lactate and 34.2 ml of 0.1 N HCl per liter). Disc electrophoresis was performed using a 7.5% polyacrylamide gel and a Tris buffer, pH 8.4. Characteristic protein patterns of each fraction were revealed by staining with Buffalo Black 10B.

LES ISOAMYLASES DU COLOSTRUM HUMAIN

Got, R., G. Bertagnolio, M. B. Pradal, and J. Frot-Coutaz, Clin. Chim. Acta **22**, 545–550 (1968) **French.**

Polyacrylamide-gel electrophoresis, carried out in a Tris–barbital buffer, pH 8.6, 0.29M, at 12 volts/cm for three hours, separated four fractions with amylase activity from human colostrum, pancreatic juice and saliva. Six different amylase isozymes were separated from colostrum, which had been chromatographed on Sephadex® G-100.

POLYMORPHISM OF γ-CASEIN IN COW'S MILK

Groves, M. L., and C. A. Kiddy, Arch. Biochem. Biophys. **126**, 188–193 (1968).

When γ-casein was subjected to disc electrophoresis in polyacrylamide gels, two γ-casein forms, designated A and B, were discovered. Disc electrophoresis was performed using the polyacrylamide system of Ornstein and David, with the exception that the gels contained 4M urea. Tris buffer, pH 8.6, was the electrolyte used. Results from the typing of milk samples from 165 cows suggested the possibility of a genetic base for the polymorphism observed.

SIZE AND CHARGE ISOMER SEPARATION AND ESTIMATION OF MOLECULAR WEIGHTS OF PROTEINS BY DISC GEL ELECTROPHORESIS

Hedrick, J. L., and A. J. Smith, Arch. Biochem. Biophys. **126**, 155–164 (1968).

A technique which uses polyacrylamide (disc) electrophoresis and which distinguishes between a size isomer of a family of proteins is described. Upon plotting the log of protein mobility relative to dye front versus concentration of gel, a family of non-parallel lines was obtained for size isomeric proteins Proteins differing in both charge and size gave non-parallel lines. The parallel lines for isomeric proteins could be extrapolated to a common point in the vicinity of 0% gel concentration, whereas non-parallel lines extrapolated to a point other than 0% concentration. The slopes of these lines were found to be related to the molecular weight.

THE PROTEIN AND LDH ISOENZYME PATTERN OF NORMAL HUMAN URINE DETERMINED BY POLYACRYLAMIDE GEL DISC ELECTROPHORESIS

Hemmingsen, L., and F. Skov, Clin. Chim. Acta **19**, 81–87 (1968).

Polyacrylamide-disc electrophoresis, modified after the method of Hemmingsen and others [Acta Ophthalmol. **45**, 359 (1967)], fractionated proteins and LDH isoenzymes in normal and pathological urine. Prealbumin, albumin, two postalbumins, a fraction having a mobility of transferrin and a γ-globulin fraction were the proteins demonstrated in normal urine. LDH_1, LDH_2 and LDH_3 were the lactate dehydrogenase isoenzymes present in normal urine; LDH_4 and LDH_5 were not present.

APPLICATION OF DEAE-SEPHADEX A-50 FOR ADDITIONAL PURIFICATION OF ANTITOXIC SERA

Hristova, G., Z. Immunitaetsforsch. Allergie Klin. Immunol. **135**, 439–448 (1968) **German.**

The electrophoretic behavior of anti-tetanus sera, which was purified by gel filtration on DEAE-Sephadex® A-50 after first being subjected to the "Diaferm 3" method of purification, was reported. Polyacrylamide-gel electrophoresis according to the technique of Davis and immunoelectrophoresis were used. The DEAE-Sephadex A-50-treated sera showed only one fraction upon polyacrylamide-gel electrophoresis. Upon immunoelectrophoresis only the preciptin arcs characteristic of 5S fragments were revealed.

HUMAN PLASMA CHOLINESTERASE ISOENZYMES

Juul, P., Clin. Chim. Acta **19**, 205–213 (1968).

Disc electrophoresis performed according to the Ornstein-Davis technique separated the cholinesterase isoenzymes of human plasma. Electrophoresis was performed at 3 mA per column for three hours. The cholinesterase isoenzymes were detected by staining the gels first with thiocholine, then with dithiooxamide. The thiocholine histochemical method proved completely reproducible, and 12 cholinesterase isoenzymes could be demonstrated.

ISOENZYME DER ALKALISCHEN PHOSPHATASE IN DER LÖSLICHEN PHASE VON RATTENLEBERZELLEN

Kaschnitz, R., J. Patsch, and M. Peterlik, Europ. J. Biochem. **5**, 51–54 (1968) **German.**

Polyacrylamide-gel electrophoresis characterized alkaline phosphatase isoenzymes isolated from rat liver bile and serum gel filtration of Sephadex® G-200. Several isoenzymes were demonstrable in bile and serum, but only one of the enzymes appeared in the soluble phase and in bile. Gel filtration on Sephadex® G-200 separated three enzymes from the soluble phase but only two were revealed on polyacrylamide gel. The enzymes of rat liver were localized in the microsomal fraction plasma membrane and the soluble phase of the cell.

VARIATIONS IN THE SMALL SUBUNITS OF DIFFERENT MYOSINS

Lacker, R. H., and C. J. Hagyard, Arch. Biochem. Biophys. **122**, 521–522 (1968).

Skeletal myosins prepared from beef, sheep, a four-month-old lamb fetus, white leghorn fowl breast and a fourteen-day-old chick embryo were characterized with polyacrylamide-gel electrophoresis. Electrophoresis was carried out in a discontinuous buffer system of 0.02M potassium phosphate, pH 7.0. Patterns of differences in the mobility and number of myosin bands were observed: rabbit cardiac myosin showed two bands, whereas skeletal myosin showed three; sheep skeletal and cardiac myosin showed three bands; and both beef skeletal myosin and chicken fetal myosin showed three bands of activity.

SOME RENAL ACTIONS OF TWO ANIONIC FRACTIONS OF PLASMA GLOBULINS

Lockett, M. F., and H. H. Siddiqui, Br. J. Pharmacol. **32**, 311–321 (1968).

Two anionic proteins were isolated from plasma by gel filtration on Sephadex® G-200 with Tris–HCl buffer, pH 3.9, with sodium chloride (100 g/l). These proteins, possessing pharmacologic action when perfused in cat kidneys, were examined for electrophoretic characteristics using polyacrylamide-disc electrophoresis. Electrophoresis was performed using the procedure of Ornstein in the Canal apparatus. A β-alanine buffer, pH 5.0, was used in the electrode compartment. Electrolytic mobility was unaltered by isolation procedures.

ESTERASES AND OTHER SOLUBLE PROTEINS OF SOME LACTIC ACID BACTERIA

Morichi, T., J. Gen. Microbiol. **53**, 405–414 (1968).

The polyacrylamide-gel electrophoresis technique of Lund [J. Gen. Microbiol. **40**, 413 (1965)] was used to separate soluble proteins of 34 strains and esterases of 134 strains of lactic acid bacteria. Similar protein patterns were obtained for three species of lactic acid streptococci (S. cremoris, S. lactis and S. diacetilatis). The Lactobacillus acidophilus and L. delbrucki strains exhibited marked differences among themselves. Leuconotics were found to fall into groups according to their physiology.

RIBOSOMAL PROTEINS FROM GOOSE RETICULOCYTES ARE NOT HISTONES

Neelin, J. M., and G. Vidall, Can. J. Biochem. Physiol. **46,** 1507–1514 (1968).

Ribosomal protein extracted from goose reticulocyte ribosomes, with either hydrochloric acid-urea or lithium chloride-urea were compared to histone material extracted with hydrochloric acid and hydrochloric acid-urea. Electrophoresis was performed in polyacrylamide disc using a discontinuous buffer of β-alanine acetate, pH 4.5. A 2.5% and 10% gel made in potassium acetate was used as stacking and running gel respectively. Starch-gel electrophoresis was performed on a 16% gel in sodium acetate buffer, pH 4.8 (μ, 0.020) at 3.5–5.0 mA/cm^2 for six hours. The ribosomal protein patterns on both polyacrylamide and starch-gel electrophoretograms were different. The amino acid composition, as well as its elution pattern on Amberlite CG-50 in 9% guanidinum chloride, was different from that of histone.

POLYACRYLAMIDE GEL ELECTROPHORESIS OF THE REPLICATIVE FORM OF INFLUENZA VIRUS RNA

Pons, M. W., and G. Hirst, Virology **1968,** 182–184.

Replicate form RNA extracted from chick fibroblast moholayer infected with WSN strain of influenza virus was shown to have five peaks. Further analysis revealed that RNA isolated by dimethyl sulfoxide denaturation of replicate form possessed a pattern much like single stranded RNA.

CHROMATOGRAPHIC AND ELECTROPHORETIC PATTERNS OF SOLUBLE PROTEINS FROM NERVOUS AND MUSCLE TISSUE IN PERIPLANETA AMERICANA

Rusca, G., P. Amaldi, and P. U. Angeletti, Brain Res. **8,** 256–259 (1968).

Soluble proteins of ganglia, brain, and muscle tissue of *Periplaneta americana* were fractionated on DEAE-cellulose columns equilibrated with a Tris-phosphate buffer, 5mM, pH 7.3. Elution of protein was achieved with sodium chloride gradient. Pooled eluates in groups of six were separated by polyacrylamide-disc electrophoresis according to the procedure of Davis and Ornstein, at pH 8.3. Chromatographic analysis revealed similarity between the tissue protein patterns; however, striking differences were observed in the electrophoretic patterns. A great amount of acidic material was found in brain extracts as compared to muscle extracts.

BIOCHEMICAL TAXONOMY OF THE DERMATOPHYTES: I. COMPARATIVE DISC ELECTROPHORESIS OF CULTURE FILTRATE PROTEINS

Schecter, Y., J. W. Landau, N. Dabrowa, and V. D. Newcomer, J. Invest. Dermatol. **51**, 165–169 (1968).

The proteins contained in water soluble extracts of filtrates of 21 dermatophyte species cultured on Sabouraud's medium were subjected to disc electrophoresis. By comparing the number of homologous fractions in each of the 21 pairs, it was possible to select two fraction groups which would distinguish between *Microsporium* and *Trichophyton*.

DARSTELLUNG DER HAPTOGLOBINTYPEN DURCH ELEKTROPHORESE IN POLYACRYLAMIDE GEL

Schleyer, F., and P. Schiable, Z. Klin. Chem. **5**, 32–34 (1968) German.

Optimum separation of serum haptoglobins was achieved using a modified disc electrophoresis technique and a discontinuous buffer system. The gels were made in a 0.03M boric acid–sodium hydroxide buffer, pH 8.9. The electrode buffer was 0.3 boric acid adjusted to pH 8.5 with sodium hydroxide. Up to 13 bands were detected in haptoglobin 2-1.

SEPARATION OF HUMAN TISSUE ALKALINE PHOSPHATASES BY ELECTRO-PHORESIS ON ACRYLAMIDE DISC GELS

Smith, I., P. J. Lighstone, and J. D. Perry, Clin. Chim. Acta **19**, 499–505 (1968).

Electrophoresis on 5% polyacrylamide gel and a continuous Tris-borate buffer, pH 9.5, separated alkaline phosphatases of liver, intestine, kidney, lung, bile, bone, cardiac and skeletal muscle tissue before and after treatment with neuraminedase. A potential of 90, 3 mA per tube (270 volts) was maintained for 30 to 40 minutes. The alkaline phosphatases were detected by staining with a solution containing β-naphthyl phosphate (2 mg/ml), and Fast Blue BB (1 mg/ml) in boric acid buffer, pH 9.5. A Joyce-Löebl Chromoscan was used to scan the gels. It was possible to distinguish among the tissues examined on the basis of mobility and number of bands obtained.

ISOLATION AND CHARACTERIZATION OF THE HEMOLYMPH LIPOPRO-TEINS OF THE AMERICAN SILKMOTH, HYALOPHORA CECROPIA

Thomas, K. K., and L. I. Gilbert, Arch. Biochem. Biophys. **127**, 512–521 (1968).

Three classes of lipoproteins isolated from the pupal hemolymph of *Hyalophora cecropia* were characterized by disc electrophoresis on polyacrylamide gels. Seventy-five percent of the lipids in total hemolymph lipoprotein were found to be associated with the HDL class. Diglyceride was found to be the major neutral lipid component in all classes.

SEPARATION OF SUBUNITS OF VETCH LEGUMIN BY CHROMATOGRAPHY ON DEAE-CELLULOSE

Vaintraub, I.A., and N. Thanh Tuen, Dokl. Biochem. Sect. (English Trans.) **180**, 147–149 (1968).

Legumins isolated from *Vicia sativa L.* seeds by zonal isoelectric deposition of Sephadex® G-50 were dissociated by treatment with urea and 4M guanidinium chloride. The dissociated protein components were characterized by DEAE-cellulose column chromatography, ultracentrifugation studies and polyacrylamide-disc electrophoresis. Initial elution of the dissociated protein from the DEAE column with Tris-citrate buffer, pH 8.0, containing 4M urea followed by 0.1M phosphoric acid, separated five components from the gemish. Polyacrylamide electrophoresis of the eluted fraction revealed a high degree of heterogeneity.

ISOZYME DEMONSTRATION TECHNIC

Wolf, R., and L. Taylor, Am. J. Clin. Pathol. **49**, 871–876 (1968).

After electrophoretic separation on polyacrylamide (disc) electrophoresis, serum amylase activity was demonstrated by using a dry starch gel on a microscope slide. The dry starch–agar gel containing substrate was incubated in contact with the polyacrylamide gel for one hour. The starch–gel slide was removed and developed for amylase by dipping in an iodine solution. Amylase activity was evidenced by a colorless area on the starch–agar slide.

DISC ELECTROPHORETIC PATTERNS OF THE POLYPEPTIDE CHAINS OF GUINEA PIG IMMUNOGLOBULINS

Aron, D., and M. E. Lamm, J. Immunol. **99**, 562–567 (1967).

Reduced and alkylated polypeptide chains of guinea pig γ_1 and γ_2 antihapten antibodies were examined by disc electrophoresis. The procedure of Davis [Ann. N. Y. Acad. Sci. **121**, 404 (1964)] as modified by Reisfeld and Small [Science **152**, 1253 (1963)] was used. L-chains yielded different banding patterns; H-chains from γ_1 and γ_2 antibodies exhibited a varied pattern at alkaline pH.

QUALITATIVE AND QUANTITATIVE ALTERATIONS IN SEROMUCOID IN MALIGNANT NEOPLASTIC DISEASE

Bacchus, H., E. R. Kennedy, and J. Blackwell, Cancer **20**, 1654–1662 (1967).

The seromucoids of normal human serum and from patients with malignant neoplastic disease were isolated by column chromatography on DEAE-cellulose using a phosphate buffer gradient. The fractions isolated were characterized by polyacrylamide-disc electrophoresis and immunoelectrophoresis. Gel electrophoresis was performed according to the procedure of Ornstein in a Tris-glycine buffer, pH 8.3. Runs were made at a current of 5 mA per tube. Immunoelectrophoresis was performed according to the procedure of Scheidegger using a Noble-agar gel and a barbital buffer, pH 8.6. A potential gradient of 250 volts (25–40 mA) was used. Eight fractions were isolated from normal serum. Characteristic hexose, hexosamine protein and hexoseamine: protein ratios were found. These parameters were altered in fractions isolated from serum of malignant neoplastics.

LOCALIZATION AND CHARACTERIZATION OF RAT AND CHICKEN HISTONES

Benson, J. L., and E. L. Triplett, Exp. Cell Res. **48**, 61–70, (1967).

Histones isolated from liver, kidney, brain and spleen of adult rats and spleen and liver histones of embryonic and adult chickens were separated by polyacrylamide-gel electrophoresis. A 7.5% polyacrylamide gel containing 8.0M urea was used. A 0.075 beta–alanine–acetic acid buffer, pH 4.3, was employed and run at a current strength of 2.5 mA/tube. Statistical analyses of proteins resolved revealed no differences that were significant. Two components were common to all preparations analyzed.

ELECTROPHORESIS OF PROTEINS OF 3 PENICILLIUM SPECIES ON ACRYLAMIDE GELS

Bent, K. J., J. Gen. Microbiol. **49**, 195–200 (1967).

Polyacrylamide disc electrophoresis performed according to the procedure of Davis and Ornstein was used to examine soluble proteins extracted from the mycelium of *P. chrysogenum, P. frequentans* and *P. griseotulvum*. The protein pattern varied with the age of mycelium. Each species was characterized by a specific protein pattern.

AN ANALYSIS BY GEL ELECTROPHORESIS OF QB-RNA COMPLEXES FORMED DURING THE LATENT PERIOD OF AN IN VITRO SYNTHESIS

Bishop, D. H., Proc. Nat. Acad. Sci. U. S. **57**, 1474–1481 (1967).

Acrylamide-gel electrophoresis, because of its high resolving power, was used to study the early fate of an input template during *in vitro* synthesis of infectious QB-RNA. QB-HS and QB-FS were separated from whole cells by the original techniques [J. Mol. Biol. **16**, 544 (1966)]. A 2.4% polyacrylamide gel was poured into a Plexiglas tube for electrophoresis. The electrophoresis buffer was 0.04M Tris–0.02M sodium acetate–0.001M sodium EDTA made pH 7.2 with 2.0 ml glacial acetic acid. A current of 5 mA was applied for 90 minutes.

ELECTROPHORETIC SEPARATION OF VIRAL NUCLEIC ACIDS ON POLY-ACRYLAMIDE GELS

Bishop, D. H. L., J. R. Claybrook, and S. Spiegleman, J. Mol. Biol. **26**, 373–387 (1967).

Polyacrylamide (disc) electrophoresis on 2.4% gel separated viral nucleic acids isolated from C^{14}-uridine labeled satellite tobacco necreosis virus. The electrophoretic run was performed in a 0.12M Tris–0.06M sodium acetate–0.003M sodium EDTA buffer, pH 7.2, for 30 minutes at 5 mA/tube and 5.5 volts/cm. Ten viral nucleic acids species and four *E. coli* RNA species were used to calibrate the system. A general relationship was exhibited between the logarithm of the molecular weight and relative electrophoretic mobility.

TWO SUBUNITS OF TROPOMYOSIN B

Bodwell, C. E., Arch. Biochem. Biophys. **122**, 246–261 (1967).

Polyacrylamide-gel electrophoresis (disc) has effectively resolved tropomyosin B into two subunits. The Arden-500 preparative apparatus was used with the discontinuous buffer system of Ornstein and Davis, pH 9.5. Electrophoresis was performed at 2.5 mA (280 volts) for 3.5 to 4 hours.

ARTIFACT PRODUCED IN DISC ELECTROPHORESIS BY AMMONIUM PER-SULFATE

Brewer, J. M., Science **156**, 256–257 (1967).

Ammonium persulfate is commonly used to polymerize acrylamide gels. The addition of 0.4 μM of persulfate/ml of gel showed heterogeneity in a yeast enolase that electrophoresed in a gel containing 8M urea. The heterogeneity of the enolase was not visible when the gel was polymerized by riboflavin and light or when 20 μM thioglycolate was added to reduce the excess ammonium persulfate. Thioglycolate is useful only if there is no cysteine or cystine which is affected. The author recommends the use of riboflavins and light for polymerization.

LACTATE DEHYDROGENASE ISOENZYMES IN HEALING BONES

Bruce, R. A., and D. S. Strachan, J. Oral. Surg. **25,** 542–549 (1967).

Tissue samples from healing mandibular rabbit bone defects were examined by disc electrophoresis after 6, 9, 12 and 15 days of healing. The electrophoresis procedure employed was essentially that described by Davis except that sample gel was omitted. Fibrous bone filled the defects after 15 days of healing. Healing rabbit bone was highest in LDH_1 LDH_2, and LDH_3.

CASEIN COMPONENTS SOLUBLE IN CHLOROFORM–METHANOL (2:1) AND ON FIFTY PERCENT AQUEOUS ETHANOL

Cerbulis, J., and J. H. Custer, J. Daily Sci. **50,** 1356–1359 (1967).

Polyacrylamide gel electrophoresis in Tris–EDTA–borate buffer, pH 8.6, was used to distinguish the mobility patterns of several casein preparations. The gel was prepared with a 7.0% acrylamide concentration in 4.5M urea. The soluble casein fractions prepared by treating skimmed milk with chloroform-methanol (2:1) and extracting skimmed milk with 50% aqueous ethanol were compared. Incorporating mercaptoethanol in the acrylamide gel did not alter the number of bands formed. Amido Black was the stain used.

A COMPARATIVE STUDY OF THE SERUM PROTEINS OF THE SPECIES OF TARICHA AND THEIR HYBRIDS

Coates, M., Evolution **21,** 130–140 (1967).

A comparative study of the serum proteins of the salamander genus *Taricha* was made using the disc electrophoretic technique of Ornstein and Davis and two-dimensional electrophoresis. The two-dimensional technique used a preliminary separation step on filter paper in 0.10M barbital buffer. The protein bands were cut from the paper lengthwise and rerun on the polyacrylamide disc. Distinct qualitative differences were found between species and subspecies, members of the same species from different geographic locations, and among members of the same population. Variations decreased significantly in the hybrid generation.

INTERRELATIONSHIPS AMONG GLYCOGEN PHOSPHORYLASE ISOZYMES

Davis, C. H., L. H. Schliselfeld, D. P. Wolf, C. A. Leavitt, and E. C. Krebs, J. Biol. Chem. **242,** 4824–4833 (1967).

Three glycogen phosphorylase b enzymes were separated from rabbit heart extracts by polyacrylamide-disc electrophoresis and column chromatography. Disc electrophoresis was achieved using a modification of the method of Ornstein and Davis. A small pore gel (5% acrylamide, 0.2% N, N'-methylenebisacrylamide buffered with pH 7.9 Tris–HCl) and a large pore gel (3% acrylamide, 0.075% N,N'-methylenebisacrylamide, buffered with pH 5.8 imidazole HCl. The column chromatographic procedure employed DEAE-cellulose to separate the enzyme from heat-treated ammonium sulfate extracts.

SEPARATION AND QUANTITATION OF LACTIC DEHYDROGENASE ISO-ENZYMES BY DISC ELECTROPHORESIS

Dietz, A. A., and T. Lubrano, Anal. Biochem. **20**, 246–257 (1967).

A modification of the disc electrophoresis system of Ornstein and Davis permitted complete separation of lactic dehydrogenase isozymes. The major modification was to omit the stacking gel and substitute 40% sucrose solution. A 5.5% gel was used for all separations as it had less background staining. The reproducibility of the modified technique was found to be within accepted limits.

AN INDIOGENIC REACTION FOR ALKALINE PHOSPHATE IN DISC ELECTRO-PHORESIS

Epstein, E., Tech. Bull. Registry Med. Technologists **37**, 270–274 (1967).

Serum alkaline phosphatase was separated by disc electrophoresis in poly-acrylamide gel and coupled with an indiogenic phosphate reagent which was a sensitive quantitative procedure adaptable to the clinical laboratory. The stock buffer was 3.0 g -2-amino–2 methyl-1,3-propanedial and 0.8 g boric acid per liter, pH 9.0 ± 0.05. This buffer gave improved results compared with the older techniques of separating alkaline phosphatase. The buffer concentration was critical to the results.

MICROELECTROPHORESIS ON POLYACRYLAMIDE GEL

Felgenhauer, K., Biochem. Biophys. Acta **133**, 165–167 (1967).

A technically simple modification of the Ornstein and Davis procedure is presented. The gels were prepared in 1 mm x 10 cm capillary tubes. The sample was introduced with 1 cm capillary after being mixed with 1 μl 50% glucose and 1 μl of Bromophenol Blue. Resolution was comparable to that obtained with the standard technique.

EFFECT OF HYDROGEN PEROXIDE ON WHEY PROTEIN NITROGEN VALUE OF HEATED SKIM MILK

Fish, N. L., and R. Mickelsen, J. Dairy Sci. **50**, 1045–1048 (1967).

Polyacrylamide-disc electrophoresis was used to study the effects of hydrogen peroxide on individual whey proteins and their interaction with other proteins. The method of Davis [Ann. N. Y. Acad. Sci. **121**, 404 (1964)] was used. A dilution of 0.3 ml of the protein sample in 0.7 ml of Tris-glycine buffer, pH 8.3, was applied to the gel. A potential of 5 mA/tube permitted separation of the protein in 1.5 hours. A 1% solution of Amido Black 10B in 7% acetic acid stained the gels in one hour. Excess dye was removed using 7% acetic acid.

EFFECT OF HYDROGEN PEROXIDE TREATMENT ON HEAT INDUCED INTERACTION OF K-CASEIN AND β-LACTOGLOBULIN

Fish, N. L., and R. Mickelsen, J. Dairy Sci. **50**, 1360–1362 (1967).

β-lactoglobulin and K-casein were heated to 85°C for 30 minutes in 0.10% hydrogen peroxide. This mixture and the heated, unheated, and hydrogen peroxide controls were then studied using polyacrylamide disc electrophoresis [J. Dairy Sci. **50**, 1045–1048 (1967)]. Treating β-lactoglobulin with hydrogen peroxide partially retarded the heat-induced interaction of the β-lactoglobulin with K-casein.

POLYACRYLAMIDE-DISC-ELECTROPHORESE: VERSUCHE ZUR ISOLIERUNG UND CHARACTERISIERUNG VON HUMAN-SERUMPROTEINEN

Geyer, E., S. Marghescu, and J. J. Müller, Klin. Wochschr. **45**, 717–721 (1967) **German.**

Preparative polyacrylamide-disc electrophoresis in a column was performed according to the technique of Davis [Ann. N.Y. Acad. Sci. **121**, 404–427 (1967)] in combination with immunodiffusion to characterize α_1 acid glycoproteins, albumins, transferrin, α_2-macroglobulins and β-lipoproteins of human sera. B_{1A} and β_{1C}-globulins migrated behind the transferrins. Prealbumins and albumins were better separated from other components when the buffer system of Lewis and Clark was used. [Anal. Biochem. **6**, 303–315 (1963)].

THE ONTOGENY OF SPERM SPECIFIC LACTATE DEHYDROGENASE IN MICE

Goldburg, E., and C. Hawtrey, J. Exp. Zool. **164**, 309–316 (1967).

The testes of sexually mature mice have a lactate dehydrogenase isozyme associated with the spermatozoan cell. This isozyme, LDH-X, differs in subunit composition from the LDH isozyme found in other tissues. It appears when spermatocytes first mature in the testes. In polyacrylamide-gel electrophoresis, LDH-X has the same mobility as LDH-5. The LDH was extracted from the testes by homogenizing it in 0.05M phosphate buffer, pH 7.0, and then disrupting the cells with a sonic vibrator. After centrifugation, the enzyme remained in the supernate. LDH isozymes were separated by electrophoresis according to the method of Ornstein and Davis. A current of 4 mA per 75 x 10 mm test tube for 30 to 40 minutes gave good separation. A Bromophenol Blue tracking dye migrated to 5 mm from the bottom of the gel in this time.

CHANGES IN MILK PROTEINS TREATED WITH HYDROGEN PEROXIDE

Grindrod, J., and T. A. Nickerson, J. Dairy Sci. **50,** 142–150 (1967).

Polyacrylamide gel was a sensitive medium to detect changes in the electrophoretic pattern of hydrogen peroxide-treated milk proteins. The gel was prepared according to the method of Davis [Ann. N. Y. Acad. Sci. **121,** 404 (1964)]. Ureamercaptoethanol was incorporated into some gels and demonstrated the rupture of heat-induced complexes of β-lactoglobulin and K-casein. The migration rates of several proteins were altered following the treatment with hydrogen peroxide.

SPECIES DIFFERENCE IN ELECTROPHORETICALLY DISTINCT FORMS OF GLUCOSE-6-PHOSPHATE DEHYDROGENASE IN SOME MAMMALS

Hori, S. H., and T. Kamada, Japan. J. Genetics **42,** 367–374 (1967).

Polyacrylamide-disc electrophoresis demonstrated two major forms of G6PD in the organs of mice, rats, hamsters and guinea pigs. The Ornstein procedure [Ann. N. Y. Acad. Sci. **121,** 321 (1964)] was used. Electrophoresis was performed for two hours at 20°C in 0.05M Tris–glycine buffer, pH 8.3, at a constant current of 3 mA per gel column.

THE ELECTROPHORESIS OF HISTONES IN POLYACRYLAMIDE GEL AND THEIR QUANTITATIVE DETERMINATION

Johns, E. W., Biochem. J. **104,** 78–82 (1967).

The separation of histones of calf-thymus and rat-liver ribosomal proteins was achieved on polyacrylamide gel (20%) prepared in 4.6N acetic acid. Electrophoresis was performed in 0.1N acetic acid at 200 volts d.c. for 4.5 hours. The gels were stained with Naphthalene Black (.5% in N-acetic acid). Quantitation of the separated bands was achieved by cutting the gel bands, dissolving them in dimethyl sulphoxide, and reading them in spectrophotometer at 600 $m\mu$ after an appropriate heating period. Results of the quantitative determinations agreed with results obtained by representative methods.

DISK-ELEKTROPHORETISCHE TRENNUNG VON ISOENZYMEN DER SAUREN PROSTATA-PHOSPHATASE

Kaschnitz, R., Z. Klin. Chem. **5,** 126–128 (1967) **German.**

The acid phosphatase from prostrate gland extracts was separated into four isoenzymes by disc electrophoresis. Separation was achieved in a 7.5% gel using a Tris–diethylbarbiturate buffer. A potential of 220 volts (4-5 mA/disc) was employed. In order to prevent thermal inactivation during electrophoresis, the temperature was maintained at 4°C by using an apparatus constructed to circulate 30% aqueous alcohol.

THE PROTEIN COMPOSITION OF HUMAN PANCREATIC JUICE

Keller, P. J., and B. J. Allan, J. Biol. Chem. **242,** 281–287 (1967).

Proteins in human pancreatic juice were separated by polyacrylamide-disc electrophoresis at pH 4.1, 2 mA per gel for three hours at 4°C. Protein zones were identified with Amido Black. Amylase, lipase, ribonuclease, deoxyribonuclease, proelastase, procarboxypeptidase A and B, chymotrypsinogen, trypsinogen, and a trypsininhibitor were identified.

THE FRACTIONATION OF HIGH-MOLECULAR-WEIGHT RIBONUCLEIC ACID BY POLYACRYLAMIDE-GEL ELECTROPHORESIS

Loening, U. E., Biochem. J. **102,** 251–257 (1967).

High molecular weight ribonucleic acid was separated on gels prepared from acrylamide and bisacrylamide (the acrylamide concentration being between 2–5%. Electrophoresis was performed in a Tris–sodium acetate–EDTA buffer, pH 7.8, for 0.5 to 3 hours. Electrophoresis in 2.2–2.6% gel revealed 45s precursor, transfer RNA and minor components. 4s and 5s RNA were separated in 5% and 7% of gels. Ribosomal RNA was excluded.

POLYACRYLAMIDE-GEL ELECTROPHORESIS ACROSS A MOLECULAR SIEVE GRADIENT

Margolis, J., and K. G. Kenrick, Nature **214,** 1334 (1967).

A linear gradient of gel was prepared for optimum molecular-sieve concentration to separate a protein mixture. A continuous gradient was maintained by 10% w/v sucrose in the acrylamide suspension and by eliminating connection currents generated by polymerization heat. Electrophoresis of serum in the direction of increasing gel concentration sharpened resolution since the advance edge was retarded more than the trailing edge.

PURIFICATION AND CHARACTERIZATION OF THE LIPASE OF PSEUDOMONAS FRAGI

Mencher, J. R., and J. A. Alford, J. Gen. Microbiol. **48,** 317–328 (1967).

The lipase of *P. fragi* (NRRL-B-25) was purified 70–100-fold through a combination of chromatography on DEAE-Sephadex® and ultra-filtration, and ammonium sulfate fractionation. Disc electrophoresis and gas liquid chromatography, density gradient centrifugation, and TLC were used to characterize the isolated protein. Polyacrylamide-disc electrophoresis was performed according to the technique of Ornstein and Davis. A 7.5% gel was used and the runs conducted at a current of 5 mA (2-3°C).

ANOMALOUS ELECTROPHORETIC PATTERN OF MILK PROTEINS

Michalak, W., J. Dairy Sci. **50**, 1319–1320 (1967).

The protein bands described by Wake and Baldwin (1.00 and 1.04) were not found in this investigation of polymorphism of milk proteins. In this study a modification of Schmidt's method [Biochim. Biophys. Acta **90**, 411 (1964)] was used. A borate buffer, pH 8.4, a temperature of 2°C and a voltage which gradually increased over the two hour run from 120 to 400 volts were the modifications. Skimmed milk was the sample rather than precipitated casein, and 2-mercaptoethanol was not added.

CRYSTALLIZATION OF HUMAN LIVER ALCOHOL DEHYDROGENASE

Mourad, N., and C. L. Woronick, Arch. Biochem. Biophys. **121**, 431–439 (1967).

Alcohol dehydrogenase was purified from human livers through a combination of ammonium sulfate fractionation, treatment with ethanol–chloroform, DEAE-cellulose, CM-cellulose chromatography, and crystallization from aqueous ethanol. Polyacrylamide-disc electrophoresis and cellulose-acetate electrophoresis were employed to ascertain purity of the preparation. The enzyme was found to be homogenous upon electrophoresis.

DISC ELECTROPHORESIS OF HUMAN AND ANIMAL SERUM LIPOPROTEINS

Narayan, K. A., Lipids **2**, 282–284 (1967).

Lipoproteins of chicken, mouse, rat, rabbit and human sera were separated by disc electrophoresis on 3.75%, 5.0% and 7.5% polyacrylamide gels. Results indicate that two to six minor lipoprotein components were present in the five species examined. The major components consist of at least one slow component close to the spacer gel. This procedure gives better resolution than paper electrophoresis, which separates two to four lipoprotein components of serum.

FRACTIONATION OF CELL MEMBRANE PROTEIN BY DISC ELECTRO-PHORESIS

Neville, D. M., Biochim. Biophys. Acta **133**, 168–170 (1967).

Disc electrophoresis performed at pH 9.4, 4.3 and 2.3 in the presence of urea separated protein of solubilized cell membrane. The protein was found to be heterogeneous; 15 bands clearly resolved, with another 10 too faint to be detected by densitometric tracings.

TURNOVER AND AGE DISTRIBUTION OF A COLLAGEN FRACTION EXTRACTABLE FROM RAT SKIN BY MERCAPTOETHYLAMINE

Nimni, M. E., K. Deshmukh, and L. A. Bavetta, Arch. Biochem. Biophys. **122**, 292–298 (1967).

Collagen material extracted from the dorsal skin of male Holtzman rats with cysteamine (0.2M) was separated into two major components. Polyacrylamide-gel (disc) electrophoresis was used. The rate of formation and degradation of the α- and β-fractions was investigated by following the labeling of hydroxyproline at different times after the injection of C^{14}-proline.

ELECTROPHORETIC SEPARATION OF THE SOLUBLE PROTEINS OF BRUCEI SUB-GROUP TRYPANOSOMES

Njogu, A., and K. Humphreys, Nature **216**, 280–282 (1967).

The soluble proteins of Brucei sub-group trypanosomes were separated by electrophoresis on paper, cellulose acetate, polyacrylamide and starch gel. Using cellulose acetate and high resolution Tris buffer, pH 8.9, at least nine protein bands were found. Disc electrophoresis yielded nine components. The results obtained with Smithies starch-gel electrophoresis method were superior to other methods as described above. In using starch gel and a discontinuous glycine–sodium hydroxide–borate buffer, pH 8.58 (8 volts/cm), 22 components were detected.

POSTPARTUM CHANGES IN MILK SERUM PROTEINS

Porter, R. M., and H. R. Conrad, J. Dairy Sci. **50**, 505–508 (1967).

Changes in the relative concentration of β-lactoglobulin α-lactalbumin and immune globulin were studied in milk serum for the first 21 days postpartum. Milk was defatted by centrifugation and then diluted for electrophoresis, 1:3 for milk of the first six days, thereafter a 1:2 dilution. A 7% polyacrylamide gel was used for disc electrophoresis according to the method of Ornstein and Davis. By ten days postpartum, a stable relationship was established in the relative concentrations of the proteins, with β-lactoglobulin being in highest concentration.

GENETIC CONTROL OF LACTATE DEHYDROGENASE ISOZYMES IN CULTURES OF LYMPHOCYTES AND GRANULOCYTES: EFFECT OF PHYTO-HEMAGGLUTININ, ACTINOMYCIN D OR PUROMYCIN

Rabinowitz, Y., and A. Dietz, Biochim. Biophys. Acta **139**, 254–264 (1967).

The isozymes of lactate dehydrogenase and malate dehydrogenase of leukocytes were separated by polyacrylamide-disc electrophoresis using a Canalco® Model 12 cell. Electrophoresis was performed at 5°C in a Tris-glycine buffer (pH 8.3) at 30 volts/cm. Two malate dehydrogenase bands were found in lymphocytes and granulocytes. It was learned that phytohemagglutinin produced a change in the proportion of H- and M- subunits synthesized by lymphocytes.

HETEROGENEITY OF CALF THYMUS HISTONES BY ELECTROPHORESIS IN POLYACRYLAMIDE GEL

Rebentish, B. A., and V. G. Debakov, Biokhimiya **32**, 169–172 (1967), cited in Biochemistry **32**, 141–144 (1967).

Histones isolated from calf thymus by homogenization in 0.14M sodium chloride and 0.01M sodium citrate and purified by column chromatography on carboxymethyl cellulose columns was characterized using polyacrylamide disc electrophoresis. The histone fractions were resolved by disc electrophoresis on a gel containing 20% acrylamide, 0.13% N, N'-methylene-bis-acrylamide, 1.67% ammonium persulfate, and 1.16% diaminopropionitrile. Electrophoresis was carried out at 5-6 mA per tube for four hours. The electrolyte used was 0.37M glycocol adjusted to pH 4.0 with glacial acetic acid. Results obtained after evaluation of the Amido Black 10B stained patterns showed gross heterogeneity of the whole histone as well as the column isolated material. Amino, acid analyses of the isolated material complete the characterization studies reported.

FRACTIONATION OF ISOLATED BACTERIAL MEMBRANES

Salton, M. R. J., M. D. Schmitt, and P. E. Trefts, Biochem. Biophys. Res. Commun. **29**, 728–733 (1967).

Cell membranes isolated by two procedures from *Micrococcus lysodeikticus*, *Sarcina lutea*, and *Bacillus subtilus* were characterized by disc electrophoresis and UV spectral studies. Disc electrophoresis was performed in either a urea-acid gel system or Tris–glycine buffer containing 8M urea and 0.005M EDTA. The acid urea gels were run at 1.5 mA per column for 150 minutes and the Tris–glycine gels were run at 2 mA per column for 100 minutes. Detection of separated protein was possible by staining with Aniline Blue Black. Marked differences in the solution protein pattern from each extraction procedure were observed.

SOME SERUM PROTEIN AND CELLULAR CONSTITUENTS OF INFLAMMATORY LESIONS

Sheldon, W. H., D. Mildvan, and J. C. Allen, Johns Hopkins Med. Bull. **121**, 113–133 (1967).

Polyacrylamide-disc electrophoresis was used to separate proteins of exudates produced by short-term injury to rabbits. A 7.5% gel, pH 9.5, prepared according to the procedure of Ornstein was used. Electrophoresis was carried out at 4.5 mA per column. The major protein fractions identified were albumin, transferrin and gamma globulin. Total protein concentration as determined by the Folin-Cliocalteau method was found to be 0.49 g % at six hours post-injury and rose to 2.9 g % 96 hours post-injury.

EFFECT OF AGE ON DEHYDROGENASE HETEROGENEITY IN THE RAT

Schmukler, M., and C. H. Barrows, J. Gerontol. **22**, 8–13 (1967).

To test the "error" theory of aging, in which aging is thought related to the production of abnormal proteins, qualitative and quantitative studies were performed on rat dehydrogenases as a function of age. Electrophoretic patterns in various organs were compared. Vertical acrylamide-gel electrophoresis by the technique of Raymond was done in 0.076M Tris–0.005M citric acid buffer, pH 8.6 (30 to 45 minutes at 300 volts with a current of 0.5–1.0 mA/strip). No qualitative differences in electrophoretic pattern were seen with aging.

LOW-pH DISC ELECTROPHORESIS OF SPINAL FLUID: CHANGES IN MULTIPLE SCLEROSIS

Shapiro, H. D., Exp. Molecular Pathol. **1**, 362 (1967).

A method is described for resolving proteins in unconcentrated spinal fluid using acrylamide-disc electrophoresis at low-pH. The gel polymer consists of acrylamide–bisacrylamide solution, 7.5 volume; TEMED buffer, 5.5 volume; persulfate catalyst solution, 2.0 volume. The gel polymerizes in the tube. Electrophoresis is performed at 2 mA for 15 minutes, then 5 mA for 120–135 minutes. Spinal fluid is mixed 3:2 with solvent. Spinal fluid of multiple sclerosis patients appears to be missing a slow-moving protein present in normal spinal fluids.

MOLECULAR WEIGHT ESTIMATION OF POLYPEPTIDE CHAINS BY ELECTROPHORESIS IN SDS POLYACRYLAMIDE GELS

Shapiro, A. L., Biochem. Biophys. Res. Commun. **28**, 815 (1967).

Anionic detergent sodium docecyl sulfate (SDS) incorporated into polyacrylamide gave a rapid estimation of protein subunits. Electrophoresis was done at 8 volts/cm for two hours in 0.1M phosphate buffer, pH 7.1, containing 0.1% SDS. The 5% polyacrylamide gel was prepared in this same buffer. Relative migration vs. log molecular weight is linear for molecular weight 15,000 to 165,000. The method was useful for proteins with isoelectric points from 4 to 11 because SDS tends to minimize charge differences.

CHANGES IN PEROXIDE-ISOZYME PATTERNS INDUCED BY VIRUS INFECTION

Solymosy, F., J. Szirmai, L. Beczner, and G. L. Farkas, Virology **32**, 117–121 (1967).

Extracts from normal leaves of *Phaseolus vulgaris* and *Nicotiana glutinosa* contained different electrophoretic peroxidase isozymes. Virus infection caused the formation of new peroxidase isozymes in both plants. Polyacrylamide-disc electrophoresis was done according to the method of Davis, 1964. The buffer conditions of Farkas and Stahlmann were used [Phytopathology **56**, 667–669 (1966)].

DEMONSTRATION OF MULTIPLE ESTERASES OF THE HUMAN DENTAL PULP AFTER ELECTROPHORESIS IN STARCH AND ACRYLAMIDE GELS

Strachan, D. S., R. Rapp, and J. K. Avery, J. Dental Res. **46**, 471 (1967).

Adult human dental pulp esterases contained in 40% sucrose homogenates, were separated by starch-gel and polyacrylamide-gel electrophoresis. Polyacrylamide-gel electrophoresis was performed for two hours at 2 mA per tube (4°C) at a pH of 8.3. The substrates used to detect enzyme activity were α-naphthyl acetate and α-naphthyl butyrate. Contrasting with results obtained on starch gels at pH 8.6, four bands were detected on polyacrylamide gel.

ELECTROPHORETIC STUDIES ON PHOSPHATASES FROM THE PANCREATIC ISLETS OF OBESE-HYPERGLYCEMIC MICE

Täljedal, I.-B., Acta Endocrinol. **55**, 153–162 (1967).

Disc electrophoresis on polyacrylamide gels was used to separate phosphatase enzymes from the pancreatic islets of American obese-hyperglycemic mice. Electrophoresis was performed at 4°C according to the procedure of Ornstein and Davis. Three sites of phosphatase activity were noted when inosine diphosphate, adenosine diphosphate, and thiamine pyrophosphate were used as substrates. Substrate specificity and ion sensitivity studies were carried out on the enzymes separated.

DISC ELECTROPHORETIC SERUM PROTEIN PATTERNS IN DIAGNOSIS

Tarnoky, A. L., and B. Dowding, Clin. Biochem. **1**, 48–65 (1967).

Disc electrophoresis in 7% acrylamide gel was carried out according to the Ornstein-Davis procedure in a buffer containing 5.7% Tris in 0.256M phosphoric acid. Electrophoresis was performed at 2 mA/gel for the first 30 minutes and 3 mA/gel thereafter. Electrophoretic patterns throughout life for normal patients and those typical of various disease entities are reported.

RIBOSOMAL PROTEINS OF ESCHERICHIA COLI: I. DEMONSTRATION OF DIFFERENT PRIMARY STRUCTURES

Traut, R. R., Proc. Nat. Acad. Sci. U.S. **57**, 1294–1301 (1967).

Acrylamide-gel electrophoresis [J. Mol. Biol. **19**, 215 (1966)] was combined with paper electrophoresis to study ribosomal proteins. Whatman® 3MM paper was spotted with the oxidized protein samples, and descending chromatography was performed in butanol–acetic acid–water (3:1:1). The papers were dried, and electrophoresis was done in 1.24% pyridine, 1.25% acetic acid, pH 4.7, for three hours at 38 volts/cm or for one hour at 70 volts/cm depending on the purity of the preparation.

ELECTROPHORETIC STUDIES OF HUMAN PLACENTAL DEHYDROGENASES

Van Bogaert, E. C., E. De Peretti, and C. A. Villee, Am. J. Obstet. Gynecol. **98**, 919–923 (1967).

Lactate, malate, isocitrate, glutamate, and glucose-6-phosphate dehydrogenase from human term placenta, early placenta and a hydatidiform mole were studied by electrophoresis on polyacrylamide gels (disc). Electrophoresis performed according to the procedure of Ornstein was adequate to separate lactate dehydrogenase, malate dehydrogenase, and glucose-6-phosphate dehydrogenase. A 4% polyacrylamide gel (lower) and a 7% acrylamide gel (upper gel) were necessary to separate TPN-specific isocitrate dehydrogenase. The gels were prepared in 56 mM imidazole aspargine buffer, pH 7.0. Eleven bands of lactate dehydrogenase were generally detected. Human t e r m placenta exhibited two malate dehydrogenase bands. Young placenta was characterized by extra glucose-6-phosphate dehydrogenase bands.

PRECIPITATION OF CAPRINE AND BOVINE CASEINS FROM ACIDIC
SOLUTIONS BY SODIUM POLYPHOSPHATE: INFLUENCE OF pH AND UREA,
UTILIZATION FOR SEPARATION OF α_2- AND K-CASEINS

Zittle, C. A., and J. H. Custer, J. Dairy Sci. **50**, 1352–1355 (1967).

Goat and bovine caseins were separated by a low concentration (4 mg/10 ml
for a 0.5% casein solution at pH 2.0) of sodium polyphosphate. In the
presence of 2M urea, K-casein was not precipitated, whereas α_s-casein was.
Vertical polyacrylamide-gel electrophoresis at pH 9.0 in 4.5M urea was
the method used to demonstrate the separation of the K- and α_2-casein.
Separation was further improved by first reducing K-casein to the monomeric
form with mercaptoethanol.

SEX ASSOCIATED QUANTITATIVE DIFFERENCES IN THE PLASMA ESTER-
ASES OF INBRED MICE

Allen, R. C., and D. J. Moore, Endocrinology **78**, 655–658 (1966).

Disc electrophoresis (8.5% separatory gel and 3.95% spacer gel) was used
to study sex-associated differences in plasma esterases of inbred mice (female
C57 BL/6). Electrophoretic separations were carried out in 0.05M Tris-glycine
buffer at constant current of 3 mA per gel tube. Esterase activity in the gels
was localized with α-naphthyl butyrate substrate contained in 0.04M Tris-
hydrochloride buffer. Fast blue RR salt was the coupling agent employed.
There were no quantitative differences in the esterase pattern of male or female
rats less than three weeks old. Esterase activity was found to be controlled
by testosterone as evidenced by results obtained from plasma of gonadectom-
ized animals.

ELECTROPHORESIS OF PITUITARY PROTEINS AFTER TREATMENT OF RATS
WITH NORETHYNODREL

Baker, B. L., and D. B. Zanott, Endocrinology **78**, 1037–1040 (1966).

Pituitary homogenates of rats ovariectomized and non-ovariectomized stimulated
with norethynodrel were subjected to electrophoresis on polyacrylamide discs.
A 7.5% gel at pH 9.5 was used according to procedure of Ornstein and
Davis. Ovariectomy reduced the prominence of the prolactin band and
accentuated the growth hormone band. Norethynodrel intensified the prolactin band in the ovariectomized but not in the non-ovariectomized animal.

INDIVIDUALITY OF GLYCOPROTEINS IN HUMAN AORTA

Berenson, G. S., B. Radhakrishnamurthy, A. F. Fishkin, H. Dessauer, and
P. Arquembourg, J. Atherosclerosis Res. **6**, 214–223 (1966).

Glycoproteins extracted from human aorta were separated by electrophoresis on 5% polyacrylamide discs using a phosphate buffer, pH 7.8, μ 0.03. The separated glycoproteins were detected by staining the gel with Amido Black and periodic acid-Schiff's reagent. A family of glycoproteins in cardiovascular connective tissue was demonstrated. Specific differences were evident in some of the individual members of the glycoproteins.

DIFFERENTIATION OF PAROTID AND PANCREATIC AMYLASE IN HUMAN SERUM

Berk, J. E., S. Hayashi, R. L. Searcy, and N. C. Hightower, Jr., Am. J. Digest. Diseases **11,** 695–701 (1966).

Amylases were isolated from saliva, pancreatic homogenates and parotid gland by gel filtration on Sephadex® G-100. The purified amylases were examined electrophoretically on polyacrylamide discs. When activity was assayed using a saccharogenic method, parotid gland and salivary amylase exhibited a mobility similar to the more anodal peak observed in normal serum. The slower moving amylase band of serum seemed to be identical to that of pancreatic tissue homogenates.

MEASUREMENT OF THE RADIOACTIVITY IN 14-CARBON AND TRITIUM-LABELED PROTEINS THAT HAVE BEEN SEPARATED BY DISC ELECTRO-PHORESIS

Boyd, J. B., Anal. Biochem. **14,** 441–454 (1966).

Replacing the water in acrylamide gel with a tolune-based scintillation solution permitted direct measurement of C^{14} and H^3. The soaking procedure involved three stages requiring a minimum total of 17 hours. Counting efficiency of C^{14} was $81\pm$ and of H^3 was 23%. A sample apparatus is described, which permits simultaneous soaking of 74 gel slices.

ELECTROPHORETIC AND IMMUNOLOGICAL COMPARISONS OF SOLUBLE ROOT PROTEINS OF MEDICAGO SATIVA L. GENOTYPES IN THE COLD HARDENED AND NON-HARDENED CONDITION

Coleman, E. A., R. J. Bula, and R. L. Davis, Plant. Physiol. **41,** 1681–1685 (1966).

Polyacrylamide-disc electrophoresis, according to the procedure of Ornstein, and immunodiffusion studies were used to compare the soluble root proteins of six *Medicago sativa L* genotypes in cold-hardened and non-hardened conditions. The cold-hardened species were found to possess a large amount of highly-charged, low molecular weight protein material. Proteins of all species were found to be anodic in migration.

PROTEIN CHANGES RELATED TO HAM PROCESSING TEMPERATURES: I. EFFECT OF TIME-TEMPERATURE ON AMOUNT AND COMPOSITION OF SOLUBLE PROTEINS

Cohen, E. H., J. Food Sci. **31,** 746–750 (1966).

The effect of different temperatures up to 165°F on protein changes in cured uncooked ham were investigated using polyacrylamide-disc electrophoresis. The technique of Ornstein and Davis was used to fractionate soluble proteins of ham heated at 120°, 140°, 150°, and 160°F. Five components were found to be relatively stable. One of these components was identified as an acid phosphatase by staining with α-naphthol phosphate substrate at pH 5.0.

PURIFICATION AND PROPERTIES OF AN ENZYME FROM BEEF LIVER WHICH CATALYZED SULPHYDRYL-DISULFIDE INTERCHANGE IN PROTEINS

De Lorenzo, F., R. F. Goldberger, E. Steers, Jr., D. Giuol, and C. B. Anfinsen, J. Biol. Chem. **241,** 1562–1567 (1966).

An enzyme which catalyzes sulphydryl-disulfide interchange in proteins containing "incorrect" disulfide bonds is described. The purified enzyme showed two bands on 7.5% polyacrylamide gel when electrophoresed with 0.2M Tris–glycine buffer, pH 8.5, at 4°C with 3.0 mA per tube. Starch-gel electrophoresis in 0.04M sodium phosphate buffer, pH 6.5, showed the same two bands. These bands were thought to be variants of the same protein since they appeared as one band after reduction and alkylation.

DISC ELECTROPHORESIS IN THE PRESENCE OF SODIUM DODECYL SULPHATE

De Vito, E., and J. A. Santome, Experientia **22,** 124–125 (1966).

Disc electrophoresis performed according to the Ornstein-Davis procedure—modified so that the gel concentration was raised to 4%, EDTA incorporated at 6 mg% and including 0.5% sodium dodecyl sulphate—was used to show the association of subunits. Insulin, aldolase and bovine growth hormone were studied. Multicomponents were observed in all three molecules when they were electrophoresed in dodecyl sulphate.

POLYACRYLAMIDE-GEL ELECTROPHORESIS OF SOYBEAN WHEY PROTEINS AND TRYPSIN INHIBITORS

Eldridge, A. C., R. L. Anderson, and W. J. Wolf, Arch. Biochem. Biophys. **115,** 495–504 (1966).

Soybean whey proteins were separated into 24 bands by polyacrylamide-gel electrophoresis in a glycine buffer, pH 9.2, containing 8M urea. Commercial preparations of trypsin inhibitors were examined by this procedure. Most preparations yielded six bands; however, one sample out of nine revealed thirteen bands.

MOLECULAR SIEVE CHROMATOGRAPHY OF PROTEINS ON GRANULATED POLYACRYLAMIDE GELS SEPARATIONS

Fawcett, J. S., and C. J. Morris, Science **1**, 9–26 (1966).

Polyacrylamide gels were successively forced through 100- and 200-mesh sieves and collected in 0.5M NaCl. The granulated gel was packed into a column to separate protein solutions by the upward flow method. Partition coefficients, K and K_{av}, were calculated. Equations relating the gel composition to the Ogstern-Laurent-Killander fundamental parameters L and r were derived. A method was also derived for estimating pore radius of the gel.

DISC ELECTROPHORESIS OF IRRADIATED TRYPSIN

Ghiron, C. A., A. F. Ghiron, and J. D. Spikes, Photochem. Photobiol. **5**, 201–205 (1966).

Trypsin inactivated by UV radiation and gamma radiation, visible radiation in the presence of sensitizing dyes and autolysis, were subjected to disc electrophoresis using a 7% gel and a beta alanine buffer, pH 5.2. Separation proceeded at 4 mA per gel column for 1.5 hours. A progressive loss in the number of bands separated was observed when inactivation was achieved with flavin-sensitized photooxidation, autolysis and treatment with UV and gamma radiation. Methylene Blue and eosin Y prevented this loss in bands. Normal trypsin exhibited four bands upon electrophoresis.

THE ELECTROPHORETIC MOVEMENT OF PROTEINS FROM VARIOUS SPECIES AS A TAXONOMIC CRITERION

Gottlieb, D., and P. M. Hepden, J. Gen. Microbiol. **44**, 95–105 (1966).

The use of polyacrylamide-gel electrophoresis for the separation of proteins extracted from various species of streptomyces is reported. Electrophoretic separation of protein extracts was performed in a Canalco® Model 12 apparatus on a 7.5% gel. Between 100–200 μg protein were applied to each column and electrophoresed in a buffer of pH 8.2–8.5 at 4 to 5 mA per tube. Electrophoresis was continued for 25 to 35 minutes. A solution of Amido Black (0.5% in 7% acetic acid) was used to stain the protein bands. A distinctive protein pattern was obtained for each strain of organism tested. Several strains of S. venezuelae and S. griseus were typed. There was more similarity between strains within one species than between strains of different species.

CHARACTERISTICS OF A PROTEINASE OF A TRICHOSPORON SPECIES ISOLATED FROM DUNGENESS CRAB MEAT

Groninger, H. S., and M. W. Eklund, Appl. Microbiol. **14,** 110–114 (1966).

The proteinase from a Trichosporon species was 170X purified by dialysis, ammonium-sulfate fractionation and gel filtration. Starch gel electrophoresis separated the enzyme into two major and two minor components. The pH 9.2 discontinuous buffer used for preparing the gel was composed of 0.05M Tris, 0.007M boric acid, and 0.0015M EDTA. A 3.3X more concentrated buffer was used for electrophoresis. A current of 25 mA (20 volts/cm) separated the protein in 1½ hours. Buffalo Blue Black NBR was the stain used.

ELECTROPHORETIC HETEROGENEITY OF β-LACTOGLOBULINS IN POLY-ACRYLAMIDE GEL

Grossmann, A., and U. Freimuth, Milchwissenschaft **21,** 344–346 (1966).

Five to eight percent polyacrylamide gel was prepared in a borate buffer, 0.03M H_3BO_3 with 7.5 x 10^{-3}M NaOH, pH 6.6. β-lactoglobulin was diluted 1:100 in this buffer and 25% sucrose added. Electrophoresis was done in a borate buffer, pH 8.2 (0.3M H_3BO_3 with 0.06M NaOH). Four to five fractions were identified. Starch-gel electrophoresis of β-lactoglobulin separated only two fractions.

STUDIES OF PROTEIN SYNTHESIS IN CULTURES OF HUMAN LYMPHO-CYTES STIMULATED BY PHYTOHAEMAGGLUTININ

Huber, H., C. Huber, F. Gabl, H. Winkler, and H. Braunsteiner, Protides Biol. Fluids, Proc. 14th Colloq. 301–306 (1966).

The nature of proteins synthesized by phytohemagglutinin-stimulated cultures of human lymphocytes was studied as to time course of protein synthesis in relation to RNA turnover, characterization of some of the proteins and their subcellular localization. Polyacrylamide-disc electrophoresis carried out according to the method of Winkler et al. revealed no quantitative differences in stimulated cells.

STUDIES ON KIDNEY TUBULOGENESIS: II. LACTIC DEHYDROGENASE ISO-ZYMES IN THE DEVELOPMENT OF MOUSE METANEPHROGENIC MESENCHYME IN VITRO

Koskimes, O., and L. Saxen, Ann. Med. Exp. Biol. Fenniae (Helsinki) **44,** 151–154 (1966).

Vertical disc electrophoresis carried out in 7.5% polyacrylamide (micro hematocrite tubes were used to cast gels) was used to characterize the LDH of minute amounts of mesenchyme tissue grown in tissue culture. No acrylamide sample gel or spacer was used. Sephadex®-sucrose buffered with 0.180M Tris buffer, pH 6.7, replaced the spacer gel. The gels were electrophoresed at 4 volts/cm for 60 to 90 minutes. Samples obtained at the beginning of culture showed three bands of activity. LDH-5 was the strongest band. After 48 hours, a shift occurred toward the faster moving isozymes. After six days of culture, the LDH pattern was the reverse of that of undifferentiated cells.

SEPARATION OF RAT ANTERIOR PITUITARY HORMONES BY POLY-ACRYLAMIDE GEL ELECTROPHORESIS

Kragt, C. L., and J. Meites, Proc. Soc. Exp. Biol. Med. **121**, 805–808 (1966).

It was possible to separate growth hormone, prolactin, TSH and LH activities in rat pituitary homogenates by polyacrylamide-gel electrophoresis. The technique of Ornstein and Davis with tetramethylethylenediamine added to the large pore gel to facilitate polymerization as done by Lewis [J. Biol. Chem. **238**, 3330 (1963)]. The 7.5% gel was prepared at pH 9.5. Fresh pituitaries were homogenized and suspended in pH 7.4 phosphate buffer, at a final concentration of 2 mg/0.1 ml buffer. Three mg of this suspension were mixed with 0.15 ml of the upper gel layer and added to each tube [62 x 5 (ID) mm]. Using a current of 2.0-2.5 mA/tube, separation was complete in about 1.0 to 1.5 hours as determined by watching the migration of a bromophenol buffer front. Tubes that were to be used to assay for the active fractions were marked to correspond to the tubes that were to be stained and immediately frozen. Amido Black dye in 6% acetic acid was the stain.

NONSPECIFIC ESTERASES IN NORMAL AND NEOPLASTIC TISSUES OF THE SYRIAN HAMSTER: A ZYMOGRAM STUDY

Kreusser, E. H., Cancer Res. **26A**, 2181–2185 (1966).

Hamster tissues were extracted by homogenizing in water (1 mg tissue/1 ml water). After centrifugation for 20 minutes at 20,000 g, the supernate could be frozen for storage. Gel electrophoresis was performed in a 0.5M sodium barbital buffer, pH 8.6, at a potential of 10 volts/cm for 2.5 hours. The gel was prepared in a 10X concentration of this buffer. The substrate used in all cases was α-Naphthyl butyrate with Blue RR salt as a diazonium coupling agent.

RESIDUAL CASEIN FRACTIONS IN RIPENED CHEESE DETERMINED BY POLYACRYLAMIDE-GEL ELECTROPHORESIS

Ledford, R. A., A. C. O'Sullivan, and K. R. Nath, J. Dairy Sci. **49**, 1098–1101 (1966).

Since specific changes in the casein fractions can be detected, polyacrylamide-gel electrophoresis is an ideal method for studying the ripening process of cheese. Cheese was prepared for electrophoresis by adding 0.8 ml distilled water and 2.0 ml 7M urea buffer to 0.2 mg cheese. Samples were warmed to 37°C and the fat layer removed. Vertical electrophoresis was performed at pH 9.0-9.1, 4.5M urea, according to the method of Thompson [J. Dairy Sci. **47**, 378 (1964)].

THE QUANTITATIVE HISTOCHEMISTRY OF A CHEMICALLY INDUCED EPENDYMOBLASTOMA: I. ENZYMES

Lehrer, G. M., H. S. Maker, and L. C. Scheinberg, J. Neurochem. **13**, 1197–1206 (1966).

Acrylamide-gel electrophoresis by the method of Davis was used to separate the lactate dehydrogenase isoenzymes in lyophilized extracts of tumor and brain tissues. Twenty-five μg of the sample were suspended in 50 μl of a 40% sucrose in 0.05M Tris–glycine buffer, pH 8.25. This buffer was also used for electrophoresis at a current of 1.0 mA/tube. The LDH was demonstrated by a colorimetric reaction at room temperature in the dark. The medium consisted of 0.1M sodium lactate, 0.01M NaDP$^+$, 0.1M Kreb's buffer (pH 7.4), 0.4 mM nitro-blue tetrazolium and 1.0mM sodium amytal. Just before use, 2.5 mg phenazine methosulphate were added to 5.0 ml of the above reagent.

ACRYLAMIDE GEL DISC ELECTROPHORESIS PATTERNS AND EXTRACTABILITY OF CHICKEN BREAST MUSCLE PROTEINS DURING POSTMORTEM AGING

Maier, G. E., and R. L. Fischer, J. Food Sci. **31**, 482–487 (1966).

Polyacrylamide-disc electrophoresis was performed on water-soluble, 1M potassium chloride-soluble and 8M urea-soluble fractions of chicken breast muscle. The technique of Ornstein and Davis was used. Two to four fast-moving faint components were found in the water-soluble extract, which increased in intensity upon postmortem aging. These components were found in the salt-soluble or urea-soluble extracts. Water-soluble proteins were less extractable at 24 hours than at 30 minutes postmortem.

DISC ELECTROPHORESIS OF RAT PLASMA LIPOPROTEINS

Narayan, K. A., H. L. Creinin, and F. A. Kummerow, J. Lipid Res. **7,** 150–157 (1966).

Prestained plasma proteins (Sudan Black B) were separated in 7.5, 5, and 3.75% polyacrylamide gels. Electrophoresis was performed according to the method of Davis and Ornstein except that the sample gel and potassium ferricyanide were omitted. Four lipoprotein components were separated.

A CHARGE DIFFERENCE BETWEEN AN INTRACELLULAR AND SECRET-ED MOUSE MYELOMA GLOBULIN

Notani, G. W., A. J. Munro, and P. M. Knopf, Biochem. Biophys. Res. Commun. **25,** 395–401 (1966).

Polyacrylamide-gel electrophoresis on a 4.5% or 7.5% gel in a Tris–TEMED buffer, pH 9.2, was used to compare intracellular and extracellular myeloma globulins of Adj-PC5 and MOPC 21 plasma cell tumors. Electrophoresis for 70 to 80 minutes at 5 mA/tube revealed differences in relative mobility, the external protein being a faster anodic migrating protein.

ELECTROPHORETIC PATTERNS OF NORMAL HUMAN SERUMS BY DISC ELECTROPHORESIS IN POLYACRYLAMIDE GEL

Pasteuka, J. V., Clin. Chim. Acta **14,** 219–226 (1966).

Two hundred μg of serum protein was resolved into 20 to 30 discrete bands using slight modifications of the Davis technique [Ann. N. Y. Acad. Sci. **121,** 404 (1964)]. The electrophoresis was done at 4-5°C, and the reservoirs were filled with undiluted stock buffer; these changes can improve resolution of the bands. Naphthol Blue Black was used as the stain. This more sensitive method of serum electrophoresis could be useful to detect early serum changes in diseases and as a tool in genetic studies.

AMINO ACID COMPOSITION OF SIX DISTINCT TYPES OF β-CASEIN

Peterson, R. F., L. W. Nauman, and D. F. Hamilton, J. Dairy Sci. **49,** 601–607 (1966).

Six genetic types of β-casein were demonstrable by electrophoresis in 7% acrylamide gel, pH 9.0 in 4.5M urea, [J. Dairy Sci. **46,** 1136 (1963)]. At 5°C, trypsin hydrolysates of casein were separated at pH 9.0 (1.0M triethylamine carbonate buffer, 2000 volts) or in acid buffer, pH 1.8, (2.5% formic acid– 9% w/v acetic acid). Whatmann® 3MM paper was the carrier. Terminal amino acid analyses of the casein types were made.

DETECTION OF NEW TYPES OF β-CASEIN BY POLYACRYLAMIDE-GEL ELECTROPHORESIS AT ACID pH: A PROPOSED NOMENCLATURE

Peterson, R. F., and F. C. Kopfler, Biochem. Biophys. Res. Commun. **22**, 388–392 (1966).

Four β-caseins were resolved when milk caseins were electrophoresed in formic acid–acetic acid buffer at pH 3.0; a fifth β-casein was typed by alkaline-gel electrophoresis. The vertical electrophoresis technique of Raymond was used to type these caseins. The stock buffer was composed of 86 ml glacial acetic acid and 25 ml of 90% formic acid in one liter of water. 15 g Cyanogum®-41 and 40.5 g urea were dissolved a total volume of 150 ml with buffer. At 25°C, 1.0 ml of TEMED and 0.35 g ammonium persulfate were added. Buffer cells were filled with 7.7% acetic acid. The samples were dissolved in stock buffer, 10N in urea. Electrophoresis was performed for 20 minutes using a 20 mA current. Then the current was raised to 100 mA with the voltage being increased to 300 volts to maintain this current. Over the next 20 hours, the current was decreased to 60 volts in order to achieve separation, which in turn made it possible to type the casein.

THE EFFECTS OF CORTISONE ON THE METABOLISM OF HISTONES AND OTHER NUCLEAR AND CYTOPLASMIC PROTEINS IN BRAIN AND LIVER

Piha, R. S., M. Cuenod, and H. Waelsch, Ann. Med. Exp. Biol. Fenniae (Helsinki) **44**, 553–562 (1966).

Polyacrylamide-disc electrophoresis was used to study the effect of administering C^{14}-cortisone acetate on the turnover and incorporation of lysine into histone fractions of brain. No effect was shown upon the electrophoretic distribution of histones.

SEROMUCOID PATTERNS OF NORMAL AND TUBERCULOUS INDIVIDUALS AS DETERMINED BY DISC ELECTROPHORESIS TECHNIQUE

Price, W., H. Harrison, and J. Molenda, Am. J. Epidemiol. **83**, 152–175 (1966).

A method for fractionating serum seromucoids by disc electrophoresis is described. Polyacrylamide-disc electrophoresis was carried out at 22–24°C at 3.75 mA/tube in a Tris-HCl buffer, pH 8.8–9.0. Sera from tuberculous and nontuberculous individuals was analyzed by this technique with reproducible results. Normal sera and tuberculous sera differed in content of heavy molecular weight suromucoids.

ELECTROPHORETIC HETEROGENEITY OF POLYPEPTIDE CHAINS OF SPECIFIC ANTIBODIES

Reisfeld, R. A., and P. A. Small, Jr., Science **152**, 1253–1255 (1966).

Polyacrylamide-disc electrophoresis was used to study the effect of admin- chains of rabbit γG-immunoglobulin and specific antibodies to haptens. Rab- bits were immunized with arsenilic acid conjugated to bovine gamma globulin; dinitrophenyl bovine gamma globulin was also used. The antibody was isolated and reduced with 7M guanidine hydrochloride. Heavy and light chains were separated on Sephadex® G-200 with 5M guanidine hydrochloride. For electro- phoresis, the pH 8.91 buffer in the upper tray consisted of 5.16 g Tris, 3.48 g glycine, 700 ml of 10M urea and water to one liter. The pH 8.07 buffer in the lower tray consisted of 14.5 g Tris, 60 ml of 1N HCl, and water to one liter. Buffers were prepared fresh just prior to use. A current of 2.5 mA per tube separated the heavy chains in three hours and the light chains in two hours. The heavy and light chains could be separated into multiple bands by this technique.

STUDY OF THE POSTERIOR PITUITARY OF NORMAL AND DEHYDRATED RATS USING DISC ELECTROPHORESIS

Rennels, M. L., Endocrinology **78**, 659–660 (1966).

Polyacrylamide (disc) electrophoresis of the posterior lobe of the pituitary of normal and dehydrated rats was performed with a 7% gel and the Canalco® Model 6 apparatus. Electrophoresis at pH 9.5 revealed the presence of a densely-staining fraction in the glands of normal rats which is absent from those of dehydrated rats.

ELECTROPHORESIS OF MAMMALIAN ALDEHYDE DEHYDROGENASE

Robbins, J. H., Arch. Biochem. Biophys. **114**, 585–592 (1966).

Bovine aldehyde dehydrogenase precipitated from bovine liver supernatant at 50 to 55% saturation with ammonium sulfate was characterized by poly- acrylamide-disc electrophoresis. Electrophoresis was performed according to the technique of Ornstein and Davis at 1.5 to 2.5 mA/tube for 1.5 hours at constant current. Dehydrogenase activity was demonstrated by the formation of blue formazan bands when a tetrazolium-phenazine methosulfate staining technique was employed. A variety of aldehydes was useful as substrates. Only one band of enzyme activity was demonstrated with any of these sub- strates. The formation of formazan bands in the absence of substrate in the same position as the lactic dehydrogenase band was attributed to "nothing" dehydrogenase.

STUDIES ON ELECTROPHORETIC HETEROGENEITY OF ISOMETRIC PLANT VIRUSES

Semancik, J. S., Virology **30**, 698–704 (1966).

Acrylamide gel disc electrophoresis demonstrated heterogeneity in purified bean pod mottle virus and cowpea yellow mosaic virus. A 3¾% large-pore gel preparation was used for electrophoresis of virus suspensions. A 7% small-pore gel in 8M urea was used for electrophoresis of extracted viral protein. Fast and slow components of the virus were separated at a pH between 4.7 and 5.5. For the large-pore system, a current of 1-2 mA/tube was applied for three to six hours; 4 mA/tube for three to five hours was used in the small-pore system.

HIGH-RESOLUTION DISC ELECTROPHORESIS OF HISTONES: I. AN IMPROVED METHOD

Shepherd, G. R., and L. R. Gurley, Anal. Biochem. **14**, 356–363 (1966).

An improved method for high-resolution polyacrylamide-gel disc electrophoretic separation of histones is described. A modification of the gel polymerization, siliconization of the tubes, and use of a discontinuous buffer to produce a steep concentration gradient were the improvements made.

THE EFFECT OF CSF PARAPROTEINS ON THE COLLOIDAL GOLD TEST

Weiss, A. H., and N. Christoff, Arch. Neurol. **14**, 100–106 (1966).

Immuno- and polyacrylamide (disc) electrophoresis were used in conjunction with the Lange colloidal gold (LCG) test to study the nature of paraproteins in the CSF. In 11 cases, positive LCG tests were found. In nine of these cases there were paraproteins which migrated in the γ-region. The paraprotein mobilities differed in the positive and negative LCG test although their amount and distribution were similar. Slower moving γ-globulins are most positive in solution and have a greater tendency to precipitate the gold. Qualitative differences in the γ-globulin can thus affect the LCG test.

ELECTROPHORESIS IN HORIZONTALEM POLYACRYLAMIDGEL

Zwisler, O., and H. Biel, Z. Klin. Chem. **4**, 58–62 (1966) **German.**

A simple technique for electrophoresis on horizontal layered non-cooled polyacrylamide gel is presented. The influence of different buffers, pH, degree of cross-linkage, running distance, and concentration of applied protein were investigated.

ELECTROFORESIS DE LAS PROTEINAS SERICAS DE COBAYOS INOCULADOS CON VIRUS JUNIN (EN PAPEL Y GEL DE POYLACRILAMIDA)

Budzko, D. B., Rev. Soc. Arg. Biol. **41**, 92–101 (1965) **Spanish.**

Serum proteins of serum from normal and *Junin* virus-infected guinea pigs were separated by paper and polyacrylamide-disc electrophoresis. Paper electrophoresis was carried out horizontally on Whatman® No. 1 paper in a Veronal-sodium Veronal buffer, pH 8.6, ionic strength 0.05. Polyacrylamide-disc electrophoresis was carried out according to the procedure of Davis and Ornstein on 7.5% polyacrylamide gel. The serum protein pattern was examined each day up to eleven days for changes. During the last days of infection, total proteins, albumin and gamma globulins were decreased; beta globulins were increased.

THE ELECTROPHORETIC BEHAVIOR OF SOME GENETICALLY CONTROLLED VARIANTS OF HUMAN SERUM ALBUMIN AT ACID pH

Cooke, K. B., Protides Biol. Fluids, Proc. 13th Colloq. 435–437 (1965).

Ten percent polyacrylamide gels were used to separate albumin varients of human serum. Electrophoresis was performed at 3 volts/cm with cooling for 24 hours. The gels used were prepared in 0.1N sodium chloride at pH 3.4 to 4.2 and sodium acetate at pH 3.4 to 4.2. Identical behavior was observed at pH 4.2 and below. Isomerization was not detected above pH 3.8; two bands were present at pH 3.8. Opposite results were obtained on cellulose-acetate membranes.

IDENTIFICATION OF THE FLUORESCENT SUBSTANCES SEPARATED BY "SMALL-VOLUME" ELECTROPHORESIS

Fischl, J., and S. Segal, Clin. Chim. Acta **12**, 349–352 (1965).

Fluorescent substances in fresh human urine were identified after separation by electrophoresis [Fischl, Clin. Chim. Acta **8**, 399 (1963)]. Formic acid–acetic acid buffer was used since it did not alter or destroy fluorescence. Ninhydrin-positive substances were also identified. Urea, uric acid, urochrome, hippuric acid, pyruvate, xanthurenic acid and amino acids were some of the substances identified. This method could be of clinical diagnostic value.

PROTEIN PATTERNS IN HUMAN PAROTID SALIVA

Gifford, G. T., and L. Yuknis, J. Chromatog. **20**, 150–153 (1965).

Using 7.5% acrylamide gel on disc electrophoresis gave good resolution in separating parotid fluid. Parotid fluids were dialyzed versus saline or water in the cold and then lyophilized. The protein concentration was adjusted to 0.4 mg/0.03ml H_2O, and 0.06 ml of sample was put on the column (5 mm i.d. tubes). A pH 8.3 Tris–glycine buffer, ionic strength 0.01, was the electrolyte; the current was 5 mA/cell for 50 to 60 minutes. Amido Black stained proteins, Sudan Black stained the lipoproteins, and the periodic acid Schiff and Alcian Blue was used to stain glycoproteins. Attempts were made to identify normal parotid electrophoretic patterns.

CHANGES IN MIXTURES OF WHEY PROTEIN AND K-CASEIN DUE TO HEAT TREATMENTS

Hartman, G. H., and A. M. Swanson, J. Dairy Sci. **48,** 1161–1167 (1965).

The high resolution of polyacrylamide-disc electrophoresis was used to separate and identify whey proteins. A 9% gel and a Tris-HCl buffer, ionic strength 0.11, pH 8.9, was used. The current was 5 mA per tube for two hours during which time a marker dye band moved 5 cm. A complex was formed when K-casein and β-lactoglobulin were heated together at 74.5°C for 30 minutes in a solution of milk salts.

ELECTROPHORETIC COMPARISON OF PITUITARY GLANDS FROM MALE AND FEMALE RATS

Jones, A. E., J. N. Fisher, U. J. Lewis, and W. P. Vanderlaan, Endocrinology **76,** 578–583 (1965).

The proteins of the anterior pituitary gland of male and female rats were examined using the method of Ornstein and Davis for disc electrophoresis on a 7.5% gel, pH 9.5. The protein bands were compared as to width and depth of staining. Male rats possessed a slowly moving band with growth hormone activity. The corresponding band appeared narrow in the female. A prolactin band was deeply stained and wide in the female. The opposite behavior was found in the male. Weanlings showed no sexual differences in protein patterns.

ACTION OF RENNIN ON K-CASEIN, THE AMINO ACID COMPOSITIONS OF THE PARA-K-CASEIN, AND GLYCOMACROPEPTIDE FRACTIONS

Kalan, E. B., and S. H. Woychick, J. Dairy Sci. **48,** 1423–1428 (1965).

A rennin reaction with K-casein giving two major components in reduced para-K-casein can be demonstrated by polyacrylamide-gel electrophoresis. Tris-buffer pH 4.5 was used to prepare gels and as the electrolyte. The amino acids of K-casein and the breakdown products were compared.

STUDIES ON THE GROWTH HORMONE OF NORMAL AND DWARF MICE

Lewis, U. J., E. V. Cheever, and W. P. VanderLaan, Endocrinology **76**, 210–215 (1965).

Pituitary extracts of normal and dwarf mice were subjected to polyacrylamide electrophoresis at pH 9.5 and 4.5. The electrophoretic separations at pH 9.5 were carried out using the procedure of Ornstein and Davis. The pH 4.5 runs were made according to a modification of Reisfeld et al. [Nature (London) **195**, 281 (1962)]. The major component seen in extracts of glands of normal mice was missing from extracts of the dwarf animals. Biological assays of the major band eluted from the gel showed it to be 100% growth hormone.

A COMPARISON OF DYES USED FOR STAINING ELECTROPHORETICALLY SEPARATED LIPOPROTEIN COMPONENTS

Narayan, K. A., and F. Kummerow, Clin. Chem. Acta **13**, 532–535 (1965).

Amido Black B was the best stain for lipoproteins separated by polyacrylamide gel electrophoresis. For unfractionated samples, either Sudan Black B or Oil Red O are more effective since they do not stain protein constituents.

SERUM PROTEIN ELECTROPHORESIS IN ACRYLAMIDE GEL: PATTERNS FROM NORMAL HUMAN SUBJECTS

Peacock, A. C., Science **147**, 1451–1453 (1965).

Raymond's vertical acrylamide-gel electrophoresis demonstrated that electrophoretic patterns of normal human serum may be divided into at least 12 discrete groups. Cold stock solutions were used to prepare to gel before each run. These solutions were: A) 20% acrylamide–380 g acrylamide–20 g N_2-N'methylenebisacrylamide, dissolved in water 2:1; B) dimethylaminopropionitrite-buffer–8 ml DMAPN–100 ml stock buffer (20 x see below)–142 ml water; C) 2% ammonium persulfate prepared fresh weekly. Gel consisted of 40 ml of A, 20 ml of B, 80 ml of C, and 20 ml water. The stock buffer contained 431 g Tris–(hydroxymethyl) aminomethane, 37 g disodium ethylene–diaminotetracetate, and 220 g boric acid to 2 liters (pH 8.3–8.4), diluted 1:20 before use. The fresh serum samples were separated at 200 volts in five hours and then stained with Amido Black.

LOCATION OF THE CARBOHYDRATE-CONTAINING FRACTION OF K-CASEIN AFTER GEL ELECTROPHORESIS

Purkayastha, R., and D. Rose, J. Dairy Sci. **48**, 1419–1422 (1965).

Casein fractions were separated by polyacrylamide electrophoresis in the presence of 8M urea and 0.04M mercaptoethanol. A Veronal buffer or Tris–

EDTA–borate, 15 or 30 mA applied for 3 to 6 hours, was used. The electrode compartments contained 0.1M sodium chloride. Proteins were stained with Amido Black, glycoproteins with periodic acid-Schiff reagents. Only one band of K-casein contained glycoprotein.

DETERMINATION OF BOVINE TRANSFERRIN TYPES BY DISK ELECTRO-PHORESIS

Rausch, W. H., T. M. Ludwick, and D. F. Weseli, J. Dairy Sci. **48,** 720–725 (1965).

Discrete transferrin bands which were easily analyzed were separated by disc electrophoresis. The method of Ornstein and Davis was adopted. In the column the upper gel was spotted with 4 μl of the sample. This was concentrated in the middle layer, the spacer gel. The lower small pore served as a sieve in which electrophoretic separation took place. The buffer was 6.0 g Tris and 28.8 g glycine per titer. Electrophoresis was complete in one hour. The procedure for preparing the gel tubes is detailed.

RIBOSOMAL PROTEIN DIFFERENCES IN A STRAIN OF E. COLI CARRYING A SUPPRESSOR OF AN OCHRE MUTATION

Reid, P. J., E. Orias, and T. K. Gartner, Biochem. Biophys. Res. Commun. **21,** 66–71 (1965).

Polyacrylamide-disc electrophoresis, using the procedure described by Leboy, Cox and Flaks [Proc. Natl. Acad. Sci. U.S. **53,** 62 (1965)], revealed that a strain of E. coli having a suppressor of an "Operator negative" mutation in lactose (Lac) operon has a different electrophoretic pattern of ribosomal protein than the nonsuppressor parent.

ELECTROFORESIS EN GEL DE POLIACRYLAMIDA DE LAS SEROPROTEINAS DE ALGUNOS PECES TELEOSTEOS

Salibian, A., Rev. Soc. Arg. Biol. **41,** 121–127 (1965) **Spanish.**

Preliminary results of separation by polyacrylamide-gel electrophoresis of serum of five species of *Pimelolodial* are presented. The technique of Ornstein and Davis was used. The protein fractions were identified by staining with Sudan Black B. A.P.A.S.-positive fraction corresponding to the human 195-gamma globulin fraction was identified.

THE INTERPRETATION OF GEL ELECTROPHORESIS

Tombs, M. P., Anal. Biochem. **13,** 121–132 (1965).

The relationship between mobility and polyacrylamide-gel concentration is mathematically described. Values do not hold over a wide concentration range.

The relationship between gel pore size and mobility is derived and fitted to electrophoretic data of various globulins.

Starch-Gel Electrophoresis

GENE DUPLICATION IN FISHES: MALATE DEHYDROGENASES OF SALMON AND TROUT

Bailey, G. S., G. T. Cocks, and A. C. Wilson, Biochem. Res. Commun. **34**, 605–612 (1969).

Starch-gel electrophoresis at pH 6, at a potential gradient of 14 volts/cm for 19 hours, was employed to examine the supernatent malate dehydrogenase of salmon and trout tissues. Mitochondrial, crude extracts and supernatent fractions were analyzed. Three forms of supernatent malate dehydrogenase were revealed. Starch-gel electrophoretic analysis of remcombinants of purified enzymes indicated that two subunits, A and B, were responsible for the enzymes.

A NEW GENETIC VARIANT OF 6-PHOSPHOGLUCONATE DEHYDROGENASE IN AUSTRALIAN ABORIGINES

Blake, N. M., and R. L. Kirk, Nature **221**, 278 (1969).

Vertical electrophoresis on an 11% starch gel was used to separate a new variant of 6-phosphogluconate dehydrogenase from erythrocytes of Australian aborigines. Electrode vessels contained 0.2M disodium hydrogen phosphate adjusted to pH 7.0 with citric acid. The starch gels were prepared in a 1:20 dilution of this solution. Separation was achieved at a potential of 3 volts/cm for 18 hours. The new variant was called "ELCHO." The phenotype was present in many tribesmen in the ELCHO island area.

LACTATE DEHYDROGENASE ELECTROPHORETIC VARIANT IN A NEW GUINEA HIGHLAND POPULATION

Blake, N. M., R. L. Kirk, E. Pryke, and P. Sinnett, Science **163**, 701–703 (1969).

Using vertical electrophoresis on a 12% starch gel, the authors demonstrated variations in the LDH-A subunit in six different erythrocyte samples. Electrophoresis was performed on a 12% gel prepared in a 0.2M phosphate–citric acid buffer, pH 7.0 (1:20 dilution). Separation proceeded at 2 volts/cm for 16 to 18 hours at 15°C. A 0.2M phosphate–citric acid, pH 7.0, was used as the electrolyte. Standard tetrazolium procedures were used to detect the enzymes.

CHANGES IN HUMAN SERUM HIGH DENSITY LIPOPROTEINS INDUCED BY DISULFIDE-EXCHANGE REAGENTS

Cohen, L., and J. Djordjevich, Proc. Soc. Exp. Biol. Med. **129**, 788–793 (1969).

High-density lipoproteins isolated from serum were freed of lipids, and the a-proteins were subjected to starch-gel electrophoresis. Three different buffers were used: aluminum lactate 8M urea, pH 3.2; aluminum lactate without urea, pH 3.2; and Tris–borate–versene, pH 8.6. The high-density lipoprotein was reacted with disulfide at pH 8.6 in the presence of urea and its reaction product examined electrophoretically. The disulfide treated apo-protein migrated the same as "normal" apo-protein on alkaline gel.

6-PHOSPHOGLUCONATE DEHYDROGENASE HEMIZYGOUS MANIFESTATIONS IN A PATIENT WITH LEUKEMIA

Faiklow, P. J., R. Lister, J. Detter, E. R. Giblett, and C. Zavala, Science **163**, 194–195 (1969).

Starch-gel electrophoresis performed according to the method of Feldes and Parr [Nature **200**, 890 (1963)] was used to study the 6-phosphogluconate dehydrogenase of 41 patients with chronic myelocytic leukemia. Two patients were found to be heterozygous (AB) and two were found to be phenotypically homozygous.

THE RELATION BETWEEN SERUM AND PLATELET ALBUMINS

Ganguly, P., Clin. Chim. Acta **23**, 514–516 (1969).

Platelet albumin, prepared by ammonium sulphate precipitation followed by gel filtration on Sephadex® G-200, was compared with serum albumin by starch-gel electrophoresis, immunologic properties, molecular weight determinations and viscosity measurements. No differences were found in these parameters under normal conditions; however, striking differences were found on starch-gel electrophoresis in 8M urea. The platelet albumin migrated more slowly than the serum albumin band. At pH 4.2, the sedimentation behavior of platelet albumin was not altered but in contrast to that of serum, which showed a sigmoid decrease.

CHARACTERIZATION OF PLACENTAL ISOENZYME OF ALKALINE PHOS-PHATASE IN HUMAN PREGNANCY SERUM

Ghosh, N. K., and W. H. Fishman, Can. J. Biochem. **47**, 147–155 (1969).

The alkaline phosphatase of human placenta and blood was isolated by standard biochemical procedures and characterized by starch-gel electrophoresis, thin-layer Sephadex®–agarose electrophoresis, sensitivity to neuraminidase and heat inactivation. Horizontal starch-gel electrophoresis was performed on a 12% gel using the discontinuous Tris-citrate lithium borate system. Electrophoresis was carried out at 150 volts (constant) for 19 hours at 4°C. Thin-layer Sephadex® agarose-gel electrophoresis was carried out on a gel made of 3% agarose in Sephadex® G-200. The buffer described above was used and incorporated 17mM magnesium chloride as the electrolyte.

HAEMOGLOBIN PRODUCTION IN MICE RECOVERING FROM RADIATION

Lord, B. I., and R. Schofield, Nature **221**, 1060–1062 (1969).

Starch-gel electrophoresis was used to separate hemoglobins extracted from blood, bone marrow, spleen and individual spleen colonies of mice that were recovering from irradiation with 850 rads of X-rays. Normal bone marrow cells were injected into a starch prepared in a 0.076M Tris–0.005M citric acid buffer at pH 8.6. This was run in a 0.3M boric acid–0.05M potassium hydroxide buffer at pH 8.2. A current of 1.6 mA/cm was maintained for 2.5 hours. It was possible to evaluate protein synthesis if Fe^{59} were injected to the animals prior to sacrificing. Marked differences were shown for responses of the tissues studied, the greatest difference being in the intensity of hemoglobin band four. A greater quantity of this hemoglobin was found in the irradiated tissues than in normal tissues.

HISTONES FROM SPERM OF SEA URCHIN ARBACIA LIXULA

Palau, J., A. Ruiz-Carrillo, and J. A. Subirana, European J. Biochem. **7**, 209–213 (1969).

Whole histones and five fractions isolated from sperm of *Arbacia lixula* were characterized by starch-gel electrophoresis. Whole histones were prepared according to a technique described by Johns [Biochem. J. **92**, 55 (1964)]. Further fractionation of whole histones was achieved by column chromatography on CM-cellulose and fractional precipitation with acetone. These isolated products were then characterized by starch-gel electrophoresis in a 0.01N hydrochloric acid buffer. In some experiments 25mM Tris–HCl was employed as a buffer. Five main histone bands were identified in whole histones. The amino acid composition of whole histones had a higher content of arginine and a lower content of acidic amino acids as compared to the calf thymus histones.

IDENTIFICATION OF HUMAN GLUCOKINASE AND SOME PROPERTIES OF THE ENZYME

Pilkis, S. J., Proc. Soc. Exp. Biol. Med. **129**, 681–684 (1969).

Glucokinase of human liver extracts was characterized by starch-gel electrophoresis, apparent molecular weight, K_M value for glucose, and its reaction toward anti-glucokinase of rat. The 105,000 g supernate of liver homogenate were characterized by four isozymes on starch-gel electrophoresis. The molecular weight as estimated by Sephadex® G-100 gel filtration was found to be 48,000. The human liver hexokinase was inhibited 85% by rat anti-hexokinase.

A COMPARATIVE STUDY OF ALKALINE LIPOLYTIC ACTIVITY IN ADIPOSE TISSUE OF VARIOUS MAMMALS

Rivello, R. C., A. Cortner, and J. D. Schnatz, Proc. Soc. Exp. Biol. Med. **130**, 232–235 (1969).

Adipose tissue esterase (alkaline lipolytic activity) contained in 0.15M KCl homogenates was separated by vertical starch-gel electrophoresis. The electrolyte used was 0.165M phosphate–citrate buffer, pH 7.0. A potential of 5 volts /cm for five hours was used for separation. This buffer in a 1:20 dilution was used to prepare the gel. The esterase activity was located by staining the sliced gel with a solution containing naphthol esters as substrate and fast 2B as coupling agent. Human adipose ALA exhibited five isozymes; no genetic variants were found in humans. Rabbit, lamb, rat, and mouse patterns were similar to that of man. The rabbit exhibited genetic variants.

STRUCTURE AND FUNCTION OF CHLOROPLAST PROTEINS: VII. RIBULOSE-1,5-DIPHOSPHATE CARBOXYLASE OF CHORELLA ELLIPSOIDEA

Sugiyama, T., C. Matsumoto, T. Akazawa, and S. Miyachi, Arch. Biochem. Biophys. **129**, 597–602 (1969).

Ribulose-1, 5-diphosphate carboxylase was isolated from autotrophically grown *Chorella ellipsoidea* by a procedure gel filtration of *Chorella* extracts on Sephadex® G-200 and DEAE-column chromatography. Starch-gel electrophoresis was carried out in a buffer system described by Mendiola and Akazawa [Biochemistry **3**, 174 (1964)] to characterize the isolated purified enzyme. Results of the electrophoretic studies revealed the similarity of the *Chorella* enzyme to spinach enzyme, and immunodiffusion studies against specific antibodies demonstrated this relationship.

ERYTHROCYTE ADENYLATE-KINASE DEFICIENCY

Szeinberg, A., S. Gavendo, and D. Cahane, Lancet **1**, 315–316 (1969).

Starch-gel electrophoresis according to the technique of Feldes and Harris [Nature (London) **209**, 261 (1966)] was used to characterize an adenylate kinase variant in erythrocytes of a dog five months old. The main component of the isozymic patterns was absent on the electrophoretogram. Only one minor band could be detected. Glucose-6-phosphate dehydrogenase was deficient, and a severe anemia was present.

THE STRUCTURE OF GOAT HEMOGLOBINS: III. HEMOGLOBIN D, A β-CHAIN VARIANT WITH ONE APPARENT AMINO ACID SUBSTITUTION (21 Asp–His)

Adams, H. R., E. M. Boyd, J. B. Wilson, A. Miller, and H. J. Huisman, Arch. Biochem. Biophys. **127**, 398–405 (1968).

Starch-gel electrophoresis of goat hemoglobin according to the method of Huisman [Advan. Clin. Chem. **6**, 231 (1963)] revealed the presence of a third type, hemoglobin D. Structural differences in the β-chain were revealed in tryptic digests of A and D-Hb. An aspartyl residue was replaced by a histidinyl residue in peptide T-3b in hemoglobin D.

PENGUIN BLOOD SERUM PROTEINS

Allison, R. G., and R. E. Feeney, Arch. Biochem. Biophys. **124**, 548–555 (1968).

The blood serum of two Antarctic penguins, the Adelie and Emperor, was studied by starch-gel electrophoresis, disc electrophoresis, immunoelectrophoresis, immunodiffusion autoradiography, gel filtrations, and ion-exchange chromatography. Horizontal starch-gel electrophoresis was performed in the discontinuous buffer of Poulik [Nature **180**, 1477 (1967)]. Disc electrophoresis was performed in a continuous buffer system (0.1M glycine NaOH, pH 8.65) without a stacking gel. Gel filtration on Sephadex® G-200 separated the sera into three main components. Strong cross-reactivity between species was demonstrated by immunoelectrophoresis.

ARYLAMIDASE OF HUMAN DUODENUM

Bahal, F. J., and G. H. Little, Clin. Chim. Acta **21**, 347–355 (1968).

Starch-gel electrophoresis characterized the arylamidase of duodenum, liver, pancreas and kidney which had been separated by column chromatography. Electrophoresis was performed in a 0.15M Tris–borate buffer, pH 8.6, for 16 hours at 4.0 volts/cm. A temperature of 40°C was maintained. The separated enzymes were detected by staining with a solution containing amino acid βNA, pH 7.3 for one hour, after which solid Fast Blue B was added.

HOMOLOGIES BETWEEN ISOENZYMES OF FISHES AND THOSE OF HIGHER VERTEBRATES

Bailey, G. S., and A. C. Wilson, J. Biol. Chem. **243**, 5843–5853 (1968).

Lactate dehydrogenase isozymes of trout liver and heart were partially puri- fied by ammonium sulfate precipitation and DEAE–cellulose column chromato- graphy. The isolated enzymes were characterized by starch-gel electrophoresis, gel filtration on Sephadex® G-200, and by reactivity with specific antisera prepared in rats. Starch-gel electrophoresis was carried out at pH 5.2, 5.8, 6.0, and 7.0 in a citrate phosphate buffer. Electrophoresis was continued for 17 to 24 hours at a voltage gradient of 13 volts/cm at 4°C. Two major groups of isoenzymes were detected in the heart and muscle group. The lactate dehydrogenase of different species of trout were characterized by differences in mobility as well as spacing of the isoenzymes.

THE MOLECULAR WEIGHT AND OTHER PROPERTIES OF ASPARTATE AMINOTRANSFERASE FROM PIG HEART MUSCLE

Banks, B. E., C. Doonan, A. J. Lawrence, and C. A. Vernon, Eur. J. Biochem. **5**, 528–539 (1968).

Soluble aspartate aminotransferase of pig heart muscle was purified by three different methods including ammonium sulfate precipitation, adsorption with calcium phosphate gel and DEAE–cellulose column chromatography. The purified enzyme was subjected to starch-gel electrophoresis according to the method of Smithies at pH 8.6, 8.0 and 7.5. At pH 8.6 the enzyme appeared to be homogeneous, at pH 8.0 the enzyme separated into four bands, and at pH 7.0 four bands with mobilities different from those found at pH 8.0 were noted. The molecular weight was estimated to be 73,200 to 77,500 from re- sults obtained from Sephadex® column chromatography.

KINETIC PROPERTIES OF RABBIT TESTICULAR LACTATE DEHYDROGENASE

Battellino, L. J., F. R. Jaime, and A. Blanco, J. Biol. Chem. **243**, 5185– 5192 (1968).

Lactic dehydrogenase X (extra testicular isoenzyme) from rabbit sperm and testes, LDH-1 from heart, and LDH-5 from liver were separated and partially purified by preparative starch-gel electrophoresis. Electrophoretic separation was possible on a 12% gel prepared in 0.03M borate buffer, pH 8.6, in the vertical cell of Boyer and Hiner. The discontinuous buffer system of Poulik, a phosphate–citrate buffer, pH 7.0, and a Tris–HCl buffer, pH 8.0, were used. A voltage of 6 volts/cm was applied for 14 hours at 4°C. K_m, optimum substrate concentrations for α-ketobutyrate, lactate, substrate inhibition, and effects of metabolites on reaction rates indicate that LDH-X is a distinct molecular form.

LACTATE DEHYDROGENASE GENES IN RODENTS

Bauer, E. W., and D. L. Pattie, Nature **218**, 341–343 (1968).

Starch-gel electrophoresis has revealed a phylogenetic distribution of a lactate dehydrogenase control gene in rodents. Lysates of rodent erythrocytes and tissues were electrophoresed in a sodium phosphate buffer (0.015M, pH 7.0 for gel, and 0.04M, pH 7.0 for electrode vessels). A potential of 2.7 volts/cm for 16 hours at 2°C was applied for separations. The separated enzymes were located by staining according to the Vesell method.

POLYMORPHISM OF ALBUMIN-LIKE PROTEINS IN THE AMERICAN TETRA-PLOID FROG ODONTOPHBYNUS AMERICANUS (SALIENTIA: CERATO-PHRYDIDAE)

Becak, W., A. R. Schwantes, and M. L. B. Schwantes, J. Exp. Zool. **168**, 473–476 (1968).

An albumin-like plasma protein of a tetraploid South American species of frog was examined. Proteins were separated by horizontal starch-gel electrophoresis with a 0.081M Tris–citrate–borate buffer, pH 8.6, at 4 volts/cm at 4°C. No allelic polymorphism at this locus was indicated in the congeneric diploid species O. *cultripes*. Five phenotypes were found in 141 frogs of the O. *americanus*.

ISOENZYME VON LACTAT-DEHYDROGENASE IN STERNALMARK- UND VENENBLUTSERUM BEI INNEREN KRANKHEITEN

Chury, Z., J. Továrek, and J. Vojková, Z. Klin. Chem. **6**, 92–95 (1968) **German.**

The lactic dehydrogenase isoenzymes in serum of bone marrow and venous blood from 44 patients and the erythrocytes and nucleated cells of bone marrow were studied by starch-gel electrophoresis. The technique of Wieme was used. Bone marrow serum contained high LDH$_3$, LDH$_4$ and LDH$_5$ isoenzymes.

STARCH GEL ELECTROPHORESIS OF LENS PROTEINS TREATED WITH UREA

Cristini, G., and L. Negroni, Exp. Eye Res. **7**, 1216–1220 (1968).

Whole lens extracts treated in urea were analyzed by starch-gel electrophoresis. A 16.9% starch gel prepared in 0.05M sodium acetate buffer, pH 5.6, and containing 7M urea run in a 0.5M sodium acetate buffer at 240 volts (15 to 20 mA) was used for separations, separation being toward the cathode. Whole lens cortex and nucleus of prebirth and postnatal lens tissue show only one boundary in the ultracentrifuge after homogenization with urea. Starch-gel electrophoresis separated a large number of fractions differing in mobility and charge. Electrophoresis of α- and β-crystallin, degraded in urea, revealed many bands.

MUSCLE LACTATE DEHYDROGENASE ISOENZYMES IN HEREDITY MYOPATHIES

Emery, A. E. H., J. Neurol. Sci. **7**, 137–148 (1968).

Starch-gel electrophoresis separated the LDH isoenzymes of fresh biopsy tissue of superficial limb muscles from patients with various kinds of muscular dystrophy. Electrophoresis was performed in a vertical direction at 4°C at 4.0 volts/cm for 18 hours. A 1 mM EDTA–25mM boric acid–45mM Tris buffer, pH 8.5, was used. A significant reduction in LDH-5 was noted in muscular dystrophy cases. The alteration of enzyme pattern was not specific for any given type of muscular dystrophy.

PEROXIDASES FROM THE EXTREME DWARF TOMATO PLANT: IDENTIFICATION, ISOLATION, AND PARTIAL PURIFICATION

Evans, J. J., Plant Physiol. **43**, 1037–1041 (1968).

The dwarf tomato shoot (*Lycopersicon esculentum*) was shown to contain 12 peroxidases when electrophoresed on starch gels in a borate buffer, pH 8.8 (5.5–6.0 volts/cm, three hours at room temperature). Four of the major isozymes were isolated and partially purified using acetone and ammonium sulfate precipitation, preparative starch-gel electrophoresis and DEAE– and CM–cellulose column chromatography. Three of these isolated peroxidases migrated toward the anode and one toward the cathode.

MYÉLOME AVEC PARAPROTÉINÉ SERIQUE γ ET ÉLIMINATION URINAIRE D'UN FRAGMENT DE γ_G DÉPOURVU DE CHAINES LEGERES

Fine, J. M., M. M. Zakin, A. Faure, and G. A. Boffa, Rev. Franc. d'Études Clin. Biol. **13**, 175–178 (1968) **French.**

A γ_G paraprotein with a sedimentation constant of 65 and a Lambda light chain Bence-Jones protein was isolated from the serum of a myeloma patient. Two anti-Fab and Fe fragments could be isolated from the urine. Paper and immunoelectrophoresis were two techniques used to aid in the identification of these proteins. Starch gel electrophoresis in formic acid buffer, pH 31, containing 8M urea was used to characterize the chains of the isolated proteins following separation on Sephadex® G-100.

DISTINCTIONS BETWEEN INTESTINAL AND PLACENTAL ISOENZYMES OF ALKALINE PHOSPHATASE

Fishman, W. H., N. R. Inglis, and N. K. Ghosh, Clin. Chim. Acta **19**, 71–79 (1968).

Horizontal starch-gel electrophoresis using the discontinuous Tris–citrate–lithium borate buffer system, containing 17mM magnesium chloride (pH 8.0), separated placenta alkaline phosphatase enzymes. A constant voltage of 150 with current of 36 mA was applied for 19 hours in order to separate the isoenzymes. The chamber buffer, pH 8.6, contained 120mM lithium hydroxide–380 mM boric acid and 17.0mM magnesium chloride.

HEAT DENATURATION OF THE OVOMUCINLYSOZYME ELECTROSTATIC COMPLEX–A SOURCE OF DAMAGE TO THE WHIPPING PROPERTIES OF PASTEURIZED EGG WHITES

Garibaldi, J. A., J. W. Donovan, J. G. Davis, and S. L. Cimino, J. Food Sci. **33**, 514–524 (1968).

The changes that occur in the whipping properties of egg whites upon pasteurization were investigated by centrifugation, whip test, viscosity, surface tension, and starch-gel electrophoresis. Starch-gel electrophoresis was performed horizontally in the discontinuous buffer system of Poulik. The 10% gels were prepared in 0.075M Tris–0.005M citric acid containing 2M urea, pH 8.5. The electrode buffer was 0.03M sodium borate–0.18M boric acid, pH 8.1. Electrophoresis was continued for 17 hours at 8 volts/cm. Changes were evident in egg white stabilized at 54° and 56°C. Decreases in live "18" material, decrease in lysozyme, and increase in amount of material remaining at the origin were noted.

ASYCHRONOUS ACTIVATION OF PARENTAL ALLELES AT THE TISSUE-SPECIFIC GENE LOCI OBSERVED ON HYBRID TROUT DURING EARLY DEVELOPMENT

Hitzeroth, H., J. Klose, S. Ohno, and U. Wolf, Biochem. Genet. **1**, 287–300 (1968).

Starch-gel electrophoresis on gels prepared in 0.03M borate buffer, pH 8.6, separated lactate dehydrogenase isoenzymes of total homogenates of hybrid trout produced from the mating of *Salmo trutta* and *Salmo irrideus*. The presence of $LDHA^2$ subunits in the whole embryo extracts was detectable from a very early stage of development. The existence of both parental alleles became evident 70 days post-hatching.

THE HEMOGLOBIN HETEROGENEITY OF THE VIRGINIA WHITE-TAILED DEER: A POSSIBLE GENETIC EXPLANATION

Huisman, T. H. J., A. M. Dozy, M. H. Blunt, and F. A. Hayes, Arch. Biochem. Biophys. **127**, 711–717 (1968).

Starch-gel electrophoresis and DEAE–Sephadex® columns were used to study the hemoglobin types of over 150 Virginia white-tailed deer. Starch-gel electrophoresis was performed at room temperature in a Tris–EDTA borate buffer, pH 9.0. The gel was prepared in a Tris–EDTA borate buffer, pH 8.1. Electrophoresis was carried overnight at 150 volts (12–15 mA). Seventeen different types of hemoglobin heterogeneity were recognized. Hemoglobin chains isolated by Sephadex® columns were subjected to starch-gel electrophoresis in urea. Subunits were identified, and a genetic explanation presented for hemoglobin heterogeneity.

A STARCH GEL ELECTROPHORETIC DEMONSTRATION OF THE EFFECT OF pH ON THE AGGREGATION OF ARGININE-RICH HISTONES

Johns, E. W., J. Chromatog. **33**, 563–565 (1968).

Electrophoresis, performed on starch gels at varying pH values (pH 2, 3, 4, 5, 6, 7) according to a modification of Johns, et al. [Biochem. J. **80**, 189 (1961)] evaluated the mobility of arginine-rich f_3 histones. As the pH value increased, the f_3 histones yielded a multiplicity of bands. The f_3 histones were the only fractions to exhibit this kind of aggregation.

ISOZYME VARIABILITY IN SPECIES OF THE GENUS DROSOPHILA: V. EJACULATORY BULB ESTERASES IN DROSOPHILA PHYLOGENY

Johnson, F. M., and S. Bealle, Biochem. Genet. **2**, 1–18 (1968).

The article describes the esterase isoenzyme patterns of the ejaculatory bulbs of some species of Drosophila. Starch-gel electrophoresis of homogenates of ejaculatory bulbs was performed at a voltage gradient of 10 to 15 volts/cm in the discontinuous buffer system of Poulik. Detection of esterase activity was possible by incubation of the starch gels in a solution of 100 ml sodium phosphate, 1.5 ml of 1% α-naphthyl acetate (w/v) in acetone–water (1:1), and 40 mg Fast Garnet GBC salt. The genus Sophophora did not have one type of esterase.

HEMOGLOBIN TYPES OF ADULT, FETAL AND NEWBORN SUBHUMAN PRIMATES: MACACA SPECIOSA

Kitchen, H., J. W. Eaton, and V. G. Stenger, Arch. Biochem. Biophys. **123**, 227–234 (1968).

Starch-gel electrophoresis using a 0.06M Tris–citrate, pH 9.0, in the

gel and 0.12M barbital buffer, pH 8.6, in the electrode vessel identified hemoglobin types of adult, newborn and juvenile animals. Two hemoglobin types
were identified in both sexes. The proportion of these hemoglobins varied
from one animal to another. The γ- and β-chains were separated by starch-
gel electrophoresis in 6M urea, pH 8.1 after procedure of Chernoff and Pettit
[Blood **24**, 750 (1964)]. Structural differences are described.

GLUCOSE 6-PHOSPHATE DEHYDROGENASE IN DROSOPHILA, A SEX-IN-FLUENCED ELECTROPHORETIC VARIANT

Komma, D. J., Biochem. Genet. **1**, 229–237 (1968).

A variant of glucose-6-phosphate dehydrogenase in *Drosophila melanogaster*
showed a different electrophoretic migration pattern on starch gels. Electrophoresis was performed in starch gels made in a buffer of 0.11M boric acid
and 0.002M EDTA pH 8.3–8.5. Before the gels were degassed, they incorporated one milligram of nicotinamide adenine dinucleotide. Electrophoresis was
carried out in a buffer of 0.105 Tris, 0.075M boric acid and 0.002M EDTA,
pH 8.5.

DIE ELEKTROPHORETISCHE BEWEGLICHKEIT DER ANTIGEN—KOMPONTEN EINIGER ADENOVIRUS-TYPEN

Lengyel, A., and W. A. K. Schmidt, Arch. Ges. Virusforsch. **24**, 419–424
(1968) **German.**

Six adenovirus strains were studied electrophoretically according to the technique of Schmidt [Arch. Ges. Virusforsch. **20**, 11 (1967)]. The hemagglutinin
(3 strains), virion (4 strains), and hexon antigen (6 strains) were examined
using HCl–Veronal–acetate buffer, pH 8.4, ionic strength 0.12. Hemagglutinin
was separated from the virion of one type 9, a type 9-15 strain and prototypes
of 15 and 19. No correlation between hemagglutinin ability and mobility was
evident.

CHARACTERIZATION OF CHOLESTEROL-BINDING GLOBULIN BY MODI-FIED ZONE ELECTROPHORESIS AND O-(DIETHYLAMINOETHYL) CELLULOSE CHROMATOGRAPHY

Ling, N. S., and T. Krasteff, Proc. Nat. Acad. Sci. U.S. **60**, 928–935
(1968).

Zone electrophoresis, at pH 8.6, as well as DEAE-cellulose chromatography
isolated and characterized serum transcholesterin. DEAE-cellulose chromatography was performed at pH 4.75 and pH 8.20 with a linear gradient of 1M
sodium chloride. The albumin, alpha and beta globulins were resolved into a
number of subfractions by this technique. Vertical zone electrophoresis on
starch at pH 8.6 resolved lipoprotein-free serum into discrete protein compon-

ents. The transcholesteria activity was localized in one of the alpha-1 and all of the alpha-2 globulin regions. Whole serum tricholesteria activity was not well separated.

STARCH GEL ELECTROPHORESIS OF MOUSE PITUITARY GONADOTRO-PHINS

Lloyd, H. M., J. Endocrinol. **40,** 313–323 (1968).

Mouse pituitary glands were frozen and homogenized in Tris–EDTA–boric acid buffer, pH 8.9 (250 ml of 0.2M Tris; 0.5 g EDTA, 10 ml of 0.4M boric acid; water to one liter). The extract was clarified by ultracentrifugation at 100,000 g for three hours. Starch gel was prepared in the above buffer. Electrophoresis was performed in boric-borate buffer, pH 8.6 (6.65 g boric acid, 10.8 g sodium tetraborate, water to one liter). The gel was held vertically for one hour for electrophoresis using 300 volts and 40 mA. It was then changed to the horizontal position for 15 hours using 120 volts and 15 mA at 20°C. The extracts were separated into 16–17 fractions which were removed from the gel by recovery electrophoresis [Clin. Chim. Acta **9,** 192 (1964)].

MULTIPLE FORMS OF ALCOHOL DEHYDROGENASE IN SACCHAROMYCES CEREVISIAE: I. PHYSIOLOGICAL CONTROL OF ADH-2 AND PROPERTIES OF ADH-2 AND ADH-4

Lutstorf, U., and R. Megnet, Arch. Biochem. Biophys. **126,** 933–944 (1968).

Vertical starch-gel electrophoresis was used to separate the alcohol dehydro-genase enzymes of S. cerevisiae grown on glucose yeast extract media at 30°C. Electrophoresis was performed in a Tris–citrate buffer, pH 8.4, at 8 volts/cm for 14 hours (4°C). The gel was prepared in the same buffer, with the concentration of starch being 11%. Five different strains were analyzed and shown to produce multiple bands of activity.

SUBUNIT STRUCTURE OF BOVINE α-CRYSTALLIN AND ALBUMINOID

Mehta, P. D., and H. Maisel, Exp. Eye Res. **7,** 265–268 (1968).

Bovine albuminoid and α-crystallin isolated from embryonic and adult lens were examined by 8M urea starch-gel electrophoresis. Starch-gel electrophoresis was performed on gels containing 8M urea in the discontinuous buffer system of Poulik. Tris–phosphate buffer gels, pH 8.6, were also used. Cellulose-acetate electrophoresis of adult and embryonic bovine lens tissue separated five bands from adult and four from embryonic lens. Urea starch-gel electrophoresis separated 13 bands from adult lens and 11 from embryonic tissue. Adult α-crystallin and albuminoid differed in migration and total number of bands separated.

PHYSICAL AND CHEMICAL STUDIES ON GLYCOPROTEINS: III. THE MICRO-HETEROGENEITY OF FETUIN, A FETAL CALF SERUM GLYCOPROTEIN

Oshiro, Y., and E. H. Eylar, Arch. Biochem. Biophys. **127,** 476–489 (1968).

Fetuin, isolated by alcohol–metal ion fractionation from calf serum, has been resolved into eight components by starch-gel electrophoresis. Electrophoresis was performed vertically in a 0.2M sodium acetate buffer, pH 4.2. A constant voltage of 200 to 210 was applied. Microheterogeneity was demonstrated by immunoelectrophoresis according to the Schmid method.

HEMOGLOBULIN POLYMORPHISM IN THE DEER MOUSE, PEROMYSCUS MANICULATUS

Rasmussen, D. I., J. N. Jensen, and R. K. Koehn, Biochem. Genet. **2,** 87–92 (1968).

Horizontal starch-gel electrophoresis on 12% starch gels revealed three hemo-globulin phenotypes in blood of natural populations of the northern Arizona deer mouse. The gels were prepared in pH 8.6 borate buffer, 0.03M, with borate buffer 0.3M used as the electrolyte. A potential gradient of 9 volts/cm was applied at room temperature. The separated hemoglobins were visible when stained with a saturated solution of Amido Black.

PROTEINASES IN HUMAN CEREBROSPINAL FLUID

Reikkinen, P. J., and U. K. Rinne, J. Neurological Science **7,** 97–106 (1968).

Human cerebrospinal fluid was collected, concentrated and fractionated by DEAE–CM–cellulose and Sephadex® G-100 chromatography. Three proteinases were identified by appropriate stain techniques. The electrophoretic mobility of the proteinases was determined on starch gel using a 0.025M McIlvaine buffer, pH 6.0. The pH optimum was 5.0, 7.1 and 8.0 respectively.

STARCH-GEL ELECTROPHORESIS OF BACTERIAL CELL-FREE EXTRACTS

Robinson, K., Lab. Pract. **17,** 196–200 (1968).

The effect of varying experimental conditions on mobility of bacterial cell-free enzymes in starch gels was investigated. Cell-free extracts of *Microbacter-ium lacticum* were separated on starch gels of different concentrations (11.2, 12.3 and 13.5%) in the discontinuous buffer of Poulik adjusted to pH 9.0, 8.6 and 8.0. Increased mobility was obtained with increased gel buffer concentra-tion. Variability of protein patterns was noted for different methods of dis-ruption.

ACTION ANTI-PROTÉNASE DE l'α_2-MACROGLOBULINE

Steinbuch, M., C. Blatrix, and F. Josso, Rev. Franc. d'Études Clin. Biol. **13**, 179–186 (1968) **French.**

Purified α_2-macroglobulin can inhibit the clotting activity of thrombin but does not alter its esterase activity. It was demonstrated by starch-gel electrophoresis that α_2-macroglobulin combines with thrombin [J. Clin. Invest. **37**, 1323 (1958)].

MALIC DEHYDROGENASES IN CORN-ROOT TIPS

Ting, I. P., Arch. Biochem. Biophys. **126**, 1–7 (1968).

Starch-gel electrophoresis characterized three malic dehydrogenase isoenzymes isolated from 2½-day-old corn-root tips. DEAE-cellulose column chromatography separated the mitochondrial and soluble enzymes from corn-root tip homogenates. Starch-gel electrophoresis subsequently characterized the mobility of the enzymes. Electrophoresis was performed in a phosphate–citrate buffer, pH 7.0, according to the procedure of Fine and Costello [Methods of Enzymology, VD. **VI**, 608 (1963)].

STUDY OF THE CHROMATIN ACIDIC PROTEINS OF RAT LIVER: HETERO-GENEITY AND COMPLEX FORMATION WITH HISTONES

Wang, T. Y., and E. W. Jones, Arch. Biochem. Biophys. **124**, 176–183 (1968).

The nonhistone chromatin acidic proteins of rat liver were fractionated on DEAE-cellulose and studied by starch-gel electrophoresis. Starch-gel electrophoresis was performed in a gel made in 0.005M Tris–citrate buffer, pH 8.6, containing 0.01M 2-mercaptoethanol. The electrode buffer was 0.3M borate buffer, pH 8.3. The amino acid analysis of the DEAE-cellulose fractions revealed that an acidic amino acid residue predominated. These results correlated with the mobility observed on the starch-gel electrophoretograms.

INHERITANCE OF FROG LACTATE DEHYDROGENASE PATTERNS AND THE PERSISTENCE OF MATERNAL ISOZYMES DURING DEVELOPMENT

Wright, D. A., and F. H. Moyer, J. Exp. Zool. **167**, 197–206 (1968).

A back-cross of a female heterozygous for the most negatively charged unit (B) of LDH was made. Results showed that the structure of the B unit of LDH is controlled by at least three alleles, that the LDH phenotype appears before hatching, and that the maternal LDH persists until eleven days after the tadpole begins feeding. Tissue homogenates were separated by starch-gel electrophoresis in citrate–phosphate buffer, pH 7.2. The LDH isozymes were identified by staining with a solution containing 0.1M lithium lactate, 0.6 mg/ml NAD, 0.25

mg/ml Nitro-BT, 0.02 mg/ml phenazine methosulfate, and 0.005M KCN in 0.25M Tris–HCl buffer, pH 7.4.

THE ONTOGENY OF LACTATE DEHYDROGENASE IN THE CHICK CORNEA

Alcorn, S., and H. Marsel, J. Exp. Zool. **165,** 309–316 (1967).

The LDH isozyme pattern changes during the development of the chick cornea. Prior to the fourteenth day of development, LDH-1-3 is predominant; in later development there is an increase in LDH-4 and 5. At hatching, LDH 2-5 stained with almost equal intensity. LDH-1 is now the less active form. Vertical starch-gel electrophoresis was used to separate the LDH isozymes. The starch gel was prepared in 0.076M Tris–citrate buffer, pH 8.6. The bridge buffer was 0.05M borate, pH 8.6. Electrophoresis was done at 6 volts/cm for 18 hours at 4°C. The gel was stained with a medium of: 45 mg NAD, 37 mg p-Nitro Blue tetrazolium salt, 3 mg phenazine methosulphate, 113 ml distilled water, 15 ml 0.5M Tris–HCl buffer, pH 7.4, 7.5 ml 0.1M potassium cyanide, 15 ml 0.5M sodium lactate pH 7.4. The gels were stained for 30 minutes.

THE HOMOGENEOUS γG-IMMUNOGLOBULIN PRODUCED BY MOUSE PLASMACYTOMA 5563 AND ITS SUBSEQUENT HETEROGENEITY IN SERUM

Awdeh, Z. L., B. A. Askonas, and A. R. Williamson, Biochem. J. **102,** 548–553 (1967).

A single molecular species of γG-immunoglobulin is produced by the mouse plasma cell tumor 5563. The heterogeneity of this globulin in tumor-bearing mice is due to changes in the charge of the protein produced in the serum environment. These changes have been reproduced in vitro. Both the chromatographic and electrophoretic characteristics were changed. Starch-gel electrophoresis was performed with 0.05M glycine buffer, pH 8.9, in the gel and 0.1M sodium borate buffer, pH 8.9, in the electrode vessels.

BISALBUMINEMIA OF THE FAST TYPE WITH A HOMOZYGATE

Bell, H. E., S. F. Nicholson, and Z. R. Thompson, Clin. Chim. Acta **15,** 247–252 (1967).

Bisalbuminemia is the condition in which two distinct albumin bands can be separated by electrophoresis at pH 8.6. This article is a report of an Indian family with bisalbuminemia of the fast type. A suspension of 77.5 g Connaught hydrolyzed starch in 620 ml of 0.03M Tris–borate, pH 8.05, was poured into a vertical column. Electrophoresis was done for three hours at 1100 volts. Cellulose-acetate electrophoresis was done in a pH 8.6 barbital buffer, ionic strength 0.05. The sera were separated in one hour at 250 volts and the strips were stained with Ponceau S. The background was cleared with 5% acetic acid.

"GALACTOSE DEHYDROGENASE", "NOTHING DEHYDROGENASE", AND ALCOHOL DEHYDROGENASE: INTERRELATION

Beutler, E., Science **156**, 1516–1517 (1967).

Similarities of electrophoretic mobility in starch gel led the author to suspect that a previously reported new pathway of galactose oxidation is actually due to an alcohol dehydrogenase. Contamination in commercial substrates by primary alcohols gave the original investigators misleading results.

THE CHARACTERIZATION OF THE WHEY PROTEINS OF GUINEA PIG MILK

Brew, K., and P. N. Campbell, Biochem. J. **102**, 258–264 (1967).

Cellulose-acetate, starch-gel, and polyacrylamide-disc electrophoresis techniques characterized the whey proteins in guinea pig milk. Cellulose-acetate electrophoresis was performed in a pH 8.6 Veronal buffer, ionic strength 0.05. Starch gel was performed in the discontinuous buffer system of Poulik [Nature **180**, 1477 (1967)]. Disc electrophoresis was performed by the method of Davis [Ann. N. Y. Acad. Sci. **121**, 404 (1965)]. The whey proteins were resolved on all three media. The relationships between bovine and guinea pig whey are noted.

MICRO-STARCH GEL ELECTROPHORESIS UNDER PETROLEUM ETHER

Brody, J. I., Clin. Chim. Acta **18**, 205 (1967).

Microelectrophoresis was performed on 13.3% hydrolyzed starch in borate buffer (pH 8.6, 0.023M H_3BO_4, 0.092M NaOH). The electrolyte was a borate buffer (pH 8.6, 0.3M, 0.06M NaOH). Microscope slides (1 x 3 inches) were used for the gel slides. Only 5 μl of sample was required. One percent Amido Black was the stain used. The center compartment of the electrophoresis chamber was filled with petroleum ether, which acts as a coolant, so that the electrophoresis could be performed at room temperature. Separation of sera was complete in 40 minutes.

HEXOKINASE ISOENZYMES IN LIVER AND ADIPOSE TISSUE OF MAN AND DOG

Brown, J., Science **155**, 205–207 (1967).

A hexokinase, with a low Michaelis constant, was found in extracts of human and dog liver but not in the rat. Vertical starch-gel electrophoresis was performed according to a modification of Katzer's procedure [Proc. Nat. Acad. Sci. U.S. **54**, 1218 (1965)]. A 0.02M Veronal buffer, pH 7.4, containing 4.5mM

EDTA and 2.3mM mercaptoethanol was used to effect separation. The hexo-kinase isoenzymes could be distinguished from isoenzymes of a liver glucokinase by substrate specific staining.

DEVELOPMENTAL AND POPULATION VARIATION IN ELECTROPHORETIC PROPERTIES OF DEHYDROGENASES, HYDROLASES AND OTHER BLOOD PROTEINS OF THE HOUSE SPARROW PASSER DOMESTICUS

Bush, F. M., Comp. Biochem. Physiol. **22**, 273–287 (1967).

Vertical starch-gel electrophoresis in gels prepared in 0.02M sodium borate buffer, pH 8.6, or 0.045M sodium barbiturate, pH 8.0, separated some blood proteins of *Passer domesticus*. Lipoprotein and eight proteins were separated from plasma; one of these enzymes was also separated from erythrocytes. Non-specific esterase, cholinesterase napthylamidase, alkaline and acid phos-phatase, β-glucuronidase and amine oxidase were found in plasma. Lactate dehydrogenase was found in plasma and erythrocytes.

GENETIC VARIANT OF HUMAN ERYTHROCYTE MALATE DEHYDROGENASE

Davidson, R. G., and J. A. Cortner, Nature **215**, 761–762 (1967).

Horizontal starch-gel electrophoresis of erythrocyte malate dehydrogenase re-vealed the existence of an inherited genetic variant. The hemolysates were surveyed for variants on horizontal starch-gel electrophoresis in a gel pre-pared in a 0.01M phosphate buffer and a 0.2M phosphate buffer in the electrophoresis trays. The pH of both buffers was 7.0. The separations were achieved at 10 volts/cm for five hours. Vertical gels were found to improve the separation of mitochondrial enzymes. These were run in a phosphate–citrate buffer, pH 7.0. Starch-block electrophoresis and DEAE·cellulose chro-matography were used to purify the enzyme.

FOUR ISOZYMIC FORMS OF A RAT PEPTIDASE RESEMBLING KALLIKREIN PURIFIED FROM THE RAT SUBMANDIBULAR GLAND

Ekfors, T. O., P. J. Reikkinen, T. Malmiharji, and V. K. Hopsu-Havu, Hoppe-Seylers Z. Physiol. Chem. **348**, 111–118 (1967).

Four isozymes of a peptidase which will hydrolyze Nα benzoyl–DL-arginine–p-nitroanilide and β-naphthylamide were purified by precipitation and chro-matographic methods from the rat submandibular gland. Starch-gel electro-phoresis and DEAE-cellulose chromatography showed the four components to be different.

THE POLYMORPHISM OF PREALBUMINS AND α_1-ANTITRYPSIN IN HUMAN SERA

Fagerhol, M. K., and C. B. Laurell, Clin. Chim. Acta **16**, 199–203 (1967).

Prealbumins of human sera have been separated by starch-gel electrophoresis in a discontinuous buffer at pH 4.5 for the starch and 9.0 in the electrode vessels. After electrophoresis, the prealbumins were electrophoresed into agarose gel containing anti-α_1-antitrypsin. Genetic polymorphism of α_1-antitrypsin was observed and a possible explanation for the precipitin pattern was proposed.

HETEROGENEITY OF AVIAN OVOMUCOIDS

Feeney, R. E., D. T. Osuga, and H. Maeda, Arch. Biochem. Biophys. **119**, 124–132 (1967).

The ovomucoids from eggs of different breeds and genetic lines of chickens were examined by starch-gel electrophoresis. Electrophoresis was carried out according to the technique of Poulik [Nature **180**, 1477 (1957)] on ovomucoids isolated with trichloroacetic acid. The discontinuous buffer system of Poulik at 7 volts/cm and 18 mA or 35 volts/cm and 75 mA, the pH 5.1 discontinuous buffer system of Lush, 17 volts/cm and 50 mA, and the pH 4.4 buffer of Wise at 10 volts/cm and 15 mA were used. The separated proteins were stained with aniline blue-black or nigrosine. The ovomucoids from eggs of two of eleven strains of chicken showed definite differences in electrophoretic patterns.

INSULIN BINDING PROTEINS IN NORMAL SERUM: QUALITATIVE AND QUANTITATIVE STUDIES IN VIVO IN CATS

Gjedde, F., Acta Physiol. Scand. **70**, 57–68 (1967).

Studies with radioactive I^{125}-labeled insulin suggest that it is bound in vivo to a fast-moving alpha-one globulin. Starch-block and starch-gel electrophoresis were used to characterize and define the globulin.

ELECTROPHORETIC AND KINETIC PROPERTIES OF RANA PIPIENS LACTATE DEHYDROGENASE ISOZYMES

Goldberg, E., and T. Wuntch, J. Exp. Zool. **165**, 101–110 (1967).

Lactate dehydrogenase of the leopard frog exists in at least three isozymes. Extracts of heart contained all three isozymes, extracts from the gastroenemius from muscle contained only one isozyme. Fresh 10% (w/v) homogenates of the tissues were prepared in 0.05M sodium phosphate buffer, pH 7.0. In addition, testes extracts were sonicated. After centrifugation for 20 minutes at 25,000 g, the LDH in the homogenates were separated by starch-gel electrophoresis. Samples of the homogenates were spotted on filter paper wicks and inserted into the gel. Electrophoresis was performed at pH 6.0, 7.0, and 8.6 according to the technique of Fine and Costello [Methods of Enzymology 6, 958–972, New York: Academic Press, 1963].

CREATINE PHOSPHOKINASE IN THYROID: ISOENZYME COMPOSITION COMPARED WITH OTHER TISSUES

Graig, F. A., Science 156, 254–255 (1967).

Starch-gel electrophoresis separated the isoenzymes of organ extracts and a commercial preparation of creatine phosphokinase (CPK) for comparison. Electrophoresis was done in a pH 8.0 borate buffer, ionic strength 0.12, for 16 hours in a 5 volt/cm current. The gel was neutralized after electrophoresis by washing with 0.2M glycine buffer, pH 6.75, for the staining reaction. The staining mixture gave a blue formazan at CPK sites; the staining solution consisted of: creatine phosphate, 7.6×10^{-3}M; ADP 1.1×10^{-3}M; Mg acetate 1.0×10^{-1}M; glucose 3.3×10^{-2}M; NAD 9.2×10^{-4}M; nitroblue tetrazolium 5.9×10^{-4}M; phenazine methosulfate 1.4×10^{-4}M; hexokinase, 4 units/100 ml; glucose–6-phosphate dehydrogenase, 420 units/100 ml, prepared in 1.0M glycine buffer.

ELECTROPHORETIC VARIANTS OF α-GLYCEROPHOSPHATE DEHYDROGENASE IN DROSOPHILA MELANOGASTER

Greth, E. H., Science 151, 1319 (1967).

Using flat-bed electrophoresis equipment made it possible to separate a major component and two slower-moving components of GDH from muscle of inbred D. melanogaster. Processing of two different hybrids revealed five visible components of GDH. The differences in electrophoretic mobility are based on these being two alleles of a genetic locus.

ON THE TISSUE SPECIFICITY AND BIOLOGICAL SIGNIFICANCE OF ALDO-LASE C IN THE CHICKEN

Herskovits, J. J., C. J. Masters, P. M. Wasserman, and N. D. Kaplan, Biochem. Biophys. Res. Commun. 26, 24–29 (1967).

Studies were made on aldolase C of the chicken, and theories on its function were proposed. The enzyme was extracted by homogenizing fresh tissue in 0.01M Tris–HCl buffer, pH 7.5, containing 0.001M EDTA and 0.01M β-mercaptoethanol. The supernate was recovered after centrifugation at 100,000 g for one hour. Starch electrophoresis was done at pH 7.0 at 4°C [Methods Enzymol. 6, 958 (1963)]. A 1.5% agar solution in 50 ml 0.05M pyrophosphate buffer, pH 8.5, was stored in the dark and the following added: 1 ml of 0.3M sodium arsenate, 2.0 ml of 0.2M sodium–1,6–diphosphate, 1 ml of DPN

(30 mg/ml), 1 ml Thiazolyl Blue (10 mg/ml), 0.25 ml phenazine methosulfate (10 mg/ml) and 0.5 ml of triosephosphate dehydrogenase (10 mg/ml). This mixture was poured over the starch block after electrophoresis and developed in the dark for three hours.

HEMOGLOBIN POLYMORPHISM IN CHIMPANZEES AND GIBBONS

Hoffman, H. A., A. S. Gottleib, and W. G. Wisecup, Science **156**, 944–945 (1967).

Starch-gel electrophoresis of carbonmonoxyhemoglobin from chimpanzees and gibbons demonstrated polymorphism. Tris–EDTA–borate gel buffer, pH 9.1, and a borate buffer, pH 8.6, were used in the electrode chambers, at a voltage gradient of 15 volts/cm for 75 minutes. Hemoglobin zones were stained with a benzidine–peroxide mixture.

A BRIEF METHOD FOR THE IDENTIFICATION OF THE α_1-ACID GLYCO-PROTEIN VARIANTS

Hunziker, K., and K. Schmid, Anal. Biochem. **20**, 495–501 (1967).

A rapid method for the identification of the α_1-acid glycoprotein variants of human serum is presented. Starch-gel electrophoresis was performed on perchloric digests in order to separate the α_1-glycoproteins. Electrophoresis was performed using a sodium acetate buffer (ionic strength 0.02, pH 5.1). A potential of 300 volts was sufficient for a gel 24 x 10 x 0.3 cm.

ALKALINE PHOSPHATASE AND LEUCINE AMINOPEPTIDASE ASSOCIATION IN PLASMA OF THE CHICKEN

Law, G. R., Science **156**, 1249–1250 (1967).

Gel electrophoresis showed two variants of leucine aminopeptidase directly associated with two genetically controlled forms of alkaline phosphatase. After electrophoresis in a 0.04M Tris buffer, pH 8.6, with boric acid, at 5 volts/cm for 16 hours, one-half of the gel was stained for leucine aminopeptidase activity, the other half for alkaline phosphatase. It was suggested that the two forms of the enzyme might be due to the presence or absence of a gene controlling the attachment of sialic acid to the enzymes.

HUMAN RED CELL PEPTIDASES

Lewis, W., and H. Harris, Nature **215**, 351–355 (1967).

Red cell peptidases were separated by electrophoresis of distilled water hemolysates on starch gels. A 0.005M Tris–maleate gel buffer and a 0.1M Tris–maleate bridge buffer, pH 7.4, were used. Separation was achieved at 5 volts/cm for 18 hours. The peptidases were located in the sliced gel by staining with a solution containing phosphate buffer, the appropriate peptide, crude *Crotalus adamenteus* venom, horse radish, peroxidase, O-dianisidine hydrochloride and manganese chloride. By using four tripeptides and twelve dipeptides, the authors were able to characterize peptidases from a great many people. A genetic-determined variant of two of these enzymes was found.

GLUTATHIONE REDUCTASE IN RED BLOOD CELLS: VARIANT ASSOCIATED WITH GOUT

Long, W. K., Science **155**, 712–713 (1967).

A variant of glutathione reductase characterized by greater electrophoretic mobility and an autosomal mode of inheritance was demonstrated in erythrocytes of a Negro population by starch-gel electrophoresis. A two-stage staining procedure for glutathione reductase is described.

HAEMOGLOBIN J~BANGKOK~: A CLINICAL HAEMATOLOGICAL AND GENETICAL STUDY

McHenry, A. E., Brit. J. Haematol. **13**, 303–309 (1967).

Hemoglobin J~BANGKOK~ was separated and identified by starch-gel electrophoresis in a pH 8.6 Tris–EDTA–borate buffer. Amido-Black was the stain. Cellulose-acetate strip electrophoresis in a pH 8.6 Veronal buffer, ionic strength 0.025, confirmed the finding. Nine members of one Thai family tested positive for this hemoglobin.

MALATE DEHYDROGENASE ISOZYMES OF THE MARINE SNAIL, ILLYANASSA OBSOLTA

Meizel, S., and C. L. Markert, Arch. Biochem. Biophys. **122**, 753–765 (1967).

Vertical starch-gel electrophoresis separated the isoenzymes of the marine snail, *Illyanassa obsoleta*. The starch gels used were 13 to 14% and were prepared in a pH 7.0 Tris–citrate buffer. The buffer, diluted 1:20, was used in the electrode vessels. Electrophoresis, carried out at a constant voltage of 6 volts/cm for 14 to 18 hours at 4°C, revealed several isoenzymes which were shown to be either mitochondrial or supernatant in nature.

PREPARATION OF STARCH GELS CONTAINING 8 MOLAR UREA

Melamod, M. D., Anal. Biochem. **19**, 187 (1967).

Urea in starch gel improves resolution of some proteins and eliminates multiple bands caused by aggregation of protein subunits. The pH ($\pm$ 0.01) and ionic strength ($\pm$ 0.002) of an 8M urea starch-gel can be maintained by using Zeocarb® 225 resin in the H+ form when preparing the mixture. The final pH is 0.8 units higher than that of the buffer used as the electrolyte.

STUDIES ON COMPOSITION AND LEAKAGE OF PROTEINS AND ESTERASES OF NORMAL RAT LIVER AND MORRIS HEPATOMA

Murray, R. K., H. Kalant, M. Guttman, and H. P. Morris, **5123** T.c., Cancer Res. **27**, 403–411 (1967).

Vertical starch-gel electrophoresis separated proteins and esterases which leaked into the media when 0.5% EDTA was added. Before electrophoresis the media were pooled and dialyzed against eight volumes of water overnight at 4°C. Clarification was done by centrifugation and the supernate was then lyophilized. Electrophoresis was performed in a 0.021M sodium borate buffer, pH 8.4, using a 10 mA current for 16 hours. Amido Black B stained a vertical slice of the gel for proteins. The method of Ecobichon [Pharmacol. **11**, 573 (1963)] was used to locate the esterases.

HETEROGENEITY OF SERUM γ-GLUTAMYL TRANSPEPTIDASE IN HEPATO-BILIARY DISEASES

Orlowski, M., and A. Szczeklik, Clin. Chim. Acta **15**, 387–391 (1967).

Horizontal starch-gel electrophoresis, performed according to the method of Smithies using a discontinuous borate buffer, separated seven isoenzymes from serum of patients with hepatobiliary diseases. The individual isoenzymes were separated by Sephadex® G-200 filtration. Their distribution patterns are discussed.

LINKAGE OF ES-1 AND ES-2 IN THE MOUSE

Papp, R. A., J. Heredity **58**, 186–188 (1967).

This is a report of the effect of Es-2^a, on the esterase pattern of tissues from the Rfm/μn mouse. Es-2^a is a silent allele which expresses no esterase. It was previously described in the serum esterase pattern of the Rfm/μn and fecal mouse. Serum or tissue extracts were applied to Whatman® No. 1 filter paper and imbedded into the starch gel for electrophoresis. The esterases were located by a diazonium dye-coupling reaction. These methods are described by Papp in an earlier paper [J. Heredity **53**, 111–114 (1962)].

ABNORMAL HAEMOGLOBINS IN IRAN OBSERVATION OF A NEW VARIANT: HAEMOGLOBIN J. IRAN (α_2 β_2 77 His → Asp)

Rahbar, S., D. Beale, W. A. Issacs, and W. A. Lehman, Brit. Med. J. **Part 1**, 674 (1967).

A new variant of hemoglobin was identified following pH 8.0 electrophoresis of globulin in starch gel in the presence of 6M urea. Electrophoresis at pH 8.9 and pH 6.5 demonstrated the absence of one histidine in the variant.

PURIFICATION AND PROPERTIES OF HUMAN TRANSFERRIN C AND A SLOW MOVING GENETIC VARIANT

Roop, W. E., J. Biol. Chem. **242,** 2507–2513 (1967).

Starch-gel and immunoelectrophoresis were used to demonstrate the homogeneity of a transferrin C and a variant of it isolated from human plasma. The purification procedure is outlined in detail. Immunoelectrophoresis was done in 1% agar using 0.025M Veronal buffer, pH 8.6. Commercial whole anti-human sera was used to form the precipitin lines. Starch-gel electrophoresis was done in the discontinuous buffer system of Poulik [J. Immunol. **82,** 502 (1959)]. Ultracentrifuge and moving-boundary electrophoresis confirmed the homogeneity of the transferrins.

TISSUE SPECIFIC ESTERASE ISOZYMES OF THE MOUSE (MUS MUSCULUS)

Ruddle, F. H., and L. Harrington, J. Exp. Zool. **166,** 51–64 (1967).

Vertical starch-gel electrophoresis was used to study tissue-specific esterase isoenzymes in 35 organs of the mouse. A discontinuous buffer system was used. The gel buffer was 0.076M Tris and 0.005M citric acid at pH 8.65. The buffer in the vessels was 0.29M boric acid and 0.5 NaOH at pH 8.0. The gel was 14% potato starch. A continuous voltage of 450 volts gave good separation in five to six hours at 4°C. The esterases were stained by the coupling-reaction products of naphthyl substrates with a diazonium salt.

THE EFFECT OF IRON ADDITION TO AVIAN EGG WHITE ON THE BEHAVIOUR OF CONALBUMIN FRACTIONS IN STARCH GEL ELECTROPHORESIS

Stratil, A., Comp. Biochem. Physiol. **22,** 227–233 (1967).

Egg white conalbumins of the type CO^{66} of the chicken *(Gallus gallus)* were separated into ten electrophoretic fractions by starch-gel electrophoresis after injection of Fe $+3$. Electrophoresis was performed in a 16% starch-gel discontinuous buffer. The buffer was prepared according to Kristjansson (Genetics **48,** (1963)]. Vertical electrophoresis was carried out for seven hours at 12 volts/cm. Horizontal separations were carried out at 300 volts for four hours. In addition, autoradiography of the separated conalbumins was used to identify Fe^{+3}-conalbumins.

AUTOSOMAL PHOSPHOGLUCONIC DEHYDROGENASE POLYMORPHISM IN THE CAT (FELIS CATUS L.)

Thuline, H. C., A. C. Morrow, D. E. Norby, and H. G. Motulsky, Science **157**, 431–432 (1967).

Horizontal starch-gel electrophoresis carried out in a Tris–EDTA–boric acid buffer separated isoenzymes of 6-phosphogluconic dehydrogenase of blood of cats. Separations were carried out at 4°C at 350 volts/cm for 18 hours. Cats showed either a single fast band, a single slow band or three bands. These findings were compatible with at least a dimeric structure for the enzyme.

PURIFICATION AND PARTIAL CHARACTERIZATION OF A BASIC PROTEIN FROM PIG BRAIN

Tomasi, L. C., and S. E. Kornguth, J. Biol. Chem. **242**, 4933–4938 (1967).

A basic protein was isolated from pig brains by chloroform–methanol (2:1) and 10% sodium chloride and dilute hydrochloric acid extraction. Increased purity of the protein was achieved by pH and ammonium sulfate fractionation, Sephadex® G-100 gel filtration and chromatography on Amberlite® IRC-50 with a 5 to 20% guanidinium chloride gradient. Starch-gel electrophoresis of the eluted fractions was performed in three different buffer systems: sodium formate (0.025M, pH 3.0) run at 3.75 volts/cm for eight hours; sodium formate (0.025M, pH 3.0 in 8M urea) run at 3.75 volts/cm; and sodium borate (0.026M, pH 8.6) run at 6.25 volts/cm for 20 hours. Polyacrylamide-gel electrophoretic separations were carried out on 15% gel in a β-alanine–acetate buffer, pH 4.3. A current of 3 mA per gel was applied for three hours. Both techniques demonstrated that the purification procedure yielded homogeneous results.

THE BEHAVIOR OF HUMAN GLOBIN IN RELATION TO ITS CONTENTS OF DENATURED PROTEINS

Vodrázka, Z., Z. Hrkal, A. Ceika, and H. Sipalova, Collection Czech. Chem. Commun. **32**, 3250–3261 (1967).

The starch-gel electrophoretic behavior, Sephadex® gel filtration characteristics, and some optical properties of human globulin are described. Electrophoresis was performed at pH 1.9, Sephadex®-gel filtration was performed at pH 1.9 and 4.8. Differences in solubility at pH 7.0 for native globin and heat-denatured globin were observed.

COMPARATIVE ELECTROPHORETIC STUDIES OF THE SEED PROTEINS OF CERTAIN SPECIES OF BRASSICA AND SINAPIS

Vaughan, J. G., and A. Waite, J. Exp. Botany **18**, 100–109 (1967).

Serologic methods and starch-gel electrophoresis characterized the saline-soluble seed proteins of *Brassica compestris* var. automnalis, *B. nigra*, *B. oleracea*, *B. campestris* var. rapa and *Sinapis alba*. Starch-gel electrophoresis was carried out using Smithies' method and a pH 8.5 borate buffer. After the electrophoretic run (3.5 to 4 hours at 4°C, 40 ma), the starch gel was sliced and stained to detect β-galactosidase and esterases using the appropriate substrate. The serologic studies were performed using adsorbed antisera and testing against powdered extract of the seeds. Serological studies indicated greater relationships among *B. campestris*, *B. oleracea* and *B. nigra* than among any one of the species of *S. alba*. Starch-gel electrophoresis revealed no differences among the species.

THE ENZYMES OF HONEY: EXAMINATION BY ION EXCHANGE CHROMATOGRAPHY, GEL FILTRATION AND STARCH-GEL ELECTROPHORESIS

White, J. W., Jr., and I. J. Kushnir, Apicult. Res. **6**, 69–89 (1967).

DEAE-cellulose column chromatography gel filtration on Sephadex® G-200 and starch-gel electrophoresis were used to characterize crude enzymes of bulk honey, comb honeys and honey produced by sugar-feeding of caged bees. Three 9-α-glycosidases were separated from crude honey preparations. Starch-gel electrophoresis using the procedure of Fielber separated 7 to 18 components.

GEL FILTRATION OF ACID CASEIN AND SKIM MILK ON SEPHADEX®

Yaguchi, M., and N. P. Tarassuk, J. Dairy Sci. **50**, 1985–1988 (1967).

The fractionation of casein and skimmed milk on Sephadex® was done to study the nature of the micelle structure. One hundred thirty mg casein were eluted from a Sephadex® G-200 column at 4°C with 0.02M sodium-phosphate pH 8.0, containing 1M NaCl. The absorbancy of the fractions at 280 mμ was determined and then the fractions were pooled into five samples. These samples (A-E) and whole casein (w) were separated by starch-gel electrophoresis at pH 8.6, in 7.0M urea, 2-mercaptoethanol using the modified method of Wake [Biochim. Biophys. Acta **90**, 441 (1964)]. Two other 130 mg samples of casein were treated in a similar manner, except that the column eluant was changed: 0.02M sodium-phosphate, pH 8.0 and pH 8.6, containing 6M urea gave better dissociation of the α_2- and β-casein units.

GEL FILTRATION OF THE WATER SOLUBLE PROTEIN FRACTION OF WHEAT FLOUR

Abbott, D. C., and J. A. Johnson, J. Food Sci. **31**, 38–47 (1966).

A combination of gel filtration and starch-gel electrophoresis was employed to characterize proteins of wheat flour. Water soluble extracts of wheat flour were filtered on Sephadex® G-25, G-50, G-75, G-100 and G-200, while distilled water or lactic acid in various concentrations were used as eluants. The electrophoresis system of Waychick et al. [Arch. Biochem. Biophys. **94**, 477 (1961)] was employed to analyze the eluted proteins. An aluminum lactate buffer containing urea was used. Electrophoresis was performed at 5 volts/cm at 4°C for 13 hours. Proteins were detected by staining with Amido 10B in 5% acetic acid. A single component was isolated by gel filtration and preparative starch-gel electrophoresis.

SOLUBILITY AND ELECTROPHORETIC BEHAVIOUR OF SOME PROTEINS OF POST-MORTEM AGED BOVINE MUSCLE

Aberle, E. D., and R. A. Merkel, J. Food Sci. **31**, 151–160 (1966).

The solubility characteristics and starch-gel electrophoresis of intracellular bovine muscle proteins were investigated during a 336-hour period of post-mortem aging. Starch-gel electrophoresis of muscle homogenates was performed in an apparatus designed by Wake and Baldwin. A 12% gel prepared in a .012M Tris–0.0018M citric acid buffer, pH 8.6, and run in a 0.1M boric acid–0.06M Tris buffer, pH 8.6, was used. A potential of 300 volts (12 mA) was maintained for nine hours. As aging proceeded, changes were evident in the starch-gel patterns. Individual zones became more distinct at 168 and 336 hours postmortem aging. A new protein band was visible at 336 hours postmortem. Ordinarily only 14 bands were visible after staining with Amido Black B.

SERUM ESTERASES OF APODEMUS SYLVATICUS AND MUS MUSCULUS

Arnason, A., and E. M. Pantelouris, Comp. Biochem. Physiol. **19**, 53–61 (1966).

Starch-gel electrophoresis of serum of *Apodemus sylvaticus* and *Mus musculus* revealed 23 esterases of the former and 19 of the latter. Separation was achieved on gels prepared in Tris–citrate buffer, pH 7.6 or pH 8.6, containing 10% borate buffer. The electrode vessels contained a pH 8.7 borate buffer. The electrophoretic run proceeded at 20 mV/cm (2–4 mA) for 2½ hours. The esterases were detected by staining the sliced gel with substrate solution containing naphthol or naphthyl ester.

MODIFIED PROCEDURE OF STARCH GEL ELECTROPHORESIS FOR β-CASEIN PHENOTYPING

Aschaffenburg, R., J. Dairy Sci. **49**, 1284–1285 (1966).

β-casein phenotyping can be done without isolation of the caseins from milk samples. The pH 1.7 buffer was prepared with 30 ml formic acid (98/100%) and 120 ml glacial acid in water to one liter. Whole milk was diluted in an equal volume of 100 g urea in 100 ml water. Gels were prepared as mentioned previously [J. Dairy Sci. **48**, 1524 (1965)] *except* the mercapto-ethanol was omitted and the urea was reduced to 24 g per 80 ml buffer. At 4°C a constant voltage of 220 volts separated the β-casein in 16 hours.

DETERMINATION OF RADIOACTIVE PROTEINS AFTER SEPARATION BY STARCH-GEL ELECTROPHORESIS

Bacci, V., and A. Viti, Gazz. Biochem. **15**, 303–313 (1966).

Autoradiography and gamma counting sensitivities were compared with protein staining to find the best quantitative method of measuring protein fractions separated by starch-gel electrophoresis. I^{131} serum proteins were separated by 5 volts/cm at 2°C. Spectophotometric determinations of Amino Black 10B stained fractions did not give the reproducibility of counting the I^{131}.

SPECIFIC MILK PROTEINS ASSOCIATED WITH RESUMPTION OF DEVELOPMENT BY QUIESCENT BLASTOCYST OF THE LACTATING RED KANGAROO

Bailey, L. F., J. Reprod. Fertility **11**, 473–475 (1966).

Starch-gel electrophoresis was used to examine milk proteins of three lactating red kangaroos (*Megaleia rufa*). The technique of Ferguson and Wallace [Nature **190**, 629 (1961)], was used. Two new specific milk proteins occurring concurrently with the development of the quiescent blastocyst following post-partum mating were detected.

SPECIES, TISSUE AND INDIVIDUAL SPECIFICITY OF LOW IONIC STRENGTH EXTRACTS OF OVIAN MUSCLE AND OTHER ORGANS REVEALED BY STARCH-GEL ELECTROPHORESIS

Baker, C. M. A., Can. J. Biochem. **44**, 853–859 (1966).

Vertical starch-gel electrophoresis was used to separate proteins soluble at low ionic strength. Various muscles and other tissues from five species of birds were examined. Electrophoresis was performed in a borate buffer, pH 8.4; potassium phosphate, pH 7.0; and Tris–EDTA–borate, pH 8.6–8.8. The separated samples were stained for esterases, lactate dehydrogenase and glutamic dehydrogenase with the appropriate substrate specific stain. When stained nonspecifically for protein with Amido Black 10B, most tissues yielded 15 bands.

QUANTITATIVE EVALUATION OF SERUM PROTEINS SEPARATED BY STARCH GEL ELECTROPHORESIS

Battistini, A., Lattante **37**, 97–135 (1966) **Italian.**

Recent advances in electrophoretic separation of serum proteins, with special emphasis on techniques, are discussed. Agar, continuous, cellulose acetate, and immunoelectrophoresis as well as quantitative methods are covered.

PREGNANCY ENZYMES AND PLACENTAL POLYMORPHISM: I. ALKALINE PHOSPHATASE

Beckman, L., G. Bjorling, and C. Christodoulou, Acta Genet. (Basel) **16**, 59–73 (1966).

Vertical starch-gel electrophoresis, using a modification of the discontinuous buffer of Ashton and Barden, was used to separate alkaline phosphatase of serum and placentas of full term pregnant women. Improved separation was possible with this technique. In addition to the three common alkaline phosphatases, four new variants were found. An electrophoretic variation between the main pregnancy bands and another slow-moving alkaline phosphatase was thought to be caused by the presence of a common polypeptide chain.

THE ISOLATION OF SULFUR-RICH FRACTIONS FROM K-CASEIN

Beeby, R., Intern. Dairy Congr. Proc. 17th, 1965, **2**, 99–103 (1966).

Casein was treated with HCl, pH 3.0. Five fractions were isolated from the sediment. Starch-gel electrophoresis further separated the five fractions into several components, some of which were common to each fraction.

LACTIC DEHYDROGENASE ISOZYMES IN LENS AND CORNEA

Bernstein, L., M. Kerrigan, and H. Maisel, Exp. Eye Res. **5**, 309–314 (1966).

Bovine and rabbit lens and cornea were analyzed for lactic dehydrogenase isozymes by starch-gel electrophoresis. Separation was achieved first on filter paper (Whatman® No. 3) in a 0.1M sodium acetate buffer for 16 hours. The area of filter paper containing protein was cut out, placed in starch gel and electrophoresed at right angles to that of paper separation. A boric acid electrode buffer and a 12.5% starch gel prepared in a pH 8.6 Tris–citrate buffer were used. A progressive loss of enzyme activity was evident in the lens of both species with increasing age. However, the isozyme loss of bovine lens was more prominent. Isozyme 1 predominates in the adult bovine lens, isozymes 3–5 in the adult rabbit lens. The cornea of adult bovine lens tissue was characterized by isozymes 3–5, rabbit cornea by enzymes 1–4.

BASIC PROTEINS IN SERRATIA MARCESCENS

Blazsek, V. A., and A. Marx, Naturwissenschaften **53**, 553–554 (1966).

Starch-gel electrophoresis showed a major fast-moving component and five slower-moving components in an acid extract of *Serratia marcscens*. Dried cells were ground in 0.25N HCl and the acid soluble material was precipitated with TCA. The precipitate was dissolved in 0.05M Tris, pH 8.6, and fractionated on a CM-cellulose column with 0.1N HCl. Starch-gel electrophoresis separated the basic proteins eluted.

LACTATE DEHYDROGENASE ISOENZYMES: A DIFFERENCE BETWEEN CUTANEOUS AND MUSCULAR NERVES

Brody, I. A., J. Neurochem. **13**, 975–978 (1966).

Starch-gel electrophoresis according to the method of Smithies in the discontinuous buffer of Poulik was used to separate the lactate dehydrogenase isoenzymes of peripheral nerves and spinal cord of adult guinea pigs and cats. 0.03M sodium chloride was added to gels and to the electrode compartment. A potential gradient of 4 volts/cm was applied for seven hours at 4°C. Upon staining the sliced gel by standard technique, it was evident that normal peripheral nerves differed in LDH isoenzyme patterns. Five main isoenzymes were found in all nerves examined. The fastest-moving isoenzyme (LDH-1) was highest in nerves supplying skeletal muscle than nerves supplying skin. Motor and sensory nerves did not exhibit significant differences in LDH activity.

GENETIC CONTROL OF SALIVARY PROTEIN COMPOSITION

Burgess, R. C., G. E. Connell, C. M. Maclaren, and T. A. Williams, J. Dental Res. 45, Suppl. **3**, 613–621 (1966).

Urea–formate starch-gel electrophoresis was used to study the proteins of human parotid and extraparotid saliva. Vertical starch-gel electrophoresis was performed on gels made in formate–urea–EDTA buffer, pH 3.7, at 4 volts/cm (30 mA–45 mA for 20 hours at 5°C). Parotid saliva showed as many as 17 bands, while extraparotid saliva showed 18 bands. Better reproducibility of protein separations could be obtained if 3 ml of parotid saliva were rejected first.

REDUCTION AND PROTEOLYTIC DEGRADATION OF IMMUNOGLOBULIN A FROM HUMAN COLOSTRUM (IgA-HC)

Cederblad, G., Acta Chem. Scand. **20**, 2349–2357 (1966).

Human colostrum (IgA–HC) was reduced with β-mercaptoethanol, and the urea–starch-gel electrophoresis pattern compared with serum IgA and IgG. Differences in migration of the H-chains of IgA–HC and IgA–S were demonstrated. Starch-gel, 8M urea in 0.05M sodium formate buffer, pH 2.8, with 0.02M β-mercaptoethanol, 24 hours, 3.5 volts/cm was used. A proteolytic degraded 3.5S component of IgA–HC did not show an Fc fragment on electrophoresis, whereas the IgG-fraction showed an Fc fragment.

STARCH GEL ELECTROPHORETIC ANALYSIS OF MATERNAL AND UMBILICAL SERUM PROTEINS

Chojnowska, I., Ginekol. Polska **37**, 701–708 (1966).

Eight of fourteen bands in mothers' sera were quantitatively evaluated after electrophoresis (150 volts for 18 hours) and staining with Amido Black 10B. The mothers' sera were compared with umbilical sera. The umbilical sera contained fewer α_2-fast globulins, β-transferrin, and β-lipoproteins than the mothers' sera. Haptoglobin and lipoprotein bands were missing in some umbilical sera.

DETECTION OF PROTEINS SYNTHESIZED DURING THE ESTABLISHMENT OF LYSOGENY WITH PHAGE P22

Cohen, L. W., and M. Levine, Virology **28**, 208–214 (1966).

The synthesis of protein at various time intervals after infection of *Salmonella typhimurium* with phage P22 was detected by a method which combines pulse labeling with S^{35}, starch-gel electrophoresis and autoradiography. Soluble proteins of infected bacteria labeled with S^{35} were separated by vertical starch-gel electrophoresis. The electrophoretic run was carried out on 15% gel made in a Tris–EDTA–Borate buffer (0.045M–.001M–0.025M), buffer pH 8.6. Electrophoresis was performed at 550 volts for 18 hours. The bridge buffer (0.11M Tris, 0.063M boric acid, 0.0025M EDTA) was used. The protein bands containing S^{35} were detected by autoradiography. Infection of S. *typhimurium* with phage C+, C_1 and C_2 caused the synthesis of four new protein bands at four minutes of infection. The synthesis of these bands persisted throughout the latent period.

FRAGMENTATION OF IMMUNOGLOBULIN DURING STORAGE

Connell, G. E., and R. H. Painter, Can. J. Biochem. **44**, 371–379 (1966).

The fragments remaining after the breakdown of γ-globulin during storage

were isolated on Sephadex® G-100 and characterized by starch-gel electrophoresis. The buffer was 8M urea–formate. Electrophoresis was done in a
vertical system for 18 hours, using 3 to 4 volts/cm [Smithies, Am. J. Human
Genet. **14**, 14 (1962)]. Double-diffusion Ouchterlony analysis was done in
1.5% agar using a pH 8.4 sodium barbital–HCl buffer, ionic strength 0.05, with
0.05% azide. The fragments isolated after storage were similar to those
present after digestion of γ-globulin by plasmin. This suggests plasmin contamination in the γ-globulin preparation.

SERUM TRANSFERRINS OF TWIN SHEEP-GOAT HYBRIDS

Cooper, D. W., L. F. Bailey, G. Alexander, and D. Williams, Australian
J. Biol. Sci. **19**, 1175–1177 (1966).

Starch-gel electrophoretic analysis of the sera of two hybrid fetuses, Australian
Merino sheep and goat (*Saanen*), for transferrin variants revealed that hybrids
have transferrin bands with mobility identical to that of both parents. Minor
transferrin bands found in other normal fetal sera were absent in the hybrid.

HEMOGLOBIN POLYMORPHISM IN MACACA NEMESTRINA

Crawford, M. H., Science **154**, 394–398 (1966).

Vertical starch-gel electrophoresis performed in a 0.9M Tris–0.5M boric acid–
0.2M EDTA separated four hemoglobin phenotypes in *Macaca nemestrina*. Pedigree studies suggested a simple codominant Mendelian type of inheritance for
three of these animals.

ELECTROPHORETIC HETEROGENEITY OF MAMMALIAN GALACTOSE DE-HYDROGENASE

Cuatrecasas, P., and S. Segal, Science **154**, 533–535 (1966).

Starch-gel electrophoresis separated electrophoretically-distinct forms of galactose dehydrogenase in various tissues of the rat. Vertical starch-gel electrophoresis was performed with 0.0053M phosphate buffer, pH 6.7, at 5 volts/cm
for 15 hours at 4°C. The isozymes were detected with a modified tetrazolium
procedure with galactose (50mM) as the substrate. Five isozyme bands
were noted, their migration being toward the cathode.

STUDIES ON THE COAT PROTEIN OF BACTERIOPHAGE φX174

Dann-Markert, A., Virology **29**, 126–132 (1966).

Starch-gel electrophoresis was one of several techniques used to study the
chemistry of the coat protein of the phage φX174. A 14% w/v starch gel was
prepared in 2 x 10^{-2}M ammonium formate containing 6M urea, final pH 3.8.
At pH 9.0 electrophoresis was done for 14 hours at 250 volts, 4°C. Twenty

mg of sample was required. Amido Black 10B was the stain used to detect proteins. The fractions were eluted from the gel and a protein determination made.

HETEROGENEITY OF INSULIN: I. ISOLATION OF A CHROMATOGRAPHI- CALLY PURIFIED HIGH POTENCY INSULIN AND SOME OF ITS PROPERTIES

Dillon, W. W., and R. G. Romans, Can. J. Biochem. **44**, 1171–1181 (1966).

Crystalline beef (zinc) insulin was separated into four components by column chromatography on Amberlite IRC-50® resin in 7M urea–0.13M sodium phosphate buffer, pH 6.05. Starch-gel electrophoresis (13.5% gel in 0.02M HCl) with 0.2M HCl as the electrolyte was used to further characterize the isolated components. The main component was not homogeneous, as shown by starch-gel electrophoresis.

HYDROLASES AND DEHYDROGENASES OF CHICKEN TISSUES

Ecobichon, D. U., Can. J. Biochem. **44**, 1277–1283 (1966).

Starch-gel electrophoresis using a 16% gel prepared in 0.01M sodium cacodylate–0.001M cacodylic acid buffer, pH 6.9, was used to characterize proteins and hydrolases of chicken tissues. The bridge buffer was 0.1M sodium cacodylate–0.01M cacodylic acid. Electrophoresis was carried out at 18-20 mA for 18 hours. Dehydrogenases were separated on 14% gels prepared with sodium citrate–phosphate buffer. Eighteen bands of esterase activity were detected with various substrates and specific inhibitors. Tissue-specific enzyme patterns were obtained for six enzymes.

RAPID RECOVERY, PRESERVATION, AND PHENOTYPING OF MILK PRO- TEINS BY A MODIFIED STARCH GEL TECHNIQUE OF SUPERIOR RESOLV- ING POWER

El-Negoumy, A. M., Anal. Biochem. **15**, 437–447 (1966).

Casein was recovered from 10 ml of whole milk by precipitation with an equal volume of 1M acetate buffer, pH 4.6, at 50°C. When the mixture was centrifuged, some of the casein went to the bottom of the tube and some was trapped in the milk fat at the top. The middle whey portion was removed and the preparation washed with distilled water. Solid urea and water to a total of 7M urea in 6 ml were added to prepare the precipitate for electrophoresis. The composition of the starch gel was: 50 g starch, 205 ml deionized distilled water, 55 ml Tris–citrate buffer [Nature **180**, 1477 (1967)], 115 g urea and 1.40 ml 2-mercaptoethanol. Sodium borate buffer was used for electrophoresis. The literature values for current and voltage were used [Biochim. Biophys. Acta **47**, 225 (1961)].

CLASSIFICATION OF HUMAN SERUM PREALBUMINS AFTER STARCH GEL ELECTROPHORESIS

Fagerhol, M. K., and M. Braend, Acta Pathol. Microbiol. Scand. **68,** 434–438 (1966).

Starch-gel electrophoresis using a modification of Poulik's discontinuous system [Nature **180,** 1477–1479 (1957)] was used to examine the prealbumins of normal serum. Two protein zones were found to migrate in front of the albumin. A third zone, P_1 proteins, migrated in front of albumin at pH 7.7 or lower. When pH 5.5 was used, more zones emerged, at least eight moving faster than albumin.

STARCH GEL ELECTROPHORESIS OF THE SERUM PROTEINS OF INFANTS WITH HEMOLYTIC DISEASE

Fenina, E. P., and N. I. Lopatina, Pediatriya **1966,** 13–15.

Umbilical blood serum of infants with hemolytic disease was compared to normal adult blood. Starch-gel electrophoresis in a borate buffer, pH 8.8, separated 12 proteins in the controls and 10 proteins in healthy infants. Sera of hemolytic infants were characterized by decreased amounts of albumin and γ-globulin. After transfusion the protein pattern was normal.

STUDIES ON THE INHERITANCE OF ELECTROPHORETIC FORMS OF TRANS-FERRIN ALBUMINS, PREALBUMINS AND PLASMA ESTERASES OF HORSES

Gahne, B., Genetics **53,** 681–694 (1966).

Sixteen transferrin, three albumin, eight prealbumin and six esterase phenotypes in the sera of 147 Saleritana horses were demonstrated by starch-gel electrophoresis. Transferrins were separated using 0.06M lithium hydroxide + 0.229M boric acid, pH 8.5, as electrolyte to run gels prepared in 1 volume of above + 5.4 volumes of Tris–citrate + 0.0085M EDTA. The prealbumins and albumins were demonstrated by preparing gels in a pH 5.4 sodium acetate EDTA buffer diluted 1:4. The gels were run in a 0.125M sodium acetate + 0.0085M EDTA solution in the cathode compartment. The starch gel used was 11.5%, and a voltage gradient of 15 volts/cm was applied for 3½ hours at pH 8.5 and for 4 hours at pH 5.4. Esterases were located by staining one of the gel slices from each pH condition with a solution containing 50 ml Tris–malate buffer (pH 6.5), 50 ml methanol, 2 ml 1% α-naphthyl acetate and 50 mg Fast Garnet GBC salt.

GENETICS OF PHYTOPATHOGENIC FUNGI: PECTOLYTIC ENZYMES OF VIRULENT AND AVIRULENT STRAINS OF THREE PHYTOPATHOGENIC PENICILLIA

Garber, E. D., and L. Beraha, Can. J. Botany **44**, 1645–1650 (1,966).

Culture filtrates from a virulent and a nonvirulent strain of three phytopatho-genic species of penicillium grown in a defined medium were electrophoresed on vertical starch gels to separate pectolytic enzymes. Gels were prepared in a 0.00935M Tris–0.005M citric acid buffer, pH 5.0 to 5.1. The separation was carried out at 2.5 volts/cm for 14 hours at 4°C. Endopolygalacturonase and exopalygalacturonase were detected. Pectin methylesterase and pectin lysase were not detectable.

TWO RARE HAPTOGLOBIN PHENOTYPES, 1-B AND 2-B, CONTAINING A PREVIOUSLY UNDESCRIBED α POLYPEPTIDE CHAIN

Giblett, E. R., I. Uchida, and L. E. Brooks, Am. J. Human Genet. **18**, 448–453 (1966).

The vertical starch-gel electrophoresis technique of Smithies [Advan. Protein Chem. **14**, 65 (1959)] was used to compare the haptoglobins of seven family members. The formate–urea gel electrophoretic technique of Smithies [Am. J. Human Genet. **14**, 14 (1962)] was used to demonstrate the haptoglobin phenotype after reduction cleavage.

CHARACTERISTICS OF A PROTEINASE OF A TRICHOSPORON SPECIES ISOLATED FROM DUNGENESS CRAB MEAT

Groninger, H. S., Jr., and M. W. Eklund, Appl. Microbiol. **14**, 110–114 (1966).

A proteinase of a *Trichosporon* species isolated from Dungeness crab meat was partially purified and characterized electrophoretically and chemically. The enzyme was purified by dialysis ammonium sulfate precipitation and gel-filtration on Sephadex® G-100. Starch-gel electrophoresis in a discontinuous buffer system according to the method of Smithies separated the proteinase into two major components. The pH optimum of the enzyme was found to be 5.8 to 6.2. The proteinase was active against casein, hemoglobin and crab meat substrates.

ADDITIONAL VARIANTS OF αS_1-CASEIN AND β-LACTOGLOBULIN IN CATTLE

Grosclaude, F., J. Pujolle, J. Gariner, and B. Rabideau-Dumas, Ann. Biol. Animale Biochim. Biophys. **6**, 215–222 (1966).

Two variants of cattle milk protein were isolated by starch gel electrophoresis.

The first variant was αS_1-casein from the Flamande breed and the second was β-lactoglobulin D from the Montbeliarde breed. The relative electrophoretic mobilities of the variants and natural proteins are discussed.

ÉTUDES SUR LA STRUCTURE DES γ_G-GLOBULINES NORMALES ET MYÉLO-MATEUSES: II. SÉPARATION ET ÉTUDE DE FRAGMENTS LIBÉRÉS AU COUR DE LA RÉDUCTION DES γ_G-GLOBULINES NORMALES AND MYELOMATEUSES

Havez, R., Pathol. Biol. (Paris) **14**, 810–815 (1966) **French.**

Reduced and alkylated γ_G-globulins were fractionated into heavy and light fragments by Sephadex® G-200. Starch-gel electrophoresis of the fractions was done at pH 3.0 and pH 8.0. The 8M urea–sodium formate buffer was at pH 3.0. A buffer consisting of 8M urea and 0.035M glycolate was used at pH 8.0. At pH 3.0, 0.2M formic acid in 0.08M soda, pH 8.2, was used for electrophoresis at pH 8.0. Amido Black was the stain.

ISOZYMES AND GENETIC CONTROL OF NADP-MALATE DEHYDROGENASE IN MICE

Henderson, N., Arch. Biochem. Biophys. **117**, 28–33 (1966).

Vertical starch-gel electrophoresis at 6–8°C and 6 to 7 volts/cm was used to study isozymes of malate dehydrogenase of mouse tissue homogenates. Borate–sodium hydroxide, 0.03M, pH 8.6, and .005M citrate, pH 5.0, buffers were used. Mouse tissues were characterized by two isozymes of NADP-dependent malate dehydrogenases. The two isozymes differ in mobility on starch gels, in tissue composition, and in subcellular location. Mitochondrial isozymes predominate in the heart, and liver tissue enzymes were found in the supernatent fraction. A genetic variant was found in an inbred strain of mice.

PYRIDINE NUCLEOTIDE-LINKED DEHYDROGENASES AND ISOZYMES OF NORMAL RAT BREAST AND GROWING AND REGRESSING BREAST CANCERS

Hershey, F. B., Cancer Res. **26**, 265–268 (1966).

Quantitative measurements were made of the enzymes in tissues. Starch-gel electrophoresis was done using 4.7 volts/cm for six hours at 4°C. The gels were prepared as follows:
1. Lactate and malate dehydrogenase gel was prepared in a Veronal buffer, pH 8.5, ionic strength 0.10.
2. Isocitric and glucose-6-phosphate dehydrogenase–Tris–citrate buffer, ionic strength 0.076 and 0.03 respectively, pH 8.65.

Borate buffer, pH 8.2, ionic strength 1.8, was used for electrophoresis. The basic incubation mixture for all enzymes was 0.15m PO_4, pH 7.5; 75 mg/l p NBT; 125 mg/l phenazine methosulfate. For LDH and MDH, the medium was 270 mg/1 NAD and 112M sodium lactate or 37.5gmM malic acid, pH 7.25. For G-6-PDH and ICDH, NADP was 38 mg/1, MgCl₂ was 5 mm and glucose-6-phosphate or sodium isocitrate was 2.7 mm.

LACTATE DEHYDROGENASES IN POIKILOTHERMS: DEFINITION OF A COMPLEX ISOZYME SYSTEM

Hochachka, P. W., Comp. Biochem. Physiol. **18**, 261–269 (1966).

Starch-gel electrophoresis performed in a Buchler-type cell, using a phosphate–citrate buffer (pH 7.0, ionic strength 0.52) was used to examine LDH activity of trout tissue homogenates. Five kinds of subunits (A–E) found generated 14 isozymes of lactic dehydrogenase in skeletal muscle of trout *(Salmo namaycush* and *S. fontinalis)*. The liver of lake trout contains five LDHs homologous to the A–B series in muscle.

INFLUENCE OF HEPARIN AND NUTRITION OF α- AND β-LIPOPROTEINS

Houtsmuller, A. J., Protides Biol. Fluids, Proc. 14th Colloq. 413–417 (1966).

Electrophoresis according to the technique of Wieme, using 0.5% Reinagar Behringwerke, pH 9.5 [Clin. Chim. Acta **9**, 497 (1964)], proved successful in separating α- and β-lipoproteins. This technique was used to evaluate the effect of the administration of i.v. heparin on lipoproteins of serum. The pre-β-lipoproteins disappeared immediately, and the α- and β- lipoproteins showed increased mobility after electrophoresis.

ELECTROPHORETIC SEPARATION OF CARBONIC ANHYDRASE AND NAPHTHYL ESTERASE ACTIVITIES

Hyyppä, M., L. K. Korhaven, and E. Kornhonen, Ann. Med. Exp. Bio. Fenniae (Helsinki) **44**, 63–66 (1966).

Starch-gel and cellulose-acetate electrophoresis were used to study the carbonic anhydrase and naphthyl esterase activity in human, bovine, rat, pig, sheep, goat and guinea pig erythrocyte hemolysate and rat kidney. Cellulose-acetate electrophoresis was performed horizontally in a 0.025M Veronal–medinal buffer, pH 7.90 or 8.15, at a potential gradient of 300 volts and 10 mA for two 50 x 90 mm strips for 90 minutes. Starch-gel electrophoresis was performed in a 0.025M medinal–hydrochloride buffer, pH 8.2, 300 volts for two 170 x 4 x 90 mm gels for 11 hours. Esterase activity was demonstrated by the technique of Markert and Hunter [J. Histochem. Cytochem. **7**, 42–49 (1959)]. The method of Waldeyer and Häusler [Acta Biol. Med. Ger. **2**, 568–569 (1959)] was used to demonstrate carbonic anhydrase activity. Separation of carbonic anhydrase from naphthyl esterase activity was possible with each of the techniques used. Erythrocyte carbonic anhydrase activity was different for each species examined.

ENZYMATIC DIFFERENTIATION OF RAT YOLK-SAC PLACENTA AS AFFECTED BY A TERATOGENIC AGENT

Johnson, E. M., and R. Spinuzzi, J. Embryol. Exp. Morphol. **16**, 271–288 (1966).

Equipment was designed specifically for the acrylamide gel electrophoresis done in these experiments. A scale drawing of this apparatus is presented. The yolk-sac and placenta were separated and the yolk-sacs cooled and homogenized in an equal volume of distilled water. All electrophoresis was performed at room temperature. Separation on starch gel was completed in four hours with 6 volts/cm, and separation in acrylamide was completed in 20 to 30 minutes with 5 mA per gel. When testing for esterases in yolk sac preparations, a pH 8.3 buffer was used in the gel and the tank. Blue RR salt and α-naphthyl acetate buffered to pH 7.19 with the above buffer as the stain. When assayed for acid phosphatase, the gel buffer was 0.02M Tris/HCl, pH 8.3, and the tank buffer was 0.03M borate, pH 8.4. Sodium α-naphthyl acid phosphate and the diazonium salt Blue RR in 0.02M sodium acetate, pH 7.5, was the stain.

ELECTROPHORETIC DETERMINATION OF ALBUMIN/GLOBULIN RATIOS IN MICE

Klasky, S., and M. S. Pickett, Nature **211**, 1298–1299 (1966).

This study compares normal, bovine serum albumin immune, hyperimmune BSA, newborn, pathogen-free, and S. aureus infected mice. Cellulose acetate with a B-2 buffer was used for the electrophoretic separation, which was run for 30 minutes. The albumin and globulin were quantitated on a Beckman® Analytrol.

ONTOGENY OF CATTLE HAEMOGLOBIN

Kleihauer, E., E. Brauchle, and G. Brandt, Nature **212**, 1272–1273 (1966).

Electrophoresis demonstrated three different types of hemoglobin appearing at various stages in the development of cattle. Starch-gel electrophoresis was done in Tris–EDTA–borate buffer at pH 9.0. Starch-block electrophoresis was performed in barbital–sodium barbital buffer, pH 8.6. Blood was drawn from the heart or extracted from the liver and diluted in 1.4% sodium chloride. Hemolysates were prepared by adding water and then shaking with carbon tetrachloride. Polypeptide chains of the hemoglobins were separated by agar-gel electrophoresis in urea–barbital buffer, pH 8.6.

254 Electrophoresis – Technical Applications

FRACTIONATION OF SERUM ALBUMIN AND GENETIC CONTROL OF TWO ALBUMIN FRACTIONS IN PIGS

Kristjansson, F. K., Genetics, **53**, 675–679 (1966).

Starch-gel electrophoresis performed on an 11.4% gel in a pH 6.3 Tris–citrate buffer was used to supply albumin. For 75 minutes 165 volts were applied, and then 350 volts until the borate boundary was 10 cm from the end of the gel. (A Tris–borate buffer was used in the electrode compartment.) Pig serum albumin was separated into 13 discrete fractions.

EFFECTS OF STARCH CONCENTRATION ON THE RESOLUTION OF SERUM PROTEINS BY GEL ELECTROPHORESIS: A SIX-SPECIES COMPARISON

Krotoski, W. A., D. C. Benjamin, and H. E. Weimer, Can. J. Biochem. **44**, 545–555 (1966).

The effects of starch concentration on the resolution of serum components by gel electrophoresis were investigated. Gels varying from 11.8 to 20.6% were used to separate serum of man, rabbit, guinea pig, rat, mouse and chicken. Vertical and two-dimensional electrophoresis proceeded in discontinuous Tris–citrate and borate buffers. An increase in the number of protein bands was observed in gels with concentrations between 13.0 and 16.0%. Species differences were observed. The maximum number of serum proteins detected were: man, 27; rabbit, 15; guinea pig, 22; rat, 21; mouse, 19; and chicken, 12.

A SPECIFIC PROTEIN DIFFERENCE IN THE MILK FROM TWO MAMMARY GLANDS OF A RED KANGAROO

Lemon, M., and L. F. Bailey, Australian J. Exp. Biol. Med. Sci. **44**, 705–708 (1966).

Starch-gel electrophoresis of the whey proteins from mammary glands at different stages of lactation demonstrated different proteins in the glands of the same animal. Evidence indicated that the lactogenic stimulus for specific whey proteins disappears before the birth of the second offspring.

STARCH GEL ELECTROPHORESIS OF SERUM PROTEINS OF HUMAN UMBILICAL BLOOD

Lewandowski, J., A. Rozalska, and A. Lyko, Am. J. Obstet. Gynecol. **96**, 1159–1163 (1966).

Serum proteins of umbilical and maternal venous blood were separated by vertical starch-gel electrophoresis. A 15% starch suspension was prepared in a pH 8.6 borate buffer, ionic strength 0.026. Electrophoresis was done in this buffer for 18 to 22 hours with 3 to 4 volts/cm. Amido Black 10B stained the strips after electrophoresis.

IMPROVED TECHNIQUE FOR SEPARATION AND IDENTIFICATION OF BOVINE BETA GLOBULINS BY STARCH GEL ELECTROPHORESIS

Makarechian, M., and W. E. Howell, Can. J. Biochem. **44**, 1089–1091 (1966).

Use of a pH 7.30 to 7.35 buffer for electrophoresis and preparation of the starch gel made possible the demonstration of six protein bands in each transferrin allele of beta globulin. The buffer used was 2.3368 g Tris (hydroxy-methyl) aminomethane and 2.4509 g cacodylic acid to 1 liter; 168 g hydrolyzed starch was dissolved in 1 liter of buffer. The composition of the buffer is critical for duplication of these experiments. Gels were stained with Amido Black 10B.

ONTOGENY OF HEMOGLOBIN IN THE CHICKEN

Manwell, C., C. M. Baker, and T. W. Betz, J. Embryol. Exp. Morphol. **16**, 65–81 (1966).

Vertical starch-gel electrophoresis was performed in a series of buffers to study the polypeptide composition and effects of chemical modification on adult and embryonic chicken hemoglobin. The buffers used were: borate (gel pH 8.4–8.5), Tris–EDTA–borate (gel pH 8.6-8.8), barbiturate (gel pH 8.4), phosphate, ionic strength 0.02 (gel pH 6.0, 6.7, and 7.3–7.5) and a lithium modification of Poulik's discontinuous Tris–citrate–borate system (gel pH 8.0). The borate and Tris–EDTA–borate buffers yielded the best resolution of multiple hemoglobin forms. The technique used is described by Manwell et al. [Proc. Nat. Acad. Sci. U.S. **49**, 496–503 (1963)].

IN VITRO HYBRIDIZATION OF LACTATE DEHYDROGENASE IN THE PRE-SENCE OF ARSENATE AND NITRATE IONS

Markert, C. L., and E. J. Massaro, Arch. Biochem. Biophys. **115**, 417–426 (1966).

Vertical starch-gel electrophoresis (14% gel) prepared and run at 10 volts/cm in the EBT buffer of Boyer was used to study the hybridization of lactate dehydrogenase. Purified lactate dehydrogenase isoenzymes (enzyme 1, 2, 3, 4, 5) were purified by chromatography on DEAE-cellulose. Hybridization was achieved by freezing and thawing the mixtures of the various enzymes. The freeze-thaw operation was performed in the presence of various concentrations of salts at $-20°C$. Freeze-thaw hybridization was bound to occur at rates of efficiency equivalent to that of the standard neutral phosphate technique in the presence of neutral $NaHAsO_4 \cdot HCl$ and in $NaNo_3$–NaOH.

CHARACTERIZATION OF CASEINS IN GEL ELECTROPHORETOGRAMS

McCabe, E. M., and J. R. Brunner, J. Dairy Sci. **49**, 1148–1149 (1966).

Starch-gel electrophoresis in the presence of urea demonstrates 15 to 20 zones of the casein complex. The effect of a low calcium concentration on the appearance of these zones is the subject of this paper. Whole precipitated casein was applied to a urea-starch gel, and this was separated by electrophoresis [Biochim. Biophys. Acta **47**, 225 (1961)]. The gel was cut into sections and soaked in deionized water to remove the urea. One section was stained with Amido Black and the others were soaked in varying concentrations (0.1–0.5M) of calcium chloride. Aggregation in the αs_2 and αs_3 was noted. Concentrations of calcium high enough to aggregate β-casein caused an opaque zone in the γ-casein region.

ZONE ELECTROPHORESIS OF β-LACTOGLOBULINS

McKenzie, H. A., and H. A. Sawyer, Nature **212**, 161–163 (1966).

Whey protein β-lactoglobulins were separated by starch-gel electrophoresis using a variety of buffers. At pH 7.5 and 8.5, the β-lactoglobulins dissociated rapidly at low concentration. Varying this concentration of NaOH and boric acid, at constant pH 8.5, showed a slower-moving second band in 0.02M buffer that disappeared in 0.3M buffer. The bands are most likely due to irreversible denaturation of the β-lactoglobulin in dilute buffer. Best results of zone electrophoresis of β-lactoglobulin are achieved when careful attention is given to the gel buffer.

SERUM PROTEIN ELECTROPHORESIS IN THE TAXONOMY OF SOME SPECIES OF THE GROUND SQUIRREL SUBGENUS SPERMOPHILUS

Nadler, C. F., and C. E. Hughes, Comp. Biochem. Physiol. **18**, 639–651 (1966).

Serum proteins of S. undulatus, S. columbians, and S. beldingi were analyzed by two-dimensional starch-gel electrophoresis. The horizontal two-dimensional technique of Poulik and Smithies [Biochem. J. **68**, 636–643 (1958)] and a Tris-discontinuous buffer system were used. The patterns were found to be generally similar but exhibited variation in nine fractions. Close relationship among all species was indicated by the patterns.

A DISTINCTIVE FRACTION OF ALKALINE PHOSPHATASE IN HEALTH AND DISEASE

Newton, M. A., J. Clin. Pathol. **19**, 491–495 (1966).

A slow-moving fraction of plasma alkaline phosphatase was found in one-fifth of the normal subjects and in some patients with parenchymal liver disease.

Starch-gel electrophoresis separated the fractions in five to six hours at 4.7 volts/cm gel. A 0.076M Tris–citrate buffer was used in a discontinuous system. Alkaline phosphatase activity was located by incubation at 37°C with 0.005M α-1-naphthyl phosphate in pH 2.0 bicarbonate buffer. α-1-Naphthol was then detected by its fluorescence in ultra-violet light. A fresh aqueous solution of diazo-o-dianisidine stained the preparation. A methanol–acetic acid–water solution was used as a fixative.

KINETICS AND ELECTROPHORETIC PROPERTIES OF THE ISOZYMES OF ASPARTATE AMINO-TRANSFERASE FROM PIG HEART

Nisselbaum, J. S., and O. Bodansky, J. Biol. Chem. **241**, 2661–2664 (1966).

Pig heart aminotransferase, isolated and purified according to a procedure which employed chromatography on CM-cellulose, was characterized using starch-gel electrophoresis. The enzyme was eluted from a CM-cellulose column with a pH 5.4 sodium–acetate buffer gradient (5mM–100mM). Starch-gel electrophoresis of the eluted material was performed in gels prepared in 5.0mM Tris–succinate buffer, pH 7.2. The electrode vessel contained 0.1M phosphate buffer, pH 7.2. The electrophoretic run was carried out at 4.0 volts/cm (10–16 mA for 18 hours at 4–6°C). The enzymes were detected by a specific staining procedure described by Schwartz, Nisselbaum and Bodansky [Am. J. Clin. Pathol. **40**, 103 (1963)]. Two enzyme fractions with cathodic mobility were detected. The kinetic results indicated that the two enzymes were different.

MULTIPLE FORMS OF LACTATE DEHYDROGENASE AND ASPARTATE AMINO TRANSFERASE IN HERRING (CLUPEA HARENGUS HARENGUS L)

Odense, P. H., T. M. Allen, and T. C. Leung, Can. J. Biochem. **44,** 1319–1326 (1966).

Vertical starch-gel electrophoresis separated lactate dehydrogenase and aspartate amino transferase from crude homogenates of herring *(Clupea harengus harengus)*. A modified Aronsson-Grönwall [Scand. J. Clin. Lab. Invest. **9,** 338 (1957)] was employed. Two hybrid forms representing mutant alleles at the β locus were observed.

AUTOGENESIS OF SERUM ESTERASE IN MUS MUSCULUS

Pantelouris, E. M., and A. Arnason, J. Embryol. Exp. Morphol. **16,** 55–64 (1966).

Two buffer systems were used to separate the esterases in mouse serum by starch-gel electrophoresis. The first buffer system was: gel buffer–10.5 g citric acid, 92.0 g Tris, water to one liter; vessel buffer–pH 8.6, 18.6 g boric acid, 4.0 g NaOH, water to one liter. The second buffer system employed was: gel buffer–pH 8.6, 0.5 citric acid, 2.7 g Tris, 100 ml vessel buffer, water to one liter; vessel buffer–pH 8.6, 11.8 g boric acid, 1.2 g NaOH, water to one liter, Hydrolyzed starch was used to prepare the gel. Electrophoresis was done for 2.5 hours at 15–20 mV/cm length and 2–4 mA/cm width of the gel plate.

ELECTROPHORESIS OF PIG AND HUMAN RENIN

Peart, W. S., G. D. Lubash, G. N. Thatcher, and G. Muriesan, Biochim. Biophys. Acta **118**, 640–643 (1966).

Purified pig and human renin migrate at different speeds in starch gel, acrylamide, and paper electrophoresis. The discontinuous buffer system of Poulik [Nature **180**, 1477 (1957)] was used for starch-gel and acrylamide electrophoresis. Paper electrophoresis was done on Whatman® 3 MM paper at a constant voltage of 220 to 230 volts for 30 to 60 minutes. A 0.5M Tris–EDTA–borate buffer, pH 9.0, and 0.4M Tris–maleate buffer, pH 7.0, were used. Bromophenol Blue was the stain. Paper electrophoresis in Tris–EDTA–borate buffer gave the highest pig renin recovery of all methods employed—60%. Human renin recovery was 25%. Electrophoresis in the cold might increase recovery of the renin.

INHERITANCE OF AN ERYTHROCYTE AND KIDNEY ESTERASE IN THE MOUSE

Popp, R., J. Heredity **57**, 197–201 (1966).

Starch-gel electrophoresis in 0.03M boric acid–sodium hydroxide buffer, pH 8.5, was used to study the pattern of inheritance of a propionyl esterase of RFM/Un, YBR/He, RF/J and RF/Lo strains of mice (inbred strains). Three phenotypic propionyl esterases were observed. An autosomal locus Es-3 controls the phenotypes.

SEPARATION OF PROTEINS AND PROTEIN SUB-UNITS

Porter, R. R., Brit. Med. Bull. **22**, 164 (1966).

Starch and acrylamide gels were used for the separation and purification of protein subunits. Dissociating agents can be incorporated into the gels at both acid and alkaline pH. Both gels can separate 20 to 25 protein components in serum. Techniques were developed for sharpening zones by using a large-pore space gel and a small-pore separation gel.

ÉTUDES SUR LA STRUCTURE DES γ_G-GLOBULINES NORMALES ET MYÉLO-MATEUSES: I. ÉTUDE DE LA RÉDUCTION DES DISULFURE PONTS DES γ_G-GLOBULINES NORMALES ET MYELOMATEUSES

Racadot-Leroy, N., Pathol. Biol. (Paris) **814**, 805–809 (1966) **French.**

The sugar and partial reduction by the technique of Fleischmann [Arch. Biochem. Biophys. Suppl. **1**, 174 (1962)] were used for reduction of the disulfide bridges of γ_G-globulins. Electrophoresis of the protein components was performed on paper in a pH 8.9 Veronal buffer. An 8M urea–sodium formate buffer (formic acid 0.05M, 0.01M), pH 3.0, was used for starch-gel electrophoresis. Better separation was achieved on starch gel.

SERUM TRANSFERRIN POLYMORPHISM IN THE DEER MOUSE

Rasmussen, D. I., and R. K. Koehn, Genetics **54**, 1353–1357 (1966).

A Tris–citrate gel buffer, pH 7.5, and a sodium borate electrode chamber buffer, pH 8.7, formed the discontinuous system used to study transferrins. A voltage of 10.0 volts/cm was applied until the borate boundary migrated 8 cm from the origin (about two hours). An iron-staining kit was used to locate the transferrin.

ELECTROPHORETIC BEHAVIOR OF H AND L CHAINS OF HUMAN SERUM AND COLOSTRUM GAMMA GLOBULIN

Rejnek, J., J. Kostka, and D. Travicek, Nature **209**, 926–928 (1966).

Improved separation of the L-chain of human serum and colostrum was achieved by starch-gel electrophoresis in 0.035M glycine and 8M urea. A fast-moving zone, which was absent in serum L-chain, was found in colostrum. The H-chains were separated into a number of zones which differed in serum and colostrum. The method of preparing the H-chain for electrophoresis is important.

OBSERVATIONS ON THE ORIGIN OF PREGNANCY-ASSOCIATED PLASMA PROTEINS

Robinson, J. C., W. T. London, and J. E. Pierce, Am. J. Obstet. Gynecol. **96**, 226–230 (1966).

The plasma protein patterns of pregnant women were compared with those of nonpregnant women, women who were receiving estrogenic or progestational compounds, and women with abnormal trophoblastic tissue. Forty-four sera were examined for pregnancy-associated phosphatase, alpha-2-globulin, and aminopeptidases using starch-gel electrophoresis. Alkaline phosphatase was observed by incubating the gel with α-naphthyl phosphate after electrophoresis. Amido Black 10B was used to stain the gel after electrophoresis to localize the alpha-2-globulin. Amidopeptidase was stained by α-bucyl–β-naphthylamide after electrophoresis.

AN EXAMINATION OF CORYNEBACTERIUM SPECIES BY GEL ELECTRO-PHORESIS

Robinson, K., J. Appl. Bacteriol. **29**, 179–184 (1966).

Cell-free homogenates of 24 species of *Corynebacterium* were examined for proteins by polyacrylamide or starch-gel electrophoresis. Electrophoresis was performed on a 7% gel prepared from Cyanogum® 41 in a Tris–citrate buffer. Separation proceeded at 8 volts/cm. Starch-gel electrophoresis was employed for separating and detecting esterase, catalase and peroxidase activity of the species. Enzymes were detected using conventional techniques for staining the gel. The genus was separated into definite subgroups using information obtained as to enzyme pattern and protein pattern. Human pathogens possessed a single catalase and no peroxidase. The animal pathogens were characterized by a double catalase, one of which possessed mobility similar to that of the human enzyme. Plant pathogens possessed a slow-moving catalase, one peroxidase, and at least one esterase with high activity.

PEAFOWL LACTATE DEHYDROGENASE: PROBLEM OF ISOENZYME IDENTI-FICATION

Rose, R. G., and A. C. Wilson, Science **153**, 1411–1413 (1966).

Extracts were made from fresh and frozen peafowl by homogenizing 1 g of tissue in 5 ml of cold 0.25M sucrose. After centrifugation the extracts were frozen for storage. Starch-gel electrophoresis was performed in several different buffers: 1) citrate–phosphate, pH 7.0, ionic strength 0.025 for 20 hours at 10°C using a gradient of 12 volts/cm; 2) Tris–borate–EDTA, pH 8.6, ionic strength 0.025; 3) glycine buffer, pH 9.7. The enzyme from the peafowl and the chicken had different mobilities at pH 7.0. Extracts of enzyme from heart and muscle migrated differently depending upon the buffer. Thermal inactivation studies indicated that peafowl breast LDH is an M_4 isoenzyme, and peafowl heart LDH is an H_4 isoenzyme.

AN IMPROVED PROCEDURE OF TWO DIMENSIONAL ELECTROPHORESIS OF SERUM PROTEINS USING 1.5M UREA-STARCH GEL AND A SYSTEM OF TEB-BUFFERS

Rösler, B., Acta Biol. Med. Ger. **17**, 371–373 (1966).

A technique of two-dimensional electrophoresis of proteins which combines electrophoresis in 1.5M urea starch gel with paper electrophoresis is described. Serum applied in a 2.7 cm band was separated in the first direction on filter paper FN 7 (VEB spezialpapierfabrik Niederschlag. Erzgebirge) at 3.5 volts/

cm for 16 hours (1.2 mA per strip). A strip 10.5 cm long was cut from the middle of the paper chromatogram and placed in the 13% starch–urea gel (1.5M) made in a Tris–EDTA buffer. Electrophoresis was then performed at 10 volts/cm in a sodium tetraborate–boric acid buffer for four hours. Proteins that diffuse readily were isolated as discrete bands. Reproducibility of starch-gel electrophoresis was improved by the addition of urea.

AMYLASE ISOZYME POLYMORPHISM IN MAIZE

Scandalios, J. G., Planta **69,** 244–248 (1966).

Starch-gel electrophoresis revealed the presence of amylase of maize in isozymic forms. Homogenates of whole seeds and seedlings prepared in saline were subjected to horizontal starch-gel electrophoresis in the discontinuous buffer system of Ashton and Braden [Australian J. Biol. Sci. **14,** 253 (1961)]. The electrophoretic run was performed at room temperature at 6 to 8 volts/cm for 3.5 hours. Activity of amylase was detected in sliced starch gel by incubating at 37°C after immersion in 0.1M phosphate buffer, pH 6.0. The gels were then washed in a solution of methanol–acetic acid–water (5:5:1) to render the gel translucent. The area of amylase, however, showed up as a transparent band. The earliest period of amylase activity appeared on the second day after germination of seeds. Adult plants showed no amylase activity.

HUMAN SERUM LIPOPROTEINS: ADVANCES IN PHYSICAL AND CHEMICAL METHODS OF ISOLATION AND ANALYSIS

Scanic, A., and J. L. Granda, Progr. Clin. Pathol. **1,** 398 (1966).

Human serum lipoproteins were classified by their density in the ultracentrifuge, which is the most practical means of quantitative separation. Electrophoresis in gel is applicable only to the high density lipoproteins which can penetrate the gel to move in the α_1-globulin region. Vertical starch electrophoresis demonstrated polymorphism of these lipoproteins. Lighter lipoproteins were poorly resolved.

PATTERNS OF ALKALINE PHOSPHATASE IN DEVELOPING DROSOPHILA

Schneiderman, H., W. J. Young, and B. Childs, Science **151,** 461–463 (1966).

Vertical starch-gel electrophoresis performed in a 0.05M Tris–HCl buffer, pH 8.6, was used to separate alkaline phosphatase enzymes in developing *Drosophila.* Homogenates of whole larvae, eggs or organs were electrophoresed at 7–9 volts/cm for two to four hours at 4°C. The alkaline phosphatase of *Drosophila* seems to be a family of enzymes whose members, except band 1, are organ specific. Band 1 was present in all stages of development.

RELATIONSHIP OF THE ELECTROPHORETIC PATTERNS OF PHOSPHOHEXO-
ISOMERASE AND GLUTAMIC-OXALACETIC TRANSAMINASE IN HUMAN
TISSUES TO THE PATTERNS IN SERUM OF PATIENTS WITH NEOPLASTIC
DISEASE

Schwartz, M. K., and O. Bodansky, Am. J. Med. **40**, 231–242 (1966).

Starch-block electrophoresis using a 0.1M barbital buffer, pH 8.5 to 8.6, separated phosphohexoisomerase and glutamic-oxalacetic transaminase of serum and tissue homogenates from cancer patients. Phosphohexoisomerase activity was cathodic in behavior and migrated with the gamma globulin fraction. Tissue GOT was distributed in two peaks, whereas serum GOT was confined to one peak, the alpha-beta region.

GENETICAL EVIDENCE FOR MATERNAL ORIGIN OF AMNIOTIC FLUID
PROTEIN

Seppälä, M., E. Ruoslahti, and T. Fallberg, Ann. Med. Exp. Biol. Fenniae (Helsinki) **44**, 6–7 (1966).

Starch-gel electrophoresis separated transferrins of maternal serum, cord serum, and amniotic fluid. A discontinuous Tris–citrate buffer at 15 volts/cm was employed for separation. The separated transferrins were detected by staining according to the method of Seppälä (Ann. Med. Exp. Biol. Fenniae (Helsinki) **43** Suppl. 4, 1–40 (1965)].

HETEROGENEITY OF HEAVY (γ) CHAIN PREPARATIONS FROM HUMAN
γG-IMMUNOGLOBULINS

Sjoquist, J., Nature **210**, 1182–1183 (1966).

After alkylation of a preparation of heavy and light chains by the method of Fleischman [Arch. Biochem. Biophys. Suppl. **1**, 174 (1967)], the chains were dissociated by dialysis against 0.1M formic acid. The mixture of chains was separated on Sephadex® G-200, A 50 in 0.1M formic acid. Light and γ-chains were separated from reduced but non-alkylated material in 0.1M formic acid containing 1 ml mercaptoethanol. Urea starch-gel electrophoresis was performed in formate buffer and in glycine buffer [Biochem. J. **90**, 278 (1964); J. Exp. Med. **113**, 861 (1961)]. The data suggested that the banding of γ-chain preparations is due to charge differences in the primary structure.

HETEROGENEITY OF H AND L CHAINS OF NORMAL AND MYELOMA
γG-GLOBULIN

Sjoquist, J., and M. W. Vaughan, Jr., J. Mol. Biol. **25**, 527–536 (1966).

Heterogeneity of the H-chain was demonstrated by glycine–urea starch electrophoresis at pH 7–8. H- and L-chains were separated on Sephadex®

G-200 in 0.1M formic acid. Electrophoresis was performed for 16 to 18 hours, 250 volts, 16 mA. Amido Black B was the stain. Reduced but non-alkylated or iodoacetic acid-alkylated H-chains from normal patients showed 18 bands; myeloma H-chains showed fewer bands.

ZONE ELECTROPHORETIC COMPARISON OF MUSCLE MYOGENS AND BLOOD PROTEINS OF ARTIFICIAL HYBRIDS OF SALMONIDAE WITH THEIR PARENTAL SPECIES

Tsuyuki, H., and E. Roberts, Fisheries Res. Board Canada **22**, 767–773 (1966).

Starch-gel electrophoresis and polyacrylamide-disc electrophoresis separated muscle myogens and blood proteins of artificial hybrids and their parent species of some fresh water members of *Salmonidae*. Starch-gel separations were achieved on gels prepared in 0.023M sodium borate buffer and electrode buffer of 0.3M sodium borate. Polyacrylamide-disc electrophoresis was performed in a Canalco® Model 12 apparatus at 2.5 mA per gel for 1.5 hours. Serum proteins were found to show a degree of uniformity.

PROTEIN FRACTIONS IN SERUMS SUBJECTED TO DEEP FREEZING AND THAWING

Utevs'ka, L. A., Ukr. Biokhim. Zh. **38**, 362–366 (1966) **Russian.**

The electrophoretic patterns of human and albino rat sera were not changed when frozen in liquid nitrogen (−19°C) and then quick thawed. The glyco-protein content was not altered. The relative composition of electrophoretic fractions was unaltered after freezing in liquid nitrogen.

TRANSFERRIN VARIANTS IN BELGIAN POPULATIONS: A STUDY USING AGAR GEL ELECTROPHORESIS

Vuylsteek, K., and R. J. Wieme, Protides Biol. Fluids, Proc. 14th Colloq. 93–96 (1966).

The serum proteins of 4,859 individuals of Belgian nationality and Flemish stock were investigated for transferrin variants using starch-gel electrophoresis. Routine electrophoretic separation was effected at 12°C at a field strength of 20 volts/cm for 40 minutes. A second technique achieved separation at 7°C and 20 volts/cm for 90 minutes. In both instances 1% Difco Noble agar was used with a pH 8.4 barbital buffer, ionic strength 0.05. Transferrin variants T_B and T_D were detected with a frequency of 1.0% and 0.2% respectively. The measurement of m_r values demonstrated the occurrence of subclasses in the T_B group.

LACTIC ACID DEHYDROGENASE ISOENZYMES OF BUFFY COAT CELLS AND ERYTHROCYTES FROM DIFFERENT SPECIES

Walter, H., F. W. Selby, and J. R. Francisco, Nature **212**, 613–614 (1966).

LDH isoenzymes from the buffy coat and RBC of different species were tested to determine if young RBC had the same LDH isoenzyme pattern as the cells of the buffy coat. One hundred units of the enzyme ($\triangle$O.D.340 of 0.001/min/ml of lysate) were electrophoresed on starch gel in 0.01M phosphate buffer, pH 7.5, for five hours in the cold at 10 volts/cm. LDH was assayed using a buffered substrate agar preparation poured over the starch gel. Buffy coat cells had more isoenzymes than the RBC of the same species; no buffy coat population lacked isoenzyme # 3.

APPLICATION OF STARCH GEL ELECTROPHORESIS TO THE SEPARATION OF EXTRACELLULAR PRODUCTS OF GROUP A STREPTOCOCCI

Winter, J., Dissertation Abstr. **B-27**, 1722 (1966).

Extracellular fluids of eleven strains of group A *Streptococcus pyrogenes*, grown on a medium containing only low molecular weight compounds, were examined by starch-gel electrophoresis using Smithies' method. The extracellular proteins were well resolved and a standardized set of conditions for separation was presented. A maximum of 11 bands was found in the strains studied. Biological assays were developed to detect streptococcal products in starch gel.

SEQUENCE OF LINKAGE BETWEEN THE PROSTHETIC GROUPS AND THE POLYPEPTIDE CHAINS OF HAEMOGLOBIN

Winterhalter, K. H., Nature **211**, 932–934 (1966).

Three classes of intermediary compounds between hemin and globulin were postulated. Molecular weights of the compounds were estimated by gel filtration. Starch-gel electrophoresis separated the intermediaries. A discontinuous system of 0.076M Tris and 0.005M citric acid buffer was used for the gel; a 0.30M boric–0.06M sodium hydroxide system was used in the buffer vessels, pH 8.6.

SIMULTANEOUS PHENOTYPING PROCEDURE FOR THE PRINCIPAL PROTEINS IN COW'S MILK

Aschaffenburg, R., J. Dairy Sci. **48**, 1524–1526 (1965).

Starch-gel electrophoresis was used in K-casein phenotyping. The gel consists of 20 g hydrolyzed starch in 80 ml buffer (60 g Tris, 8 g EDTA, 4.6 g H_3BO_3 per liter). The gel was added to the buffer and heated until the starch grains disrupted. Forty grams urea were quickly added and the mixture heated to boiling. The mixture was transferred to a 1-liter Büchner flask, reheated, and evacuated until there were no air bubbles. Three drops of 2-mercaptoethanol were added. The gel was poured into troughs that were previously treated with silicone grease. The filled troughs were covered with a thin sheet of polyethene and pressed down with a cover plate to exude excess fluid. A heavy weight was then placed on this plate and the gels dried overnight. For electrophoresis the buffer was $0.3H_3BO_3$ and 0.1M NaOH; the constant voltage was 300 volts. Separation was complete in 4.5 to 5 hours in the cold. Amido Black 10B was the stain. This technique is not suitable for detection of the α-lactalbumin variants found in the milk of Zebu cattle. The simultaneous classification of variants of K-casein, α_{s1}-casein, β-casein, and β-lactoglobulin can be made in all milk samples if one drop of 2-mercaptoethanol is added to each 0.5 ml milk sample.

STARCH-GEL ELECTROPHORESIS OF ERYTHROCYTE STROMA

Azen, E. A., and O. Smithies, J. Lab. Clin. Med. **65,** 340–349 (1965).

Freeze dried proteins of erythrocyte membranes were separated by starch-gel electrophoresis. Vertical electrophoresis was performed on urea–sodium formate gels and urea–aluminum lactate gels. The urea–formate gel was run at potential gradient of 3.5 volts/cm (20–25 mA) for 20 hours at pH 3.7. The aluminum lactate gel was run at a pH 4.0 for 20 hours, 3.5 volts/cm (10–15 mA). Reproducible separation patterns of the membranes freed of lipids and hemoglobin were obtained. The bridge buffers incorporated mercapethanol.

HETEROGENEITY OF CALF INTESTINAL ALKALINE PHOSPHATASE

Behal, F. J., and M. Center, Arch. Biochem. Biophys. **110,** 500–505 (1965).

Calf intestinal alkaline-phosphatase activity was resolved into several chromatographically distinct enzyme components with DEAE-cellulose column chromatography. By subjecting portions of the column eluate to vertical starch-gel electrophoresis, the presence of multiple enzyme components with distinct electrophoretic mobilities was also demonstrated. Study of the relative activities of these enzymes on several phosphomonoester substrates revealed additional differentiating characteristics among the isolated enzyme components. A similarity was found, however, in the sensitivities of the enzyme components towards heat.

HYBRIDIZATION OF GLUCOSE-6-PHOSPHATE DEHYDROGENASE FROM RAT AND HUMAN ERYTHROCYTES

Beutler, E., and Z. Collins, Science **150,** 1306–1307 (1965).

Glucose-6-phosphate dehydrogenase from rats and humans was partially purified and was dissociated by removing trophosphopyridine nucleotide. A hybrid enzyme could be formed before reactivation by mixing enzymes from two species with triphosphopyridine nucleotide at 25°C. Vertical starch-gel electrophoresis of reactivated human, rat, and hybrid enzyme showed that the hybrid band was located between the fast-moving rat dehydrogenase and the slow-moving type B human enzyme.

EFFECT OF DENERVATION ON THE LACTATE DEHYDROGENASE ISOZYMES OF SKELETAL MUSCLE

Brody, E., Nature **205,** 196 (1965).

Starch-gel electrophoresis of lactate dehydrogenase isozymes of denerved skeletal muscle of guinea pigs indicated that a reversion to immature lactate dehydrogenase occurs with denervation. There was a decrease in the amount of fast-moving isozymes and an increase in the intensity of the slow-moving isozymes.

SEPARATION OF ACID POLYSACCHARIDES BY STARCH GEL ELECTROPHORESIS

Brookhart, J. H., J. Chromatog. **20,** 191–193 (1965).

Horizontal starch-gel electrophoresis in a 13% gel prepared in a 0.01 citric acid buffer, pH 3.5, was used to separate five acid polysaccharides. The electrolyte buffer was 0.1M citric acid buffer, pH 3.5. Separation was carried out at 10 mA for eight hours at room temperature. Heparin sulfate, hyaluronic acid sulfate, hyaluronic acid, chondroitin sulfate B, and chondroitin sulfate A were separated in quantities as small at 50 μg. The separated compounds were revealed as oval red spots after staining with Toludine Blue O and destaining the gel background with acetone–glacial acid (24:1).

A SYSTEM OF MICROSTARCH-GEL ELECTROPHORESIS AS APPLIED TO THE SEPARATION OF MILK PROTEINS

Copius-Peereboom, J. W., Neth. Milk Daily J. **19,** 234–244 (1965).

Whey proteins were separated into nine bands by a micromethod using starch–urea gels on microscope slides. A Veronal buffer, pH 8.6, and a potential gradient of 170 volts (4 mA per slide) was applied for five hours at 4°C. 2-Mercaptoethanol in the starch gel helped separate the K-casein complex.

A NEW MEDIUM FOR HEMOGLOBIN ELECTROPHORESIS

Cronenberger, J. H., Am. J. Med. Technol. **31,** 9–15 (1965).

An improved method of hemoglobin electrophoresis is described. The method employs a mixture of starch (11.5 g/100 cc buffer) and agar (2 g). Electrophoresis was performed vertically in a discontinuous buffer system as described by Smithies. Better separation was achieved on the mixture than either component yielded alone.

ATYPICAL γ_{1A}-GLOBULIN WITH THE ELECTROPHORETIC PROPERTIES OF AN α_2-GLOBULIN OCCURRING IN MULTIPLE MYELOMA

Cummings, N. A., and E. C. Franklin, J. Lab. Clin. Med. **65,** 8–17 (1965).

An abnormal α_2-globulin isolated from serum of a paraproteinemia was characterized by physiochemical as well as chemical means. The abnormal globulin was isolated by starch-zone electrophoresis as described by Kunkel ("Zone Electrophoresis," in Methods of Biochemical Analysis, New York: Academic Press, 1964). The isolated globulin was found to be genically related to γ_{1A}-globulin but not to the 7S or 19S γ-globulin.

THE BEHAVIOUR OF SOME PROTEINS OF NORMAL RAT SERUM STUDIED BY GEL FILTRATION AND ELECTROPHORESIS

Dolezalová, V., Z. Brada, and A. Kocent, Biochim. Biophys. Acta **107,** 294–306 (1965).

Rat serum proteins were separated into three fractions by chromatography on Sephadex® G-200. A Tris–HCl buffer, pH 8.0, 1.0M in NaCl was used to chromatograph the serum. Starch-gel electrophoresis of these three fractions, performed according to the Smithies method, yielded 17 zones.

PHOSPHATASE MUTANTS IN ASPERGILLUS NIDULANS

Dorn, G., Science **150,** 1183–1184 (1965).

Electrophoretic and genetic studies indicate that the phosphatase system in *Aspergillus nidulans* is distinguished by a large variety of phosphatase mutants and the large number of genetic loci affecting the enzyme activity. Enzyme patterns were analyzed at room temperature by horizontal starch-gel electrophoresis at a potential difference of 5 volts/cm.

RAPID RECOVERY PRESERVATION, AND PHENOTYPING OF MILK PRO-TEINS BY A STARCH GEL TECHNIQUE OF SUPERIOR RESOLVING POWER

El-Negoumy, A. M., J. Dairy Sci. **48**, 784 (1965).

A reproducible technique suitable for large-scale screening of milk samples is described. The casein in 10 ml milk was precipitated by adding 10 ml acetate buffer, pH 4.60, at 50°C. The whey was filtered, exhaustively dialyzed, lyophilyzed, and dissolved in 2 ml 7M urea. Whatman® No. 3 strips were spotted with the protein solution and air-dried for preservation. The gel was composed of: 50 ml Tris–citrate buffer, 190 ml water, 44 g hydrolyzed starch, 105 g urea, and 1.40 ml mercaptoethanol. Sodium borate buffer, pH 8.60, was used in electrophoresis. A 1 x 5 mm insert of the protein-spotted filter paper was embedded into the gel for electrophoresis. A potential of 250 volts for 20 hours at 15°C separated the proteins into 18 to 20 bands. One hundred milk samples can be processed for electrophoresis in three hours. Protein-spotted papers showed no change in electrophoretic pattern after two years storage.

SERUM PREALBUMIN: POLYMORPHISM IN MAN

Fagerhol, M. K., and M. Braend, Science **149**, 986–987 (1965).

Four hundred twenty-one sera were analyzed by starch-gel electrophoresis, gel buffer, pH 4.95, for examination of the prealbumin bands. Based on a study of 31 members in two families, a genetic theory of three codominant alleles corresponding to the migratory rates of the prealbumin bands is proposed.

RED CELL ACID PHOSPHATASE: RACIAL DISTRIBUTION AND REPORT OF A NEW PHENOTYPE

Giblett, E. R., and N. M. Scott, Am. J. Human Genet. **17**, 425–432 (1965).

A new low-frequency acid phosphatase phenotype is described. Using a pH 5.0 formic acid–NaOH buffer (10.2 ml 90% formic acid and 9.25 g NaOH per liter), the usual phenotypes and the new phenotype of acid phosphatase were seen after starch-gel electrophoresis. Separation was distinct in 16 to 20 hours at 4 to 6 volts/cm at 6°C. This method is an improvement over the succinate–Tris–borate method of separation, since bands are better separated. A family was studied to trace the inheritance of the new phenotype.

A REPORT ON THE PRESENCE OF A THIRD α-COMPONENT IN CALF AND RAT SKIN TROPOCOLLAGENS

Heidrich, H.-G., and L. K. Wynston, Hoppe-Seylers Z. Physiol. Chem. **342**, 166–169 (1965) **German.**

A new type of α-component separated from calf and rat-skin collagen by chromatography on carboxymethyl cellulose was characterized by starch-gel electrophoresis, molecular weight determination, and amino acid composition. The component was found to be very difficult to separate from the α_1 component. Its molecular weight was 95,000 as determined by ultracentrifuge data. Eighteen amino acid residues were found.

STUDIES ON THE CLASSIFICATION OF THE ENZYMES HYDROLYZING ESTERFORM DRUGS IN LIVER MICROSOMES

Iwatsubo, K., Jap. J. Pharmacol. **15**, 244–256 (1965).

The esterases in rodent liver microsomes were extracted with sodium deoxycholate and fractionated either by chromatography or electrophoresis. A potato starch electrophoresis column was prepared in Veronal buffer. Electrophoresis was carried out at 300 volts and 23–25 mA for 30 to 36 hours in the cold (4°C). Then the fluid in the tube was slowly displaced with Veronal buffer and 0.7 ml fractions collected for assay.

ELECTROPHORETIC CHARACTERIZATION OF HUMAN DEHYDROGENASES

Katz, A. M., and W. Kalow, Can. J. Biochem. **43**, 1653–1659 (1965).

Vertical starch-gel electrophoresis performed in a phosphate–citrate buffer, pH 7.0, separated the isoenzymes of lactate, malate, and isocitric dehydrogenase of human skeletal muscle, heart and liver. Electrophoresis was carried out at 4.5 volts/cm for 18 hours at 4°C. Zones of isozyme activity were located with tetrazolium procedures employing the appropriate substrate.

X-LINKED 6-PHOSPHOGLUCONATE DEHYDROGENASE IN DROSOPHILA: SUBUNIT ASSOCIATION

Kazazian, H. H., Science **150**, 1601–1602 (1965).

6-Phosphogluconate dehydrogenase (PGD) subunits were dissociated and reassociated in vitro. An intermediate form of the heterozygous phenotype was produced by reassociation of subunits derived from the two types of PGD from homozygous flies. The phenotypes were identified by starch-gel electrophoresis.

HETEROGENEITY OF LIVER ALCOHOL DEHYDROGENASE ON STARCH GEL ELECTROPHORESIS

Mckinley-Mckee, J. S., and D. W. Moss, Biochem. J. **96,** 583–587 (1965).

Horizontal starch-gel electrophoresis according to the method of Smithies [Biochem. J. **61,** 635 (1955)] was performed on alcohol dehydrogenase purified from horse liver. Several buffer systems were employed. Upon electrophoresis in a pH 7.1 phosphate buffer, four bands of activity were found. Similar results were obtained using Tris–phosphate and Tris–acetate buffers, pH 7.1. When Tris–chloride, pH 7.1, and borate buffer were used, an extra band (staining faintly) was observed. Alteration of the electrophoretic patterns was possible by adding oxidized or reduced coenzymes, isobutyramide, metal ions or metal-chelating agents to the buffers.

SEX-LINKAGE OF ERYTHROCYTE GLUCOSE-6-PHOSPHATE DEHYDROGENASE IN TWO SPECIES OF WILD HARES

Ohno, S., and I. Gustavsson, Science **150,** 1737–1738 (1965).

Vertical starch-gel electrophoresis, using pH 8.6 borate buffer or a pH 7.0 phosphate buffer has successfully separated glucose-6-phosphate dehydrogenase of each of two wild European hares. Electrophoresis continued for 15 hours at 4°C with a potential gradient of 4 volts/cm. The sex-linkage of this enzyme was tested through reciprocal hybrids of the two species (L. europaeus and L. timidus). Each male hybrid had a single band of activity identical to that of the mother. Coexistence of both parental types of this enzyme was found in the female hybrids.

CHARACTERIZATION OF EFFORT PROTEINURIA

Poortmans, J., Protides Biol. Fluids, Proc. 12th Colloq. 327–329 (1965).

Gel and starch electrophoresis demonstrated two types of proteinuria in 22 male athletes. Prealbumin, α_1-, α_2, β_1-, β_2-, γ_1-, and γ_2-globulin fractions characterized the first type. In the second type of proteinuria, the prealbumin was primarily orosomucoid and was more pronounced; the α- and β-globulin were the same, but other protein fractions differed. Bence-Jones proteins were demonstrated by immunoelectrophoresis.

CONFIRMATION OF ASSOCIATION BETWEEN ABO BLOOD GROUPS AND SALIVARY ABH SECRETOR PHENOTYPES AND ELECTROPHORETIC PATTERNS OF SERUM ALKALINE PHOSPHATASE

Price Evans, D. A., J. Med. Genet. (London) **2,** 126–127 (1965).

Electrophoresis of saliva samples on a 12.1% starch-gel prepared in an

0.05M Tris–HCl buffer, pH 8.8, separated the alkaline phosphatase enzyme. Separation was carried out at 20 mA, 260 volts for 3.5 hours. The electrode buffer was 0.30M Tris–HCl, pH 8.6. Electrophoresis revealed only one serum alkaline phosphatase band in blood group A and the group of nonsecretors of ABH. Two serum alkaline phosphatase bands were found in persons with blood groups O or B who secreted A, B, or H salivary factors.

MINOR PROTEASES IN THE STOMACH OF THE PIG

Ryle, A. P., Biochem. J. **98**, 485–487 (1965).

Starch-gel electrophoresis of proteases from the gastric contents of pigs stomachs revealed the presence of pepsins C and D, as well as pepsin E. Electrophoresis was performed at pH 3.2 in a 0.05M citrate buffer at 25 volts/cm for 22 hours. The presence of the peptidases on the starch gel was revealed by staining with a warm solution of 5% gelatin in 0.05 hydrochloric acid containing Congo Red. Pepsin B was indicated by a high ratio of peptidase activity when tested against acetyl-L-phenylalnyl–L–diiodotyrosine and hemoglobin.

FRACTIONATION OF MILK PROTEINS, ESPECIALLY OF MILK PRODUCED ON A PROTEIN-FREE FEED OF UREA AND AMMONIUM SALT AS THE SOLE SOURCE OF NITROGEN

Syvaoja, E. L., and A. I. Vitanen, Enzymologia **29**, 205–220 (1965).

Paper, starch-gel and polyacrylamide-gel electrophoresis and gel filtration were used to study the protein patterns of milk from cows on restricted diets. No differences were found in the proteins from protein-free fed cows and the normal-fed controls. Amino acid composition of the proteins was the same.

SEX-LINKAGE OF GLUCOSE-6-PHOSPHATE DEHYDROGENASE IN THE HORSE AND DONKEY

Trujillo, J. M., Science **144**, 1603–1604 (1965).

Different electrophoretic patterns of red cell glucose-6-phosphate dehydrogenase were found in hemolysates from the horse and donkey. Ascending starch-gel electrophoresis with a gradient of 4–5 volts/cm at 2–4°C gave good separation in 16 to 18 hours. The buffer was EDTA, boric acid and Tris, pH 8.0. Findings suggested an X-linkage of inheritance in male reciprocal hybrids and the existence of isozymes within the RBC system.

COMPARATIVE ZONE ELECTROPHEROGRAMS OF MUSCLE MYOGENS AND BLOOD HEMOGLOBINS OF MARINE AND FRESH WATER VERTEBRATES AND THEIR APPLICATION TO BIOCHEMICAL SYSTEMATICS

Tsuyuki, H., E. Roberts, and W. E. J. Vanstone, Fisheries Res. Board Canada **22**, 203–213 (1965).

The heterogeneity of the muscle myogens and hemoglobins of 50 species of fishes representing specimens from three of the four living classes of fishes of the superclass Gnathostomata were separated on starch gels. Electrophoresis was performed on starch gels prepared in 0.023M sodium borate buffer, pH 8.5 in order to separate myogens. Hemoglobins were separated using the same buffer with 0.1% EDTA added to the gel buffer. The myogen patterns were found to be quite different, as no two patterns were alike in 46 species examined. There was not any correlation observed between complexity of patterns and increasing order. The hemoglobin patterns were shown to exhibit diverse heterogeneity.

DIAGNOSTIC APPLICATION OF LACTATE DEHYDROGENASE ISOZYMES

Vessell, E. S., Exp. Med. Surg. Suppl. **23**, 10–19 (1965).

Starch-gel electrophoresis of LDH can readily be used as a diagnostic tool in most laboratories because five isozymes can be identified. The relative levels of the isozymes can be correlated with different diseased tissues. Methodological sources of error are discussed, and illustrated examples and 57 references are given.

POLYMORPHISM OF HUMAN LACTATE DEHYDROGENASE ISOZYMES

Vessell, E. S., Science **148**, 1103–1105 (1965).

Hemolysates subjected to vertical starch-gel electrophoresis in a discontinuous Tris–borate buffer system (12 volts/cm^2 at 4°C for three hours) demonstrated variants at either the a- or b- genetic locus in 4 out of 1200 individuals screened. Family pedigree suggests autosomal codominant inheritance of the variant phenotype.

LYTIC ENZYMES OF SPORANGIUM SPECIES: ISOLATION AND ENZYMATIC PROPERTIES OF THE α- AND β-LYTIC PROTEASES

Whitaker, D. R., Can. J. Biochem. **43**, 1935–1954 (1965).

Two lytic proteases were isolated from culture filtrates of Sporangium species by adsorption on Amberlite® CG50 and elution with citrate buffer containing a gradient of sodium citrate. Starch-gel electrophoresis of the isolated enzymes at pH 8.0 in a Tris buffer showed that the α-enzyme migrated faster than egg white lysozyme, while the β-enzyme migrated slightly slower. Both enzymes were found to hydrolyze casein, although the α-enzyme was more reactive.

SOME PROPERTIES OF PHOTOOXIDIZED K-CASEIN

Zittle, C. A., J. Dairy Sci. **48**, 1149–1153 (1965).

Photooxidation reduced the electrophoretic mobility of K-casein on starch gel [J. Dairy Sci. **47**, 1052 (1964)]. The mobility of photooxidized K-casein was increased on cellulose acetate [J. Dairy Sci. **45**, 747 (1962)].

Starch (Block) Electrophoresis

UTILISATION DE LA MÉTHODE D'ÉLECTROPHORÈSE A TROIS PARAMÈTRES CONSTANTS DE WIEME POUR LA CARACTÉRISATION SIMULTANEÉ DES HAPTOGLOBINES ET DES TRANSFERRINES EN GEL D'AMIDOS

De Boissezon, J. F., H. Vergnes, and C. Decea, Rev. Franc. d'Études Clin. Biol. **12**, 86–90 (1967) **French.**

The Weime electrophoresis apparatus (diagrammed in this paper) and Poulik's discontinuous buffer system were used to separate serum and identify the proteins, haptoglobin and transferrins present. The borate buffer of Smithies at pH 8.6 was used in the chamber, and the Tris–citrate buffer of Poulik, pH 8.7, was used in the starch gel. A 60 mA current and a 200 volt potential separated the serum in 2½ hours. The starch block was sliced in three: proteins were stained with Amido Black, haptoglobins with benzidine, and transferrin was identified by autoradiography.

ANTIGENS OF BRUCELLA ABORTUS: I. CHEMICAL AND IMMUNOELEC-TROPHORETIC CHARACTERIZATION

Hinsdill, R. D., and D. T. Berman, J. Bacteriol. **93**, 544–549 (1967).

High-titer anti-Brucella cow serum was used to precipitate Brucella antigens separated by electrophoresis. A variety of purification methods were tried to obtain the maximum number of antigens. Powdered potato starch was used to prepare a starch block. Tris (hydroxylmethyl) aminomethane–acetic acid buffer, pH 8.2, ionic strength 0.02, was set up in a continuous flow system [See Bodman, Chromatographic and Electrophoretic Techniques, Vol. 2. New York:

Interscience Publishers, Inc., (1960)]. Complete separation of all antigenic components was not achieved. The fractions separated can be used to study the biological activity of the organisms.

MULTIPLICITY OF LEUCINAMIDE-SPLITTING ENZYMES IN NORMAL AND HEPATITIS SERA

Nakagawa, S., and H. Tsuji, Clin. Chim. Acta **13**, 155–160 (1966).

Starch-block electrophoresis was employed to separate leucinamide-splitting enzymes in normal and hepatitis sera as well as tissue homogenates. The electrophoretic separation was carried out according to Kunkel's procedure [Kunkel, in Methods of Biochemical Analysis, D. Glick, ed., New York: Interscience, (1954)]. After electrophoresis, the separated enzymes were eluted with saline solution, and enzymatic activity was estimated with leucyl-β-naphthylamide as the substrate. Normal serum contained two LNAase, one migrating in the α_1-globulin-albumin region, the other remaining at the origin. The LNA protein of hepatitis sera migrated to both sides of the origin.

NATURE OF HAEMOGLOBIN$_{\text{MOLDENBURG}}$

Pik, C., Nature **210**, 1182 (1966).

A detailed chemical investigation was done to find the amino acid replacement in Hb$_{\text{MOLDENBURG}}$. The Hb$_M$ fraction was separated from the hemolysate by starch-block electrophoresis after the hemoglobin was oxidized. A phosphate buffer (0.025M Na$_2$H PO$_4$ x 2H$_2$O and 0.025M NaH$_2$ PO$_4$ x 2H$_2$O), pH 7.0, was used. This purified Hb$_M$ was digested with trypsin for preparation of a peptide map. High-voltage paper electrophoresis at pH 6.4 (water–pyridine 9:1, acidified with acetic acid) was performed. An unusual peptide was present in Hb$_M$. This peptide was further separated by high-voltage electrophoresis at pH 3.6 (water–pyridine–acetic acid–n-butanol, 7.6:1:7:2). Further studies suggested that a histidine was replaced by a tyrosine in this variant of Hb.

Thin-Layer Electrophoresis

THE USE OF SEPHADEX® G-25 IN THIN-LAYER ELECTROPHORESIS AND CHROMATOELECTROPHORESIS OF AMINO AND LOW MOLECULAR WEIGHT PEPTIDES

Chudzik, J., and A. Klein, J. Chromatog. **36**, 262–264 (1968).

Thin-layer chromato-electrophoresis on Sephadex® G-25 successfully separated a mixture of 12 amino acids and 8 low molecular weight peptides. Chromatography of the mixture was achieved on paper with descending development in a solvent system composed of pyridine—methanol—water (1:20:5). The paper chromatogram was then covered with a 0.5 mm-thick layer of Sephadex® G-25 and electrophoresed in acetic acid—formic acid—water (3:1:135), pH 1.9. A voltage of 600 was sufficient for a 30 x 2 mm paper strip.

SCREENING FOR DISORDERS OF AMINO ACID METABOLISM BY THIN-LAYER HIGH VOLTAGE ELECTROPHORESIS

Farrelly, R. O., and W. B. Walkins, Clin. Chim. Acta **20**, 291–294 (1968).

High-voltage electrophoresis on thin-layer plates composed of cellulose—silica gel (5:2) separated amino acids in urine and serum. A buffer containing 25 ml of 98% formic acid and 87 ml glacial acetic acid per liter was used. Electrophoresis was carried out at 27 volts/cm for 45 minutes at room temperature. Fourteen amino acids were possible in one direction.

THE ESTIMATION OF VANILMANDELIC ACID (VMA) IN SERUM AND URINE BY A RADIOACTIVE C^{14} LABELING TECHNIQUE

O'Gorman, L. P., Clin. Chim. Acta **19**, 485–492 (1968).

A method which allows one to estimate vanilmandelic acid (VMA) in urine and serum is described. VMA was separated from other phenolic substances by electrophoresis on cellulose thin-layer plates in a pH 3.9 buffer of 3.0 ml pyridine $+$ 16.0 ml glacial acetic acid $+$ 991 ml H_2O. A constant potential of 500 volts was employed. The VMA isolated by TLE was acetylated with $[I-C^{14}]$ acetic anhydride, and reaction products were counted for radioactivity content. Recovery rates of 84% through the entire procedure, which requires several extraction steps, was 84%.

ELECTROPHORESIS OF PEPTIDES ON THIN LAYERS OF SILICA GEL

Sargant, J. R., and B. P. Vadlamudi, Anal. Biochem. **25**, 583–587 (1968).

The authors describe a technique for simultaneously fingerprinting two proteins on thin layers of silica gel. Two-dimensional development of tryptic digests of proteins was achieved on thin layers of silica gel bonded to flexible polyester (Chromatogram sheets Type 6061 Kodak Company). Each sample was applied to the 20 x 20 cm sheets so that the origins approximated 9 cm from each of the sides. After wetting with a solution of pyridine—acetic acid—water (25:1:225), electrophoresis was carried out in a horizontal direction in pyridine acetate buffer, pH 6.5, at 300 volts for two hours, 10 (mA). Chromatography was then performed in the second dimension in butanol—acetic acid—water

(3:1:1). Quantities as low as 100–200 μg of proteins would be used in the procedure and were detectable by spraying with cadmium-ninhydrin spray.

TERMINAL AMINO GROUP DETERMINATION, PEPTIDE MAPPING, BIOLOGICAL AND IMMUNOLOGICAL ACTIVITY OF AN INSULIN-LIKE PROTEIN AND THE AMINO ACID COMPOSITIONS OF ITS TWO S-SULFOPOLYPEPTIDE CHAINS

Yip, C. C., Arch. Biochem. Biophys. **127**, 741–748 (1968).

Thin-layer electrophoresis on silica gel G, following chromatography in a solvent system [methanol–chloroform–conc. ammonium hydroxide (2:2:1)], characterized a protein found in several insulin (bovine) preparations. TLE performed in a buffer system of pyridine–acetic acid–water (1:10:289), pH 3.6, has been used to determine amino acids and to map peptides released upon hydrolysis with formic acid.

DÜNNSCHICHTELECTROPHORESE IN ACRYLAMIDE UND STARKEGEL AUF MIKROSKOPISCHEN OBJEKTTRÄGERN

Klein, U. E., Clin. Chim. Acta **16**, 163–168 (1967) **German.**

A technique which lends itself to horizontal or vertical one- or two-dimensional microelectrophoresis was presented. Starch or acrylamide gels were used. The apparatus was constructed of Plexiglas, and the sample compartments could contain 2.5 to 20 μl of material.

EINFACHER TEST AUF PHÄOCHROMOCYTOME SOWIE NEUROBLASTOME UND DÜNNDARMCARCINOIDE: BESTIMMUNG DER 3-METHOXY-4-HYDROXYMANDELSÄURE, 3-METHOXY-4-HYDROXYPHENYLESSIGSÄURE UND 5-HYDROXYINDOLESSIGSÄURE IM URIN MITTLES CHROMATOELEKTROPHORESE

Lachhein, L., and J. Schutz, Clin. Chim. Acta **15**, 429–434 (1967).

Vanilmandelic acid, homovanilic acid and 5-hydroxyindoleacetic acid were separated from unextracted urine by chromato-electrophoresis. Separation was achieved on paper, and electrophoresis proceeded for 30 minutes.

THIN LAYER ELECTROPHORESIS OF ALKALI IONS ON SALTS OF HETEROPOLYACIDS

Lesigang Buchtela, M., and K. Buchtela, Mikrochim. Acta **6**, 1164–1167 (1967).

The thin-layer electrophoretic separation of the alkali metal ions Cs^+, Rb^+, K^+ and N^+ on ammonium phosphododecamolybdate oxine germanododecamolybdate and oxine germadodecamolybdate carriers was described. Ammonium nitrate 0.5M, pH 7.6, was used as the electrolyte.

THIN-LAYER PEPTIDE MAPPING

Ballieux, R. E., T. Sehens, and N. A. J. Mul, Protides Biol. Fluids, Proc. 14th Colloq. 527–529 (1966).

Tryptic digests of human IgG were separated by thin-layer chromato-electrophoresis. Thin layers of Kieselgel G (28.5g) and amylopectin (1.5g in 65 ml H_2O) were prepared to a height of 0.2 mm. Chromatography was performed in a direction perpendicular to the direction of application in a solvent system composed of propanol–ethanol–conc. ammonia–water (20:60:2:2) for seven to eight hours. Electrophoresis was then performed in a second direction for 50 minutes at a voltage of 1000 volts. A buffer of pyridine–acetic acid–water (1:10:489), pH 3.5, was used. When sprayed with ninhydrin, 45 ninhydrin positive spots were detected. Good resolution and reproducibility characterized the technique.

SEPARATION AND ESTIMATION OF AMINO ACIDS IN CRUDE PLANT EXTRACTS BY THIN-LAYER ELECTROPHORESIS AND CHROMATOGRAPHY

Bieleski, R. L., and N. A. Turner, Anal. Biochem. **17**, 278–293 (1966).

Fresh or frozen tissue amino acids extracted with methanol–chloroform–water (12:5:3) have been separated, without further purification, by a combination of thin-layer electrophoresis and chromatography. A mixed layer of cellulose–silica gel was the medium used. The buffer used for electrophoretic separation was 17 ml 90% formic acid + 57 ml acetic acid per liter, pH 2.0. Electrophoresis was performed at 55 volts/cm for 20 minutes (20 x 20 cm plate). Chromatography was then performed twice in methyl ethyl ketone–pyridine–water–acetic acid (70:15:15:2). A colorimetric method for the quantitative estimation of the separated amino acids was presented.

THIN-LAYER ELECTROPHORESIS: III. STUDIES ON THE DISSIPATION OF HEAT

Criddle, W. J., J. Chromatog. **24**, 112–116 (1966).

In order to separate amino acids and peptides, thin-layer electrophoresis is usually performed at high voltages, thus creating a great deal of excessive heat. The Frigistor® thermoelectric cooling device is an effective means of dissipating heat produced in the range of 25–50 volts/cm. Kieselguhr G, silica gel G, and alumina G thin-layers were tested. Borax, 0.05M, was the electrolyte. The use of the Frigistor® cooler extended the length of time that the current could be applied for electrophoresis.

THIN-LAYER ELECTROPHORESIS OF SERUM LIPOPROTEINS

Reissel, P. K., L. M. Hagopian, and F. T. Hatch, J. Lipid Res. **7,** 551–557 (1966).

A method for the semiquantitative separation of serum lipoproteins is presented. Serum samples were electrophoresed on starch granules (potato starch) spread on 20 x 20 m glass plates. Electrophoresis was performed at 600 volts (18 to 20 mA) for two hours. A pH 8.6 barbital buffer, ionic strength 0.1, containing 0.001M EDTA was used. The separated lipoproteins were located by staining with Oil Red O. This method was coupled with TLC and was found to possess advantages over paper electrophoresis.

SEPARATION OF URINARY AMINO ACIDS BY THIN-LAYER HIGH VOLTAGE ELECTROPHORESIS AND CHROMATOGRAPHY

Troughton, W. D., Am. J. Clin. Pathol. **36,** 139–143 (1966).

A mixture of 10 g cellulose, 4 g silica gel and 8.0 ml water was prepared for thin-layer plates. The plates were poured and washed overnight in a continuously ascending buffer of 2.5% formic acid, 8.7% acidic acid, and distilled water; final pH 1.9. This buffer was also used for electrophoresis. Twenty to forty μl of urine were applied and subjected to a current of 20–30 mA (50 volts/cm) for 30 minutes. The plates were dried at 37°C; distilled water ascended the plates to condense the spots. The dried plate was chromatographed in butanol–acetic acid–water (5:1:4, the top phase), phenol–water (80:20 w/v) or methylethyl ketone–acetic acid–water (14:6:5). Spots were stained with 0.5% ninhydrin in acetone.

FUNDAMENTAL ASPECTS IN THE PRACTICE OF THIN-LAYER ELECTRO-PHORESIS

Criddle, W. J., Proc. Soc. Anal. Chem. Conf. Nottingham, England **1965,** 135–140.

Alterations in electrophoresis (0.05M aqueous $NaBH_4$ electrolyte) due to heat generation and layer thickness of Kieselguhr G are described. A load of .021 w/cm^2 is the tolerated current for short term experiments with no danger of excessive heat. Lyophilization of plates overcomes the excessive migration often due to drying at elevated temperatures.

THIN-FILM ELECTROPHORESIS: II. FREEZE-DRYING OF ELECTROPHORETO-GRAMS

Criddle, W. J., J. Moody, and J. D. R. Thomas, J. Chromatog. **18,** 530–534 (1965).

The paper describes an apparatus constructed of ordinary glassware for use in freeze-drying of thin-films after electrophoresis. The apparatus is useful in overcoming zone migration which occurs during drying of thin-film electrophoretograms.

GIBBERELLINE: VI. MITT DIE DÜNNSCHICHTELECTROPHORESIS VON GIBBERELLINEN

Schneider, G., G. Sembdner, and K. Schreiber, J. Chromatog. **19,** 358–363 (1965) **German.**

Thin-layer electrophoresis performed on thin layers of silica gel G was used to study the mobility of gibberellins A_1, A_3-A_9. A Theorell-Stenhagen buffer with pH ranges of 2–12, ionic strength 0.05–0.06, was used. Electrophoresis was carried out at 4 to 6 volts/cm for five hours.

Miscellaneous Methods

DETERMINATION OF SERUM HAPTOGLOBIN BY ELECTROPHORESIS ON PLASTER OF PARIS

Affonso, A., Clin. Chim. Acta **12,** 466–467 (1968).

Serum haptoglobin was separated on 9 x 4 cm strips, 2 mm thick, using a pH 8.6 borate buffer. Separation was achieved at a potential gradient of 11.0 volts/cm for three hours. Free hemoglobin migrated behind the haptoglobin; both were well-defined bands when stained by the benzidine stain.

PREPARATIVE ELECTROPHORESIS OF SERUM PROTEINS ON PLASTER OF PARIS

Affonso, A., and E. Affonso, Clin. Chim. Acta **21,** 315–319 (1968).

A procedure utilizing 22 x 24 cm plaster of Paris strips ($2CaSo_4$, H_2O) for the electrophoretic separation and isolation of serum proteins was presented. The 3 mm-thick strips of solidified plaster of Paris were soaked for ½ hour in Veronal buffer, pH 8.6. Electrophoresis was then performed at 14 volts/cm for six hours in the Veronal buffer in a 38 x 38 cm electrophoresis chamber. After electrophoresis was completed, it was possible to break off a longitudinal strip of the media stain and fix to locate the separated bands. This stained

ANTIBACTERIAL ACTION OF PMN LYSOSOMAL CATIONIC PROTEINS RESOLVED BY DENSITY GRADIENT ELECTROPHORESIS

Zeya, H. I., J. K. Spitznagel, and J. H. Schwab, Proc. Soc. Exp. Biol. Med. **121**, 250–253 (1966).

Ascending electrophoresis was performed in a sucrose gradient column with a pH 4.0 acetate buffer, ionic strength 0.01, at 700 volts, 16 mA at 22°C. The lysosomal preparation was from inflammatory PMN of the rabbit peritoneum. The anode was filled with 50% sucrose, and the other vessel was filled with buffer. The gradient was formed by mixing and gravity flow. The fractions were assayed for anti-bacterial activity and identified using cellulose-acetate electrophoresis. The control enzymes were lysozyme, ribonuclease, deoxyribonuclease, acid phosphatase, and β-glucuronidase.

High-Voltage Electrophoresis

CHEMICAL TYPING OF IMMUNOGLOBULINS

Frangione, B., C. Milstein, and E. C. Franklin, Nature **221**, 149–151 (1969).

High-voltage electrophoresis combined with radioautography separated and characterized tryptic digests of partially reduced and C^{14}- iodoacetate alkylated myeloma proteins. Electrophoresis was performed on Whatman® 3MM paper for one hour at pH 3.5. A potential of 60 volts/cm was applied. Myeloma proteins containing each of the four types of γ-chains gave a distinctive autoradioautographic pattern for peptides of interchain disulphide bridges. Only a few bands were found in all 14 types of γ-globulin. It was possible to characterize the light chains with this technique during that same run.

ELECTROPHORETIC AND CHROMATOGRAPHIC SEPARATION AND FLUOR-OMETRIC ANALYSIS OF POLYNUCLEAR PHENOLS: APPLICATION TO AIR POLLUTION

Abstracts of 155 meeting of ACS, National Meeting U-33 (1968).

A number of polynuclear phenols were separated by thin-layer chromatography and low voltage and high voltage electrophoresis. The location of the separated compounds with OphthaLdehyde and 3-methyl-2-benzeothiozolinone hydrazone was described. Coupling this technique with fluorometric analysis of hydrolysis products, eight phenols were detected in coal tar.

ELECTROPHORESIS IN ZWITTERIONIC BUFFERS

Frigerio, N. A., and L. K. Kleiman, Abstracts of 156 meeting of ACS, National Meeting p-B-279 (1968).

The feasibility of high voltage gradients for electrophoretic separations by using Zwitterionic buffer solutions is discussed. The use of Zwitterionic buffers avoids the increased heat generation, diffusion, and thermal convection encountered with conventional buffers.

STARKE-GEL-HOCHSPANN UNGSELEKTROPHORESE BEI KLINISCHCHEM-ISCHEN SERUMUNTERSUCHUNGEN

Lange, V., Z. Klin. Chem. **5**, 168–175 (1968) **German.**

High voltage starch-gel electrophoresis was used to investigate pathological traits in serum of 140 persons. The gels were prepared in a 0.015M borate–0.006M sodium hydroxide buffer, pH 8.4. The gels were run at 18 volts/cm. The electrode buffer was 0.15M boric acid and 0.03M sodium hydroxide. The separated protein bands were revealed by staining with Amido Black. As many as 19 fractions were revealed in normal serum. Hepatocirrhosis serum was characterized by weak or absent haptoglobins and prealbumin combined with strong γ-globulins.

ELECTROPHORESIS ON CARRIERS AT HIGH ELECTRIC FIELD STRENGTH. SEPARATION OF AMINO ACIDS

Badzio, T., and T. Pompowki, Chem. Anal. **12**, 409–416 (1967).

Hydroxyproline, glutamic acid, methionine, leucine, aspartic acid, analine, glycine, lysine, and histidine were separated n an electrolyte composed of glacial acetic acid:formic acid (75 ml: 20.5 ml diluted to 1 liter, pH 1.9). A voltage of 410 volts/cm (paper strip 23 cm long) was applied for seven minutes. The construction of the apparatus, which will tolerate up to 450 volts, is described. It is possible to cool on both sides with pressure of 120 g/cm$_3$ applied to the top plate.

SELECTIVE PURIFICATION OF PHOSPHOSERINE PEPTIDES BY DIAGONAL ELECTROPHORESIS

Milstein, C. P., Nature **215**, 1190–1191 (1967).

A pepsin hydrolysate of ovalbumin was spotted on Whatman® No. 1 chromatography paper for electrophoresis at pH 6.5 with 53 volts/cm for one hour. The strip was then treated with alkaline phosphatase in 0.2 ammonium car-

acetic acid on 4.5M urea. One molar acetic acid was used in the electrode vessels at pH 9.0. The gel was either 7% or 10% acrylamide gel. Paper electrophoresis was done in 2.5% formic acid–9% (w/v) acetic acid buffer at 2000 volts for two hours. A variation in the histidine pattern was best seen using paper electrophoresis at pH 9.0 in 1.0M triethylamine carbonate buffer for two hours at 2000 volts.

GEGENSTROMIONOPHORESE: I. PRINZIP UND THEORETISCHE

Preetz, W., Talanta **13**, 1649–1660 (1966) **German.**

The techniques by which separation of ions with the same charge and different mobilities are separated are discussed. This technique is known as counter-current distribution. Ions to be separated form stationary zones, depending on mobility, between the point of entry of the counter-current into the tube and the point of exit, if the counter-current and anode solutions have a mobility greater than the ions to be separated. By the same token, the effluent and cathode solution must contain anions of lower mobility. The principle can be applied to the separation of cations as well.

BIOLOGICAL ACTIVITY OF BOVINE BRAIN CONSTITUENTS

Shaw, R. K., B. H. Chandler, and W. C. Stewart, Can. J. Physiol. Pharmacol. **44,** 69–76 (1966).

Preliminary experiments were done to assess the biological activity in 89 bovine brain extracts, which were fractionated by electrophoresis and chromatography. Definite correlations between peptides and biological activity could not be made at this time.

UN NOUVEAU SUPPORT POUR DÉS SÉPARATIONS ÉLECTROPHORETIQUES: LE GEL MIXTE D'ACRYLAMIDE-AGAROSE

Uriel, J., and J. Berges, Compt. Rend. Sci. (Paris) **C262**, 164–167 (1966) **French.**

A support medium for electrophoresis was prepared with 5% acrylamide gel and 0.7% agarose. The purpose of these experiments was to find a medium with the ease of preparation of agar gel and the excellent resolution properties of polyacrylamide gel. The buffer solution, pH 8.2, for preparation of the gel was: 6 g Tris, 15 g glycine, 0.36 g boric acid and 800 ml water. Final concentration was 0.2M. For electrophoresis the same buffer at pH 8.7 was used. The gel was prepared by dissolving 15 g agarose in 100 ml of boiling buffer which was maintained in a water bath at 55°C. Ten g acrylamide 0.26 g N,N'-methylene–bisacrylamide was placed in 100 ml buffer at 55°C. Just before use, 160 mg ammonium persulfate and 0.12 ml N,N,N',N'-tetramethylethylenediamine was added to the acrylamide preparation, and then this mixture was rapidly mixed with the agarose preparation. After 10 to 15 minutes incubation, the plates were poured. Electrophoresis was performed at 4°C. This method was applicable to the separation of a wide variety of biological substances.

PRÄPARATIVE ZONENELEKTROPHORPHORESE IM GIBSBLOCK

Böckemuller, W., and P. Kaiser, J. Chromatog. **18,** 86–99 (1965) **German.**

An apparatus was constructed for use in preparative electrophoresis employing pure gypsum as the stabilizing medium. Gypsum medium is especially suited to protein separation because it does not adsorb proteins and because electroosmosis is very minor. It was possible to fractionate 1 g of substance in one operation. Care must be taken to avoid use of gypsum with substances that form soluble sulfates or calcium salts.

PURIFICATION AND PROPERTIES OF SULFATED SIALOPOLYSACCHARIDES ISOLATED FROM PIG COLONIC MUCOSA

Inoue, S., and F. Yosizawa, Arch. Biochem. Biophys. **117,** 257–265 (1965).

Mucopolysaccharides obtained from pronase digests of mucin were separated and partially purified by fractionation on DEAE–Sephadex® A-50. Zone electrophoresis on GEON (a copolymer of vinylacetate and vinylchloride) and paper electrophoresis were used to characterize the fractions recovered. Electrophoresis on GEON was carried out in a 1.5 x 5.0 x 45 cm plastic tray using a 0.05M Tris–acetate buffer, pH 6.0 (GEON was suspended and poured into the tray). The run was for seven hours at 16–18 mA and 1.1 RV in the cold. Paper electrophoresis was performed on paper strips 17 x 40 cm at 12–13 mA (200–800 volts) for three to four hours. An acetate buffer, pH 2–4, 0.1M, citrate buffer, pH 2–3, or 0.05 Tris–acetate buffer, ph 6–7, were used. Additional chemical analyses were also employed to characterize the isolated material. DEAE–Sephadex® chromatography separated the material into two groups, one being sulfated sialopolysaccharide and the other a uronic containing mucopolysaccharide.

STUDIES ON β-PAROTIN: II. PURIFICATION OF β-PAROTIN

Ito, Y., S. Okabe, and S. Namba, Endocrinol. Japon. **12,** 69–77 (1965).

Zone electrophoresis on starch was the superior method for purification of acetone extracted β-parotin. Potato starch was washed in 0.3% HCl, water and acetone, and dried before using. The starch was suspended in a pH 8.6 Veronal buffer, ionic strength 0.1, and poured into the cell. An electric field

IDENTIFICATION AND GENETIC CONTROL OF TWO RABBIT LOW-DENSITY LIPOPROTEIN ALLOTYPES

Albers, J. J., and S. Dray, Biochem. Genetics **2**, 25–35 (1968).

Immunoelectrophoresis on 1% agarose in a sodium barbital buffer, pH 8.4, revealed the presence of two allotypic specifications in the low-density lipoprotein fractions isolated from rabbit serum. Immunodiffusion studies on agarose distinguished the variants.

ISOLEMENT ET ÉTUDE DES LIPOPROTÉINES SÉRIQUES ANORMALES AU COURS DES ICTÈRES PAR RÉTENTION APRÈS FLOCULATION PAR LE POLYVINYLPYRROLIDONE

Burstein, M., and J. Caroli, Rev. Franc. d'Etudes. Clin. Biol. **13**, 387–391 (1968) **French.**

Zone electrophoresis in Veronal buffer, pH 8.6, was performed to study alterations in the mobility of serum lipoproteins after reaction with anti-β-lipoprotein antibody. Micro-immunoelectrophoresis was done in gels and paper electrophoresis was also done in further studies. Light Green was the protein stain, Sudan Black was the lipoprotein stain.

ANTIGENS OF THE LENS OF XENOPUS LAEVIS

Campbell, J. C., R. M. Clayton, and D. E. S. Truman, Exp. Eye Res. **7**, 4–10 (1968).

The antigens of the lens of *Xenopus laevis* were investigated by electrophoresis, immunoelectrophoresis and diffusion against heterologous antiserum prepared in rabbits. At least 11 soluble protein fractions were demonstrated by cellulose-acetate electrophoresis in a 0.05M Tris-borate buffer, pH 8.6, run for 1.75 hours at 20 volts/cm. Immunoelectrophoresis according to the technique of Scheidegger in the high resolution buffer of Aronsson and Gronwall, pH 8.9, showed at least 22 antigens present.

STUDIES ON THE CHARACTERIZATION OF SOYBEAN PROTEINS BY IMMUNOELECTROPHORESIS

Catsimpoolas, N., E. Leuthner, and E. W. Meyer, Arch. Biochem. Biophys. **127**, 338–345 (1968).

Soybean proteins were separated into at least 12 antigenetically distinct components by immunoelectrophoresis in agar gel. Immunoelectrophoresis was carried out according to the procedure of Grabar and Williams as modified by Scheidegger on microscope slides. The gel was made of 1% Ionagar No. 2 (oxoid) in a Tris-barbital buffer, pH 8.8, ionic strength 0.05. A potential gradient of 5 mA per microscope slide was used. The antigenic components were detected with antiserum prepared by injecting isolated soybean protein into New Zealand white rabbits. Monospecific antiserum detected two components, 11S and 7S.

RADIOIMMUNOELECTROPHORETIC ANALYSIS OF THYROXINE-BINDING PROTEINS

Miyai, K., K. F. Itoh, H. Abe, and Y. Kumahara, Clin. Chim. Acta **22,** 341–347 (1968).

A technique described as radioimmunoelectrophoresis was used to analyze serum-binding globulins. Normal serum was mixed with low concentrations of purified I^{131}-thyroxine and incubated at room temperature for one hour or at 4°C overnight. Immunoelectrophoresis was performed by using a modification of the technique of Grabar and Williams. A pH 7.4 phosphate buffer, ionic strength 0.05, and a 1% agar plate were used separately. Separation was achieved at 2.5 to 3.5 volts/cm for 2.0 to 2.5 hours. Antiserum was added to the proper well and allowed to diffuse for 24 to 48 hours. Radioautography was then performed on the washed plate. Using this technique, a line of identity was found in TBG-deficient serum which was not found in normal serum. Five distinct TBG components were identified in normal serum.

THE DISTRIBUTION OF SPECIFIC ANTIBODY AMONG THE IMMUNO-GLOBULINS IN WHEY FROM THE LOCALLY IMMUNIZED GLAND

Outteridge, P. M., D. D. S. Mackenzie, and A. K. Lascelles, Arch. Biochem. Biophys. **126,** 105–110 (1968).

Immunoelectrophoresis techniques were used to identify antibodies of serum samples and whey samples isolated by gel filtration on Sephadex® G-200 and DEZE-cellulose columns. Immunoelectrophoresis was performed using microscope slides coated with Noble Agar (Difco) and LBK® electrophoresis cell. A Veronal buffer, pH 8.6 was employed. The immunoelectrophoretic patterns were developed with antisera prepared in rabbits.

ANALYSIS OF THE PROTEINS IN HUMAN SEMINAL PLASMA

Quinlivan, W. L. G., Arch. Biochem. Biophys. **127,** 680–687 (1968).

Cellulose-acetate electrophoresis, immunoelectrophoresis, and disc-gel electrophoresis have been used to analyze the proteins of human seminal plasma. Cellulose-acetate electrophoresis was performed in a Beckman MicroZone® system using a barbital buffer (0.1M, pH 8.6). A potential gradient of 250 volts for 20 minutes at room temperature was used. Immunoelectrophoresis was performed in a Shandon Universal Electrophoresis Apparatus on glass slides coated with agarose. An Owens modified barbitone buffer, pH 8.6, and a constant 50 mA current equivalent to 20 volts/cm for 4.5 hours were employed. Disc electrophoresis was performed according to the method of Davis [J. Ann.

Polyacrylamide (Disc) Electrophoresis

ELECTROPHORETIC SEPARATION OF MULTIPLE FORMS OF PARTICLE AS-SOCIATED ACID PHOSPHATASE

Allen, J. M., and J. Gockerman, Ann. N.Y. Acad. Sci. **121**, Art. 2, 616–633 (1964).

The demonstration by acrylamide-gel electrophoresis of the multiple acid phosphatases associated with particles and their differential release in rat liver mitochondrial-lysosomal fractions is described. Quantitative and electrophoretic examination indicated that both components of acid phosphatase were concentrated in the mitochondrial-lysosomal fraction, and it was concluded that two types of acid phosphatase, which differ in the nature of their binding to lysosomal structure as well as in their electrophoretic properties, may reside in lysosomal particles. The paper was presented at a conference on "Gel Electrophoresis," held in 1963 by the New York Academy of Sciences.

SOME ASPECTS OF PREPARATIVE ELECTROPHORESIS ON POLYACRYLA-MIDE GEL: APPLICATION TO BOVINE SERUM ALBUMIN

Altschul, A. M., W. J. Evans, W. B. Carney, E. J. McCourtney, and H. D. Brown, Life Sciences **3**, 611–615 (1964).

Some difficulties with polyacrylamide gel as the medium for high-resolution preparative electrophoresis reported by earlier workers were overcome by the authors by control of the physical parameters involved in the preparation of the column and in the mechanical design of associated equipment. Application of the technique to bovine serum albumin is described.

PROTEIN PHENOTYPING BY DIRECT POLYACRYLAMIDE-GEL ELECTRO-PHORESIS OF WHOLE MILK

Aschaffenburg, R., Biochem. Biophys. Acta **82**, 188–191 (1964).

Evidence of the existence of genetic variants of the principal protein consti-

MICROIMMUNOELECTROPHORESE AUF CELLULOSE ACETEGEL

Lomanto, B., and C. Vergani, Clin. Chim. Acta **15,** 169–171 (1967).

A method for performing immunoelectrophoresis on cellulose acetate was described. The optimum pH for electrophoresis was found to be 6.5–8.2. Veronal buffer, pH 8.2, ionic strength 0.05, was used routinely. A potential gradient of 160 volts, 0.5mA per cm, was employed to achieve separation.

QUANTITATIVE IMMUNOASSAY BY DISC ELECTROPHORESIS

Louis-Ferdinand, R., and W. F. Blatt, Clin. Chim. Acta **16,** 259–266 (1967).

By using polyacrylamide-gel electrophoresis, it was possible to quantitatively determine the extend of reaction of antigen to antibody. The antibody (anti-horse albumin) was mixed with a large-pore gel. Migration of albumin was retarded because of the combination of antigen and antibody. Quantitation was achieved by comparing the densitometric evaluation of stained protein bands of antibody-reacted material with a system in which the antibody was omitted. This technique was found to be valid only if the antibody was capable of fixing the antigen over a wide range of antigen concentration.

ELECTROIMMUNODIFFUSION (EID): A SIMPLE, RAPID METHOD FOR QUANTITATION OF IMMUNOGLOBULINS IN DILUTE BIOLOGICAL FLUIDS

Merrill, D., T. F. Hartley, and H. N. Claman, J. Lab. Clin. Med. **69,** 151–159 (1967).

A technique is described for diffusing antigen under an electric field into a layer of agar containing specific antiserum. The Ab-Ag precipitate length is proportional to the concentration of antigen and the duration of electrophoresis. Cerebrospinal fluid, a biological fluid with low γ-globulin concentration, was used as a test system. Microslides were prepared using 1% agar in a pH 8.6 Veronal buffer, ionic strength 0.05, which contained anti-γ-G. Four μL of CSF were applied to a well at the end of the slide. Electrophoresis was performed for 30 to 90 minutes at 150 volts in a pH 8.6 Veronal buffer, ionic strength 0.1. The lower limits of the technique are: γG 0.2 mg%; γA 0.5 mg%; γM 0.8 mg%. Normal CSF contained 1.0 to 3.1 mg/100 ml of γG.

EVALUATION OF LIGHT CHAIN ANTIGEN BINDING BY RADIOIMMUNO-ELECTROPHORESIS

Minden, P., H. Grey, and R. S. Farr, J. Immunol. **99,** 590–595 (1967).

Radioimmunoelectrophoresis demonstrated specific antigen binding by isolated light chain antibody. Anti-BSA was purified by the procedure of Uliky et al. [Immunochemistry **1,** 219 (1964)]. Electrophoresis of the purified preparation was carried out according to the procedure of Yagi. After slides were washed, I^{131} BSA was added to the wells, and allowed to diffuse for 24 hours, and radioautography was performed. Radioactive arcs and specific binding of I^{131} were demonstrated.

STUDIES IN SARCOIDOSIS: III. SERUM PROTEINS IN CASES WITH CONCOMITANT ERYTHEMA NODOSUM

Norberg, R., Acta Med. Scand. **181,** 101–114 (1967).

The proteins in 29 sera from patients with active erythema nodosum were quantitated by electrophoresis [Bottiger, Clin. Chim. Acta **5,** 664 (1960)]. Proteins, hexoses, hexosamines, and sialic acids in the electrophoretic fractions were also quantitated. Immunoelectrophoresis using commercial rabbit antihuman sera was performed to try to distinguish differences between patients and the controls. A pH 8.6 Veronal buffer, ionic strength 0.01, was used. Transferrin, the complement complex proteins, γA and γM immunoglobulins, and rapidly migrating γG migrated between the β- and the γ-globulin of erythema patients.

OPTIMAL CONDITIONS FOR IMMUNOELECTROPHORESIS OF SOME SOLUBLE BOVINE LENS ANTIGEN

Rathbun, W. B., M. A. Morrison, and R. M. Fusaro, Exp. Eye Res. **6,** 267–272 (1967).

The conventional techniques for the immunoelectrophoretic separation of bovine α-crystallin, β-crystallin and other lens proteins are evaluated. Conventional pH values, temperature and specific conductance values were found to be inadequate. Tris-phosphate buffer, pH 6.0–7.5 (specific conductance equivalent 0.025M KCl at $4°$C), gave improved results. Optimum conditions were different for each protein studied.

CORRELATION BETWEEN NET CHARGE OF ANTIGENS AND ELECTROPHORETIC MOBILITY OF IMMUNOGLOBULIN M ANTIBODIES

Robbins, J., Nature **213,** 1013–1014 (1967).

The net charge of the antigen influenced the electrophoretic mobility of IgM antibodies. A 0.05 sodium barbital buffer, pH 8.6, was used for electrophoresis and preparation of the 1.5% agar. A current of 220 volts, 31 mA, separated the proteins on 18 slides in 100 minutes. Rabbit sera was the sample. Goat antirabbit antisera precipitated the separated rabbit serum proteins. Acidic antigens tested were diphtheria toxoid and DNP-(T,G)-A-L; basic antigens were lysozyme, ribonuclease, uridine-A-L, and DNP-poly-L-lysine.

patients without structural changes in the nervous system. The application of this method to pathological fluids and the differences are also discussed.

POLYACRYLAMIDE GEL ELECTROPHORESIS OF SPINAL-FLUID PROTEINS

Evans, J. H., and D. T. Quick, Clin. Chem. **12**, 28–36 (1964).

Polyacrylamide-gel (disc) electrophoresis on vertical gels (7.5%) separated proteins of 23 control samples of CSF. CSF was concentrated by pressure dialysis against 0.01 phosphate buffer and electrophoresed at 3.5 mA per column of gel in a Tris-glycine buffer, pH 8.3. Fifteen bands were regularly observed, exclusive of the aphtoglobin band.

FAST FINGERPRINTING ON THIN LAYERS WITH POLYACRYLAMIDE AS DIAPHRAGM

Stegemann, H., and B. Lerch, Anal. Biochem. **9**, 417–422 (1964).

The authors describe an apparatus for electrophoresis and for fingerprinting (chromatography as a second run in the rectangular direction) on thin layers or paper. Directly connected to the thin layer or paper are the electrode vessels containing polyacrylamide as the diaphragm. With a relatively simple equipment, 60 volts/cm are applied and an electrophoresis is finished in one hour, a fingerprint in three to four hours.

IDENTIFICATION OF CATALASE FOLLOWING ELECTROPHORESIS ON ACRYLAMIDE GELS

Baumgarten, A., Blood **22**, No. 4, 466–471 (1963).

The use of acrylamide gels for the electrophoretic separation of plasma proteins and hemoglobins gave sharper bands and improved resolution as compared with starch gels, facilitated photometric scanning with their transparency, is easy to prepare, and can be stored for long periods before use. The author developed a micromethod of serum protein electrophoresis suitable for the rapid typing of haptoglobins, using acrylamide gels. This technique has been used in the investigation of electrophoretic mobility of hemoglobins, during which it was sometimes advantageous to intensify the color of the hemoglobin band using the staining procedure previously adopted for haptoglobins. Whenever this was done, there could be seen a narrow, completely clear band, approximately midway between the hemoglobin A band and the origin, whereas the rest of the gel assumed a reddish tint. This paper describes investigation into the activity associated with the clear zone, and presents a method for identifying catalase in acrylamide gels.

MICROMETHOD OF HAPTOGLOBIN TYPING USING ACRYLAMIDE GELS

Baumgarten, A., Nature **199**, No. 4892, 490–491 (1963).

A micromethod which utilized electrophoresis in acrylamide gels on microscope slides was proven more effective than the usual starch-gel electrophoresis techniques in the typing of haptoglobin in human sera. The micromethod can be completed in 50 minutes, offers better resolution of haptoglobin types, and requires only 0.1 μl of serum for examination. The technique is described in this paper.

A RAPID METHOD OF DISC ELECTROPHORESIS

Broome, J., Nature **199**, No. 4889, 179–180 (1963).

This paper describes and evaluates the results of a method of disc electrophoresis which obtains its results in about 70 minutes rather than in the two hours required by the original disc electrophoresis technique. In addition to speed, this method enables electropherograms of the plasma proteins (as anions) to be obtained from capillary blood, and excellent resolution is attainable.

SERUM PROTEIN ELECTROPHORESIS IN ACRYLAMIDE GEL: IV. TECHNICAL CONSIDERATIONS

Ferris, T. G., R. E. Easterling, and R. E. Budd, Am. J. Med. Technol. **29**, 163–168 (1963).

Several years of investigation by the authors on the use of acrylamide gel as a supporting medium for protein electrophoresis in a vertical cell revealed that the results of this technique were similar, if not identical, to those obtained with starch gel. After establishing basic operating parameters, certain unanticipated difficulties, some due to physical causes and others to matters of technique, were encountered which adversely affected the quality of electrophoretic separation. These difficulties are discussed and solutions to them are offered.

ELECTROPHORESIS OF SERUM PROTEINS IN ACRYLAMIDE GEL: II. AN IMPROVED APPARATUS AND TECHNIC FOR THE REMOVAL OF BACKGROUND STAIN, BY MEANS OF AN ELECTRIC CURRENT

Ferris, T. G., R. E. Easterling, and R. E. Budd, Am. J. Clin. Pathol. **33**, 193–197 (1963).

Problems had been encountered in the use of acrylamide gel as a medium for the electrophoretic separation of serum proteins because of the necessity of a prolonged washing to remove the background stain from the gel. The use of an electric current to remove the background stain also was found to have

ISOLATION AND CHARACTERIZATION OF TWO ANTIGENS OF CORYNE-BACTERIUM HOFMANNII

Banach, T. M., and R. Z. Hawirko, J. Bacteriol. **92**, 1304–1309 (1966).

Two seriologically-active substances extracted from sonic extracts of C. *hofmanni* were partially purified by column chromatography on DEAE-cellulose and by Sephadex® G-200 gel filtration. The active components isolated were characterized by immunodiffusion, immunoelectrophoresis and polyacrylamide-disc electrophoresis. Glutamic acid, aspartic acid, alanine, glycine, valine and leucine were the main amino acids present in antigen A.

SIMPLIFIED TECHNIQUE OF IMMUNOELECTROPHORETIC ASSAY OF HUMAN INTRINSIC FACTOR ON ACRYLAMIDE GEL

Bardhan, K. D., Gut **7**, 566– 568 (1966).

Electrophoresis was performed in 16-cm test tubes using 11 cm of the following gel mixture: 1 part water : 1 part solution A : 2 parts solution B : 4 parts solution C. The solutions were: (A) 48 ml 1N HCl; 36.6 g 2-amino-2-(hydroxymethyl)-1, 3-propanediol (Tris); 0.23 ml N.N.N^1, N^1-tetramethylethylenediamine; 100 ml water; (B) 28 g acrylamide; 0.74 g N,N^1-methylenebisacrylamide; 100 ml water; (C) 14 g ammonium persulfate; 100 ml water. Gastric juice was neutralized with 1.0N NaOH and filtered. One ml of the gastric juice was the sample placed on the gel. Tris-glycine stock buffer (6 g Tris–28.8 g glycine to 1 liter with water) was diluted 1:10 (pH 8.6) for the buffer. Electrophoresis was performed for 3½ to 4½ hours at 3.75–4 mA per tube.

ANTIGENIC STRUCTURE OF BRUCELLA SUIS SPHEROPLASTS

Baughn, R. E., and B. A. Freeman, J. Bacteriol. **92**, 1298–1303 (1966).

Immunoelectrophoresis was used to quantitate the antigens of normal cells of *Brucella suis* and of spheroplasts induced by penicillin and/or glycine. The supporting medium consisted of a 1% solution of Difco purified agar in 0.075M barbituric buffer, pH 8.6, pipetted in 2-ml quantities onto microscope slides. A 0.05M barbituric buffer, pH 8.6, was used in the electrode vessels. Electrophoresis was performed with a total of 150 volts for 100 minutes for eight slides. Glycine spheroplasts retained only six antigens, and penicillin spheroplasts eight, as contrasted with the normal bacteria which had at least thirteen antigens.

DENATURATION PROCESS IN ALBUMINS ISOLATED FROM THE SERUM OF HEALTHY WOMEN AND FROM UMBILICAL BLOOD SERUM OF THEIR NEWBORNS

Chojnowska, I., Ginekol. Polska **37**, 933–934 (1966).

Albumins were separated by TCA and ethanol precipitation. On starch gel and immunoelectrophoresis, maternal blood albumin showed two fractions while umbilical serum showed one component. Umbilical cord serum albumin was more easily denatured by heat and urea than was maternal albumin.

STRAIN VARIATIONS OF MOUSE 7Sγ_1 GLOBULINS

Coe, J. E., Immunochemistry **3**, 427–432 (1966).

Immunoelectrophoresis performed on glass slides coated with 2% ion agar in a pH 8.2 barbital buffer (ionic strength 0.04) demonstrated strain differences in the electrophoretic mobility of 7Sγ_1 globulin of mice. Samples were electrophoresed for 60 to 90 minutes at 6 volts/cm. Antiserum was added to the trough and allowed to diffuse for 24 to 48 hours at 4°C. Autoradiographs were obtained by adding HEA^{131}I to the trough after precipitin lines were developed with the antiserum. There seemed to be variation in mobility, which was probably hereditary and was not accompanied by antigenic dissimilarity.

LACK OF GAMMA-A IMMUNOGLOBULINS IN SERUM OF PATIENTS WITH STEATORRHEA

Crabbé, P. A., and J. F. Heremans, Gut **7**, 119–127 (1966).

The serum of three patients with steatorrhea was examined by immunoelectrophoresis according to the technique of Scheidegger. With the use of a multivalent horse antiserum against whole human serum and specific antiserum (rabbit) against the three human immunoglobulins, the authors demonstrate a deficiency of γA-immunoglobulins. Histochemical studies of the jejunal mucosal biopsy showed these patients to be almost completely devoid of plasma cells containing γA and γM-immunoglobulins in the lamina propria.

IMMUNOCHEMICAL STUDY OF PARAINFLUENZA VIRUS (TYPE 2) IN AMNION CELLS

De Vaux St. Cyr, C., and C. Howe, J. Bacteriol. **91**, 1911–1916 (1966).

Immunoelectrophoresis demonstrated that, in addition to hemagglutinin and enzyme, there are at least three infection-specific antigens that appear intracellularly during virus development. Electrophoresis of infected and control amnion cells was done in a 0.06M Veronal buffer, pH 8.2. Standard microscope slides coated with agar were used. Separation was carried out at

seven bands were obtained. The four principal bands had relative mobilities of 1.00, 0.83, 0.53, and 0.06. The glycine-1-C^{14} was isolated as DNP-glycine and the specific activity in each band determined. Band IV incorporated the largest amount of C^{14}-glycine.

ACRYLAMIDE-GEL ELECTROPHORESIS OF HEMOGLOBINS

Nakamichi, M., and S. Raymond, Clin. Chem. **9**, No. 2, 135–145 (1963).

A method is described which offers increased speed, reproducibility, and precision measurements from minor components in the analysis of hemoglobin mixtures using acrylamide-gel electrophoresis. Measurements on the unstained gel are made at 525 and at 420 mμ which is then stained with Amido Black to detect and measure non-hemoglobin components. The major component of normal human adult hemoglobin is separated in the gel into two components, A_1 and A_3, which appear in varying relative amounts depending on the source of the specimen. The normal minor component A_2 also appears, and, after staining the pattern with a protein dye such as Amido Black 10B, additional components which have a constant migration ratio as opposed to hemoglobin A_1 and appear to be nonheme-containing intraerythrocytic proteins, appear in the electrophoresis patterns of hemolysates from carefully washed red cells.

ACRYLAMIDE-GEL ELECTROPHORESIS OF SERUM AND SKELETAL MUSCLE EXTRACT

Namba, T., A. H. Wolintz, and D. Grob, Federation Proc. **22**, No. 2, Part I, 236, Sec. 453 (1963).

Electrophoretic patterns of proteins of human and rabbit sera and skeletal muscle extracts, using acrylamide gel as a supporting medium, were sharper, and more fractions were obtained than with any other zone electrophoresis method. The experiments, correlating each fraction with that of filter paper and starch-gel electrophoresis, and the findings are described. It was concluded that acrylamide gel is the best medium for the electrophoretic separation of serum and muscle proteins.

A CONVENIENT METHOD OF CASTING POLYACRYLAMIDE GELS

Tsuyuki, H., Anal. Biochem. **6**, 203–205 (1963).

The author describes a method which permits the casting of polyacrylamide gel, a supporting medium for zone electrophoresis, under a variety of conditions. The original method of casting the gel by floating a sheet of plastic on the surface of the monomer solution proved not to be completely satisfactory. Specially designed equipment is not needed because any type of casting tray, including that used for immunoelectrophoresis, may be used. The gel is cast under an atmosphere of nitrogen.

DISC ELECTROPHORESIS STUDY OF SERUM PROTEINS FROM PATIENTS WITH MULTIPLE MYELOMA AND MACROGLOBULINEMIA

Zingale, S. B., C. A. Mattioli, H. D. Bohner, and M. P. Bueno, Blood **22**, No. 2, 152–164 (1963).

The authors applied a new version of disc electrophoresis technique (Ornstein and Davis) to the study of the abnormal proteins from cases of multiple myeloma and macroglobulinemia. Sera from 31 patients with plasma cell myeloma and seven with macroglobulinemia were analyzed and of the 31 myeloma cases, 24 were found to have only one abnormal component. In spite of apparent homogeneity in paper electrophoresis, several abnormal components were seen in the remaining cases. Disc electrophoresis is simple to perform and provides a procedure of high resolution. It can be used for protein dissociation experiments with urea and sulfhydryl liberating compounds and is a valuable procedure for the diagnosis of macroglobulinemia. It does not, however, seem to demonstrate all the myeloma protein abnormalities revealed by starch-gel electrophoresis.

SYNTHETIC ACRYLIC GEL: A NEW MEDIUM FOR THE STUDY OF IMMUNE PRECIPITATES

Antoine, B., Science **138**, 977–978 (1962).

The use of a synthetic gel, Cyanogum®, as a medium for the double diffusion of antigen and antibody to form immune precipitates and a method for polymerization of the gel which avoids the use of the β-dimethylaminopropionitrile are described in this report. Among the practical advantages of the physicochemical properties of photocatalyzed Cyanogum® gel in the analysis of specific immunological precipitation techniques are the use of any buffer at any pH; most probable neutrality for stains or reagents; remarkable resolution of the immunological precipitates, visible microscopically; possibility of direct dark-background examination of the precipitates, so that their kinetics can be studied and they can be photographed before dyeing and staining. Cyanogum® gels also seem especially suitable for the immunological study of complex and unstable antigenic products, such as tissue homogenates.

SEPARATION OF HAEMOGLOBINS LEPORE AND H FROM A AND A₂ IN STARCH OR ACRYLAMIDE GELS BUFFERED WITH TRIS-E.D.T.A.

Curtain, C. C., J. Clin. Pathol. **15**, 288–289 (1962).

The use of Tris-EDTA buffer at low ionic strength gave an excellent resolution of hemoglobin Lepore in starch gels, and had the advantage of a much narrower zone on which to apply the samples than was possible on paper. It was also possible to scan photometrically without staining after clarification with anhydrous glycerol. The procedure is described.

done in ion agar using a Veronal buffer, pH 8.4, ionic strength 0.5M, for 2½ hours in a 4 mA current. The beta 1 b arc (hemopexin) decreased in size as the patient improved. The IgM arc was shorter in these patients than in controls. If the patient became worse, the IgM arc could disappear.

IMMUNOELECTROPHORETIC STUDIES ON RELATIONSHIPS BETWEEN PROTEINS OF PORCINE COLOSTRUM, MILK AND BLOOD SERUM

Karlsson, B. W., Acta Pathol. Microbiol. Scand. **67**, 83–101 (1966).

Agar-gel electrophoresis was performed on whole milk, colostrum and blood serum to characterize and compare the proteins present. Antisera to these samples was prepared in hyperimmunized rabbits. Causes of the occurrence of serum proteins appearing in milk and colostrum are speculated.

IMMUNOELECTROPHORETIC STUDY OF BIOLOGICAL FALSE-POSITIVE SERUM REACTIONS FOR SYPHILIS

Kiraly, K., Acta Dermato-Venereol. **46**, 506–510 (1966).

Paper electrophoresis in Michaelis buffer, pH 8.6, was done on sera from patients with a biological false positive to the *Treponema pallidum* immobilization test. The micro method of Scheidegger [Intern. Arch. Allergy Appl. Immunol. **7**, 103 (1955)] was used for immunoelectrophoresis at a potential gradient of 3.5 volts/cm. Horse or rabbit anti-human sera were used to form the precipitin lines. A hyperactive state of the immune system, characterized by increase level of γM and γG and decrease in albumin, was frequently found in biologically false-positive sera.

DERMATITIS HERPETIFORMIS AND HERPES GESTATIONSIS: ANALYSIS OF A γ_A AND γ_M SERUM PROTEIN BY IMMUNOELECTROPHORESIS

Kjartannson, S., R. M. Fusaro, and W. C. Peterson, Jr., J. Invest. Dermatol. **46**, 480–483 (1966).

When serum from patients with *dermatitis herpetiformis, herpes gestationsis,* and *pemphigus vulgaris* was examined immunoelectrophoretically, an increase in the γ_A globulin fraction of serum was observed. Antihuman serum-horse serum, antihuman-rabbit serum, antiherpes dermatitis-rabbit serum, and anti-dermatitis herpetiformis rabbit serum were used to demonstrate the increases. The immunoelectrophoretic method of Lim and Fusaro [J. Invest. Dermatol. **39**, 303 (1962)] was used.

IMMUNOELECTROPHORETIC ANALYSIS IN ACRODERMATITIS CHRONICA ATROPHICANS

Kraus, Z., and F. Mateja, Acta Dermato-Venereol. **46**, 217–223 (1966).

A pH 8.6 sodium barbital Veronal buffer, ionic strength 0.05, was used for immunoelectrophoresis of serum from nine treated and ten untreated acrodermatitis chronica atrophicans patients. The 1.5% agar solution was spread on 5 x 5 cm disposable slides and a voltage of 30 volts was applied. Horse anti-human sera formed precipitin lines with the separated sera. Amido Black 10B was the stain. An increase in all immunoglobulins was found in 12 patients; an increase in 2M globulins was found in seven.

THE OVERTAKING OR REACTION ELECTROPHORESIS AS A QUANTITATIVE IMMUNOLOGICAL METHOD

Lang, N., Protides Biol. Fluids, Proc. 14th Colloq. 517–526 (1966).

The technique of overtaking or reaction electrophoresis performed in a barbitone acetate buffer, pH 8.6, ionic strength $\mu0.1$, is described. The faster migrating component was applied to the electrophoresis strip behind the slower substance. During electrophoresis the fast migrating substance can catch up with the slower substance and react. If one or both of the substances were labeled, it was possible to determine quantitatively the amount of reacting substances. Only a few μl or μg of reacting components were needed to perform the technique.

COMPARATIVE BIOCHEMICAL STUDIES OF MILKS: III. IMMUNOELECTROPHORETIC COMPARISONS OF MILK PROTEINS OF THE ARTIODACTYLA

Lyster, R. I., R. Jenness, N. I. Phillips, and R. E. Sloan, Comp. Biochem. Physiol. **17**, 967–971 (1966).

The milk proteins of 19 species of artiodactyls were studied by immunoelectrophoresis using rabbit antisera from *Bos taurus*. Electrophoresis was performed on lantern slide cover glasses coated with a 1% agar solution. A 0.05M Veronal buffer, pH 8.6, with 10 mM EDTA to improve resolution of the casein was used for electrophoresis. The milk proteins of both ruminant and non-ruminant species reacted with anti-cow serum albumin and anti-cow casein sera.

IMMUNOELECTROPHORETIC EVALUATIONS OF PRESERVED BLOOD PROTEINS

Mackiewicz, U., and J. Pech, Farm. Polska, **22**, 271–273 (1966).

Blood was stored at 4°C. Plasma was prepared by centrifuging, and then compared to fresh citrated plasma. Tests were done on the day of collection

medium has the advantages of being easy to prepare, colorless, transparent, nonfluorescent, chemically inert, insoluble in all histochemical reagents so far tested, and strong enough to permit involved manipulation.

REACTIONS OF SERUM PROTEINS WITH A SULFHYDRYL COMPOUND DURING ELECTROPHORESIS

Ressler, N., and D. Rolandson, Clin. Chim. Acta **6**, 433–441 (1961).

Reactions of serum proteins with a sulfhydryl compound, during fluid film electrophoresis, were investigated by dissolving 2-mercaptoethanol throughout the buffer medium, thus keeping constant the concentration of the reacting sulfhydryl compound during the protein migration. An immunoelectrophoretic method, in which anti-normal human antiserum was dissolved throughout the buffer, was also used to demonstrate the components. The procedures followed and the results obtained are described and discussed.

ELECTROPHORETIC SEPARATION OF SERUM PROTEINS ON ACRYLA-MIDE GEL

Ott, W., Med. Welt. **1960**, 2697–2700.

Polyacrylamide gel electrophoresis prepared according to the method of Raymond and Weintraub separated serum proteins. When a pH 8.5 barbiturate–sodium acetate–hydrochloride buffer was used, as many as 20 protein bands were seen. The gamma globulin was divided into several fractions. Two-dimensional separations were achieved by first separating the proteins on paper and then transferring the paper pattern to the gel for further separation.

PREPARATION AND PROPERTIES OF ACRYLAMIDE GEL FOR USE IN ELECTROPHORESIS

Raymond, S., and Yi-Ju Wang, Anal. Biochem. **1**, 391–396 (1960).

Reasons for considering acrylamide gel as a replacement for starch gel were given with a description of its preparation and properties, and suitable apparatus for preparing the gel and carrying out the electrophoresis. Acrylamide is a water-soluble monomer which dissolves in buffers over a wide range of pH and on polymerization forms transparent flexible stable insoluble gels suitable for gel electrophoresis. During electrophoresis the gel must be adequately protected from evaporation.

PROPERTIES OF CORN SEEDLING GLUTAMATE DEHYDROGENASE

Gil' Manov, M. K., V. I. Yakovleva, and V. L. Kretovich, Dokl. Biochem. Sect. (English Trans.) **175**, 233–235 (1967).

L-glutamate dehydrogenase of corn root seedlings was examined for heterogeneity using polyacrylamide-disc electrophoresis according to the procedure of Ornstein and Davis [Ann. N. Y. Acad. Sci. **121**, 321 (1964)] as modified by Safonou et al. [Dan **167**, 1168 (1966)]. Electrophoresis was carried out for one hour at a current strength of 2.5 to 4 mA/tube. The separated enzymes were detected by staining the gels with a substrate solution and deposition of formazon. Only one band having enzymic activity was revealed by disc electrophoresis.

Agar–Gel
Electrophoresis

SEPARATION OF MICROSOMAL RNA INTO FIVE BANDS DURING AGAR ELECTROPHORESIS

Bachvaroff, R., and P. R. B. McMaster, Science **143**, 1177–1179 (1964).

This report describes investigations in which agar-gel electrophoresis separated microsomal RNA from rabbit livers and lymph nodes into five major bands, and simple agar electrophoresis divided the 33S and 19S peaks of microsomal RNA from sucrose-gradient zone centrifugation into two bands. The electrophoretic method may be used both as an analytical or as a preparative tool.

STUDIES ON THE HUMAN DANDRUFF ALLERGEN: II. PRESENCE OF THE ALLERGEN IN NORMAL HUMAN SWEAT

Berrens, L., and E. Young, Dermatologica **128**, 287–296 (1964).

In order to establish the presence of minor quantities of allergen in normal human sweat, the sweat proteins were subjected to electrophoretic analysis. The results are reported in this paper. Electrophoresis in agar gel separated the specific glycoproteins contained in normal sweat and paper electrophoresis indicated the presence of traces of the human dandruff allergen which could be proved by skin tests in atopic patients. The significance of the correlation found between the skin reactions to purified human dandruff and to the sweat

for 40 to 50 minutes, Bromophenol Blue stain). Fifteen individual proteins were identified: seven were derived from serum, the others were unique to urine.

DIFFERENTIAL STAINING OF CERULOPLASMIN AS AN AID TO INTER-PRETATION IN IMMUNOELECTROPHORESIS

Schen, R. J., and M. Rabinovitz, Clin. Chim. Acta **13,** 537–538 (1966).

Prussian Blue was used as a selective stain for the arc of ceruloplasmin which separates in the α_2-protein region in electrophoresis of serum. After electrophoresis, the excess proteins were eluted and the slide dried at low temperature. The slide was immersed for ten minutes at 37°C in a 0.08% solution of ferrous sulfate in 0.1M acetate buffer, pH 5.7. The slide was washed in water and immersed in a 1:1 solution of 2% potassium ferrocyanide and 0.2N HCl. The slide was again washed in water and dried. The usual protein stains may be used as a counterstain.

THE METHOD OF SIMULTANEOUS ELECTROPHORESIS OF ANTISERUM AND ANTIGEN (IMMUNOOSMOPHORESIS) APPLIED TO LENS ANTIGENS

Swanborn, P. L., Exp. Eye Res. **5,** 302–308 (1966).

Vertebrate α-crystallins were investigated using the technique of immunoosmophoresis. Antigen and antibody were electrophoresed simultaneously and met to form precipitates in the zone of contact. The technique was carried out on microscope slides coated with 1.5% Noble Agar in Veronal buffer (ionic strength 0.05, pH 8.4). Electrophoresis was performed at 20 volts/cm at 8°C for 15 minutes. A pre-α-crystallin group was found only in mammalian species.

ALLOTYPIC SPECIFICITIES IN PIGS

Trávníček, J., Folia Microbiol. **11,** 11–13 (1966).

The formation of isoantibodies in the pig was investigated by the injection of gamma globulin isolated from pig serum in 50 mg doses. A total of 500 mg gamma globulin was given each animal. Antibodies formed were characterized by immunoelectrophoresis and classified as gamma M globulins. At least two allotypic specificities of gamma M globulin in pigs was demonstrated.

IMMUNE REACTIONS IN ELDERLY HUMANS

Virag, S., and L. Kochar, Zh. Microbiol. Epidermol. Immunobiol. **43,** 99–103 (1966).

Elderly humans and rabbits revealed a decreased response to tetanus toxoid as compared to younger subjects. The γ-globulin level was measured by radioimmunoelectrophoresis using I^{131}-labeled antibody.

EMBRYO-SPECIFIC SERUM PROTEINS IN THE CHICK

Weller, E. M., Texas Rep. Biol. Med. **24,** 164–172 (1966).

Immunoelectrophoresis on 1% lonagar-coated lantern slides using a discontinuous Veronal acetate buffer was used to characterize the proteins of chickens. The Ionagar was prepared in 0.06M sodium barbital and 0.007 barbituric acid, pH 8.2, ionic strength 0.05. The electrode compartment consisted of 0.04M sodium barbital and 0.06M sodium acetate, pH 7.8, ionic strength 0.1. Separation was achieved at 20 mA for two hours at room temperature. Antiserum prepared in rabbits was used to detect specific proteins. Results obtained with heterologous and homogeneous antisera revealed existence of adult-like as well as embryo-specific proteins.

IMMUNOELECTROPHORETIC ANALYSIS OF CYTOPLASMIC PROTEINS OF NEUROSPORA CRASSA

Williams, C. A., and E. L. Tatum, J. Gen. Microbiol. **44,** 59–68 (1966).

Immunoelectrophoretic separation of cytoplasmic proteins of N. crassa was achieved on microscope slides by the procedure as described by Graber and Williams [Biochem. Biophys. Acta **17,** 67 (1955)]. Electrophoresis was for 90 minutes at 5 volts/cm in a Veronal buffer (ionic strength 0.038, pH 8.2). Catalase and other nonspecific proteins were identified according to the method of Uriel (Methods in ImmunoChemistry Immunol **2,** Ed. by M. S. Chase and C. A. Williams, Ch. 14, New York: Academic Press). Antisera prepared by immunization of rabbits with N. crassa strain 37401-116-3 and strain 17-3A (wild) were used to detect species-specific proteins. The technique revealed more than 30 antigenic components in cytoplasmic proteins. In addition a phyletic relationship within the class ascomycetes was detectable from results obtained with antisera prepared against N. crassa.

PROTEIN-IRON LEGANDS AND THE TRANSPORT OF IRON IN HUMAN MILK. IN VITRO RESEARCH USING RADIOACTIVE IRON AND AUTO-RADIOGRAPHS

Ambrosino, C., A. Ponzone, and G. Papa, Minerva Med. **56,** 4165–4167 (1965).

The iron-binding capacity of defatted human milk was analyzed by electrophoresis. Two iron-binding components were found in the β-lactoglobulin fraction, one in the α-globulin fraction, and another in an immunoglobulin.

of proteins by agar-gel electrophoresis, the authors successfully modified the use of a piece of equipment which had been used for paper strip electrophoresis in their laboratory. Proper modification of this procedure for electrical potential and buffer pH expedites pilot studies of new problems without necessitating large expenditures of time or money because the instrument is already used in many laboratories. The procedure is described in this paper.

ELECTROPHORETIC STUDIES OF THYROXINE AND TRIIODOTHYRONINE BINDING BY THE SERA OF GUINEA PIGS IMMUNIZED AGAINST THYROGLOBULIN

Premachandra, B. N., A. K. Ray, Y. Hirata, and H. T. Blumenthal, Endocrinology **73**, No. 2, 135–144 (1963).

Both agar and paper electrophoresis were used to study the distribution of radiothyroxine and radiotriiodothyronine in the electrophoretic pattern of normal and thyroglobulin-immune guinea pig sera. The significance of the binding of thyroid hormones by thyroglobulin-immune sera is discussed.

EFFECT OF THYROGLOBULIN IMMUNIZATION ON THYROID FUNCTION AND MORPHOLOGY IN THE GUINEA PIG

Premachandra, B. N., A. K. Ray, and H. T. Blumenthal, Endocrinology **73**, No. 2, 145–154 (1963).

The authors discuss their study of the effect of thyroglobulin immunization on thyroxine secretion rate and peripheral utilization of thyroid hormones and their study of the morphologic changes in the thyroid, adrenals and hypophysis. Also presented are possible explanations for the anomalous situation of decreased peripheral utilization of thyroid hormones and decreased TSH secretion. Agar and paper electrophoresis were used in the study.

ÉLECTROPHORÈSE EN GÉLOSE DES PROTÉINES TISSULAIRES DE LA SOURIS

Sassen, A., F. Kennes, and J. R. Maisin, Protides Biol. Fluids, Proc. 11th Colloq., 442–445 (1963) **French**.

Microelectrophoresis on agar gel was used to study the soluble proteins of mice tissues, namely liver, lymph nodes and spleen. An electrophoretic pattern of 24 fractions in the liver, 16 in the lymph nodes and 15 in the spleen was yielded by a microtechnique suitable for a small amount of tissues. Abnormal mobility of tissue albumin was observed.

A NEW VARIANT OF HUMAN FETAL HEMOGLOBIN: Hb F$_{ROMA}$

Silvestroni, E., and I. Bianco, Blood **22**, No. 5, 545–553 (1963).

A new type of abnormal fetal hemoglobin, identified from the cord blood hemolysate of a healthy, newborn girl in Rome and designated as Hb F$_{Roma}$, is described. It has an electrophoretic mobility at alkaline pH identical to that of Hb Bart's and a spectrum in the UV of the fetal type; however, it is composed of normal alpha chains and altered gamma chains. It was present at birth in the portion of 17 percent but disappeared completely during the fifth month of life. Various methods, including starch-block electrophoresis and agar-gel electrophoresis, were used to study the umbilical cord blood from which the abnormal hemoglobin was first identified.

MICRO-ELECTROPHORESIS OF MUCOPOLYSACCHARIDES ON AGAROSE GEL

Van Arkel, C., R. E. Ballieux, and F. L. J. Jordan, J. Chromatog. **11**, 421–423 (1963).

A method, based on micro agar-gel electrophoresis, and using only microgram amounts of material, was developed for the separation and analysis of mucopolysaccharides. In order to eliminate the metachromatical staining, a sulfate-free agar (agarose) was prepared. This microelectrophoretic technique is described.

AN INTEGRATED PROCEDURE FOR AGAR GEL ELECTROPHORESIS

Wieme, R. J., Protides Biol. Fluids, Proc. 11th Colloq., 397–400 (1963).

Quantitation by direct photometry, the final step in a procedure of agar-gel electrophoresis with petroleum ether acting as a coolant and sealer which can also be carried out at constant low temperature, is described.

LACTATE DEHYDROGENASE ISOZYMES: LABILITY AT LOW TEMPERATURES

Zonday, H. A., Science **142**, 965-967 (1963).

Agar-gel electrophoresis on microscope slides indicated that changes can arise in LDH isozyme pattern when tissue homogenates are stored at $-20°C$. The change in isozyme pattern is most pronounced when electrophoresis is run in a pH 9.1 barbital buffer. Barbital (pH 7.0), Tris-HCl (pH 7.9 and 7.0) and phosphate (pH 7.4) buffers were also tried. When nicotinamide adeninedinucleotide is added to the tissue homogenates before freezing, no alteration in the isozyme patterns is seen.

THE QUANTITATION OF HUMAN SERUM ALBUMIN BY IMMUNOELEC- TROPHORESIS

Griffiths, B. W., B. G. Sparkes, and L. Greenberg, Can. J. Biochem. Physiol. **43**, 1915–1917 (1965).

A method for analysis of HSA which depends upon detecting areas of antigen excess, equivalence, and antibody excess is described. A pH 8.6 Veronal buffer, ionic strength 0.05, was the electrolyte in agar-gel electrophoresis; 250 volts was applied for one hour. Goat antisera was applied and the bands allowed to develop for 20 hours. The antisera was titrated for estimation of the amount of HSA.

DIFFERENTIATION OF LEUKOCYTIC FIBRINOLYTIC ENZYMES FROM PLAS- MIN BY THE USE OF PLASMATIC PROTEOLYTIC INHIBITORS

Hermann, G., and P. A. Miescher, Intern. Arch. Allergy Appl. Immunol. **27**, 346–354 (1965).

Immunoelectrophoresis performed on agar plates into which fibrinogen had been incorporated was used to distinguish plasmin from fibrinogen. A pH 8.2 barbital buffer, ionic strength 0.05, contained 0.0082N calcium chloride. Extracts of normal human leukocytes and lymphocytes from patients with lymphocytic leukemia were shown to contain fibrinolytic enzymes. The mobility of these enzymes was in the alpha-1 to beta-2 region.

IMMUNOCHEMISCHE BEFUNDE BEI HEPATITIS EPIDEMICA: II. DIE IM- MUNOLOGIE DER β-LIPOPROTEINE

Kellen, V. J., Acta Hepato-Splenol. **12**, 228–231 (1965) **German.**

Qualitative and quantitative differences in the β-lipoproteins were found by performing immunoelectrophoresis on the serum of patients with hepatitis and normal controls. Agar electrophoresis was performed in sodium barbital buffer, pH 8.6, ionic strength 0.05, with a voltage of 5 volts/cm for 100 minutes. Commercial antihuman sera was used to form the precipitin lines.

MERCURIC BROMOPHENOL BLUE STAINING OF PRECIPITIN LINES IN AGAR

LaVelle, A., Stain Technol. **40**, 347–349 (1965).

Mercuric Bromophenol Blue is an excellent stain for antibody-antigen precipi- tin lines in agar. The precipitin-carrying slides were incubated overnight in saline at 4°C and then washed in saline for two hours at 25°C. Soaking in saline stopped the precipitin reaction. The slides were rinsed with water for ten minutes and then dried overnight at 37°C. Filter paper should not be used for drying. The precipitin lines were fixed in 95% ethanol and hydrated for five minutes in water. The slides were stained for ten minutes in the MBB mixture which consisted of 10 mg $HgCl_2$, 0.1 mg Bromophenol Blue; 100 ml 95% ethanol. The slides were rinsed through 95% and absolute alcohol and xylene before mounting with resin.

ISOLATION AND PARTIAL CHARACTERIZATION OF TRACE PROTEINS AND IMMUNOGLOBULIN G FROM CEREBROSPINAL FLUID

Link, H., J. Neurol. Neurosurg. Psychiat. **28**, 552–559 (1965).

The finding of β-trace protein and γ-trace proteins from human cerebrospinal fluid was confirmed. DEAE cellulose and Sephadex® chromatography were used to further study these low molecular weight proteins. An isolation procedure is described. Agar-gel electrophoresis was performed in a 0.9% agar solution in a continuous sodium Veronal buffer, pH 8.4, ionic strength 0.05. Electrophoresis was done for 25 minutes at 200 volts. Slides were stained with Amido Black DB. Starch-gel electrophoresis was done in 0.035M glycine buffer, pH 8.8, containing 8M urea. The electrode vessels contained 0.3M boric acid — 0.06N NaOH, pH 8.2.

IMMUNOELEKTROPHORETISCHE DETEKTION UND EIGENSCHAFTEN DER OSCENDOGENEN SERUM-LIPOPROTEINLIPASE

Losticky, C., Hoppe-Seylers Z. Physiol. Chem. **342**, 13–19 (1965) **German.**

Agar-gel electrophoresis and immunoelectrophoresis on agar gel, using a pH 8.3 Veronal-citrate buffer, μ .05, were used to demonstrate plasma lipase activity to Tween-60 in normal plasma and serum. Lipoprotein lipase was shown to be the cause of this activity through the use of the inhibitors protamine sulfate, heparin and sodium desoxycholate.

AN ALBUMIN AS γ-GLOBULIN CHARACTERISTIC OF BOVINE CEREBRO-SPINAL FLUID

Macpherson, C. F. C., and M. Saffran, J. Immunol. **95**, 629–634 (1965).

Agar-gel immunoelectrophoresis by the method of Scheidegger [Intern. Arch. Allergy Appl. Immunol. **7**, 103 (1955)] was used to identify an albumin and γ-globulin in bovine cerebrospinal fluid. High titer rabbit antibovine-CSF sera, adsorbed with guinea pig anti-CSF sera, was used to form the precipitin lines.

UBER DIE RADIOIMMUNOLOGISCHE BESTIMMUNG VON INSULIN IM BLUT

Melani, F., H. Ditschuneit, K. M. Bartlett, H. Friedrich, and E. F. Pfeiffer, Klin. Wochschr. **43**, 1000–1007 (1965) **German.**

The authors describe various methods for isolating antibody-bound I^{131} insulin.

AGAR GEL ELECTROPHORESIS FOR THE DETERMINATION OF ISOZYMES OF LACTIC AND MALIC DEHYDROGENASE

Yakulis, V. J., C. W. Gibson, and P. Heller, Am. J. Clin. Pathol. **38,** No. 4, 378–382 (1962).

A simple technique, based on agar-gel electrophoresis and the staining of the electropherogram with a histological stain, for the semiquantitative determination of isozymes of lactic and malic dehydrogenase (MD) is described. Consistent isozyme patterns which had been described previously were confirmed and the clinical usefulness of the isozymic pattern in the serum was illustrated by clinical examples. Tissues with predominance of anodic fractions found to include erythrocytes were kidney and heart. The cathodic fractions were most prominent in muscle and liver. The malic hydrogenase isoenzyme pattern with the described technique was not sufficiently tissue specific for diagnostic usefulness.

AN APPARATUS FOR AGAR GEL ELECTROPHORESIS

Osborn, E. C., Clin. Chim. Acta **6,** 743–745 (1961).

An apparatus evolved for the investigation by agar-gel electrophoretic techniques of proteins concerned in blood coagulation is described. It operates at a temperature of approximately 22° with a current flow of approximately 120 mA and with stainless steel dividing sheets of only 0.55 mm thickness enabling the central trough to obtain five fractions per cm of the supporting agar.

ELECTROPHORETIC SEPARATION OF PROTEINS ON EMULSIONS OF ADSORBENTS

Ostrowski, W., Clin. Chim. Acta **6,** 38–43 (1961).

A method is presented for the electrophoretic separation of proteins on adsorbent emulsions in low concentration agar gel which can be used for analytical or preparative purposes. Hyflo-super cel as the solid phase in concentrations of one to five percent gave the best results. Examples of resolution of protein mixtures using this technique are described.

COMPARATIVE STUDIES OF HUMAN-BRAIN AND BLOOD PROTEINS: AN ELECTROPHORETIC AND SEROLOGIC INVESTIGATION

Schalekamp, M.A.D.H., and M.P.A. Kuyken, Protides Biol. Fluids, Proc. 9th Colloq., 243–248 (1961).

Both agar electrophoresis and immunoelectrophoresis were employed in the separation of the soluble proteins of human brain. Antisera against brain as well as against serum and hemoglobin was used in order to determine whether some proteins in brain extracts are serologically different from blood proteins. The materials and methods used in the investigation are described.

DEMONSTRATION OF ANTIGEN-ANTIBODY REACTION WITH A MONO-BACTERIAL CULTURE OF ENTAMOEBA HISTOLYTICA IN AGAR-GEL AND ON CELLULOSE ACETATE MEMBRANE

Siddiqui, W. A., J. Parasitol **47**, 371-372 (1961).

A simple and reliable method for preparing an antigen extract of E. *histolytica* (*DKB* strain growing with *Clostridium perfringens*); a method for the production of *anti-E. histolytica* serum; and a demonstration of antigen-antibody reaction by two independent methods, slide-gel diffusion precipitin test and the cellulose acetate paper technique, are described in this paper.

ELECTROPHORETIC MOBILITY AND DETECTION OF HAEMOLYTIC ACTIVITY OF STREPTOLYSINS O AND S IN AGAR-GEL

Stock, A. H., and J. Uriel, Nature **192**, No. 4801, 435–436 (1961).

A procedure is described for the direct demonstration of the electrophoretic mobility of streptolysins O and S after electrophoresis in agar gel. The two lysins were easy to differentiate because streptolysin O migrated very slowly while streptolysin S exhibited rapid mobility.

THE HAEMOGLOBIN OF FOETAL BLOOD

Butler, E. A., F. V. Flynn, and E. R. Huehns, Clin. Chim. Acta **5**, 571–576 (1960).

Specimens of blood obtained from 26 human fetuses, 26 newborn infants and 6 normal adults were studied by alkali denaturation, resin chromatography, and agar- and starch-gel electrophoresis in order to demonstrate the existence of the embryonic hemoglobin as the largest component in the red cells of the young human embryo. The embryonic hemoglobin was shown to differ from fetal hemoglobin (HbF) in its electrophoretic mobility in agar at pH 6.2 as well as by its rate of alkali denaturation, but on electrophoresis at alkaline pH it had a mobility similar to that of adult hemoglobin (HbA) and HbF. No major hemoglobin component which differed from HbA and HbF was found in any fetus older than nine weeks.

STUDIES ON AN ABNORMAL HEMOGLOBIN CAUSING HEREDITARY CONGENITAL CYANOSIS

Hansen, H. A., O. R. Jagenburg, and B. G. Johansson, Acta Paediat. **49**, 503–511 (1960).

A study was made of a family with hereditary cyanosis, and the anomaly was

contained 5% phosphoric acid and the lower one contained 5% ethylene-
diamine solution. Electrofocusing was performed at a starting current of
5 mA per gel for one hour at 150 volts. The current drops during the run
to 0.5 mA per gel. Most of the separated protein bands could be visualized by
placing the gel in 12% trichloroacetic acid solution.

ISOLATION OF THE PI 4.5 SOYBEAN TRYPSIN INHIBITOR BY ISOELECTRIC FOCUSING

Catsimpoolas, N., C. Ekenstam, and E. W. Meyer, Biochim. Biophys.
Acta **175**, 76–81 (1969).

Soybean trypsin inhibitor was isolated in a pure state using isoelectric focusing in the region pH 3.0–pH 10.0. An LBK 8102® electrofocusing column with
a capacity of 440 ml column was used. Carrier ampholyte Ampholyne® was
used to prepare density gradients. Electrofocusing for 16 hours at a potential
of 500 volts (10°C) was sufficient to separate the inhibitor. Polyacrylamide-
disc electrophoresis was used to demonstrate the purity of the isolated material.

DEMONSTRATION OF THE SUBUNIT STRUCTURE OF ALPHA-CRYSTALLIN BY ISOELECTRIC FOCUSING

Bloemendal, H., and J. G. G. Shoenmakers, Sci. Tools **15**, 6–7 (1968).

Bovine α-crystallin was separated by electrofocusing in a sucrose gradient
containing 1% ampholytes (pH 5–8) and 6M urea. Electrophoresis was performed for two hours at 500 volts and subsequently at 800 volts for 48 hours.
Fractions were collected and analyzed by polyacrylamide-gel electrophoresis.
Considerable heterogeneity of the fractions separated was evident.

ISOELECTRIC FOCUSING IN POLYACRYLAMIDE GELS

Dale, G., Lancet **1**, 847–848 (1968).

Isoelectric focusing takes place when a potential is applied in an electrolyte
system in which the pH steadily increases from the anode to the cathode.
Thus, proteins separate as each accumulates in the region of its isoelectric
point. A method is described for stabilizing the pH gradient for polyacrylamide-
disc electrophoresis. The results demonstrated separation of 40 bands in
human serum and more LDH isoenzymes that can be seen with conventional
techniques.

ISOELECTRIC FOCUSING OF PLANT PIGMENTS

Jonsson, M., Sci. Tools **15**, 2–6 (1968).

The pigments of raw saps of red beet, bilberry, black currant and strawberry
were separated by isoelectric focusing in a pH gradient in a region of 3–10.

The gradients were created with Carrier ampholytes (Ampholyne®). An electrolysis column of 110 ml volume was used. Electrolysis was performed at 700 volts for the Carrier ampholyte gradients, 200–500 volts for the acid gradients for four to six hours at 4°C. Separation was excellent.

THE ISOELECTRIC FRACTIONATION OF HEN'S EGG OVOTRANSFERRIN

Wenn, R. V., and J. Williams, Biochem. J. **108**, 69–74 (1968).

Six protein fractions were separated from ovotransferrin, half saturated with iron, by isoelectric focusing. Ampholyte solutions and the apparatus of LBK® company were used. The ampholyte solutions had a pH range of 5 to 8 and a current strength of 500 volts for 24 hours. Starch-gel electrophoresis was performed on the fractions. Fractions 1, 3, and 5 showed mobility similar to that of the slower, major component. Fractions 2, 4, and 6 possessed fast mobility. Amino acid analyses of peak 5 (pH_I 5.78) and peak 6 (pH_I 5.62) showed all the amino acids as the unfractionated material except tryptophan and an amide. Labeling of ovotransferrin with 59_{Fe} and then subjecting it to isoelectric fractionation revealed that two major peaks of 59_{Fe} activity were present, one at pH_I 5.2 and pH_I 5.5.

ISOELECTRIC FRACTIONATION, ANALYSIS AND CHARACTERIZATION OF AMPHOLYTES IN NATURAL PH GRADIENTS

Vesterberg, O., and H. Svenson, Acta Chem. Scand. **20**, 820–834 (1966).

Myoglobulins with isoelectric point (pl) differences of only 0.06 units were separated by electrophoresis using a mixture of low-molecular aliphatic poly-amino–polycarboxylic acids as an ampholyte carrier (LBK–Produkter AB, Stockholm–Bromma). The ampholyte was easily dialyzable, colorless, a buffer, and non-light-absorbing at 280 mm. Zone breadth of myoglobulin separation satisfied a theoretically derived equation. A mathematical definition of resolving power was also derived.

Moving-Boundary Electrophoresis

MOVING BOUNDARY ELECTROPHORESIS OF FOOD STABILIZERS

Hidalgo, J., and P. M. T. Hansen, J. Food Sci. **33**, 7–11 (1968).

Moving-boundary electrophoresis using the Perkin–Elmer Model 38 A ap-

AGAR-GEL DIFFUSION AND IMMUNOELECTROPHORETIC ANALYSIS

Grabar, P., Ann. N.Y. Acad. Sci. **69**, 591–607 (1957).

This paper was given at the conference on "Immunology and Cancer," held in 1957 by the New York Academy of Sciences in collaboration with the National Institutes of Health, Allergy and Immunology Study Section, and the National Cancer Institute. It presents some historical background, practical applications, advantages, and limitations of agar-gel diffusion. A method called immuno-electrophoretic analysis, which is a combination of electrophoresis in a gel with double diffusion, as in Ouchterlony's technique, was developed to overcome some of the difficulties encountered in the gel-diffusion methods. This technique is described, its advantages are reported, and results obtained by its use are discussed. A micromodification of immunoelectrophoretic analysis, which is useful when only small amounts of substances are available and rapid results are desired, is described.

LES RÉACTIONS DE CARACTÉRISATION DES CONSTITUANTS DES LIQUIDES BIOLOGIQUES APRÈS ÉLECTROPHORÈSE EN GÉLOSE

Uriel, J., Protides Biol. Fluids, Proc. 5th Colloq., 17–23 (1957) **French.**

The principle of general or specific characterization techniques of the constituents of biologic fluids according to agar electrophoresis or immunoelectrophoretic analysis is presented and the advantages and disadvantages of this method, in comparison to other fields of migration in zone electrophoresis (paper, starch, etc.) are discussed. The possibilities of the application of the method to fundamental or applied research in protein analysis are also considered.

Starch-Gel Electrophoresis

NEW ELECTROPHORETIC PROTEIN ZONE IN PREGNANCY

Afonso, J. F., and R. R. De Alvarez, with N. G. Farnham, I. E. Thompson, and B. Wiederkehr, Am. J. of Obstet. and Gynecol. **89**, 204–214 (1964).

A protein zone in the alpha$_2$ globulin range among haptoglobins has been identified in pregnant women, through Smithies' starch-gel electrophoresis technique, and is assumed to occur in about 95 percent of normal pregnancies. The purpose of this study was to establish the time of the appearance of this protein zone, known as the "pregnancy zone," during the course of pregnancy and to explore its nature and significance. A total of 853 samples of blood, drawn from 480 randomly selected normal pregnant women, were submitted to vertical starch-gel electrophoresis, and the results are summarized in this report.

SERUM ALBUMIN POLYMORPHISM IN CATTLE

Ashton, G. C., Genetics **50**, No. 6, 1421–1426 (1964).

Five cattle serum albumin phenotypes resolved by starch-gel electrophoresis had been described, and it had been suggested that this variation represented serum albumin polymorphism controlled by a locus with three alleles. The author, using starch-gel electrophoresis, made further observations on albumin phenotypes in British and Zebu beef and dairy cattle in Australia, and his work is discussed in this report. Matings of 460 beef cattle showed that albumin polymorphism in cattle is controlled by two autosomal codominant alleles Alb^A and Alb^B. Alb^B is frequently found in Zebu beef and dairy cattle, but it is apparently absent in European cattle. It was noticed that the appearance of the phenotypes changes on storage and at acid pH.

ARYLESTERASES

Augustinsson, Klas-Bertil, J. Histochem. Cytochem. **12**, 744–747 (1964).

This paper defines arylesterases (ArE), a term used for esterases which preferentially split aromatic esters (esters of phenol, naphthol, and indole), and discusses their occurrence, separation from other esterases, specificity and other properties, normal variation in plasma arylesterase activity and genetic control of plasma arylesterase activity. The best-known ArE's are those present in mammalian blood sera. In this study various electrophoretic techniques were used to separate the various esterase types: human plasma (serum) by two-dimensional electrophoresis; mouse plasma by starch-gel electrophoresis; mammalian plasmata by cellulose column electrophoresis; and esterase zones by immunoelectrophoresis.

A CARBONIC ANHYDRASE VARIANT IN THE BABOON

Barnicot, N. A., C. Jolly, E. R. Huehns, J. Moor-Jankowski, Nature **202**, No. 4928, 198–199 (1964).

Red-cell hemolysates from baboons were examined by starch-gel electrophoresis, using the discontinuous Tris-borate system. Individual variations were noticed in the main non-hemoglobin protein (NHP) band after staining the starch gels with Amido Black. Details of the investigation are described in this paper. It was concluded that the variable main non-hemoglobin protein component in these baboons was carbonic anhydrase, and the existence of three phenotypes, S, F and S+F, is strongly suggestive of a variation determined by a single pair of alleles, though there was no evidence that it was genetically determined.

ELECTROPHORESIS OF SERUM HAPTOGLOBINS AND ERYTHROCYTE CATALASE USING CAPILLARY BLOOD

Baur, E. W., Clin. Chim. Acta **9**, 252–253 (1964).

A technique is described for the direct use of whole blood as an electrophoretic substrate for the thin-layer starch-gel electrophoretic determination of serum haptoglobin and red blood cell catalase types. Capillary blood, absorbed into filter paper strips, may be applied freshly to the gel surface or may be stored at −20°C for later use. Blood dried on filter paper strips may be exposed to room temperature for several days without significant interference to subsequent electrophoretic results.

HETEROGENEITY OF THE INHERITED GROUP-SPECIFIC COMPONENT OF HUMAN SERUM

Bearn, A. G., F. D. Kitchin, and B. H. Bowman, J. of Exp. Med. **120**, 83-91 (1964).

Conventional vertical starch-gel electrophoresis with a lithium borate buffer system and prolonged immunoelectrophoresis were used to demonstrate the heterogeneity of the group-specific (Gc) components in normal human serum. A protein component was found to migrate ahead of the main band in both Gc 1-1 and Gc 2-2 phenotypes, and immunological evidence indicated that the faster migrating band contains Gc specificity. The possibility that the two electrophoretically distinct Gc components share a common peptide chain is discussed.

STUDIES ON FOETAL HAEMOGLOBIN: VI. THALASSAEMIA-LIKE CONDITIONS IN BRITISH FAMILIES

Beaven, G. H., B. L. Stevens, M. J. Ellis, J. C. White, L. Bernstock, P. Masters, and T. Stapleton, Brit. J. Haematol. **10**, 1–14 (1964).

A study was made on the occurrence of a thalassaemia-like anemia in some members of five families of purely Anglo-Saxon stock. Routine hematological investigations were carried out, plus some special investigations including paper electrophoresis. An estimation of the minor component Hb-A_2 was made by electrophoresis in starch block and by densitometry of the separated zones in starch-gel electrophoretic patterns after staining and impregnation with glycerol to render the gel transparent. Both paper and starch-gel electrophoresis excluded in all cases the presence of Hb-H. The clinical and hematological findings are described and discussed. The results of the study support the view that various characteristics of heterozygous thalassaemia can occur separately and may be subject to genetic dissociation.

GENETIC VARIATIONS OF PHOSPHATASES IN LARVAE OF DROSOPHILA MELANOGASTER

Beckman, L., and F. M. Johnson, Nature **201**, No. 4916, 321 (1964).

With improvements in protein separation techniques, such as starch-gel electrophoresis, a large number of protein polymorphisms in different animal groups have been described. Electrophoretic variations in various enzymes were able to be examined by combining starch-gel electrophoresis with different enzyme staining methods. This paper discusses a genetic phosphatase variation found in larvae of *D. melanogaster* by starch-gel electrophoresis.

VARIATIONS IN LARVAL ALKALINE PHOSPHATASE CONTROLLED BY APH ALLELES IN DROSOPHILA MELANOGASTER

Beckman, L., and F. M. Johnson, Genetics **49**, 829–835 (1964).

An electrophoretic variation in the larval alkaline phosphatase of *Drosophila melanogaster* is reported in this paper. Findings show that a pair of co-dominant alleles control the inheritance of this enzyme variation, and that the locus (*Aph*-alkaline phosphatase) is present on the third chromosome. In addition to the parental enzymes, heterozygotes show a hybrid enzyme of intermediate electrophoretic mobility. Starch-gel electrophoresis was used in the study.

ISOZYME STUDIES OF SOME HUMAN CELL LINES

Beckman, L., and J. D. Regan, Acta Pathol. Microbiol. Scand. **62**, 567–574 (1964).

Starch-gel electrophoresis was used in these isozyme studies on some continuous human cell lines and primary cells. Electrophoretic variations in esterase, alkaline phosphatase, leucine amino-peptidase and catalase were examined in some human cell lines (HeLa, WISH amnion and RA amnion), the WI-38 strain, primary amnion and in serum.

CATALASE HYBRID ENZYMES IN MAIZE

Beckman, L., J. G. Scandalios, and J. L. Brewbaker, Science **146**, 1174–1175 (1964).

Two electrophoretic variants of catalase found in maize endosperm are under genetic control, and the heterozygote shows three hybrid enzymes with mobilities intermediate between the parental enzymes. Maize catalase may exist as a tetramer, therefore, and the hybrid enzymes may be formed by random association of two catalase monomers. Details of the investigation are reported.

GENETICS OF LEUCINE AMINOPEPTIDASE ISOZYMES IN MAIZE

Beckman, L., J. G. Scandalios, and J. L. Brewbaker, Genetics **50**, No. 5, Part 1, 899-904 (1964).

Multiple molecular forms of enzymes had been demonstrated in a large variety of organisms by a combination of starch-gel electrophoresis and various enzyme staining techniques. One of the molecular forms of an enzyme had been defined as an isozyme. The genetic control of leucine aminopeptidase (LAP) isozymes in maize is the subject of this report. By means of starch-gel electrophoresis, four different molecular forms of leucine aminopeptidase were found in maize endosperm (zones A, B, C and D). The results show that the LAP A and LAP D enzymes have electrophoretic variations, that each enzyme is controlled by a pair of alleles acting without dominance, and that the two loci, LpA and LpD are linked.

THE FRACTIONATION OF α-HISTONES FROM CHICKEN ERYTHROCYTE NUCLEI: I. THE EFFECT OF ETHANOL CONCENTRATION AND pH ON THE FRACTIONAL PRECIPITATION OF α-HISTONES

Bellair, J. T., and C. M. Mauritzen, Australian J. Biol. Sci. **17**, 990–1000 (1964).

A group of low molecular weight, lysine-rich α-histones from chicken erythrocyte nuclei was isolated and shown by starch-gel electrophoresis to contain 13 components. The materials and methods used in the study are described and the findings are discussed.

AMYLOID: II. STARCH GEL ELECTROPHORETIC ANALYSIS OF SOME PROTEINS EXTRACTED FROM AMYLOID

Benditt, E. P., and N. Eriksen, Arch. Pathol. **78**, No. 4, 325–330 (1964).

Continuing study to determine the nature of amyloid is discussed in this paper. When subjected to starch-gel electrophoresis, proteins extracted by strong urea from washed tissue of patients with amyloidosis secondary to several diseases or associated with familial Mediterranean fever were shown to have common major electrophoretic components. There may be present in hearts of primary amyloidosis one component in common with the secondary amyloid, but primary amyloid is much more difficult to dissolve and does not regularly give the components observed when the secondary amyloid is extracted. What constituent(s) comprise the fibrils now known to be part of amyloid is still in question.

PLASMA PROTEINS AND HAEMOGLOBINS OF THE AFRICAN ELEPHANT AND THE HYRAX

Buettner-Janusch, J., and V. Buettner-Janusch, Nature **201**, No. 4918, 510–511 (1964).

Starch-gel electrophoresis was performed on the plasma and hemoglobin of whole blood from an adult hyrax and an adult elephant, with human hemoglobin A added to the plasma samples to demonstrate haptoglobin bands. Results of the study, discussed in this paper, reconfirm previous findings indicating a close serological relationship between the hyrax and the elephant.

DISTRIBUTION AND FATE OF SOLUBLE PROTEINS LABELLED WITH IODINE-131 FROM RAT AND RABBIT LIVERS

Bocci, V., Nature **203**, No. 4947, 888–889 (1964).

The distribution and fate of liver I^{131}-proteins in relation to the chloramine T and iodine monochloride (IC1) procedures are described in this paper. Included in the investigation was vertical starch-gel electrophoresis performed with a continuous buffer system, and a contact autoradiogram was taken with x-ray film on the bottom slice of the gel. In this test, more I^{131}-proteins remained near to or at the origin when chloramine T amounts were higher, caused perhaps by the sieving effect of the gel on a modified protein molecular size, as suggested by the appearance of opalescence.

EFFECT OF OXIDIZED GLUTATHIONE ON HUMAN RED CELL ACID PHOSPHATASES

Bottini, E., and G. Modiano, Biochem. Biophys. Res. Commun. **17**, No. 3, 260–264 (1964).

This paper describes experiments performed to clarify the relationship between the presence of oxidized glutathione in the sample and the formation of fast moving bands of acid phosphatases, and an investigation of the correlation between the formation of these bands and the resulting loss of total enzyme activity. Starch-gel electrophoresis performed on the acid phosphatases of human red blood cells revealed five distinct patterns of fairly common occurrence and proved them to be an additional example of true genetical polymorphism, depending upon an autosomal locus with multiple allelism.

FOETAL AND NEONATAL HAEMOGLOBINS IN SHEEP AND GOATS

Breathnach, C., Quart. J. Exp. Physiol. **49**, 277–289 (1964).

The transition of hemoglobin from fetal to adult in fetal and newborn lambs and kids was studied, using as indices differences in alkaline denaturation rate, electrophoretic mobility and ease of elution from fixed red cells. In both species fetal hemoglobin was found exclusively before and at birth, then adult hemoglobin appeared in the first week, amounted to half the total by the third week and completely replaced the fetal pigment at the end of two months. Vertical starch-gel electrophoresis, in five to ten percent solution, was used to determine the mobility of the carboxyhemoglobin. The methods and results are evaluated.

A RENIN-LIKE ENZYME IN NORMAL HUMAN URINIE

Brown, J. J., A. F. Lever, D. L. Davies, A. M. Lloyd, J. I. S. Robertson, and M. Tree, Lancet **2**, No. 7362, 709–710 (1964).

The finding of a renin-like enzyme in extracts of normal human urine is reported in this paper, and its characterization and the possible significance of its presence in urine are discussed. Starch-gel electrophoresis was used in the analysis.

ACTIN: A COMPARATIVE STUDY

Carsten, M. E., and A. M. Katz, Biochim. Biophys. Acta **90**, 534–541 (1964).

The sedimentation coefficients, starch-gel electrophoresis patterns, amino acid composition, and peptide maps of actin preparations from beef, pig, lamb, chicken, frog, fish and pecten were compared. No differences were detected in the actin preparations within the group of mammals and the single avian species studied, but variations were found when these preparations were compared with actin from single species of amphibia, fish and mollusca. The results of this study seem to indicate that there are variations in actin when a wide enough range of the evolutionary scale is considered, and that these variations decrease with closer relationship of the species.

SOME NOTES ON THE STARCH GEL ELECTROPHORESIS OF HEMOGLOBINS

Chernoff, A. I., and N. Pettit, J. Lab. Clin. Med. **63**, No. 2, 290–296 (1964).

The application of a buffer system containing lithium salts to the starch-gel electrophoresis of hemoglobins and the effect of methemoglobin on hemoglobin electrophoresis are discussed by the authors. In the starch-gel electrophoresis of hemoglobin solutions carried out in two different buffer systems of approximately the same pH, the Tris-citrate buffer was noted to provide sharp definition of the more slowly moving hemoglobins at pH 8.7, but was inadequate for the separation of Hgb F from Hgb A. It was also noted that methemoglobin forms of various hemoglobin types moved more slowly than the oxy, reduced, CO or cyanmethemoglobin compounds. In a lithium buffer system, pH 8.04, electrophoresis provided sharp definitions of hemoglobins with a mobility equal to or greater than that of Hgb S, permitted a sharp distinction of Hgb F from Hgb A, and demonstrated an even more striking displacement cathodally of the methemoglobin forms of the various hemoglobin types.

ALPHA-CHAIN OF HUMAN HEMOGLOBIN: OCCURRENCE IN VIVO

Fahey, J. L., Science **143**, 588–589 (1964).

A minor hemoglobin component, apparently representing uncombined α-chains, was detected in the hemolysates of persons with inherited β-chain deficiency. Horizontal starch-gel electrophoresis of centrifuged cell lysate was performed in a discontinuous barbital-TEB buffer, pH 8.2, for six to seven hours in a water-cooled system. The gels were stained with a benzidine stain.

CONTRIBUTION OF γ-GLOBULIN SUBUNITS TO ELECTROPHORETIC HETEROGENEITY: IDENTIFICATION OF A DISTINCTIVE GROUP OF 6.6S γ-MYELOMA PROTEINS

Fahey, J. L., Int. J. Immunochem. **1**, 121–131 (1964).

The electrophoretic heterogeneity of 6.6S γ-globulins was investigated by starch-gel electrophoresis in a 0.05M glycine buffer, pH 8.8. The subunits obtained by reduction, alkylation and papain digestion were characterized. The relative mobility of H-chains and S-pieces corresponded to that of the intact proteins.

A SIMPLE MICROELECTROPHORETIC METHOD USING STARCH-AGAR GEL

Hase, T., Clin. Chem. **10**, No. 1, 62–68 (1964).

A simple microelectrophoretic method of protein analysis using a mixture of starch and agar is described and discussed in this paper. This method is easily standardized, results show consistency and reproducibility, and protein patterns obtained show higher resolution than those of paper or agar-gel electrophoresis and simulate that of starch-gel electrophoresis. Further advantages of this technique are that the developed protein patterns are suitable for quantitative analysis by appropriate electrophometric or UV light scanning devices and can be preserved as permanent records.

318   Electrophoresis – Technical Applications

HEMOGLOBIN POLYMORPHISM: ITS RELATION TO SICKLING OF ERY-
THROCYTES IN WHITE-TAILED DEER

Kitchen, H., F. W. Putnam, and W. J. Taylor, Science **144**, 1234–1239
(1964).

Variants of hemoglobin in the white-tailed deer were separated electrophoretically and identified with RBC sickling. Multiple hemoglobins were best demonstrated by starch-gel electrophoresis in a discontinuous EDTA-Tris borate
buffer, 0.06M, pH 9.0, and by agar-gel electrophoresis in sodium citrate–
citric acid buffer, 0.03M, pH 6.0. In starch-gel electrophoresis a 0.12M barbital
buffer, pH 8.6, was used in the vessels (10 volts/cm, 60 minutes at 10°C).
Separation in agar gel was complete in 1.5 hours, 15 volts/cm at 10°C. The
hemoglobins were in the carbon monoxide form. Buffalo Black was the
stain used.

HUMAN ERYTHROCYTE LACTATE DEHYDROGENASE: FOUR GENETICALLY
DETERMINED VARIANTS

Kraus, A. P., and G. L. Neely, Jr., Science **145**, 595–597 (1964).

Four variants of human erythrocyte LDH are described. Hemolysates were
fractionated by vertical starch-gel electrophoresis at 4°C with Tris-EDTA buffer,
pH 8.6, in a current of 8 volts/cm for 14 to 16 hours. Following the pedigree
of the variants, it appeared that inheritance was an autosomal codominant
characteristic. Eight persons from a sample of 940 possessed an LDH-variant,
suggesting that variant enzyme patterns are more common than previously
suspected.

ACID PHOSPHATASES OF HUMAN RED CELLS: PREDICTED PHENOTYPE
CONFORMS TO A GENETIC HYPOTHESIS

Lai, L., Science **148**, 1187–1188 (1964).

Starch-gel electrophoresis of acid phosphatases from human red cells was
used to demonstrate the phenotype of the enzyme. It was demonstrated that
the phenotypes are determined by three alleles at an autosomal locus. The
polymorphism of this enzyme is a useful tool in studying population genetics
and genetic control of enzyme structure.

AN ELECTROPHORETIC METHOD FOR RECOVERY OF PROTEINS FROM
STARCH GEL

Lloyd, H. M., and J. D. Meares, Clin. Chim. Acta **9**, 194–196 (1964).

A method devised for the simultaneous recovery of proteins from portions of
a starch gel two cm thick is described. The dimensions of the gel may be

modified as required. The method was found to be time-saving and highly efficient.

AUTOSOMALLY DETERMINED POLYMORPHISM OF GLUCOSE-6-PHOS-PHATE DEHYDROGENASE IN PEROMYSEUS

Shaw, C. R., Science **148**, 1099–1100 (1964).

Glucose-6-phosphate dehydrogenase commonly occurs in two forms in most tissues of the deer mouse, *Peromyscus maniculatus*. Vertical starch-gel electrophoresis at a gradient of 8 volts/cm at 40°C in 0.5M Tris–EDTA–borate buffer, pH 8.0, gave excellent separation in 18 hours. Polymorphism of this enzyme in the deer mouse appears to be controlled by an autosomal gene.

ISOZYME PATTERNS OF THE HUMAN GASTROINTESTINAL TRACT IN THE NORMAL STATE AND IN NONTROPICAL SPRUE

Weiser, M. M., R. J. Bolt, and H. M. Pollard, J. Lab. Clin. Med. **63**, No. 4, 656–665 (1964).

In a study of isozyme patterns of human intestinal mucosa homogenates, normal patterns were established and compared to those of patients with gluten-sensitive nontropical sprue. With small bowel biopsy specimens and surgical specimens, the supernatant of centrifugal mucosal homogenates were subjected to vertical starch-gel electrophoresis, and histochemical techniques were employed to demonstrate protein, nonspecific esterase, cholinesterase, lactic hydrogenase, and alkaline phosphatase. Findings showed that various parts of the gastrointestinal tract could be differentiated by esterase isozyme patterns and that the cholinesterase of the gut migrated in a different manner from cholinesterase of plasma, but that there was no differentiating pattern for intestinal lactic hydrogenase isozymes. Four bands of alkaline phosphatase isozymes, found only in the small intestine, were found, with the third band paralleling that of purified calf intestinal alkaline phosphatase, and the fourth band paralleling that of serum alkaline phosphatase. Patients with gluten-sensitive nontropical sprue showed normal isozyme patterns in the mucosal homogenates.

LACTATE DEHYDROGENASE IN PIGEON TESTES: GENETIC CONTROL BY THREE LOCI

Zinkham, W. H., Science **144**, 1353–1354 (1964).

Homogenates of pigeon testes show three types of lactate dehydrogenase isozyme patterns on starch-gel electrophoresis. Results of dissociation and recombination experiments indicated that LDH synthesis in pigeon testes is controlled by three genetic loci. Before electrophoresis, the isozymes were dissociated with 0.5M sodium chloride and 0.1M phosphate.

VARIATIONS IN THE ELECTROPHORETICALLY SEPARATED ACID PHOS-PHATASES OF TETRAHYMENA

Allen, S. L., M. S. Misch, and B. M. Morrison, J. Histochem. Cytochem. 11, 706–719 (1963).

Electrophoresis in starch gels of pH 7.5 separated the acid phosphatases of variety 1 of the ciliated protozoan, Tetrahymena pyriformis into 17 zones. These separated acid phosphatases can be distinguished genetically by their chemical properties, by their behavior under different growth conditions, and by their intracellular localization, and investigation of these characteristics suggest that the acid phosphatases are a family of enzymes of varying degrees of relationship.

POLYMORPHISM IN THE SERUM POST-ALBUMINS OF CATTLE

Ashton, G. C., Nature **198**, No. 4885, 1117–1118 (1963).

Three polymorphic serum protein systems, transferrins, thread proteins, and slow-α-globulins, had been described in cattle, and a fourth system involving the post albumins had been found. Its investigation is the subject of this paper. Starch-gel electrophoresis in a modified discontinuous borate electrolyte, Tris-citric buffer system, using the horizontal technique was utilized to reveal the polymorphic post-albumins and the results are described.

A STUDY OF HEMOGLOBIN DIFFERENTIATION IN RANA CATESBEIANA

Baglioni, C., and C. E. Sparks, Develop. Biol. **8**, 272–285 (1963).

Starch-gel electrophoresis of the hemoglobins of tadpole, metamorphosing tadpole, and frog (Rana catesbeiana) revealed the presence of three hemo-globins in tadpoles and four different ones in frogs. Both tadpole and frog hemoglobin are present in the metamorphosing tadpoles. The hemoglobins, which were isolated and characterized by fingerprinting and by electrophoresis in urea-containing starch gels, were composed of two types of peptide chains like mammalian hemoglobins. The materials and methods used in the study are described.

ABNORMAL HUMAN HEMOGLOBINS: IX. CHEMISTRY OF HEMOGLOBIN J BALTIMORE

Baglioni, C., and D. J. Weatherall, Biochim. Biophys. Acta **78**, 637–643 (1963).

A human abnormal hemoglobin, obtained from an 18-year-old Negro male, with the electrophoretic mobility of hemoglobin J, was isolated and studied, and the amino acid substitution in this abnormal hemoglobin was investigated. An aspartic acid residue was found to substitute in this hemoglobin J the glycine residue present in position 16 of the β-peptide chain. Because there are reasons to believe that this hemoglobin J is different from other hemoglobin J's, it has been designated J$_{\text{BALTIMORE}}$. Starch-block electrophoresis at pH 8.6 was used to separate the hemoglobin.

STARCH GEL ELECTROPHORESIS OF RAT HYPOPHYSIS IN RELATION TO PROLACTIN ACTIVITY

Baker, B. L., R. H. Clark, and R. L. Hunter, Proc. Soc. Exp. Biol. Med. **114**, 251–255 (1963).

A slowly-moving protein band, revealed by starch-gel electrophoresis of hypophysial suspensions, was eliminated by combined ovariectomy, thyroidectomy and parathyroidectomy, and was not restored by estrogen therapy. The prominence of a rapidly moving band (J) was similarly reduced by these operations, but it was restored by estrogen therapy. Prolactin activity was concentrated in band J and in the adjacent portion of the column. Details of the study are discussed.

A METHOD FOR APPLYING HAEMOLYSATES ON STARCH BLOCKS FOR ELECTROPHORESIS

Barkhan, P., and M. E. Stevenson, J. Clin. Pathol. **16**, 386–387 (1963).

Electrophoresis of hemolysates on a starch block is an excellent method for the separation and quantitation of the various normal and abnormal hemoglobins. It is a useful method for isolating a particular hemoglobin for further analysis because relatively large amounts of hemoglobin solutions can be applied. Application of the sample in a narrow band is important for good electrophoretic resolution on the block, and a simple method for applying the hemolysate is described.

THIN-LAYER STARCH-GEL ELECTROPHORESIS AND PLASTIFICATION METHOD

Baur, E., J. Lab. Clin. Med. **61**, No. 1, 166–173 (1963).

A method is described for reducing zonal profile distortion to an acceptable degree during electrophoresis, for solving the problems of fragility, and for obtaining durable pherograms. A 1 mm thick starch-gel layer was pressed onto a supporting sheet of cellulose or cellophane by a simple technique, and the coated sheet was freely suspended in a paper electrophoresis for electrochromatographic fractionation, stained, and plasticized. This method of plasticizing can be combined with protein as well as enzyme stains, although some of the latter have to be modified to stabilize colors. It is possible to separate normal adult hemoglobin A$_1$ into two subfractions by this method of electrophoresis.

GLYCOPROTEINS AND GLYCOPETIDES IN NORMAL HUMAN URINE

Blix, G., Protides Biol. Fluid, Proc. 11th Colloq., 303–307 (1963).

An attempt was made ro isolate uniform substances from the mixture of glycoproteins of normal human urine. The techniques employed in the investigation, which included zone electrophoretic fractionation of the urinary colloid fractions designated A, B and C, and the results obtained are presented.

LACTATE DEHYDROGENASE VARIANT FROM HUMAN BLOOD: EVIDENCE FOR MOLECULAR SUBUNITS

Boyer, S. H., Science **141**, 642–643 (1963).

Vertical starch-gel electrophoresis (4°C, 4.5 volts/cm) for 16 hours in ethylenediaminotetraacetic acid–boric acid–Tris (EBT) buffer or borate buffer demonstrated a human variant of erythrocyte LDH. The variant LDH had five LDH-1, four LDH-2, three LDH-3 and two LDH-4 components. This suggested that LDH isozymes are tetramers formed from various combinations of two types of subunits.

HAEMOGLOBINS OF TWO ELEPHANT SHREWS

Buettner-Janusch, J., and V. Buettner-Janusch, Nature **199**, No. 4896, 918–919 (1963).

As part of an investigation of the evolution of the hemoglobin molecule among the Primates, the authors studied the hemoglobin of two East African elephant shrews, *Petrodromus sultani* and *Rhynchocyon chrysopygus,* a species of Macroscelididae believed to represent a link between Primates and Insectivora. Blood was obtained by heart puncture of tranquillized shrews and vertical starch-gel electrophoresis was carried out on the hemolysates. Results of the study are discussed.

STARCH-GEL ELECTROPHORESIS OF URINARY ALKALINE PHOSPHATASE

Butterworth, P. J., E. Pitkanen, and D. W. Moss, Biochem. J. **88**, 19–20 (1963).

The alkaline phosphatase content of urine specimens from healthy individuals and patients with varying renal and non-renal diseases was determined. In the study, horizontal starch-gel electrophoresis of the urinary phosphatase was carried out in a discontinuous buffer system at pH 8.7, revealing several distinct patterns. The results are discussed.

LACTATE DEHYDROGENASE ISOENZYMES IN SKIN

Carr, A., and A. W. Skillen, Brit. J. Dermatol. **75,** 331–343 (1963).

This paper describes work using starch-gel electrophoresis to investigate differences in the patterns of lactate dehydrogenase (LDH) isoenzymes in epidermal and dermal tissue extracts. The zones of enzyme activity detected were termed LD-1, LD-2, LD-3, LD-4, and LD-5. The subscripts one to five denoted increasing electrophoretic mobility in the direction of the anode. Most epidermal activity appeared in LD-1 or LD-2, and also some in LD-3, depending on the total enzyme concentration in the sample. LD-2, LD-3, and LD-4, with a trace of activity in LD-5, were contained in the dermis. In defining the effect of serum proteins on these patterns, it was seen that with the addition of serum proteins the dermal pattern remained unaltered, but in the epidermal pattern there was a shift in activity from the LD-2 to the LD position. The presence of γ-globulins was shown to be associated in some way with the altered pattern.

ACID MUCOPOLYSACCHARIDES OF NORMAL HUMAN URINE

Di Ferrante, N., J. Lab. Clin. Med. **61,** No. 4, 633–641 (1963).

A study to isolate and characterize the different acid mucopolysaccharides present in normal human urine is reported in this paper. The clear supernatant fluid from the preparation of the crude extract of urinary acid mucopolysaccharides was submitted to zone electrophoresis on starch, using 45 x 10 cm blocks and barbital buffer, pH 8.6, 0.05. Separation into three fractions with progressively increasing sulfur content was achieved in total urinary acid mucopolysaccharides by chromatography on ECTEOLA ion-exchanger cellulose. Analyses of the material obtained in each fraction indicated the presence of small amounts of chondroitin sulfate B and of products of partial and complete desulfation of chondroitin sulfate A or C in normal urine. Incubation of total urinary acid mucopolysaccharides with testicular hyaluronidase produced depolymerization of 76 percent of the original material, and analyses of the 24 percent residue confirmed the presence in normal urine of chondroitin sulfate B and indicated the possible presence of small amounts of hepartin sulfate. Results of the study demonstrate the complexity of the mixture of acid mucopolysaccharides present in normal human urine.

STUDIES ON NORMAL AND PATHOLOGICAL γ_{1A}-GLOBULINS AND THEIR SUBUNITS

Heremans, J. F., A. O. Carbonara, G. Mancini, and R. Lontie, Protides Biol. Fluids, Proc. 11th Colloq., 45–55 (1963).

Normal γ_{1A}-globulin and paraproteins of the γ_{1A}-class, which consist of two types of polypeptide chains, were separated after reduction and alkylation of the native molecules by starch-gel electrophoresis on Sephadex® G-100 in formate buffer (pH 3.1) containing 8M urea.

PROTEINS AND ISOZYMES OF ESTERASES AND CHOLINESTERASES FROM SERA OF DIFFERENT SPECIES

Hess, A. R., R. W. Angel, K. D. Barron, and J. Bernsohn, Clin. Chim. Acta **8**, 656–667 (1963).

Starch-gel electrophoresis was used in the separation of human, cat, rabbit, monkey and rat serum proteins, and in the determination of their cholinesterase, esterase, and proteolytic activities. The techniques and results are discussed.

NORMAL AND ABNORMAL HUMAN HEMOGLOBINS

Huisman, T. H. J., Advan. Clin. Chem. **6**, 231–361 (1963).

In this paper the author discusses the minor protein components of the red blood cell, and the chemical structure, classification, and genetic, clinical and some physiological aspects of human hemoglobin types. The methods, including several electrophoretic techniques, used in his hematological studies are described and commented on.

Zone electrophoresis on paper, employed for the examination of hemoglobin variants, is widely used because of its low cost and simplicity. Two procedures, one using a pressure-plate apparatus and a second using an apparatus with freely hanging strips, are recommended. Because the migration of protein on paper is largely influenced by electroendosmotic and adsorption effects, this method has its limitations. Zone electrophoresis on starch, used in the discovery of the minor hemoglobin component Hb-A$_2$, is used to separate Hb-A$_2$ and Hb-Lepore fractions from the major hemoglobin component, and also lends itself well to preparative work since the colored proteins can easily be extracted. Zone electrophoresis in starch gel, used for the separation of serum proteins, has superior resolution of the protein components. A slightly modified version of the zone electrophoresis in agar gel method described by Wieme for the separation of serum proteins was found to have several advantages. Other electrophoretic techniques using cellulose acetate, carboxymethylcellulose gel, and polyacrylamide gels as supporting media are also discussed.

ÉTUDES DU SÉRUM DE CANARD

Kaminski, M., and E. Gajos, Protides Biol. Fluids, Proc. 11th Colloq., 137–141 (1963) **French**.

Part one concerns starch-gel and agar-gel electrophoretic analysis. An examination of 400 samples of duck sera by starch-gel and agar-gel electrophoresis revealed important individual variations (unrelated to race, sex or age) consisting of wide changes in electrophoretic mobility and concentration of constituents as judged from the intensity of staining. Also discussed in this paper is the performance of some specific reactions for detecting lipids, peroxydases, transferrin, and ceruloplasmin.

Part two concerns the electrophoretic study of enzymatic activity. The localization of the esterase activity in the region of α-globulin and two zones corresponding to neither proteins nor lipids as revealed by Amido Black or Sudan Black were observed in agar electrophoresis. The main zone in starch gel was just behind albumin in the region not stained by Amido Black. More or less esterase activity was shown by the individual samples.

ANWENDUNG VON STÄRKEGEL- UND POLYACRYLAMIDGEL-ELEKTROPHORESE AUF DIE TRENNUNG VON EISENHATIGEM FERRITIN IN FRAKTIONEN VON VERSCHIEDENER MOLEKÜLGRÖSSE

Kapp, R., A. Vogt and G. Maass, Protides Biol. Fluids, Proc. 11th Colloq., 430–434 (1963) **German.**

Iron-containing ferritin was separated by starch-gel and polyacrylamide-gel electrophoresis into at least three distinct bands, probably representing ferritin molecules of different size, which were stained with Amido Black 10B and showed the typical Prussian blue reaction. The ferritin, prepared from horse spleen and freed from apoferritin by ultracentrifugation, was homogeneous in cellulose-acetate electrophoresis and serologically homogeneous according to immunoelectrophoresis.

A SIMPLE MICRO STARCH GEL ELECTROPHORESIS METHOD APPLICABLE TO IMMUNODIFFUSION

Korngold, L., Anal. Biochem. **6**, 47–53 (1963).

A simple two-step method of electrophoresis, combining immunodiffusion and micro starch-gel electrophoresis, is described and discussed. The advantages of this technique are that it is on a microscale and is done on a very thin starch-gel layer; several layers may be analyzed simultaneously; no cutting of the gel is required; a complete immunoelectrophoretic pattern can be obtained; and the same specimen can be developed by two antisera in adjacent trenches.

A RAPID MICROMETHOD FOR STARCH-GEL ELECTROPHORESIS

Marsh, C. L., C. R. Jolliff and L. C. Payne, Federation Proc. **22**, No. 2, Part 1, 236, Sec. 452 (1963).

A modification of Smithies' conventional starch-gel electrophoresis technique is described. It is more rapid and more adaptable to routine clinical usage, and the resolution of serum or plasma proteins is excellent and compares favorably with gel patterns done in the conventional manner. The completed patterns of hemoglobin and haptoglobin typing by this micromethod are superior to similar patterns on paper, agar gel, or starch block.

QUANTITATIVE STARCH GEL ELECTROPHORESIS

Matsui, K., and K. Yaeno, Anal. Biochem. **6**, 491–503 (1963).

Starch-gel electrophoresis, heretofore, had been limited almost entirely to qualitative analysis despite attempts to use it quantitatively. The authors have modified, for quantitative analysis, the methods of sample insertion and staining in Smithies' starch-gel electrophoresis, and their method is useful for the estimation of the apparent distribution of components which penetrate into the gel. This report examines the reproducibility of the methods and the conditions for treatment of the starch gel, measures the apparent percentages of the components of rat serum, and discusses the limitations and possible errors of the method.

UREA-STARCH GEL IMMUNOELECTROPHORESIS

Poulik, M. D., Protides Biol. Fluids, Proc. 11th Colloq., 385–388 (1963).

Procedures for studying the immunological properties of the structural subunits of proteins are outlined in this paper. Electrophoresis in starch gel containing urea was found to be a potent method of separating the structural subunits of proteins.

THIN-LAYER STARCH GEL ELECTROPHORESIS ON GLASS SLIDES

Ramsey, H. A., Anal. Biochem. **5**, 83–91 (1963).

The flexibility of starch-gel electrophoresis in the separation of proteins was extended as an analytic tool by pouring the gel in a thin layer on glass slides. This modification of the starch-gel electrophoretic technique is described in this paper. Electrophoresis in a thin layer has advantages over the use of a thicker gel in that it lessens the effect of electrodecantation during the entry of migrating proteins into the gel, and higher voltage gradients can be applied across a thin layer because heat dissipation is more rapid per unit volume of gel. With both the reduced concentration of starch in the gel and the higher voltage gradient, the rate of protein migration is greatly increased, thus only approximately two hours is required for electrophoresis. An example of the use of this technique by the author in the separation of five esterases in the duodenal contents of a young calf is illustrated and evaluated.

FACTORS INFLUENCING THE ELECTROPHORETIC MIGRATION OF LACTIC DEHYDROGENASE ISOZYMES

Ressler, N., J. Schulz, and R. R. Joseph, J. Lab. Clin. Med. **62**, No. 4, 571–578 (1963).

It was found that different electrophoretic methods or conditions may produce inconsistent results in the determination of lactic dehydrogenase (LDH) isozymes. Investigation of these inconsistencies by means of starch-gel and dilute-agar electrophoresis showed that, in buffers of increasing dilution, the relatively more basic isozymes, especially 5, tend to migrate toward the anode instead of the cathode. This effect was apparently due to the migration of these isozymes in association with anionic substances when the buffer was sufficiently dilute. Anionic components involved could be either bound to the isozyme in the hydrogenate or could be present in the electrophoretic medium, though the former possibility is unlikely because re-electrophoresis of isozyme 5, purified under conditions which would separate it from anionic components, continued to show dependence upon ionic strength. Interactions with the isozymes were increased when the concentration of agar or starch gel was increased, and when the isozymes were separated in an agar plate which had been subjected to preliminary electrophoresis before sample application, the effect of lower buffer concentration was largely eliminated. Therefore, interactions between the isozymes and a component or components of the electrophoretic media were most consistent with these results. Findings of this study indicate that the use of buffers of sufficiently high concentration to eliminate the interactions described may be important in avoiding incorrect interpretations of LDH isozyme patterns.

PREVENTION OF BACKGROUND STAINING OF STARCH GELS: PRELIMINARY TREATMENT OF THE STARCH WITH STREPTOMYCES GRISEUS PROTEASE

Robinson, J. C., and J. E. Pierce, Am. J. Clin. Pathol. **40**, No. 6, 588–590 (1963).

Proteins appear as dark blue bands on a light blue background when starch gels are stained with Amido Black 10B after electrophoresis. A preliminary treatment of the starch with Pronase, a protease obtained from *Streptomyces griseus*, can prevent this background staining. The materials and methods used and the results are described, and the procedure commented on.

STUDIES OF ABNORMAL HEMOGLOBINS: I. PROCEDURE FOR QUANTITATION OF HEMOGLOBINS SEPARATED BY MEANS OF STARCH GEL ELECTROPHORESIS

Sunderman, F. W., Jr., Am. J. Clin. Pathol. **40**, No. 3, 227–238 (1963).

In an effort to resolve the difficulty of obtaining quantitative measurements of hemoglobins separated by starch-gel electrophoresis, a systematic evaluation of the published techniques for such measurements was made. After experimentation with these techniques, a spectrophometric method for the quantitation of hemoglobins separated by starch-gel electrophoresis was formulated and is described in this paper. Using this method, the standard deviation of replicate measurements of the percentage of hemoglobin A$_2$ was ±0.19 percent and the coefficient of variation of replicate determinations was 6.2 percent. The standard deviation and range of value of hemoglobin A$_2$ in blood samples taken from 25 male medical students and 10 persons with thalassemia trait are discussed. A revised classification of human hemoglobins, based on the electrophoretic mobilities of the various hemoglobins at pH 8.6, relative to the mobilities of hemoglobins S, A, and H, is described. Also outlined is a systematic approach to the identification of abnormal hemoglobins, which involves measurements of the electrophoretic and chromatographic mobilities of the hemoglobins, in addition to a number of ancillary analytic procedures.

MULTIPLE ELUTION OF PROTEIN ZONES FROM STARCH GEL ELECTROPHORETICALLY

Tsuyuki, H., Anal. Biochem. **6**, 205–209 (1963).

Since the introduction of starch gel as a supporting medium for zone electrophoresis of proteins, the full realization of its potentiality has been impeded by inadequate methods of elution of the separated compounds, though several different methods had been tried. In this communication, the author presents a technique for quantitatively isolating large numbers of protein zones simultaneously from the entire migration field of starch gel, and also describes a slightly different type of gel casting tray.

PURIFICATION AND SOME OF THE PROPERTIES OF α_s-CASEIN AND κ-CASEIN

Zittle, C. A., and J. H. Custer, J. Dairy Sci. **46**, 1183–1188 (1963).

Starch-gel-urea electrophoresis, which provided an effective means for estimating the purity of caseins, was used to judge the freedom of α_s-casein and κ-casein from major contaminants. The experimental procedures and the results of the purification of these caseins are described and discussed, and their electrophoretic mobility, pH of minimal solubility salt-free (isoionic point), and with 0.1M NaCl (isoelectric point), ultraviolet absorption, and interaction with calcium ions are reported.

SEPARATION OF HEMOGLOBINS LEPORE AND H FROM A AND A₂ IN STARCH OR ACRYLAMIDE GELS BUFFERED WITH TRIS-EDTA

Curtain, C. C., J. Clin. Pathol. **15**, 288–289 (1962).

Hemoglobins Lepore, A, and A_2 were separated using 'starch gels or acrylamide gels buffered with Tris–EDTA, pH 8.6. A potential of 8 volts/cm for six hours was used to effect separation. Hemoglobin Lepore was found to migrate between hemoglobin A and A_2.

PROPERTIES OF ALKALINE-PHOSPHATASE FRACTIONS SEPARATED BY STARCH GEL ELECTROPHORESIS

Moss, D. W., and E. J. King, Biochem. J. **84**, 192–195 (1962).

A number of active alkaline-phosphatase fractions were revealed in the starch-gel electrophoresis of concentrated butan-1-01 extracts of human bone, liver, kidney and small intestine. The apparent K_M values (substrate, β-naphthyl phosphate) of the individual enzyme bands from these tissues were determined. Also studied were other properties of the bands, namely pH optimum, activation by Mg^{2+} ions and inactivation by heating at 55° and pH 7. Although there are some differences between phosphatases from different tissues, the phosphatase fractions from a given organ are similar in these operations and in K_M values. The significance of these findings is reported.

A CONVENIENT APPARATUS FOR VERTICAL GEL ELECTROPHORESIS

Raymond, S., Clin. Chem. **8**, No. 5, 455–470 (1962).

An apparatus incorporating gel mold, electrode chambers, and direct-contact cooling plates was designed to utilize the superior resolving power of vertical-gel electrophoresis, either of starch or acrylamide gel, for analysis of serum, hemoglobin, and other protein mixtures. Because it eliminates several steps in preparing and handling the gel, electrophoretic analyses can be completed in two to four hours. Either rigid (starch) or flexible (acrylamide) gels may be used on it. The apparatus is described and examples of serum and hemoglobin electrophoretic patterns obtained are given.

COMPARATIVE QUANTITATIVE ANALYSIS OF NORMAL AND PATHO-LOGICAL SERA BY ELECTROPHORESIS IN STARCH GEL AND CELLULOSE ACETATE

Rubinstein, H. J., I. T. Oliver, and C. J. Brackenridge, Clin. Chim. Acta **7**, 65–72 (1962).

The measurement of the serum proteins in normal and abnormal conditions after separation by electrophoresis in both starch gel and on cellulose acetate, and the difficulties of establishing a normal range for the starch gel values are discussed in this paper. An electrophoretic method for differentiating macroglobulins from paraproteins and very raised γ-globulin, found to be more reliable than visual comparison of electrophoretic patterns, is described. Cryoglobulins are more apt to be suspected by cellulose-acetate electrophoresis than in starch-gel electrophoresis. Both electrophoretic methods were equally reliable for diagnostic purposes and gave similar information for the patients with myelomatosis, idiopathic paraproteinaemia, and a miscellaneous group with raised γ-globulins.

MOLECULAR SIZE AND STARCH-GEL ELECTROPHORESIS

Smithies, O., Arch. Biochem. Biophys. Suppl. **1**, 125–131 (1962).

In an investigation of the effect of changes in the concentration of starch on the migration of proteins during starch-gel electrophoresis, it was found that the migration (m) is proportional to the reciprocal of the starch concentration (1/s) over the range of concentrations tested. While the retardation coefficient of a given ion, defined as $(\triangle m/m)/\triangle(1/s)$, is dependent on the size of the ion, it is independent of its charge or the duration of the electrophoresis. By determining the retardation coefficients of components in complex mixtures, a measure of their relative sizes can be obtained. The methods of investigation and the results are described and evaluated.

THE GLYCOPROTEINS OF SERUM

West, C. D., and R. Hong, J. Pediat. **60**, No. 3, 430–464 (1962).

Some of the information about serum globulins, which had been obtained over the past decade, was gathered together and viewed from the chemical and physiologic standpoints. Only the glycoproteins are discussed in this review which covers separately both of the approaches to the study of serum glycoproteins. The first is the chemical approach, by which the glycoproteins are characterized and quantitated either by their differential solubility in certain acids or by determination of specific carbohydrates incorporated in their structure. The second approach, utilizing techniques of starch-gel electrophoresis and immunoelectrophoretic analysis, aims at the identification of protein entities. Some of the 30 protein entities which have been identified in serum by electrophoresis are discussed. Conclusions indicate that the level of a single protein will be of little diagnostic value, but that the determination of several to reveal an overall pattern might give information of diagnostic significance.

COMPARISON OF THE SERUM PROTEIN FRACTIONS OF THE DEVELOP-ING CHICKEN EMBRYO BY THE TECHNIQUE OF STARCH GEL ELECTRO-PHORESIS

Amin, A., Nature **192**, No. 4800, 356–357 (1961).

The pooled sera of five different age-groups of chick embryo, incubated for 10, 13, 16, 19, and 21 days, were examined by starch-gel electrophoresis performed on a vertical tray for 18 hours, using a pH 8.6 borate buffer and a field strength of four volts/cm. The samples were run simultaneously on the same gel. Results of the investigation are reported.

AN INVESTIGATION OF CLOSELY RELATED GAMMA-MYELOMA PRO-TEINS AND NORMAL MOUSE GAMMA-GLOBULIN BY PARTIAL ENZYMIC DEGRADATION AND STARCH GEL ELECTROPHORESIS

Askonas, B. A., and J. L. Fahey, Nature **190**, No. 4780, 980-983 (1961).

By an analysis on starch-gel electrophoresis of the serum of C_3H mice bearing one of the transplantable plasma cell tumors (x5563), the high concentration of γ-myeloma protein present was found to consist of five distinct components which are distinguishable immunologically and have the same molecular size, but varied electric charge. Partial degradation of the proteins with papain and cysteine was employed in an attempt to discover a basis for the difference between these closely related γ-myeloma proteins produced by a single plasma cell tumor and possibly to throw light on the complexity of normal γ-globulin.

STARCH GEL ELECTROPHORESIS OF BRAIN ESTERASES

Barron, K. D., J. I. Bernsohn, A. Hess, J. Histochem. Cytochem. **9**, 656–660 (1961).

Nine esterases which hydrolyze α-naphthyl and naphthol AS-acetate, plus an enzyme (presumably pseudocholinesterase) acting against butyrylthiocholine iodine located at the site of sample insertion, were demonstrated by starch-gel electrophoresis in rat brain homogenates. One esterase was identified as acetylcholinesterase; the nature of the others were undefined. The methods used in the investigation are described and the results are discussed in relation to similar findings in human white matter.

RESOLUTION OF PATHOLOGICAL HUMAN GAMMA-GLOBULINS BY STARCH-GEL ELECTROPHORESIS

Binette, J. P., and K. Schmid, Nature **192**, No. 4804, 732–737 (1961).

The use of starch-gel electrophoresis for the resolution of γ-globulins of certain pathological sera into several distinct bands is described and the principal results are presented in this paper. The investigation demonstrated that this technique has an advantage over paper electrophoresis in that certain abnormal γ-globulins can be revealed into several bands.

THE OCCURRENCE OF POST-GAMMA PROTEIN IN URINE: A NEW PROTEIN ABNORMALITY

Butler, E. A., and F. V. Flynn, J. Clin. Pathol. **14**, 172–178 (1961).

A new urine protein fraction was recognized as a result of the study of the urine proteins of 223 individuals by starch-gel electrophoresis. The new fraction, on starch-gel electrophoresis at alkaline pH, moves to a position appreciably nearer the cathode than the slowest-moving γ-globulin in normal serum. The urine of 46 patients with clinical proteinuria, including 19 with the Fanconi syndrome and nine with multiple myeloma, revealed a small amount of post-γ protein. The nature, origin, and clinical significance of post-γ protein are discussed.

SEPARATION OF HUMAN HEMOGLOBINS BY STARCH-GEL ZONE ELECTROPHORESIS

Engle, R. L., Jr., A. Markey, J. H. Pert, and K. R. Woods, Clin. Chim. Acta **6**, 136–141 (1961).

The use of starch-gel electrophoresis for the separation of normal and abnormal human hemoglobins and for the determination of the relative mobilities of the different components was compared to the use of other methods. It was found to have high resolution and to be capable of detecting five protein fractions in normal red cell hemolysates. Fetal hemoglobin was readily separable from normal adult hemoglobin, and, although quantitative measurements were not made, the A_2 hemoglobin elevations were demonstrated in some patients with thalassemia minor.

STARCH GEL ELECTROPHORESIS OF SERUM PROTEINS AND URINARY PROTEINS FROM PATIENTS WITH MULTIPLE MYELOMA, MACROGLOBULINEMIA, AND OTHER FORMS OF DYSPROTEINEMIA: DEMONSTRATION OF DIVERSITY AND COMPLEXITY OF PROTEIN ALTERATIONS IN 72 CASES

Engle, R. L., K. R. Woods, G. B. Castillo, and J. H. Pert, J. Lab. Clin. Med. **58**, 1–22 (1961).

A study is described in which starch-gel electrophoresis was carried out on the proteins in the blood and urine of 56 patients with multiple myeloma, three patients possibly in the early stages of multiple myeloma, five with macroglobulinemia, and eight others with abnormal peaks in the electrophoretic patterns. High resolution of protein components and a more detailed assessment of protein alterations than can be attained by other methods were given by this method of electrophoresis. The results of the study are presented and are compared with those obtained by other methods of electrophoresis, particularly filter paper, and the advantages and disadvantages of the starch-gel method are discussed. Starch gel has no great advantage over filter paper for routine diagnostic purposes, but has some important advantages as a research tool in the study of multiple myeloma and related diseases.

STARCH-GEL ELECTROPHORESIS OF RAT-SERUM PROTEINS

Espinosa, E., Biochim. Biophys. Acta **48**, 445–451 (1961).

In an attempt to correlate the multiple components of rat serum proteins as separated by starch-gel electrophoresis and the classical fractions obtained by paper electrophoresis, serum proteins separated by paper electrophoresis were run on starch and vice versa. The experimentation is described and the differences in behavior of corresponding human and rat fractions are discussed. Paper and starch-gel electrophoresis were used to identify rat-serum fractions separated by DEAE-cellulose chromatography.

STARCH-GEL ELECTROPHORESIS OF ANTERIOR PITUITARY HORMONES

Ferguson, K. A., and A. L. C. Wallace, Nature **190**, No. 629–630 (1961).

Analyses of various standard preparations of anterior pituitary hormones by starch-gel electrophoresis are reported, together with some evidence of the distribution of hormonal activities. A modification of the discontinuous buffer system of Poulik [Nature *180, 1477* (1957)] gave better separation of pituitary proteins than the original method of Smithies.

INHOMOGENEITIES IN ALKALI-RESISTANT HEMOGLOBIN: DEMONSTRATION OF ZONE ELECTROPHORETIC DIFFERENCES USING A CATIONIC DETERGENT ELECTROLYTE

Hoerman, K. C., Blood **17**, 409–417 (1961).

Normal adult hemoglobin, hemoglobin of a patient with thalassemia and cord blood hemoglobin were submitted concomitantly to electrophoresis in six gels, in order to demonstrate the differences between alkali-resistant hemoglobin of the new-born and of patients with severe thalassemia. Three of the gels contained a cationic active detergent electrolyte, trimethyloctadecyl ammonium chloride (TMOD); three did not. Selected alteration in the migration rates of alkali-resistant hemoglobin was caused by the addition of the TMOD to the starch gels before electrophoresis, while other red blood cell pigments were not similarly affected. Results of investigations with this method showed electrophoretic inhomogeneity of cord blood and thalassemia hemoglobin, revealed new components of cord blood hemoglobin, and demonstrated a new component in the hemoglobin of cases of severe thalassemia.

THE ESTERASES OF MOUSE BLOOD

Hunter, R. L., and D. S. Strachan, Ann. N.Y. Acad. Sci. **94**, 861–867 (1961).

Using the zymogram method for analyzing the blood of mice, nine esterases in plasma and two in red blood corpuscles were demonstrated, and the effect of six substrates, one activator and five inhibitors upon these esterases was determined. Use of starch gel as a medium for electrophoretic separation greatly increased the capacity for resolving proteins. The esterase-active proteins were located in numbered bands on the starch-gel column and were located by histochemical methods. Results of the study, described in this report, indicate that the multiple esterases present in mouse blood cannot be readily classified into subgroups containing common properties.

CONTACT PRINTS OF STARCH GEL ELECTROPHORESIS PATTERNS

Johns, E. W., J. Chromatog. **5**, 91–92 (1961).

Problems in obtaining a satisfactory laboratory record of separations obtained by starch-gel electrophoresis were encountered because photography of every gel was time-consuming and tedious, drawing could not easily reproduce the intensity of the bands, and dehydrating the gels in benzyl alcohol made them rigid and brittle. A technique is described which leaves the gels flexible and transparent so that a satisfactory contact print can be made. The stained gels are soaked in ethyl alcohol followed by a mixture of benzyl alcohol and glycerol (2:1 v/v), then are pressed onto photographic paper. Further details of the technique are described. Although the contact prints are not suitable for reproduction, they do afford rapid permanent records.

FORMS OF ENZYMES IN INSECT DEVELOPMENT

Laufer, H., Ann. N.Y. Acad. Sci. **94**, 825–835 (1961).

The changes in morphology during insect development that result from endocrine interactions, reflected in changes in the protein constitution of various tissues, were observed by zone electrophoresis in starch gels in Cecropia *(Hyalophora cecropia)* and Cynthia *(Samia cynthia)* silk moths at different stages of the life cycle. This study led to an inquiry into the catalytic activities of these proteins as a possible clue to their normal function. These activities included esterases, phosphatases, carbohydrases, lipases, sulfatases, and chymotrypsins, all of which appear as multiple bands in starch-gel electrophoresis as do dehydrogenating enzymes (malic dehydrogenase, α-glycerophosphate dehydrogenase, and lactic dehydrogenase). The existence of isozymes among the hydrolic and dehydrogenating enzymes that occur in cell-free blood preparations was indicated in a study reported in this paper.

MULTIPLE PEROXIDASES IN CORN

McCune, D. C., Ann. N.Y. Acad. Sci. **94**, 723–730 (1961).

Of six peroxidase active fractions obtained from corn-leaf sheath preparations by starch electrophoresis, the four major fractions differed in substrate specificity and in the effect of 2,4-dichlorophenol and $MnCl_2$ upon their IAA oxidase activity. The study of the substrate specificities of the individual peroxidases of the corn leaf sheath and the changes in the peroxidase complement of this tissue occurring with growth, dwarfism, and the application of gibberellic acid are presented in this paper.

DETECTION OF CERULOPLASMIN AFTER ZONE ELECTROPHORESIS

Owen, J. A., and H. Smith, Clin. Chim. Acta **6**, 441–444 (1961).

A dianisidine reagent for the detection of hemoglobin and hemoglobin-haptoglobin complexes after zone electrophoresis had been reported and a modification of this procedure, described in this paper, increased the sensitivity of the stain in detecting the haptoglobin-hemoglobin complexes. When applied to starch-gel electrophoresis strips of normal human serum, this modified procedure demonstrates a previously undetected reactive zone lying about two-thirds of the way between the trailing edge of the albumin zone and the transferrin zone. Experiments were carried out to identify the component responsible for staining in the zone and the results obtained gave evidence that it was ceruloplasmin.

MOLECULAR VARIATION IN SIMILAR ENZYMES FROM DIFFERENT SPECIES

Paul, J., and P. F. Fottrell, Ann N.Y. Acad. Sci. **94**, 668–677 (1961).

A study of the electrophoresis behavior of esterases, phosphatases, peroxidases, and catalases, obtained from a variety of animal species by starch-gel electrophoresis in a discontinuous buffer system, is reported in this paper. Differences between species were almost invariably demonstrated, which seemed to indicate that structural variations in similar proteins from one species to another may be the rule. The significance of the conclusion that many examples of isozymes may be local manifestations of this is discussed in relation to molecular evolution.

TISSUE-SPECIFIC AND SPECIES-SPECIFIC ESTERASES

Paul, J., and P. Fottrell, Biochem. J. **78**, 418–424 (1961).

This paper describes the separation of tissue esterases of a variety of organs in several species by starch-gel electrophoresis in order to investigate the amount of variation among proteins with the same enzyme activity. Different species exhibit entirely different and highly reproducible patterns by this technique. The significance of the conclusion drawn from these studies is that an enormous number of proteins with the same functional activity can exist.

THE CHARACTERIZATION OF PROTEINS BY ELECTRODIALYSIS IN STARCH GELS

Pierce, J. G., and C. A. Free, Biochim. Biophys. Acta **48**, 436–444 (1961).

A procedure is given which uses the starch-gel electrophoresis of Smithies [Biochem. J. *61, 629* (1955)] in a microelectrodialysis technique with cellophane membranes implanted in the gels perpendicular to the direction of movement of a protein in an electrical field. Results indicate that this technique is useful for approximating molecular weight and for studies of homogeneity and association-dissociation relationships.

APPLICATION OF STARCH GEL ELECTROPHORESIS IN UREA TO THE STUDY OF STRUCTURAL UNITS OF PROTEINS

Poulik, M. D., and G. M. Edelman, Protides Biol. Fluids, Proc. 9th Colloq., 126–132 (1961).

Examples of the range and usefulness of starch-gel electrophoresis in urea in the study of γ-globulin are presented and the application of this technique to the proteins in the γ-globulin family is emphasized although the patterns of other proteins are also described in order to illustrate certain aspects of the subject.

ELECTROPHORETIC PATTERNS OF LYMPH NODE PROTEINS IN HYBRID F₁ MICE IRRADIATED AND TREATED WITH MYELOID AND LYMPHOID CELLS FROM PARENTAL STRAIN

Schwarzmann, V., G. Mathe, and J. L. Amiel, Nature **189**, 1025–1026 (1961).

A study was made to determine whether graft immunization in lymph nodes will result in a modification of the electrophoretic pattern of the lymph node proteins as investigated by starch-gel electrophoresis. Results of the study, reported in this paper, indicated that it is not unreasonable to assume that the band shown in starch gel whenever an immunity reaction takes place in the lymph nodes, together with the striking increase in number of hyperbasophilic cytoplasma cells, might correspond to the production of the antibody.

STARCH GEL ELECTROPHORESIS OF HEN EGG WHITE, OVIDUCT WHITE, YOLK, OVA AND SERUM PROTEINS

Steven, F., Nature **192**, No. 4806, 972 (1961).

Starch-gel electrophoresis, using gels containing 15–17 percent w/v starch prepared from BDH soluble starch, was applied in the demonstration of genetic polymorphism in proteins in the white of a hen egg. Three components were detected. Results of a comparison of the thin, inner thin and thick layers of hen egg white proteins with those of the viscous fluid obtained from the magnum region of the oviducts of laying hens are presented. The protein patterns of hen serum, egg yolk and ova isolated from the ovary also are given for comparative purposes.

ASSURANCE OF THERMOSTATIC CONTROL DURING ELECTROPHORETIC EXPERIMENTS

Wieme, R. J., Nature **190**, No. 4778, 806–807 (1961).

If absolute electrophoretic mobilities are to be measured, precise thermostatic control is of paramount importance, especially in the use of starch gel or cyanogum gel as the supporting medium, for the diameters of the gel pores, which largely determine migration in those media, depend on thermal agitation of the pore-forming bridges. Application of the principle of positive control of temperature is discussed.

STARCH ELECTROPHORESIS: III. STARCH GEL ELECTROPHORESIS

Bloemendal, H., J. Chromatog. **3**, 509–519 (1960).

Variations in the procedure of electrophoresis using starch gel as the supporting medium are presented in this review which includes a description of the apparatus used, the electrodes, the preparation of the gel, the insertion of the sample, the electrophoresis, the influence of temperature, the location of substance, the quantitative evaluation of the strip and the elution. Applications of starch-gel electrophoresis to serum proteins, other proteins, enzymes, hormones and non-protein substances as carried out and reported by several scholars are also reviewed. In summary, it was concluded that starch-gel electrophoresis is a powerful analytical tool, chiefly for the study of proteins, though the success of the method depends to a great extent on the experience of the one performing it.

STARCH-GEL ELECTROPHORESIS OF WHEAT PROTEINS

Elton, G. A. H., and J. A. D. Ewart, Nature **187**, No. 4737, 600–601 (1960).

Some preliminary experiments in which starch-gel electrophoresis was used to examine the proteins cf the gluten complex of wheat are reported. The complete resolution of eight bands was obtained, demonstrating the value of starch-gel electrophoresis as a method for investigating wheat proteins.

TRANSPARENT STARCH GELS: PREPARATION, OPTICAL PROPERTIES AND APPLICATION TO HAEMOGLOBIN CHARACTERIZATION

Gratzer, W. B., and G. H. Beaven, Clin. Chim. Acta **5**, 577–582 (1960).

A simple method is given for obtaining clear starch gels, by matching the refractive index of the starch with glycerol which replaces the water in the continuous phase. Spectroscopic characterization and some applications, with particular reference to the estimation of hemoglobin-A_2, are outlined, and modified staining procedures for protein zones in cleared starch gel are discussed.

PROPERTIES AND INHERITANCE OF THE NEW FAST HEMOGLOBIN TYPE FOUND IN UMBILICAL CORD BLOOD SAMPLES OF NEGRO BABIES

Huisman, T. H. J., Clin. Chim. Acta **5**, 709–718 (1960).

A new fast minor Hb component (called "Augusta I") was discovered by starch-gel electrophoresis in cord blood samples of Negro babies, present next to Hb-F, Hb-A, and Hb-S, and was determined to be in the amount of about three percent of the total hemoglobin by carboxymethyl-cellulose chromatography. A detailed description of the properties of the new Hb component and proof of its structure are presented in this paper. Electrophoretic and chromatographic studies of both cord blood hemoglobin and hemoglobin of family members indicated that in addition to the new component found only in cord blood hemoglobin, very small quantities of the abnormal Hb-H was present in both cord blood and in blood samples of some family members.

LOCALIZATION OF LEUCINE AMINOPEPTIDASE IN SERUM AND BODY FLUIDS BY STARCH GEL ELECTROPHORESIS

Kowlessar, O. D., L. J. Haeffner, and M. Sleisenger, J. Clin. Invest. **39**, 671–675 (1960).

The determination by starch-gel electrophoresis of the electrophoretic mobility of leucine aminopeptidase (LAP) of normal serum, sera of patients with pancreatic and hepatobiliary disease, pleural fluid and pancreatic extract is reported in this paper. A single peak of activity in normal serum identical with that of bile and pancreatic extract, located in the postalbumin fraction or in the fast α_2 and postalbumin areas, and two additional peaks of LAP activity in the sera of patients with parenchymal or metastatic liver disease and with common bile duct obstruction, were found. These peaks disappear with recovery of liver disease or relief of obstruction. Unless there is common duct obstruction or metastasis to the liver, electrophoretic mobility of serum LAP in patients with carcinoma of the pancreas is normal.

A SPECIES COMPARISON OF SERUM PROTEINS AND ENZYMES BY STARCH GEL ELECTROPHORESIS

Lawrence, L. H., P. J. Melnick, and H. E. Weimer, Soc. Exp. Biol. Med., Proc. **105**, 572–575 (1960).

The occurrence and distribution of esterase, colinesterase, aminopeptidase, cytochrome oxidase, beta-glucuronidase, acid and alkaline phosphatase, and ceruloplasmin oxidase in human sera (30 adult Caucasian males) and in sera from six species of experimental adult male animals (three monkeys, three dogs, eight rabbits, eight guinea pigs, eight rats, and ten mice) were studied by starch-gel electrophoresis and appropriate staining techniques. Succenic, lactic, beta-hydroxybutyric and glutamic dehydrogenases in human serum were also studied and serum protein patterns were included in the investigation to provide a broader basis of comparison and for orientation. Species differences, with respect to the presence, number, mobility, and concentration of components exhibiting various enzymatic activities, were observed.

ELECTROPHORETIC SEPARATION OF CERTAIN IN VITRO AND IN VIVO REACTIONS OF RABBIT ANTI-MOUSE ERYTHROCYTE SERUM

Lee, Chang-Ling, T. Takahashi, and I. Davidsohn, Nature **187**, No. 4732, 157–158 (1960).

Rabbit anti-mouse erythrocyte serum was used for the quantitative study, under comparable conditions, of in vitro and in vivo immunological reactions. This antiserum provided an excellent model for the study because of its strongly hemolytic and agglutinating properties for, and its combining ability with, mouse erythrocytes, and its ready ability to produce anemia in mice, a species especially suitable for this purpose. Starch-zone electrophoresis was used to separate the component responsible for the anemia from the saline agglutinin and hemolysin in the rabbit anti-mouse erythrocyte serum.

THE USE OF UREA-STARCH-GEL ELECTROPHORESIS IN STUDIES OF REDUCTIVE CLEAVAGE OF AN α_2-MACROGLOBULIN

Poulik, M. D., Biochim, Biophys. Acta **44**, 390–393 (1960).

The effects of reductive cleavage with mercaptoethanol in urea on the protein of high molecular weight belonging to the α_2-globulins, variously known as "heat-labile glycoprotein," the "alpha 2 macroglobulin," and "slow alpha 2 globulin," are described in this report. Starch-gel electrophoresis was used for the separation and demonstration of products obtained by reductive cleavage with mercaptoethanol in urea and after stabilization of the -SH groups by addition of iodacetamide.

DEMONSTRATION OF SMALL COMPONENTS IN RED CELL HAEMOLYSATES BY STARCH-GEL ELECTROPHORESIS

Fessas, P., and N. Mastrokalos, Nature **183**, 1261–1262 (1959).

The application of starch-gel electrophoresis to the study of human hemoglobin is discussed. A phosphate buffer (pH 6.5 $\Gamma/2$ 0.033) in the electrode vessels was used. Although this technique was not yet suitable for accurate, quantitative use, it had the advantages of affording great resolution, showing small components, and demonstrating minor components of red cell hemolysates not shown by other electrophoretic procedures.

STARCH GEL ELECTROPHORESIS STUDIES ON ABNORMAL PROTEINS IN MYELOMA AND MACROGLOBULINAEMIA

Fine, J. M., and R. Creyssel, Nature **183**, 392 (1959).

Sera from 28 patients with multiple myeloma, 13 patients with macroglobulinaemia and 4 patients with atypical dysproteinaemia (in which characterization of abnormal serum components had previously been accomplished by free-boundary electrophoresis, analytical ultracentrifuging, agar gel and immuno-electrophoresis) were further studied by starch-gel electrophoresis.

INABILITY OF HAPTOGLOBIN TO BIND MYOGLOBIN

Javid, J., D. S. Fischer, and T. H. Spaet, Blood **14**, 683–687 (1959).

Old and current concepts of the renal handling of extracorpuscular plasma hemoglobin (Hgb) are briefly reviewed and studies on the starch-gel electrophoretic behavior of myoglobin (Mb) alone and Mb in serum are presented in this paper. Results of the investigations suggest that Mb is not bound by haptoglobin (Hp), which is known to bind Hgb, and that this explains the low renal threshold of Mb as compared to that of Hgb.

IDIOPATHIC PAROXYSMAL MYOGLOBINURIA: REPORT OF A CASE WITH STUDIES ON SERUM HAPTOGLOBIN LEVELS

Javid, J., H. I. Horowitz, A. R. Sanders, and T. H. Spaet, Arch. Internal Med. **104**, 628–633 (1959).

Studies on the serum haptoglobin level during active myoglobin excretion in an 18-year-old male with idiopathic paroxysmal myoglobinuria are reported in this paper. The estimation of serum haptoglobin levels as a test for differentiating myoglobinuria from hemoglobinuria and its physiologic and diagnostic implications are presented. Samples of serum and urine were subjected to starch-gel electrophoresis at five volts/cm for ten hours.

THE DISTRIBUTION OF ESTERASES IN MOUSE TISSUES

Markert, C. L., and R. L. Hunter, J. Histochem. Cytochem. **7**, 42–49 (1959).

An investigation of the esterases in 32 organs and tissues of the mouse, separated by starch-gel electrophoresis and compared using α-naphthyl butyrate as substrate, is presented along with a study of the effect of esterine sulfate at 10^{-4}M on the esterases of selected tissue. The method of analyzing the esterases is known as the zymogram technique. The acquired information was correlated with what is known about the distribution of esterases in the mouse and other mammals as revealed by techniques involving the use of tissue sections. The methods used are described and the findings are discussed and evaluated.

STARCH-GEL IMMUNOELECTROPHORESIS

Poulik, M. D., J. Immunol. **82**, 502–515 (1959).

An immunoelectrophoretic method based upon one- and two-dimensional zone electrophoresis, using a discontinuous buffer system in order to increase the resolving power of the starch-gel electrophoresis and the sensitivity of the antigen-antibody reactions in agar gel, is described in detail and a comparison between this method and the method of Grabar and Williams is discussed. Two-dimensional zone immunoelectrophoresis resulted in a demonstration of several albumin-like proteins in a nephrotic serum and the localization of an acidic α-1-glycoprotein (orosomucoid) in urine. The one-dimensional technique demonstrated immunologic specificity of crystalline transferrin (β-1-globulin) after interaction with diphtheria toxin and the heterogeneity of crystalline diphtheria toxin and highly purified tetanus toxoid.

AN IMPROVED PROCEDURE FOR STARCH-GEL ELECTROPHORESIS: FURTHER VARIATIONS IN THE SERUM PROTEINS OF NORMAL INDIVIDUALS

Smithies, O., Biochem. J. **71**, 585–587 (1959).

An improved vertical starch-gel electrophoretic technique, which permits the sample to be introduced into the gel without the use of any supporting substance thereby greatly improving the resolving power and reproducibility of the method, is described in this paper. Results of investigations with this technique suggest that genetic factors are involved in the variations, in different individuals, of the serum proteins which migrate immediately behind albumin.

PRELIMINARY STUDIES ON QUANTITATIVE ZONE ELECTROPHORESIS IN STARCH GEL

Pert, J. H., R. E. Engle, K. R. Woods, and M. H. Sleisenger, J. Lab. Clin. Med. **54**, 572–584 (1959).

This paper reports on a method for the standardization of the buffer and soluble starch used in making the gel for zone electrophoresis using potato starch gel as the supporting medium, controlled semiautomatic staining procedures using continuously decolorized solvent for elimination of excess dye, and the use of reflection densitometry for quantitative estimation of each protein fraction. A more easily attained reproducibility and a greater resolution in the separation of fractions result. This method holds considerable promise for further studies; a survey of 100 normal and 400 pathologic sera indicated that while there are many normal serum types, each with complex patterns, the changes in disease are sufficiently marked. Rapid and sensitive results comparable with or better than those obtained by other methods were obtained when zone electrophoresis in starch gel was employed in testing the homogeneity of several purified proteins.

DIE HÄMOLYMPHE-PROTEINE EINIGER INSEKTENARTEN IM LAUFE DER LARVENENTWICKLUNG (STÄRKEGEL-ELEKTROPHORESE)

Denucé, J. M., Protides Biol. Fluids, Proc. 6th Colloq., 49–52 (1958) **German**.

The protein components of larval blood of *Bombyx mori* (silk worm) and *Galleria mellonella* (wax moth) were separated by zone electrophoresis using starch gel and paper as the supporting medium, and the results obtained from the two techniques are compared. A larger number of fractions were obtained in starch-gel electrophoresis.

FRACTIONATION OF BENCE-JONES PROTEIN BY STARCH GEL ELECTRO-PHORESIS

Flynn, F. V., and E. A. Snow, J. Clin. Pathol. **11**, 334–342 (1958).

The abnormal proteins of multiple myeloma, Bence-Jones protein, behave as a single entity on paper electrophoresis, but can often be resolved into two or more components in starch-gel electrophoresis. The findings of electrophoretic investigations on both paper and starch gel of 13 urines from cases of multiple myeloma are reported in this paper. It was found that, when exposed to alterations in light and temperature, sterile culture media undergo changes which cause considerable variations in the redox curves of the sterile media used as controls. The possible cause of these changes and their importance in avoiding misinterpretation of experimental results and drawing false conclusions are discussed.

HETEROGENEITY OF SERUM MYELOMA GLOBULINS

Owen, J. A., C. Got, and H. J. Silberman, Clin. Chim. Acta **3**, 605–607 (1958).

It was found that starch-gel electrophoresis of myeloma sera resolved several abnormal components in each serum. In the study described in this paper, sera from 18 patients with myelomatosis were examined by electrophoresis in starch gel. The presence of multiple abnormal serum components in myelomatosis may be of significance in regard to the pathogenesis of the condition, and the components may consist of molecules of a single protein in various states of aggregation or they may consist of distinct protein species. The molecular dimensions are important in determining the rate of electrophoretic migration through starch gel.

ZONE ELECTROPHORESIS OF CEREBROSPINAL FLUID PROTEINS IN STARCH GEL

Pert, J. H., and H. Kutt, Soc. Exp. Biol. Med., Proc. **99**, 181–185 (1958).

A method is presented for the separation of cerebrospinal fluid proteins by starch-gel electrophoresis, including a discussion of its advantages and limitations. Ten to twelve protein fractions, which correspond to serum proteins in the same individual but are present in different relative concentrations, were observed in normal spinal fluid. Considerable alterations of protein patterns were noted in a number of cerebrospinal fluids from patients with neurologic diseases, and examples of the type of abnormalities which may be encountered are illustrated. Protein fractions designated "fast gamma" and third and fourth prealbumins, which had not yet been seen in serum or in normal spinal fluid, were found in some of these pathologic specimens. Also described is a technique for concurrently staining lipoproteins in cerebrospinal fluid and serum.

ISOLATION OF THE LETHAL FACTOR OF DIPHTHERIA TOXIN BY ELECTROPHORESIS IN STARCH GEL

Poulik, M. D., and E. Poulik, Nature **181**, 354–355 (1958).

Although highly purified toxins of Corynebacterium diphtheriae were determined by filter-paper electrophoresis to be apparently homogeneous, electrophoresis and immunoelectrophoresis in starch gel demonstrated the complex nature of such toxins by revealing the presence of two major components in addition to many minor ones. Diphtheria toxin, lethal in very high dilutions for guinea pigs and assumed to be one of the two major components which is the cause of death, was produced on a tryptic digest of beef muscle medium and contained 162Lf/mgm (0.084 mgm total N/mgm; 0.069 P-N/mgm). This material was subjected to starch-gel electrophoresis in a discontinuous system of buffers (Tris-citrate in the starch gel and sodium hydroxide/boric acid in the electrode vessels) which resolved the two major components and assured their purity. Solutions of the toxin were injected into guinea pigs to determine the mode of action of diphtheria toxin. Results of the experimentation showed that more than one lethal factor is present in the toxins grown in culture media, and that each of these is again a mixture of proteins, presumably with different enzymatic activities.

LOCALIZATION OF THE THYROXINE-BINDING PROTEIN OF SERUM BY STARCH GEL ELECTROPHORESIS

Rich, C., and A. G. Bearn, Endocrinology **62**, 687–689 (1958).

Starch-gel electrophoresis was used in an attempt to isolate and characterize the thyroxine-binding protein (TBP) of human serum. Preliminary observations of the electrophoretic mobility of TBP in starch gel are presented in this report. A much greater purification of thyroxine-binding protein can be obtained by starch-gel electrophoresis than by conventional electrophoretic methods.

A SIMPLE METHOD FOR MAKING STARCH-GEL ELECTROPHORETIC STRIPS TRANSPARENT

Vesselinovitch, S. D., Nature **182**, No. 4636, 664–665 (1958).

A simple and rapid method is described for obtaining transparency of the strips in starch-gel electrophoresis for the direct quantitative evaluation of the separated proteins. The means of evaluating these starch-gel strips photometrically and of preserving them for future reference also are described.

SERUM PROTEIN DIFFERENCES IN CATTLE BY STARCH GEL ELECTROPHORESIS

Ashton, G. C., Nature **180**, 917–919 (1957).

As a continuation of an investigation of the genetical differences obtained with cattle serum proteins by starch-gel electrophoresis, the distribution of the phenotypes obtained from cattle serum by starch-gel electrophoresis was studied. Details of this study and the results of 150 matings are reported. Two-dimensional electrophoresis, first in agar and then in starch gel, determined the nature of the components observed in one-dimensional starch-gel electrophoresis. The β_2-globulin zone gave four, five or six components depending on the serum type, and the various combinations of these components allowed recognition of five β_2-globulin phenotypes. Results of the investigation suggest the possibility of genetic control on the β_2-globulin pattern.

HISTOCHEMICAL DEMONSTRATION OF ENZYMES SEPARATED BY ZONE ELECTROPHORESIS IN STARCH GELS

Hunter, R. L., and C. L. Markert, Science **125**, 1294–1295 (1957).

Zone electrophoresis in starch gels, a rapid, simple, dependable method of high-resolving power for separating complex mixtures of proteins, was combined with histochemical techniques for locating and identifying enzymes in tissue sections in order to analyze the enzymatic composition of biological material. Application of this method to the study of mouse liver is described and the results are discussed. The method was also used in locating tyrosinases in an investigation of the distinctive properties of tyrosinases indicated by zymograms of aqueous extracts of mushrooms *(Psalliota)*, potatoes, and mouse melanoma, and in separating the esterase-active proteins from mouse liver and other organs.

STARCH ELECTROPHORESIS OF ANIMAL SERA

Latner, A. L., and A. H. Zaki, Nature **180**, 1366–1367 (1957).

A starch-gel electrophoresis technique, used by Ashton [Nature *179*, 824 (1957)] to investigate the serum protein patterns of the cow, dog, horse, and pig, was used by the authors for the study of vitamin B_{12} binding by serum proteins. The sera of various animal species was also examined for comparative purposes. Human and various animal sera were run concurrently on the same gel, and the results showed very little correspondence in the relative intensity of the two species and not always correspondence in the position of the bands.

REQUIREMENTS FOR OPTIMAL RESOLVING POWER AND REPRODUCIBILITY IN PROTEIN FRACTIONATION BY STARCH GEL ELECTROPHORESIS

Pert, J. H., M. H. Sleisenger, K. R. Woods, and R. L. Engle, Jr., Clin. Res. Proc. **5**, 156–157 (1957).

A starch-gel zone electrophoresis technique, described in this paper, was used in the separation of normal sera into 15 to 20 protein fractions, and rigid control at each step of the procedure was found to be necessary for adequate reproducibility with optimal protein fractionation. The particular potato starch used is important, the duration of both the acetone-HCl hydrolysis and the aqueous hydrolysis must not vary more than five percent, the acetone temperature must be kept within 0.2°C at 36°C, and the gel buffer pH must be within $\pm$ 0.0005 molar boric acid. Differences in normal sera and an analysis of a number of purified proteins by ultracentrifugal, immunologic, and electrophoretic methods are also discussed.

STARCH GEL ELECTROPHORESIS IN A DISCONTINUOUS SYSTEM OF BUFFERS

Poulik, M. D., Nature **180**, 1477–1479 (1957).

The application of starch-gel electrophoresis in a boric acid buffer (0.03M boric acid and 0.012M sodium hydroxide) to the study of separated and eluated fractions of the toxin of *Corynebacterium diphtheriae* in tissue cultures and in animals resulted in unsatisfactory resolutions of the proteins. A mixture of Tris-(hydroxymethyl)-aminomethane-citrate (Tris-citrate) used in conjunction with a boric acid buffer proved to be a more suitable system, and the preliminary results obtained from its use in the starch-gel electrophoresis of toxins, enzymes and human sera are briefly reported.

MULTIPLE MYELOMATOSIS AND MACROGLOBULINEMIA: DIFFERENTIATION BY STARCH GEL ELECTROPHORESIS

Silberman, H. J., Lancet **2**, 26–27 (1957).

Multiple myelomatosis and macroglobulinemia are difficult to distinguish for their clinical features are not always clear-cut, Bence-Jones proteins may be present in the urine of both conditions, the abnormal serum globulins cannot be distinguished by paper electrophoresis, and the Sia water-test is unreliable. Electrophoresis of the abnormal serum globulins from two cases of myelomatosis and two of macroglobulinemia, using starch gel as the supporting medium, indicated a striking difference in their mobilities, however.

VARIATIONS IN HUMAN SERUM β-GLOBULINS

Smithies, O., Nature **180**, 1482–1483 (1957).

Variations in human serum β-globulins, which were noticed when sera were being investigated by the one-dimensional starch-gel electrophoretic method, are reported. These variations, however, can be established with certainty only when the two-dimensional electrophoretic method is used. Investigations on serum from a young white female and a young Negro female using two-dimensional electrophoresis are presented. The presence of β-globulin D was established with certainty in serum from two female Negroes (out of 22 males and 27 females tested) and in serum from four female and one male Australian aborigines (out of 17 males and 6 females tested), but was not demonstrated in the serum from any of several hundred Canadians (largely of European ancestry) studied mainly by one-dimensional starch-gel electrophoresis. The findings of these studies suggest the existence of considerable racial differences in the frequency of the occurrence of β-globulin D in serum and indicate that its presence is probably genetically controlled.

TESTS OF PURITY OF DIPHTHERIA TOXINS BY ELECTROPHORESIS IN STARCH GEL

Poulik, M. D., Nature **177**, No. 4517, 982–983 (1956).

Highly concentrated and purified diphtheria toxins and toxoids as well as tetanus toxoids were separated into several components by electrophoresis in starch gel. The procedure is described and the results from a few of the materials tested are presented.

TWO-DIMENSIONAL ELECTROPHORESIS OF SERUM PROTEINS

Smithies, O., and M. D. Poulik, Nature **177**, No. 4518, 1033 (1956).

Results which can be obtained from two-dimensional electrophoresis are illustrated in this report. This method is a combination at right angles of starch-gel electrophoresis and electrophoresis on filter paper.

ZONE ELECTROPHORESIS IN STARCH GELS: GROUP VARIATIONS IN THE SERUM PROTEINS OF NORMAL HUMAN ADULTS

Smithies, O., Biochem. J. **61**, 629–641 (1955).

A method of zone electrophoresis using starch gel as the supporting medium, in which protein detection is achieved by staining, the possible reasons for its high resolving power, and its limitations are described. As little as 0.02 ml of the sample can be used in this technique which is very well adapted to comparing closely related samples. Use of this method for the electrophoresis of normal human sera makes possible the demonstration of several previous undescribed non-dialysable components, and excludes electrophoresis anomalies, products of the clotting process, and inadvertent hemolysis as sources of the new components. Investigation of human serum proteins by this method is presented and discussed.

Starch–Block Electrophoresis

THE POLYPEPTIDE CHAINS OF HAEMOGLOBIN-A₂ AND HAEMOGLOBIN-G₂

THE POLYPEPTIDE CHAINS OF HAEMOGLOBIN-A_2 AND HAEMOGLOBIN-G_2

Huehns, E. R., and E. M. Shooter, J. Mol. Biol. **3**, 257–262 (1961).

Dissociation and recombination experiments using Hb-A_2 with Hb-H and Hb-G_{Br}

are described confirming that Hb-A$_2$ has the molecular composition $\alpha^A{}_2 S^A{}_2 2$. Starch-block electrophoresis was used to isolate Hb-G$_{Br}$ and Hb-H from the appropriate hemolysate and to analyze Hb-A$_2$, free from nonheme protein and prepared by chromatography. With Hb-H, Hb-A was formed and with Hb-G$_{Br}$, which has abnormal $\alpha^G{}_2$ subunits, a new zone slower than Hb-A$_2$ appeared on starch-gel electrophoresis. The methods and materials used in the hemoglobin solutions, in the isolation of individual hemoglobins and in the hybridisation experiments are described and the results are discussed.

STARCH BLOCK ELECTROPHORESIS OF HEMOGLOBIN

Pearson, H. A., and W. Mcfarland, U.S. Armed Forces Med. J. **10**, 693–700 (1959).

The technique of starch-block electrophoresis of hemoglobins, found to be of particular value in its sensitivity in detecting minor hemoglobin components such as hemoglobin A$_2$, is described. The hemoglobin A$_2$ fraction accounts for 1.5 to 3.1 percent of the total hemoglobin in normal persons, but in persons with thalassemia the fraction is elevated above normal levels, thus providing an objective and relatively specific laboratory test for the diagnosis of the condition. Starch-block electrophoresis of hemoglobin is suggested to be an indicated test of the evaluation of any iron refractory, hypochromic anemia. Three case reports are given.

ELECTROPHORESIS OF FLUORESCENT ANTIBODY

Curtain, C. C., Nature **182**, 1305–1306 (1958).

A physicochemical study was made on fluorescein-protein conjugates in order to gain more information about the chemistry of non-specific staining. Attempts at separating antibody from the non-specific staining activity by electrophoresis with starch blocks and powdered cellulose was unsatisfactory but it was possible to fractionate the fluorescein-globulin conjugate by electrophoresis convection. The investigation and the results are described and discussed.

ISOLATION AND DESCRIPTION OF A FEW PROPERTIES OF THE β_{2A}-GLOBULIN OF HUMAN SERUM

Heremans, J. F., M.-Th. Heremans, and H. E. Schultze, Protides Biol. Fluids, Proc. 6th Colloq., 166–172 (1958).

β_{2A}-globulin was obtained from normal human serum and plasma by a combination of zinc sulfate precipitation, ammonium sulfate precipitation, and starch-block preparative electrophoresis. The preparation was shown by immunoelectrophoretic tests to be identical with native β_{2A} serum protein. The various properties of β_{2A}-globulin are described.

SKIN SENSITIZING ACTIVITY OF GLOBULIN FRACTIONS FROM RABBIT IMMUNE SERUMS

Aladjem, F., W. R. MacLaren, and D. H. Campbell, Science **125,** 692–693 (1957).

An investigation is reported which was undertaken to determine whether skin sensitizing activity of rabbit immune serums is associated with the same globulin component as precipitating antibody, or whether, analogous to skin-sensitizing activity of human serums, the skin-sensitizing activity of rabbit immune serums might not also be associated with other than gamma globulin components. The methods used in the study, which include starch electrophoresis performed in barbital buffer at pH 8.6, $\mu 0.1$, are described and the findings discussed.

Paper Electrophoresis

SEPARATION OF INORGANIC IONS IN FUSED SALTS BY MEANS OF CHROMATOGRAPHY AND ELECTROPHORESIS ON GLASS FIBER PAPER: III. EFFECT OF WATER, OXYGEN AND SUPPORT ON THE MIGRATION OF INORGANIC IONS DISSOLVED IN THE LiCl-KCl EUTECTIC AT 450°

Alberti, G., S. Allulli, and G. Modugno, J. Chromatog. **15,** 420–427 (1964).

Results of a study of the chromatographic and electrophoretic behavior of different metal ions dissolved in LiCl-KCl eutectic are reported. Also discussed are the chromatographic and electrophoretic behavior of some inorganic ions and the electroendosmotic effect in fused chlorides, employing glass fiber paper as support.

STUDIES ON CAMEL HEMOGLOBIN

Banerjee, S., and A. S. Bhown, Biochim. Biophys. Acta **86,** 502–510 (1964).

In order to explain the differential mobility of camel hemoglobin, the amino acid composition of the products of partial acid hydrolysis of camel hemoglobin, the N-terminal analysis of each product and total number of acid and basic groups present in camel hemoglobin were studied. Cow and human hemoglobins also were studied for comparison. Paper electrophoresis was used in the study.

MACROMOLECULAR PROPERTIES AND BIOLOGICAL ACTIVITY OF HEPARIN: III. PAPER ELECTROPHORETIC STUDIES OF HISTAMINE BINDING

Barlow, G. H., Biochim. Biophys. Acta **83**, 120–122 (1964).

Experimentation using electrophoresis was performed to ascertain whether a correlation exists between the anticoagulant activity of heparin and its ability to bind histimine. The electrophoretic studies confirm that heparin can bind histimine and further show some interesting properties of this interacting system.

A STUDY OF THE BLOOD GLUCOSE, SERUM TRANSAMINASE, AND ELECTROPHORETIC PATTERNS OF DOGS WITH INFECTIOUS CANINE HEPATITIS

Beckett, S. D., M. J. Burns, and C. H. Clark, Am. J. Vet. Res. **25**, 1186–1190 (1964).

Dogs with and without acute infectious canine hepatitis were studied to determine the serum protein patterns, transaminase levels, and blood glucose levels. All electrophoretic work was done by paper electrophoresis. Methods and results are discussed.

A CONTRIBUTION TO THE TECHNIC OF PAPER ELECTROPHORESIS

Bendezky, K. M., and R. K. Shadmanov, Biokhimiya **29**, Vol. 3, 439–444 (1964) **Russian.**

The regularities formulated by Macheboeuf et al., which offered an explanation of the difference in the partition effect of some migrating components imposed on different places of the paper strip, were followed up in paper electrophoresis of proteins and amino acids with evaporation. The conditions of evaporation electrophoresis in the EMIB and EFA apparatuses were elucidated by a mathematical interpretation of migration in the electric field.

FIBRINOLYSIS AND THE PLASMA PROTEIN ELECTROPHORETIC PATTERN

Bielawiec, M., and I. M. Nilsson, Scand. J. Lab. Clin. Invest. **16**, 513–520 (1964).

Electrophoresis of plasmas with high fibrinolytic activity revealed no signs of fibrinogen, but the presence of fibrinogen was shown on re-electrophoresis after the samples had stood for 24 hours at room temperature. The fibrinogen spot also returned in plasma samples drawn 60 minutes after the injection of nicotinic acid or 20 minutes after the venous stasis. The influence of fibrinolysis on the plasma protein pattern was studied in an attempt to explain these electrophoretic phenomena. In the paper electrophoretic pattern of plasma with increased fibrinolytic activity (urokinase-activated human plasma and plasma from individuals in whom fibrinolytic activity had been induced by injection of nicotinic acid or by venous stasis), no fibrinogen spot was found because of the low fibrinogen level due to the plasma fibrinogen level being normal. The disappearance of the fibrinogen spot was prevented by the addition of ∈-ACA to the electrophoretic buffer and by the storage of the fibrinolytically active plasma for 24 hours at room temperature. Findings showed that the local dissolution of fibrin by activators and/or plasmin on the paper, promoted by the absence of inhibitors and the adsorption of the activators to the fibrin, caused the disappearance of the fibrin spot.

FURTHER INVESTIGATIONS ON COMPLEX FORMATION IN VITRO BE-TWEEN AORTIC MUCOPOLYSACCHARIDES AND β-LIPOPROTEINS

Bihari-Varga, M., J. Gergely, and S. Gerö, J. Atherosclerosis Res. **4**, 106–109 (1964).

A demonstration in vitro of the complex formation of serum β-lipoproteins and aortic mucopolysaccharides (MPS), previously established by paper electrophoretic and turbidimetric methods, supported the hypothesis that the attractive forces between these two substances may play a role in atherogenesis. Interpretation of the paper electrophoretic results was complicated by the strong adsorption of β-lipoproteins to paper; but these adsorptive forces were absent from agar-gel systems. This paper describes a reinvestigation of the pathogenesis of atherosclerosis using agar electrophoresis.

(I^{131}) ALBUMIN TURNOVER AND LOSS OF PROTEIN INTO THE SPUTUM IN CHRONIC BRONCHITIS

Bonomo, L., and A. D'Addabbo, Clin. Chim. Acta **10**, 214–222 (1964).

Previous investigations had shown that protein fractions identical to those of the blood serum are present in the sputum of patients with various respiratory diseases and that minimal amounts of all blood serum fractions pass into the sputum of both normal subjects and patients with bronchial inflammation. An evaluation was made on the [I^{131}] albumin turnover and the loss of protein into the sputum of patients with chronic bronchitis. The material and methods used and the results of this study are described in this paper. A clear albumin band was revealed by paper electrophoresis and autoradiography in the electrophoretic strips of the bronchitic patients' sputum. The labelled albumin in the electrophoretic strips of the controls, however, could be demonstrated only by autoradiography. The investigation indicated that the changes in the albumin are connected with the persistent loss of albumin through the bronchial tree of patients with chronic bronchitis.

STUDIES ON HAPTOGLOBIN AND HAEMOPEXIN IN THE PLASMA OF CATTLE

Bremner, K. C., Australian J. Exp. Biol. Med. Sci. **42**, 643–656 (1964).

Studies were made on plasma samples from three healthy calves to determine if the nature of the bovine heme-binding proteins might be of value in veterinary clinical medicine as a sensitive indicator of mild hemolytic anemias. Filter-paper electrophoresis was used to compare the heme-binding proteins with properties similar to human haptoglobin and hemopexin. It was concluded that simultaneous absence of both hemopexin and haptoglobin from bovine plasma during the hemolytic crisis of babesiosis was indicative of pathological hemolytic processes.

PAPIERELEKTROPHORETISCHE UNTERSUCHUNGEN VON RATTENSEREN BEI EXPERIMENTELLER DIÄTETISCHER LEBERNEKROSE

Brenner, G., and H. Kremer, Z. Klin. Chem. **2**, 57–59 (1964) **German.**

Paper electrophoresis was used to measure the blood protein fractions in rat sera. A study of the changes found in the sera of rats with alimentary liver necrosis, compared statistically with values from healthy rats of the same litter, revealed a fall in albumin to about 30 percent of the norm and a significant increase of α_1-, α_2-, β_1- and β_2-globulins compared with the norm was noted. The similarity of these changes in rat sera to blood protein changes in human eclampia is briefly discussed.

ELECTROPHORETIC BEHAVIOR OF THE VESICULAR BILE OF SOME ANIMALS

Chevrier, J. P., and P. V. Creac'h, Proces-Verbaux Seances Soc. Sci. Phys. Nat. Bordeaux 247–262 (1963–1964).

Bile of the ox, pig, hen and plaice were dialyzed and electrophoresed (pH 8.6). Six protein fractions were separated, independent of sex or age of the species. Aberrant spots were found although the animals were in good health.

THIN FILM ELECTROPHORESIS: I. THE ELECTROPHORETIC BEHAVIOR OF COAL-TAR FOOD COLOURS ON PAPER AND THIN FILMS

Criddle, W. J., G. J. Moody, and J. D. R. Thomas, J. Chromatog. **16**, 350–359 (1964).

The electrophoretic behavior of the coal-tar food colors permitted in the United Kingdom, with the exception of Oil Yellow GG, Oil Yellow XP, Naphthol Yellow S, and Ponceau 3R, was determined by electrophoresis in six different electrolytes of widely varying pH values on the thin-film materials Kieselguhr, alumina G, silica gel G, and Whatman® No. 1 filter paper. A high degree of resolution was obtained using the thin-film supports. The experimental procedure and the results are presented and discussed.

SERUM PROTEIN-BOUND CARBOHYDRATES BY PAPER ELECTROPHORESIS

Green, Morris N., Lucas L. Kulczycki, Sergio I. Magalini, and Harry Shwachman, with technical assistance of Mila Crul, J. Lab. Clin. Med. **63**, No. 3, 416–424 (1964).

Paper electrophoresis was used to determine the distribution of protein-bound carbohydrates in the sera of 143 patients with cystic fibrosis, using Green's modification of the periodic acid-Schiff method to stain the paper strips. While the protein alpha-1 level underwent only a slight increase, the carbohydrates bound to the alpha-1 globulin fractions showed a marked and progressive increase as the clinical condition of the patient deteriorated. With increasing severity of the disease, there was a twofold increase in the gamma globulin protein fraction, but the carbohydrates bound to this fraction remained essentially unchanged. No significant changes were observed in the remaining electrophoretically separated serum fractions.

COMPLEXES BETWEEN POLYHYDROXY-COMPOUNDS AND INORGANIC OXY-ACIDS: VI. PAPER ELECTROPHORESIS IN STANNATE SOLUTION

Lees, E. M., and H. Weigel, J. Chromatog. **16**, 360–364 (1964).

The formation of complexes with several acyclic and cyclic polyols by stannate was shown by paper electrophoresis in stannate solution. The investigation and the results are described and the paper electrophoretic mobilities of the polyols are discussed from the viewpoint of the conformations of the polyols and the structure of the stannate ion.

A SIMPLE CHEMICAL TEST FOR DISTINGUISHING MYELOMA GLOBULINS FROM MACROGLOBULINS

Saifer, Abraham, J. Lab. Clin. Med. **63**, No. 6, 1054–1060 (1964).

The difficulty of distinguishing with certainty a multiple myeloma peak from a macroglobulin peak by means of paper electrophoresis in the absence of an analytical ultracentrifuge can be readily resolved by the addition of a Rivanol solution to the serum and by comparison of the electrophoretic pattern obtained with the supernatant fraction with the original serum. Sixty to one hundred percent of the protein remaining in the supernatant fraction in cases of multiple myeloma will constitute the abnormal peak, while Rivanol will reduce the amount of the macroglobulin peak to nearly zero levels. Electrophoretic and comparative ultracentrifugal data to support this conclusion are presented for five cases of macroglobulinemia and eight cases of multiple myeloma.

THE EFFECT OF DILUTION ON THE ACTIVITY OF AMYLASE AND ITS RE- LATION TO THE EFFECT OF ELECTROPHORESIS

Theodor, E., and D. Birnbaum, J. Lab. Clin. Med. **63**, No. 5, 879–884 (1964).

In this paper, the authors present evidence that serum-amylase activity increased after paper electrophoresis. Investigation showed that after mere dilution, both human serum amylase and hog pancreatic amylase showed a considerable increase in specific activity, though not to the same degree as after electro-phoresis. It was also found that the barbitol buffer used in electrophoresis would by itself change the color reaction of the amylocastic method used in their experiments, thus causing a spurious increase in amylase activity, resulting apparently from a rise in pH. The observed rise in activity seemed to be caused by the serum being diluted in electrophoresis and the pH of the eluate from the paper strips being raised by the alkaline buffer in which it was conducted.

ELECTROMIGRATION IN FUSED SALTS IN THE STUDY OF COMPLEX IONS

Bailey, R. A., and A. Steger, J. Chromatog. **11**, 122–123 (1963).

A technique of paper electromigration, carried out in strips of Whatman® GF/A glass fiber paper, 30 cm long and 1 cm wide, was used to show the presence of a number of transition-metal thiocyanate complexes in solution in molten KCNS, and to compare the behavior of these metals with that of molten $LiNo_3$-KNO_3 eutectic mixture where essentially no complexing is expected.

THE ANTIGENIC COMPOSITION OF RAT SERUM: AN ELECTROPHORETIC STUDY

Benjamin, David C., and Henry E. Weimer, J. Immunol. **91**, No. 3, 331– 338 (1963).

Electrophoretic methods of moving boundary, filter paper, polyvinylchloride block, starch gel, immunoelectrophoresis and starch-gel immunoelectrophoresis were used to analyze the pooled serums from normal, adult, male Sprague-Dawley rats. Results showed immunoelectrophoresis to be the most sensitive single procedure. This technique was used to observe 25 antigenic components of rat serum and a standard system of nomenclature was proposed. Also demonstrated in this report are the antigenic heterogeneity of fractions sepa-rated by other procedures.

DIE DIAGNOSTISCHE BEDEUTUNG DER LIPID-ELEKTROPHORESE FÜR DIE BEURTEILUNG VON STÖRUNGEN DES LIPID-STOFFWECHSELS

Berg, G., Z. Klin. Chem. **1**, 48–51 (1963) **German.**

Compared to paper electrophoresis, lipid electrophoresis (thin-layer electrophoresis on starch), used to demonstrate lipid components of serum fractions and the α_1-lipoproteins in the albumin band, has the advantage that larger amounts of serum can be analyzed and the fractions of low lipid content can be shown. This method was used to study cases of acute hepatitis and the transportation of natural fat after the elimination of light lipoproteins by ultracentrifugation in cases of essential hyperlipaemia, chronic hepatitis, fatty liver epithelium and nephrosis.

THE PROTEOLYTIC ENZYMES OF ASPERGILLUS ORYZAE: I. METHODS FOR THE ESTIMATION AND ISOLATION OF THE PROTEOLYTIC ENZYMES

Bergkvist, R., Acta Chem. Scand. **17**, No. 6, 1521–1540 (1963).

Methods for the separation and estimation of proteolytic enzymes of moulds are presented. Three proteolytic enzymes were isolated and purified from *Aspergillus oryzae* both with laboratory techniques and with a method usable on a larger scale. The enzyme preparation was subjected to electrophoresis under a variety of conditions with paper electrophoresis being used to follow the distribution of activity among the different protein fractions during the course of fractionations. Although the amounts of material able to be purified by curtain electrophoresis were small, they were adequate to produce enough material for preliminary characterization of the proteases by analytical methods. In the one-dimensional paper electrophoresis all runs were carried out with an electophoresis apparatus of the horizontal moist-chamber type. The best results were obtained after preceding zone-sharpening, and Whatman® No. 1 filter papers gave the most satisfactory and reproducible separations. The continuous flow paper electrophoresis cell used was the Beckman/Spinco® model CP.

THE EFFECT OF AMINO ACIDS ON THE PAPER ELECTROPHORESIS OF THYROXINE AND TRIIODOTHYRONINE

Bird, R., Clin. Chim. Acta **8**, 936–942 (1963).

Thyroxine migrates towards the anode to a position close to that of human serum thyroxine-binding protein (TBP) when submitted to paper electrophoresis in barbital buffer at pH 8.6 in the presence of either an ultrafiltrate of trypsin-digested serum or of phenylalanine or other amino acids, but at pH 7.6 it hardly moves from the origin. Triiodothyronine remains near the origin at both pH 8.6 and pH 7.6 under the same conditions. A method of testing for the presence of TBP in biological materials, based on a study of the electrophoretic mobility of thyroxine and triiodothyronine in barbital buffer at pH 8.6 and pH 7.6, is suggested. Materials, methods and results of the study are presented in this paper.

ELECTROPHORETIC METHOD FOR OBTAINING NUCLEOTIDES LABELLED WITH RADIOACTIVE CARBON

Borkowski, T., H. Brzuszkiewicz, and H. Berbéc, J. Chromatog. **12,** 229–235 (1963).

The conditions for separating adenosine monophosphate (AMP), guanosine monophosphate (GMP), cytidine monophosphate (CMP) and uridine monophosphate (UMP) by continuous electrophoresis on Whatman® 3MM filter paper in acid buffer at pH 3.5 are presented. The purity of the ribonucleotides of RNA, obtained by this method, was tested. In the investigations reported, the four mononucleotides were produced by this method on a laboratory scale and good results were obtained for the production of nucleotides from biosynthetically labelled RNA of E. coli grown on a synthetic medium in the presence of $C^{14}O_2$. Results showed that the radioactive mononucleotides were labelled in purine and pyrimidine bases only and that the specific activity of the four mononucleotides was similar.

ZUR QUANTITATIVEN BESTIMMUNG VON AMINOSÄUREN UNTER ANWENDUNG DER HOCHSPANNUNGSELEKTROPHORESE

Braun, L., Biochem. Z. **339,** 8–12 (1963) **German.**

A high-voltage electrophoresis method, which differs from the usual procedures in that the color complexes of the amino acid-ninhydrin- compounds are not prepared by spraying the paper with a copper reagent, but rather with an alcoholic solution of copper nitrate during an elution, was developed for the quantitative determination of the amino acids in serum. The new method, described in this paper, appears to give better reproducible results.

STUDIES ON SERUM PROTEIN CHANGES AND ORGAN DYE CONCENTRATIONS IN TRYPAN BLUE CARCINOGENESIS

Brown, D. V., L. M. Norlind, A. Adamovics, and A. Bowen, Proc. Soc. Exp. Biol. Med. **114,** 290–293 (1963).

Alterations in serum proteins after one dye injection were compared with those found later in animals developing dye induced neoplasms, and the concentration of protein bound dye in the serum and reticuloendothelial organs was evaluated and related to the degree of immediate cellular response and the eventual development of neoplasms. Paper electrophoresis in a Spinco® inverted V-type cell was used to measure serum protein fractions. This study, made with Wistar strain rats, is discussed by the authors.

AN INVESTIGATION OF THE INTERACTIONS BETWEEN MILK PROTEINS AND TEA POLYPHENOLS

Brown, P. J., and W. B. Wright, J. Chromatog. 11, 504–514 (1963).

Starch-column, membrane filter, and paper electrophoresis were used in an investigation to determine whether the reduction in the astringent taste of a black tea infusion after an addition of milk to the infusion could be attributed to some interaction between milk proteins and the black tea polyphenols. Paper electrophoresis was employed to separate the colored tea compounds; starch grain column electrophoresis achieved a certain separation of the components of the protein system present in skim milk. Cellulose-acetate electrophoresis identified the milk proteins in the resulting fractions.

A SIMPLE AND RAPID METHOD FOR THE PAPER ELECTROPHORETIC DETERMINATION OF URINARY PROTEINS

Collens, R., H. Meyers, and K. Lange, Clin. Chem. 9, No. 3, 330–333 (1963).

A simple, rapid and accurate method for the determination of paper electrophoretic patterns of urinary proteins, using polyethylene glycol (Carbowax® 20M) for the concentration of the urine, is described. When added to a protein-free urine and subjected to this method, the electrophoretic patterns of sera with normal and abnormal constituents did not change. Insight may be given into the physical type of permeability disturbance in the kidneys of a patient by use of electrophoretic determination of urinary proteins as a routine procedure.

A SIMPLE LOW-VOLTAGE PAPER ELECTROPHORETIC METHOD FOR THE DETERMINATION OF URINARY VANILMANDELIC ACID (VMA)

Eichhorn, F., and A. Rutenberg, Clin. Chem. 9, No. 5, 615–619 (1963).

A simple and accurate method for the determination of urinary 3-methoxy-4-hydroxy-mandelic acid (vanilmandelic acid VMA) which applies the new principle of two solutions paper electrophoresis for better separation of amino acids, proteins, and other biologic mixtures, is described. VMA was separated from all other phenolic acids in approximately 0.05 ml untreated urine by using low-voltage paper electrophoresis with different buffer concentrations for the anode and the cathode, and the separated VMA was then determined colorimetrically in the eluate. Using this procedure, it was found that values of excreted VMA in 24-hour normal urine and those from subjects with pheochromocytoma differed considerably.

ACTH ANTIBODIES AND THEIR USE FOR A RADIOIMMUNOASSAY FOR ACTH

Felber, J. P., Experientia **19,** 227–229 (1963).

Guinea pigs were immunized with ACTH and the sera separated by paper electrophoresis to search for ACTH antibodies. Electrophoresis was done in pH 8.6 Veronal buffer, ionic strength 0.1, for 15 hours at 4°C at a constant voltage of 4 volts/cm. I^{131} labeled ACTH in excess was added to the sera before electrophoresis and incubated 24 hours at 0°C. The I^{131} ACTH remained at the origin when incubated with normal serum and moved with γ-globulin in ACTH immune serum. The ACTH antibody is non-precipitating.

TRICHROME, A NEW STAIN FOR ELECTROPHORESIS

Fischl, J., and J. Gabor, Clin. Chim. Acta **8,** 330–331 (1963).

The various dyes used in paper electrophoresis were absorbed at different rates by the protein molecules of the separated fractions, resulting in a wide variation of the normal values obtained by the different techniques. In order to facilitate the comparison and correlation of the different reported values, references were made to values obtained by free electrophoresis, and to obtain the true globulin concentration, correction factors ranging from 1 to 1.6 were introduced. By combining three different dyes into one stain, the authors found that the correlation to free values became so close that no correction factor was necessary. This new stain, called Trichrome, is described and evaluated in this report.

PROTEIN ELECTROPHORETIC ANALYSES OF SERUM OF CANCER PATIENTS

Graham, W. D., Clin. Chem. **9,** No. 5, 582–593 (1963).

Studies of the blood serum proteins of cancer patients, using paper electrophoretic separations to find some change which would be useful in the diagnosis or evaluation of treatment of the disease, were of little value because the rather nonspecific changes obtained occurred too late in the course of the disease. By including nonionic surfactants in the buffer, more extensive fractionation of the serum proteins was permitted. Because this increased detail was believed to be useful in the diagnosis of cancer, further studies with this method were made. Sera from 253 cancer patients, from 273 patients free of organic disease, and from 785 ambulatory patients suffering from a wide variety of non-neoplastic diseases were analyzed electrophoretically using this procedure, and the results are discussed in this paper.

QUANTITATIVE ANALYSES ON DOG AND HUMAN HEPATIC BILE

Hardwicke, J., K. J. Baker, J. G. Rankin, and R. Preisig, Protides Biol. Fluids, Proc. 11th Colloq., 264–268 (1963).

Qualitative and quantitative data on the proteins in dog and human bile are presented in this paper. Paper electrophoresis and agar-gel electrophoresis, both at pH 8.7, were used in the study.

QUANTITATIVE MEASUREMENT OF INDIVIDUAL AND TOTAL FREE AMINO ACIDS IN URINE: RAPID METHOD EMPLOYING HIGH-VOLTAGE PAPER ELECTROPHORESIS AND DIRECT DENSITOMETRY AND ITS APPLICATION TO THE URINARY EXCRETIONS OF AMINO ACIDS IN NORMAL SUBJECTS

Mabry, C. C., and W. R. Todd, J. Lab. Clin. Med. **61**, No. 1, 146–157 (1963).

A quantitative method based on partitioning amino acids on paper by high-voltage electrophoresis, development of color, and subsequent measurement of their intensity by direct densitometry was developed for the determination of amino acids in urine for clinical purposes. This technique is more precise than conventional paper chromatography, simpler than column chromatography, and faster than either, and preliminary experience has shown it to be suitable for clinical purposes. It yields data for the major amino acids found in urine that is comparable to the more elaborate column chromatographic method, and the data obtained from the analysis of a large number of human urine specimens are reported as micromoles amino acids excreted per day. High-voltage paper electrophoresis permits rapid partition of amino acids and quantitative measurement by subsequent direct densitometry with a recording, integrating densitometer. The analytic range on paper is 7 to 86 mμ moles amino acid with a sensitivity of 4 mμ moles.

ELECTROPHORESIS OF LIPOPROTEINS USING PRE-STAINED SERUM

Ribeiro, L., and H. J. McDonald, J. Chromatog. **10**, 443–449 (1963).

Presented in this paper is a simple, inexpensive and time-saving method for the paper electrophoretic determination of serum lipoproteins, pre-stained with a saturated amount of one volume of the dye solution to 10 volumes of serum. Electrophoresis was performed on Macherey and Nagel® No. 2214 ff filter paper in a pH 8.6 Veronal buffer, ionic strength 0.05, and a potential gradient of 8 volts/cm used for a period of two hours, with the results showing a reproducibility within ± 3.0 percent for each individual fraction. The conditions for prestaining are discussed with consideration given to the influence of light and other factors.

ÜBER DIE SICHBARMACHUNG VON EIWEISSFRAKTIONEN NACH PAPIER-
ELEKTROPHORESE DURCH IHRE REDUKTIONSWIRKUNG (REDOXOGRAMM)

Wachter, H., Protides Biol. Fluids, Proc. 11th Colloq., 364–367 (1963)
German.

A simple procedure is described for the detection of reduction activity in serum proteins after zone electrophoresis in which a reagent consisting of potassium permanganate and sulfuric acid is reduced to the brownish manganese dioxyde by some serum proteins. The pattern of this redoxogram is described.

SERUM PROTEIN AND GLUCOPROTEIN CONCENTRATIONS IN NEW-BORNS AND INFANTS

Böttiger, L. E., and G. Sterky, Scand. J. Clin. Lab. Invest. **14**, Suppl. 64, 39–44 (1962).

Serum protein fractions and protein-bound hexoses, hexosamines, and sialic acids were analyzed in 19 cord blood samples from full-term babies and in 36 serum samples from infants of different ages. Paper electrophoresis, one of the methods used in the analysis, was carried out in a barbital buffer at pH 8.6 and ionic strength 0.1, and the strips were stained with Bromophenol Blue, cut and the concentration of the dye measured spectrophotometrically after elution. Glucoproteins, total protein, albumin, and alpha and beta globulins were low at birth, but increased rapidly during the first six months. Gamma globulins showed an initial decrease.

SERUM PROTEINS AND GLUCOPROTEINS IN NORMAL SCHOOLCHILDREN

Böttiger, L. E., and G. Sterky, Acta Med. Scand. **172**, 339–350 (1962).

Analyses were made of the total protein, protein electrophoretic pattern, and protein-bound hexoses, hexosamines, and sialic acids of the sera of 105 children, aged seven to twenty years. Of special importance was the fact that even a very small increase in ESR will give considerable rise in serum glucoprotein values. Paper electrophoresis was performed in a barbital buffer of pH 8.6 and ionic strength 0.1. Other procedures used in the study and the results are reported.

ÜBER EINE SPEZIFISCHE EIGENSCHAFT DER SERUMPLASMOCYTOMLIPIDE BEI AUSFÄLLUNG MIT DEXTRANSULFALLÖSUNG

Keller-Bacoka, M., Protides Biol. Fluids, Proc. 10th Colloq., 151–153 (1962) **German.**

The results of experimentation of the lipids of 15 specimens of sera from patients with plasmocytoma, precipitated with dextran sulphate and calcium chloride solutions, are reported. One-dimensional paper-strip electrophoresis was used to examine the supernatant fluids after centrifugation and the native sera for lipids.

STRUCTURAL AND IMMUNOCHEMICAL RELATIONSHIPS AMONG BENCE-JONES PROTEINS

Putnam, F. W., S. Migita, and C. W. Easley, Protides Biol. Fluids, Proc. 10th Colloq., 93–107 (1962).

An examination was made on Bence-Jones proteins from 30 patients with multiple myeloma by various physicochemical and immunological methods to investigate possible structural relationships to each other and to normal human γ-globulin. In the ultracentrifuge and in free electrophoresis most of the proteins exhibited a single boundary; in paper electrophoresis a single band was observed; and in starch-gel electrophoresis multiple bands that gave a single precipitan line were usually produced.

DETERMINATION OF URINARY 3-METHOXY-4-HYDROXY MANDELIC (VANILMANDELIC) ACID BY ELECTROPHORESIS AT LOW pH

Randrup, A., Scand. J. Clin. Lab. Invest. **14**, 262–266 (1962).

A modification of the high-voltage paper electrophoretic method of V. Studnitz [Scand. J. Clin. Lab. Invest. *12*, Suppl. 48 (1960)] for the determination of 3-methoxy-4-hydroxy mandelic (vanilmandelic) acid in urine is described and evaluated. The electrophoresis is performed at low pH and in the modified method, which was found to be advantageous when compared to the original technique.

THYROXINE-SERUM PROTEIN COMPLEXES: TWO-DIMENSIONAL GEL AND PAPER ELECTROPHORESIS STUDIES

Blumberg, B. S., L. Farer, J. E. Rall, and J. Robbins, Endocrinology **68**, 25–35 (1961).

Two-dimensional (paper-paper and paper-gel) electrophoresis of human serum containing labeled thyroxine was performed with various buffers in an attempt to identify the various thyroxine binding fractions. Starch-gel electrophoresis was used for the second run of the experiment. Increases or decreases of thyroxine binding by one or more of the protein species was observed in abnormal sera which was also studied to assist in the identification.

Electrophoresis – Technical Applications

A COMPARISON OF THE PAPER AND STARCH BLOCK ELECTROPHORETIC METHODS FOR DETERMINATION OF A$_2$ HEMOGLOBIN

Hilgartner, M. W., M. E. Erlandson, B. S. Walden, and C. H. Smith, Am. J. Clin. Pathol. **35**, No. 1, 26–30 (1961).

A study was made, using blood specimens from a group of adults known to be heterozygous for thalassemia, to correlate the results obtained by the starch block and by the paper techniques of electrophoresis in order to evaluate the validity of the more simple paper method. Using both techniques and comparing them, it was found that the values for A$_2$ hemoglobin are statistically valid levels for normal persons and those heterozygous for thalassemia. The significantly higher values of the paper method and the diagnostic value of the A$_2$ hemoglobin levels for thalassemia trait are discussed.

A NEW ABNORMAL HUMAN HEMOGLOBIN Hb: ZURICH

Huisman, T. H. J., B. Horton, and M. T. Bridges, Clin. Chim. Acta **6**, 347-355 (1961).

An abnormal slow moving hemoglobin, characterized by its electrophoretic and chromatographic mobilities was discovered in a white family of purely Swiss descent and tentatively named Hb–Zurich. The component, which is alkali non-resistant and, in the reduced state, possesses the same solubility in concentrated phosphate solutions as found for Hb-A, was found to be different from the known abnormal hemoglobin types. The investigation, which included paper, starch-block and starch-gel electrophoresis for the characterization of the abnormal hemoglobin, is described and the genetic implications of the findings are discussed.

IDENTIFICATION OF SUGARS IN URINE BY ELECTROPHORESIS ON PAPER

Jusic, D., and M. Fiser-Herman, Clin. Chim. Acta **6**, 472–474 (1961).

Electrophoresis of sugars on Whatman® No. 1 filter paper was proposed for the identification of unknown melliturias in clinical laboratories because it has greater speed than chromatography, which had been used extensively for this purpose. The most frequently encountered sugars in non-diabetic urines were lactose, galactose, glucose, arabinose and xylose, and within a few hours glucosuria, glactosuria, lactosuria, fructosuria and pentosurias could easily be identified. Neither purification nor concentration of the urine was necessary, for the salts and organic constituents of urine do not inhibit the migration of sugars.

PAPER ELECTROPHORESIS OF SERA FROM MAN AND EXPERIMENTAL ANIMALS INFECTED WITH VARIOUS HELMINTHS

Kagan, I. G., and C. G. Goodchild, J. Parasitol. **47**, 373–377 (1961).

Paper electrophoresis was employed in the analysis of the distribution of serum proteins obtained from 66 human sera representing 26 sera found serologically positive for trichinosis, 23 sera positive for echinococcosis, 16 sera positive for schistosomiasis, and a Versatol standard, and of sera from experimental hosts (mice, hamsters, rats, rabbits and guinea pigs) infected with *Schistosoma mansoni* and with *Trichinella* larvae (rats and rabbits). The Biuret method determined the total protein changes for human trichinosis, echinococcosis, and schistosomiasis.

THE PHYSIOLOGY AND CLINICAL SIGNIFICANCE OF HAPTOGLOBIN

Kauder, E., and A. M. Maurer, J. Pediat. **59**, 286–293 (1961).

This paper briefly reviews the biologic properties and functions of a plasma protein, haptoglobin, and discusses the application of plasma haptoglobin determination to various clinical conditions, stressing some of the limitations of this laboratory technique. The four methods by which haptoglobin may be demonstrated or quantitated—paper electrophoresis, titrimetric analyses, starch-gel electrophoresis and immunoelectrophoresis—are presented. Normal haptoglobin values are given and changes in plasma haptoglobin level in hemolysis, myoglobinuria, liver disease, infectious mononucleosis, malignancy and inflammation, tissue proliferation, or destruction, which may be helpful in the diagnosis and evaluation of patient progress, are described.

DETERMINATION OF URINARY 3-METHOXY-4-HYDROXYMANDELIC ACID (VANILMANDELIC ACID) BY PAPER ELECTROPHORESIS

Klein, D., and J. M. Chernaik, Clin. Chem. **7**, No. 3, 257–264 (1961).

A procedure for the determination of 3-methoxy-4-hydroxy-mandelic acid (also called vanilmandelic acid VMA) in urine was developed by combining conditions for hydrolysis and extraction of urine with a modification of the Studnitz-Hansen [Scand. J. Clin. Lab. Invest. **11**, 101 (1959)] electrophoresis and elution technique. Paper electrophoresis at a low constant current is employed for separation from other phenolic acids. Results showed this method to be useful as a confirmatory test for pheochromocytoma, in which elevated urinary levels of vanilmandelic acid are found.

MULTI-DIMENSIONAL STUDY OF PLASMA PROTEINS: COLUMN CHROMA-TOGRAPHY STARCH GEL AND IMMUNO-ELECTROPHORESIS

Lawrence, S. H., and D. C. Benjamin, Clin. Chim. Acta **6**, 398–402 (1961).

The effluent fractions, resulting from the separation of normal human plasma into its chromatographic components using a DEAE cellulose anion exchanger, were secondarily characterized by paper electrophoresis, starch-gel electrophoresis, immunoelectrophoresis, and enzyme stains. The results of the investigation are discussed.

ELECTROPHORETIC BEHAVIOUR OF RAT SERUM AMYLASE

McGeachin, R. L., and B. A. Potter, Nature **189**, 751 (1961).

The proteins of sera obtained from both normal rats and those suffering severe liver damage from treatment with N-nitrosodimethylamine were separated by paper electrophoresis. Results of the investigation, which is described in this paper, indicate that the distribution pattern of serum amylase may vary from one species to another, and that the distribution pattern found in the rat is definitely different from the one found in man.

ELECTROPHORETIC SUBFRACTIONATION OF HUMAN SERUM LIPOPRO-TEINS BY USE OF TEB BUFFER

Sonnino, F. R., and P. P. Gazzaniga, Clin. Chim. Acta **6**, 295–297 (1961).

A buffer consisting of Tris (hydroxymethyl)-aminomethane, and EDTA acid and boric acid (TEB), prepared according to Aronsson and Grönwall and used in the paper electrophoretic separation of normal human serum lipoproteins, had the advantages over the usual buffers of permitting a component corresponding to α_2-globulins to be detected, the less dispersed fractions to be more precisely evaluated because of their smaller adsorption on paper, and a greater quantity of serum to be applied to the strips. In some human serums with a β/α ratio increase, a doubling of the β-lipoproteins peak was observed.

ELECTROPHORETIC SEPARATION OF CONJUGATED BILRUBIN ON PAPER

Talafant, E., Nature **192**, No. 4806, 972–973 (1961).

The use of 0.02M phosphate buffer of pH 6.8 for the electrophoresis of bile pigments on paper is discussed and compared to the use of sodium acetate-acetic acid buffer of pH 5.2, barbital buffer of pH 8.6, and buffers containing pyridine acetate. The electrophoretic separation of preparations of Bilrubin, even those devoid of bile salts, was not possible in mere aqueous buffer solutions but could be achieved by the use of buffers containing pyridine acetate.

THE DETERMINATION OF 3-METHOXY 4-HYDROXY MANDELIC ACID IN URINE

Woiwod, A. J., and R. Knight, J. Clin. Pathol. **14**, 502–504 (1961).

A simple and accurate method for determining 3-methoxy 4-hydroxy mandelic acid, which is suitable for rapid routine investigations with large numbers of urine samples, is described. It can be performed without the use of special techniques such as paper chromatography, column chromatography or high-voltage electrophoresis.

A METHOD FOR THE DETERMINATION OF 3-METHOXY-4-HYDROXYMAN-DELIC ACID (VANILMANDELIC ACID) FOR THE DIAGNOSIS OF PHEO-CHROMOCYTOMA

Sunderman, F. W., Jr., P. D. Cleveland, N. C. Law, and F. W. Sunderman, Am. J. Clin. Pathol. **34**, No. 3, 293–312 (1960).

A colorimetric method for the measurement of 3-methoxy-4-hydroxymandelic (vanilmandelic acid), involving the removal of interfering constituents from acidified urine by adsorption upon activated magnesium silicate, ferricyanide oxidation of vanilmandelic acid to vanillin, and photometric measurements of the color developed with vanillin by the addition of indole and phosphoric acid, is described and discussed. Also discussed are the comparative merits of measurements of vanilmandelic acid and catecholamines, with reference to their clinical significance, relative precision of analysis, and adaptability for use in clinical laboratories. That the determination of vanilmandelic acid has several advantages over the estimation of catecholamines for the diagnosis of pheochromocytoma is suggested by this comparison. This method is felt to be an improvement over the usual complex and laborious methods used involving paper chromatography, paper electrophoresis, and radioisotopic techniques.

STUDIES OF THE SERUM PROTEINS: V. CAUSES FOR DISCREPANCIES IN FRACTIONATIONS OF SERUM PROTEINS BY PAPER ELECTROPHORESIS

Sunderman, F. W., Jr., and F. W. Sunderman, Am. J. Clin. Pathol. **33**, No. 5, 369–399 (1960).

In a nationwide survey of electrophoretic fractionations of serum proteins performed in clinical laboratories, large discrepancies were observed in the results of the fractionations. Two of the most prevalent techniques for paper electrophoresis, the Grassman-Hannig and the Spinco® procedures, were compared in order to elucidate some of the causes for the observed discrepancies. Findings from this comparison are described and discussed and eight practical proposals are advanced to improve the accuracy and promote uniformity of electrophoretic fractionations of the serum proteins when applied routinely for diagnostic purposes.

ELECTROPHORETIC PROTEIN AND POLYSACCHARIDE PATTERNS IN TUBERCULOSIS AND AMYLOIDOSIS

Aronsson, T., A. Grönwall, and E. Lausing, Clin. Chim. Acta **4**, 124–126 (1959).

Improved methods of paper electrophoretic separation and PAS-staining of the protein-bound carbohydrates, described in this paper, were used for analysing a number of blood sera from cases of tuberculosis and of tuberculosis complicated with amyloidosis. The electrophoretic serum patterns were essentially the same in tuberculosis with and without amyloidosis alterations, but they were more marked in amyloidosis.

FAMILIAL INCREASE IN THE THYROXINE-BINDING SITES IN SERUM ALPHA GLOBULIN

Beierwaltes, W. H., and J. Robbins, J. Clin. Invest. **38**, 1683–1688 (1959).

Because the serum protein-bound iodine (PBI) was found elevated in euthyroid states associated with an increase of thyroxine-binding capacity of plasma thyroxine-binding α-globulin (TBP) due to known causes, a search was made for individuals with unexplained elevation of the serum PBI. Reported in this paper is a study made on the occurrence of an elevated PBI and TBP capacity in an otherwise normal adult male and in one of his three children. Reverse-flow paper electrophoresis, carried out in 0.1M ammonium carbonate as well as in a barbital buffer, was used in the measurement of the thyroxine-binding capacity of TBP.

A COMPARISON OF THE SERUM PROTEINS OF NORMAL RATS WITH THOSE OF RATS BEARING LIVER TUMOURS

Campbell, P. N., B. A. Kernot, and I. M. Roitt, Biochem. J. **71**, 155–159 (1959).

A study comparing the composition of the serum proteins of normal rats and those bearing liver tumors by paper electrophoresis, zone electrophoresis and immunological methods is described in this paper.

A DISCONTINUOUS BUFFER SYSTEM FOR PAPER ELECTROPHORESIS OF HUMAN HEMOGLOBINS

Goldberg, C. A. J., Clin. Chem. **5**, No. 5, 446–451 (1959).

A method for the paper electrophoresis of hemoglobins in a discontinuous buffer system is described in which Tris-EDTA-borate buffer 0.12M, pH 9.1, was applied to the paper and barbital buffer 0.06M, pH 8.6, was used in the buffer vessels. This discontinuous buffer system enables the detection of hemoglobin A₂ in small samples of hemolysate, resolution of abnormal hemoglobins is superior to that in barbital or Tris-EDTA-borate buffer alone, and the system is of special value in the examination of aged samples of hemoglobin. Reasons for the behavior of the hemoglobin in this buffer system are discussed.

ACTION OF ENZYMES IN PRESENCE OF CERTAIN HORMONES: II. URINE DEOXYRIBONUCLEASES

Hakim, A., Clin. Chim. Acta **4**, 484–493 (1959).

The study of deoxyribonucleases excreted into urine, using lyophilized dried urine powder as a starting material, is reported in this paper. Deoxyribonucleases are found in free (active) and combined (inactive) form, and also as acid deoxyribonuclease with optimum pH activity at 4.60, requiring magnesium activation, and as neutral deoxyribonuclease, with optimum pH activity at 7.17 and not requiring magnesium for activation. It was found that the daily and monthly variations of the free and combined deoxyribonuclease were parallel with the variation of urinary estrogenic hormones. The techniques, including paper electrophoresis, used in the experimentation and the results are described.

ACID MUCOPOLYSACCHARIDES OF NORMAL URINE

Heremans, J. F., J. P. Vaerman, and M. Th. Heremans, Nature **183**, 1606 (1959).

At least one acid mucopolysaccharide, which behaved chromatographically and electrophoretically like chondroitan sulfate A, had been found to occur in normal urine, but separation and staining had been unsatisfactory. A one percent solution of Alcian Blue used in a 9:1 mixture of acetic acid and water (5' minutes), followed by alternate washings with acetic acid and tap water, proved to be a convenient and highly specific staining method for acid mucopolysaccharides. Very sharp separation of three different acid mucopolysaccharides which migrated ahead of the albumin peak was achieved by electrophoresis on Whatman® No. 1 filter paper in a Veronal-acetate buffer of pH 8.6 and ionic strength 0.1, using 3 volts/cm for 12 hours.

THE DISTRIBUTION OF EXTRACORPUSCULAR HEMOGLOBIN IN CIRCULATING PLASMA

Lathem, W., and W. E. Worley, J. Clin. Invest. **38**, 474–483 (1959).

A method for the partition and quantitation of plasma hemoglobin was developed utilizing paper electrophoretic techniques. The plasmahemoglobin was separated electrophoretically into a free and a protein-bound component, and these fractions were identified (when benzidine and hydrogen peroxide were used for staining) by the characteristic positional relationships on paper; they were quantified by photometric analysis of the stained filter paper strips. Using these techniques, a quantitative study was made of the distribution and transport of intravenously administered hemoglobin in human subjects.

"MYELOMA" SERUM ELECTROPHORETIC PATTERNS IN CONDITIONS OTHER THAN MYELOMATOSIS

Owen, J. A., W. R. Pitney, and J. F. O'Dea, J. Clin. Pathol. **12**, 344–350 (1959).

The case histories of ten patients are presented where clinical findings and laboratory evidence do not support a diagnosis of myelomatosis, but whose serum electrophoretic patterns are of the myeloma type. The techniques of paper electrophoresis, starch-gel electrophoresis, ultracentrifugal analysis and precipitin reactions used in the investigation are described, and the significance of the findings is discussed. The conclusion was reached that while discrete abnormal electrophoretic components in serum are usually associated with myelomatosis and lymphoma, they are occasionally present in a variety of other conditions.

PAPER ELECTROPHORESIS: PRINCIPLES AND TECHNIQUES

Peeters, H., Advan. Clin. Chem. **2**, 1–134 (1959).

This report on paper electrophoresis contains a review of one-dimensional methods including general conditions of the electrophoretic separation, the run, quantitation and results; a description of the general problems, general conditions, and the run in the two-dimensional methods; and discussion of the new developments in both the one- and two-dimensional methods.

STUDIES ON THE SERUM PROTEINS: IV. THE DYE-BINDING OF PURIFIED SERUM PROTEINS SEPARATED BY CONTINUOUS-FLOW ELECTROPHORESIS

Sunderman, F. W., Jr., and F. W. Sunderman, Clin. Chem. **5**, No. 3, 171–185 (1959).

Differences in the relative affinities of electrophoretic fractions for the protein stains, one of the important variables that influence quantitative estimations of serum proteins by paper electrophoresis, are discussed. The paper also cites pertinent reviews of quantitative aspects of paper electrophoresis and presents a brief resumé of studies of albumin-trailing and of the deviations from the Beer-Lambert law.

CLINICAL OBSERVATIONS ON SERUM GLOBULIN THYROXINE-BINDING CAPACITY USING A SIMPLIFIED TECHNIQUE

Tanaka, S., and P. Starr, J. Clin. Endocrinol. Metab. **19**, 84–91 (1959).

A relatively simple and reproducible paper electrophoretic technique was used to measure the thyroxine-binding capacity of the serum in normal subjects, in pregnant women, in cases of hepatic cirrhosis, hypothyroidism and hyperthyroidism, and in euthyroid subjects with subnormal and elevated levels of serum protein-bound iodine (PBI).

ELECTROPHORETIC PROTEIN AND POLYSACCHARIDE PATTERNS IN TUBERCULOSIS AND AMYLOIDOSIS

Aronsson, T., A. Grönwall, and E. Lausing, Protides Biol. Fluids, Proc. 6th Colloq., 233–235 (1958).

Alterations of the electrophoretic serum patterns are essentially the same in cases of tuberculosis with and without amyloidosis, but are more marked in amyloidosis. Improvements in the methods for electrophoretic separation on paper and PAS-staining of the protein-bound carbohydrates are described.

MULTIPLE HAEMOGLOBINS IN THE HORSE

Bangham, A. D., and H. Lehmann. Nature **181**, 267–268 (1958).

Hemoglobin from 65 horses was examined by paper electrophoresis at alkaline pH and two fractions were found in 64, and only one, the fast-moving component, in one. The one horse in which only one hemoglobin was found was re-investigated by the use of starch or agar zone electrophoresis and two hemoglobin fractions were revealed with the slow to fast fraction ratio being 0.25:1. By using thicker paper, and more hemoglobin, paper electrophoresis also demonstrated the slow fraction. This investigation showed that there were two fractions present in all 65 horses, and that this distribution strongly indicated the presence of multiple hemoglobins in the horse rather than that of two allelomorphs. A similar study was made on 14 donkeys, 19 mules, and 3 jennets and the results are given.

SIMPLIFIED RAPID METHOD FOR THE FIXATION OF PAPER ELECTROPHORETOGRAMS

Connerty, H. V., A. R. Briggs, and E. H. Eaton, Jr., Am. J. Clin. Pathol. **30**, 343–344 (1958).

Two traditional methods used in the combined fixation and staining process to make visible the serum proteins separated by paper electrophoresis were found unsatisfactory by the authors. A simpler and more rapid method is described which eliminates the use of heat and employs a means of fixation superior to that afforded by ethyl alcohol saturated with mercuric chloride.

EFFECT OF THE BORATE ION IN BUFFERS ON THE ELECTROPHORESIS OF RAT SERUM

Garbers, C. F., and F. J. Joubert, Nature **182**, 520–531 (1958).

A further investigation of the electrophoretic mobility of certain protein components of human, bovine and rat sera altered in buffers containing borate was carried out by the authors. They studied the transport of vitamin A alcohol in rat serum to investigate the effect of borate buffers, and the results are reported in this paper. Separation of rat serum proteins was achieved by electrophoresis on partially acetylated cellulose powder columns and by paper electrophoretic techniques.

AN IMPROVED PERIODIC ACID FUCHSIN SULFITE STAINING METHOD FOR EVALUATION OF GLYCOPROTEINS

McGuckin, W. F. and B. F. McKenzie, Clin. Chem. **4**, No. 6, 476–483 (1958).

This paper gives the modification of the Köiw and Grönwall procedure [Scand. J. Clin. Lab. Invest. **4**, 244 (1952)] for the staining of glycoproteins separated by paper electrophoresis, adapted to papers of the Whatman® 3MM class. The results are comparable to those reported by direct chemical determination of hexose in the individual protein fractions. Satisfactory glycoprotein distribution patterns were obtained when this method was applied to the analysis of several body fluids of normal persons and patients with certain disease entities.

CROSSING PAPER ELECTROPHORESIS FOR THE DETECTION OF IMMUNE REACTIONS

Nakamura, S., and T. Ueta, Nature **182**, 875 (1958).

A simplification and improvement of the procedure of Grassman and Hübner [Naturwiss. **40**, 272 (1953)] for the detection of loosely bound molecular compounds has been developed merely by making use of ordinary paper electrophoresis. The antigen-antibody reaction was demonstrated on filter paper by this method, using rabbit antiserum (anti-bovine serum) and the antigen (bovine serum).

DETECTION OF HAEMOGLOBIN, HAEMOGLOBIN-HAPTOGLOBIN COM-PLEXES AND OTHER SUBSTANCES WITH PEROXIDASE ACTIVITY AFTER ZONE ELECTROPHORESIS

Owen, J. A., H. J. Silberman, and C. Got, Nature **182**, No. 4646, 1373 (1958).

A number of substances were compared in a search for a more satisfactory peroxidase color reaction in the study of haptoglobins of human serum. In each case, the reagents were prepared by dissolving 100 mgm of the test substance in 70 ml of ethanol and adding 10 ml of 1.5M acetate buffer (pH 4.7) and 18 ml of water, then 2 ml of hydrogen peroxide (100 vol.) was added imme-diately before use. Following electrophoresis, the paper or starch-gel strips were placed in the reagent for 15 minutes, then were washed in three changes of water. Results indicated that o-dianisidine was the most satisfactory reagent. Most of the experiments were carried out with hemoglobin or with hemoglobin-haptoglobin complexes. The dianisidine reagent was found to be satisfactory also in the detection of peroxidase activity in extracts of horse-radish *(Coch-learia armoracia* L.) and other plants.

COMPARATIVE STUDIES OF THE PROTEIN FRACTIONS FROM HUMAN GASTRO-INTESTINAL JUICES WITH PAPER ELECTROPHORESIS COMBINED WITH VARIOUS DETECTION METHODS

Verschure, J. C. M., Protides Biol. Fluids, Proc. 6th Colloq., 194–201 (1958).

Conditions were studied for the optimal routine separation of the proteins in saliva, gastric juice, bile and pancreatic juice by paper electrophoresis, and from an experiment with more than 5000 electrophoresis diagrams a method was developed in which the proteins were labelled with ^{131}I before paper elec-trophoresis. The method is described in detail and the results of its application to saliva, gastric juice and pancreatic juice are reported. A standard method was worked out, and resulting patterns that were sufficiently constant to give normal values are demonstrated for saliva.

IMPROVED SEPARATION OF SERUM PROTEINS IN PAPER ELECTROPHO-RESIS—A NEW ELECTROPHORESIS BUFFER

Aronsson, T., and A. Grönwall, Scand. J. Clin. Lab. Invest. **9**, 338–341 (1957).

Described in this paper is an electrophoresis buffer consisting of Tris EDTA and boric acid. Paper electrophoresis using this buffer separates serum proteins into nine fractions: prealbumin, albumin, three α-components, three β-com-ponents and γ-globulin.

THE ENZYMIC HYDROLYSIS OF β-GLUCOSIDES

Crook, E. M., and B. A. Stone, Biochem. J. **65,** 1–12 (1957).

This paper presents the techniques used and the results obtained in experiments aimed at understanding the course of cellulose breakdown in preparations of cellulolytic enzymes which also contained cellobiase activity. Paper electrophoresis of oligosaccharides was performed as part of the investigation.

URINARY EXCRETION OF ACID MUCOPOLYSACCHARIDES BY PATIENTS WITH RHEUMATOID ARTHRITIS

Di Ferrante, N., J. Clin. Invest. **36,** 1516–1520 (1957).

Data on the urinary excretion of acid mucopolysaccharides by patients with rheumatoid arthritis and the identity of the urinary acid mucopolysaccharides excreted in this disease are presented and discussed. The daily urinary excretion of acid mucopolysaccharides, found to be higher in patients with active, untreated rheumatoid arthritis than in normal individuals, was decreased by the administration of sodium silicate to those patients. The urinary acid mucopolysaccharides from normal individuals and those from patients with rheumatoid arthritis were indicated to be a mixture of chondroitinsulfate and hyaluronate by paper chromatography and paper electrophoresis, analytical and enzymatic data.

PAPER ELECTROPHORESIS OF AVIAN AND MAMMALIAN HEMOGLOBINS

Saha, A., R. Dutta, and J. Ghosh, Science **125,** 447–448 (1957).

An investigation of the electrophoretic behavior of the hemoglobins of birds for comparison with hemoglobins of certain mammalian species is described, in which the hemoglobins of the pigeon *(Columba livia),* Duck *(Anas),* guinea fowl *(Numida melagris* Linn.) and chick *(Gallus gallus)* and of man (one normal and two cases of Hb-E-thalassemia), cow, goat and rabbit were studied in an LKB paper electrophoresis apparatus using barbiturate buffer of pH 8.6, ionic strength 0.05. Only one component (Hb-A) was observed in the blood of mammals, including the normal human adult, while the blood of the different birds investigated showed two hemoglobin components. Although none of the hemoglobin components of avian blood is identical with the mammalian hemoglobins, component 2 of avian hemoglobin appears to be identical with hemoglobin E, which is the special hemoglobin component present in the blood of patients with hemoglobin E-thalassemia.

OBSERVATIONS ON A FAST-MOVING PROTEIN IN AVIAN MALARIAL SERUM

Schinazi, L. A., Science **125**, 695–697 (1957).

Filter-paper electrophoresis revealed the occurrence of a marked qualitative change in the serum protein patterns of pigeons infected with the 1P1-1 strain of *Plasmodium relictum*—namely, the appearance of a new component possessing an electrophoretic mobility greater than that of albumin. Paper electrophoresis requires an extremely small sample for analysis, an important advantage of the moving-boundary method of Tiselius, and is therefore ideally suited to the serial examination of the blood proteins of small laboratory animals without material interference with the usual course of an induced infection. The investigation is described and the findings are reported.

ELECTROPHORESIS OF PLASMA PROTEINS IN THE PARAKEET

Wall, R. L., and H. G. Schlumberger, Science **125**, 993–994 (1957).

During a study of the effects of spontaneous or transplanted pituitary tumors of the shell parakeet *(Melopsittacus undulatus)*, the plasma electrophoretic pattern of these birds was investigated using Whatman® 3MM filter paper on a conventional vertical principle electrophoretic cell utilizing Veronal buffer at pH 8.6, μ 0.05 with five percent glycerine. The method of investigation and the findings are reported.

DIE PROTEINANALYSE DES SCHWEIZERISCHEN STANDARDTROCKEN-SERUM

Wunderly, Ch., and V. Bustamante, Protides Biol. Fluids, Proc. 5th Colloq., 92–98 (1957).

The protein composition of the standard-dryserum (Berne), after being measured 40 times by paper electrophoresis in the Spinco-Beckman apparatus, proved to be equal to the mean values found in 30 normal human sera. The statistical calculation of the σ-values, attained by this largely automatic technique of paper electrophoresis, demonstrates the regularity of the results.

BUFFER COMPOSITION IN PAPER ELECTROPHORESIS: CONSIDERATIONS ON ITS INFLUENCE, WITH SPECIAL REFERENCE TO THE INTERACTION BETWEEN SMALL IONS AND PROTEINS

Laurell, C. B., S. Laurell, and N. Skoog, Clin. Chem. **2**, No. 2, 99-111 (1956).

An apparatus for paper electrophoresis according to the moist-chamber principle and a simple method for quantitative evaluation of the different electrophoretic fractions together with the limits for the normal variation of the different serum fractions are presented. Stressed in this paper is the influence of the composition of the buffer on the separation of proteins. The β-fraction separates into two distinct fractions, β, and $\beta_2{}^{Ca}$, as a result of the addition of small amounts of calcium ions. The phenomenon of the decrease in the mobility of the β-lipoproteins in the presence of calcium and the refractive contribution of the lipids in the β-lipoproteins are illustrated by moving-boundary electrophoresis. The calcium ion does not change the mobility of the main β-component containing carbohydrate or of the transferrin.

A MODIFIED METHOD OF TWO-DIMENSIONAL ZONE ELECTROPHORESIS APPLIED TO MUCOPROTEINS IN SERUM AND URINE

Markham, R. L., Nature **177**, No. 4499, 125–126 (1956).

A technique of two-dimensional zone electrophoresis on paper was described by Durrum [J. Coll. Sci., **6**, 274 (1951)], in which the analysis in the first buffer is transferred on a strip of paper to the surface of a sheet of paper soaked in the second buffer. A modification of this method, using one buffer, is described.

CONTROL OF THE STAINING PROCEDURE AFTER PAPER ELECTROPHORESIS

Wunderly, Ch., Nature **177**, No. 4508, 586 (1956).

The staining procedure, followed by removing the surplus stain, constitutes sources of error when quantitative techniques are attempted for the analysis of serum proteins separated by electrophoresis on filter paper. In order to get more satisfactory reproducibility, 0.02 ml of a 0.05 percent solution of poly-ethylhafen in water was applied to each paper strip after electrophoresis but before staining. The advantage of this polybase, $-[-CH_2-CH_2-NH_2-]_n-$, is that it is stained by Naphthalene Black 12B as well as Sudan Black, and therefore it can be used equally well as a reference substance for protein staining and lipoprotein staining. Application of the procedure is described.

THE HUMAN HEMOGLOBINS IN HEALTH AND DISEASE

Chernoff, A. I., New Eng. J. Med. **253**, 322–331 (1955).

A short review of some of the more important differentiating characteristics of The human hemoglobins covered include fetal, sickle-cell, Hgb C, Hgb D, Hgb E, cerning the various diseases due to these abnormal pigments are presented. The human hemoglobins covered include fetal, sickle-cell, Hgb C, Hgb D, Hgb E, Hgb G, Hgb H, and several others that do not correspond completely to these types. The alkali denaturation procedure for fetal hemoglobin and the technique of paper electrophoresis, two simple procedures which may be utilized for the identification of most known hemoglobin types in addition to the more special-ized methods of solubility determinations, spectrophotometry, and moving-boundary electrophoresis, are described.

URINARY EXCRETION OF ACID MUCOPOLYSACCHARIDES BY DIABETIC PATIENTS

Craddock, J. G., and G. P. Kerby, J. Lab. Clin. Med. **46**, 193–198 (1955).

Investigations on the vascular and renal complications of diabetes, which are increasingly apparent with the use of insulin, indicated that the complicating lesions reflect the abnormal metabolism of polysaccharides. Paper electrophoresis demonstrated a progressive rise in the total serum protein-bound polysaccharides as diabetic complications ensued, evident before clinical renal complications were demonstrable. The overall findings suggested that the source of the hyaline material deposited in the glomeruli and elsewhere may be a polysaccharide present in the blood of the diabetic patient in abnormal amounts. It was also suggested that most of the characteristic changes in diabetes millitus may be interpreted in terms of a disturbance in the metabolism (or structural organization) of a simple polysaccharide (glycogen) or of more complex polysaccharides (mucopolysaccharides). This paper reports a quantitative study of the urinary acid mucopolysaccharides in diabetes with vascular and renal complications.

PAPER ELECTROPHORESIS AS A QUANTITATIVE METHOD: SERUM PROTEINS

Jencks, W. P., M. R. Jetton, and E. L. Durrum, Biochem. J. **60**, 205–215 (1955).

Four main problems were encountered in the separation and quantitative estimation of serum by paper electrophoresis: albumin 'tailing' due to irreversible adsorption of albumin on paper during its migration, variation in dye binding capacity with the amount and area of application of protein, fading and incomplete elution of the commonly used dyes Bromophenol Blue and Amido Black, and the lack of a linear relationship of dye concentration to photocell response in the direct scanning of dye on paper; and there was widespread disagreement as to the magnitude and usefulness of 'correction factors' for the conversion of results of paper electrophoresis to those of moving boundary electrophoresis. These and related problems were studied and led to the development of a procedure for the paper electrophoretic analysis of serum proteins, described in this paper. Also presented are the normal values for serum protein fractions measured by this technique and the results of a study of the reproducibility of this method under conditions of routine use.

ELECTROPHORETIC BEHAVIOUR IN FILTER PAPER AND MOLECULAR WEIGHT OF INSULIN

Sluyterman, L. A. AE, Biochim. Biophys. Acta **17**, 169–176 (1955).

Preparations of partially acetylated insulin, analysed by paper electrophoresis, were found to travel in a well-defined band in the electropherograms if acetic acid-water (1:2 v/v) is used as a buffer. The results indicated the molecular weight of insulin to be 6,000. An improvement in the amino-nitrogen determination of insulin is also described.

THE ELECTROPHORESIS OF SERUM AND OTHER BODY FLUIDS IN FILTER PAPER

Griffiths, L. L., J. Lab. Clin. Med. **41**, 188–198 (1953).

A technique of paper electrophoresis is described, in which Whatman® drop reaction and Whatman 3MM papers were found to be the ideal supporting mediums. This method requires more simple apparatus than that needed in previously published techniques, it can be carried out with relatively unskilled assistance, and, when analyzing body fluids, both the serum and urine can be run at the same time in cases where the urine contains protein.

MICRO-ELECTROPHORESIS OF PROTEIN ON FILTER-PAPER

Flynn, F. V., and P. deMayo, Lancet **2**, 235–239 (1951).

The techniques of protein electrophoresis on filter paper, a method of potential value in clinical studies of the proteins of serum and other body fluids because of its simplicity, cheapness, suitability for multiple analyses, and requirement for only minute quantities of serum (0.015-0.160), are discussed. Application of qualitative and quantitative methods and the results obtained are described in detail.

STUDIES ON THE MUCOPROTEINS OF HUMAN PLASMA: I. DETERMINATION AND ISOLATION

Winzler, R. J., A. W. Devor, J. W. Mehl, and I. M. Smyth, J. Clin. Invest. **27**, 609–616 (1948).

The determination, isolation, and chemical characterization of mucoproteins from normal human plasma, reported in this paper, was undertaken as part of a study of their physiological significance, especially in relation to cancer. Electrophoretic studies on filter paper indicated the presence of three components, all with low isoelectric points. Higher than normal amounts of mucoprotein were found in plasma from cancer patients.

Microelectrophoresis

ELECTROPHORETIC STUDIES OF TURBIDITY REMOVAL WITH FERRIC SULFATE

Block, A. P., and J. V. Walters, J. Am. Water Works Assoc. **56,** 99–110 (1964).

A research project carried out to determine the relationships between electrophoretic mobility and the other measurable variables involved in the removal of turbidity by coagulation with ferric sulfate is reported in this paper. The microelectrophoretic technique used in the study appeared to be useful for the prediction of the pH zone of most effective coagulation of natural turbid water.

ANALYSIS OF INDIVIDUAL RABBIT OLFACTORY BULB NEURON RESPONSES TO THE MICROELECTROPHORESIS OF ACETYLCHOLINE NOREPINEPHRINE, AND SEROTONIN SYNERGISTS AND ANTAGONISTS

Bloom, F. E., E. Costa, and G. C. Salmoiraghi, J. Pharmacol. Exp. Therap. **146,** 16–23 (1964).

An investigation of the reduction of spontaneous discharge rate of olfactory neurons by microelectrophoretic administration of acetylcholine (ACh), norepinephrine (NE), and serotonin (5-HT) to individual nerve cells of unanesthetized, young, adult, pigmented rabbits is described. The objective of the study was to establish certain pharmacological points of reference from which characterization of olfactory bulb inhibitory transmitters might be indicated. The use of five-barreled glass micropipette electrodes in the microelectrophoresis of desired drugs and amines into the immediate environment of a neuron, while simultaneously recording its activity extracellularly, excludes many of the problems of the response interpretation related to blood-brain barrier permeability. It also helps rule out drug effects due to fluctuations in systemic blood pressure, to changes in the composition of the extracellular fluid, or to primary drug effects upon neurons many synapses removed.

MICROELECTROPHORETIC STUDIES OF ADRENERGIC MECHANISMS OF RABBIT OLFACTORY NEURONS

Bloom, F. E., R. von Baumgarten, A. P. Oliver, E. Costa, G. C. Salmoiraghi, Life Sciences **3**, 131–136 (1964).

The technique of microelectrophoresis can overcome the impediments arising from problems of blood-brain barrier penetration, atraumatic localization of drug within the desired portion of brain, and exclusion of indirect drug effects. Preliminary data suggesting participation of norepinephrine in synaptic events in the olfactory bulb are presented in this paper with experimentation carried out on rabbits decerebrated electrolytically. Previous studies had shown that acetylcholine (ACh), norepinephrine (NE), and serotonin amines stored by neurons in rabbit brain decreased the spontaneous firing of olfactory neurons when applied microelectrophoretically to them through five-barreled glass micropipette electrodes.

MICROELECTROPHORESIS OF HISTONES

Schonne, E., Protides Biol. Fluids, Proc. 11th Colloq., 368–370 (1963).

The heterogeneity of histones had been demonstrated by several techniques, involving differential extraction or precipitation, chromatography, sedimentation and moving-boundary, paper, cellulose-acetate, agar-gel, agarose-gel, and starch-gel electrophoresis, but various problems were encountered in each. Use of a new substrate, acrylamide gel, effected greater resolution because this gel is inert, mechanically strong, transparent, stable against temperature and pH changes, and free of electro-osmotic flow. Histone electrophoresis on acrylamide gel is described.

DESIGN AND USE OF A REFINED MICROELECTROPHORESIS UNIT

Grunbaum, B. W., and P. L. Kirk, Anal. Chem. **35**, 564–566 (1960).

An improved, precise microelectrophoresis apparatus, in which 0.01- to 0.1- μl can be subjected simultaneously to electrophoresis, is described. An unusually high degree of reproducibility between patterns is produced by accurate positioning of precisely cut, multistrip papers. Ink dye, proteins of blood serum, and ferritin solution were used to test and illustrate the quality of the separations.

CONTROL OF LECITHINASE ACTIVITY BY THE ELECTROPHORETIC CHARGE ON ITS SUBSTRATE SURFACE

Bangham, A. D., and R.M.C. Dawson, Nature **182**, No. 4645, 1292–1293 (1958).

Lecithin particles were examined by microelectrophoresis in order to determine the sign and magnitude of their electrophoretic charge. Materials and methods used in the investigation are described and the findings are discussed.

Apparatus — Instrumentation — Quantitation

EINE NEUE ELEKTROPHORESE-APPARATUR NACH DEM TISELIUS-PRINZIP

Wiedemann, E., Protides Biol. Fluids, Proc. 10th Colloq., 298–305 (1962) **German.**

An electrophoresis apparatus according to the Tiselius principle was described for analytical and preparative investigations as well as diffusion measurements in scientific research work. The overall length of the instrument could be reduced to 2.2 m, while at the same time doubling its capacity. A novel straight-lined optical system could achieve this progress even with a slight increase of the resolving power (in comparison with a standard instrument of at least 4.5 m in length). Catadioptic systems with considerably shortened focal distances and greater apertures which enable two cell sets to be used simultaneously and without displacement for exposure replaced the usual collimator lenses.

MICROLECTROPHORESIS ON CELLULOSE ACETATE MEMBRANES

,Grunbaum, B. W., P. L. Kirk, and W. A. Atchley, Anal. Chem. **32,** No. 10, 1361–1362 (1960).

A new microelectrophoresis apparatus with cellulose acetate as the supporting medium separates serum protein fractions with a speed and clarity far superior to conventional filter paper techniques. The procedure is described.

APPLICATION OF DIRECT PHOTOGRAPHIC PHOTOMETRY TO PREPARATIVE ELECTROPHORESIS

Ressler, N., Nature **182,** No. 4633, 463–464 (1958).

A procedure is described for the direct photographic photometry of serum proteins separated and isolated by a method of electrophoresis which uses, as a medium, a thin film of buffer solution. The technique described may prove to be of value for various purposes because it allows the course of the proteins to be followed during electrophoresis and the individual components to be isolated without continuous sectioning or analysis along the migration path.

DETERMINATION OF SERUM-PROTEIN FRACTIONS BY ZONE ELECTRO-PHORESIS ON PAPER AND DIRECT REFLECTION PHOTOMETRY

Owen, J. A., Analyst **81**, 26–37 (1956).

Dyed serum protein-fractions obtained by zone electrophoresis on paper were evaluated by the use of direct reflection photometry. The proteins in a series of normal and pathological sera were examined by paper electrophoresis and the results and their reproducibilities, obtained by elution of the dye and by direct photometry, were compared. These investigations are described and the validity of the results and those obtained by other electrophoretic methods are discussed.

Techniques

A RAPID TECHNIQUE FOR THE SEPARATION AND QUANTITATION OF FREE ALPHA-KETO ACIDS IN COMPLEX SOLUTIONS

Berry, S. A., and J. N. Campbell, Anal. Biochem. **8**, 495–502 (1964).

A rapid and simple technique for the separation and quantitation of free α-keto acids, involving the electrophoretic separation of the free acids and the use of a 2,4-dinitrophenylhydrozine (DNPH) spray for the location of the acids, is described in this paper. Derivations of the DNPH are formed *in situ*. This technique may be used for the quantitative recovery of α-keto acids from complex biological fluids such as bacterial culture media, urine and blood, and requires no initial deproteinization, desalting or extraction steps.

FORCED-FLOW ELECTROPHORESIS—INSTRUMENTATION AND RESULTS

Bier, M., Federation Proc. **22**, No. 2, Part I, 236, Sec. 454 (1963).

A variety of problems, relating to instrumentation, technique, and colloidal stability of the processed fluid, were encountered in the practical application of forced-flow electrophoresis. Study was directed towards a better under-standing of the processes occurring in the cell, and the information gained was incorporated into the design of the equipment. The primary advantage of forced-flow electrophoresis is that it can be applied to a number of problems not usually associated with electrophoresis, such as selective plasmapheresis (*in vivo* fractionation of whole blood), concentration of dilute solutions of bacteriophages and other biologically active materials, filtration of slurries, etc.

CAPACITY OF THYROXINE-BINDING GLOBULIN TO BIND TRIIODOTHYRO-NINE AND THYROXINE IN MATERNAL AND CORD BLOOD

Michener, W. M., W. N. Tauxe, and A. B. Hayles, Pediatrics **29**, No. 3, 369–375 (1962).

Normal values were established for the measurement of thyroid function using the erythrocytic uptake of I^{131}-labeled triiodothyronine and the thyroxine-binding capacity of the interalpha globulin in a study of paried maternal and cord blood samples collected at the time of delivery. The cord blood, which showed an increase in erythrocytic uptake of labeled hormone as compared to maternal blood, apparently binds exogeneous triiodothyronine in a different manner than it does exogeneous thyroxine. Reverse-flow electrophoresis was used to determine the thyroxine-binding capacity of the interalpha globulin (TBG).

SUDAN BLACK B IN ETHYL ACETATE-PROPYLENE GLYCOL FOR PRE-STAINING LIPOPROTEINS

McDonald, H. J., and J. Q. Kissane, Anal. Biochem. **1**, 178–179 (1960).

The preparation and use of Sudan Black B in ethylene and propylene glycols in the prestaining of lipoproteins for electrophoresis in various stabilized media had been reported. This paper presents a new solubilizing procedure which greatly increases the concentration of Sudan Black B and results in a more intensely stained lipoprotein pattern on the resultant ionograms or paper chromatograms.

ACRYLAMIDE GEL AS A SUPPORTING MEDIUM FOR ZONE ELECTRO-PHORESIS

Raymond, S., and L. Weintraub, Science **130**, 711 (1959).

A stable, flexible transparent gel which is useful in zone electrophoresis is formed by acrylamide polymerized in buffer solutions. A commercially available, synthetic gelling agent, Cyanogum® 41, which has many advantages over previously reported agents for electrophoresis, is described in this paper. Allowing the gel to dry out completely is a convenient way of preserving the electrophoretic patterns, whether stained or not. The gel shrinks uniformly in all dimensions, producing a thin, flexible, transparent celluloid-like film which preserves the original pattern relationships, and this film can be re-hydrated to its original dimensions by soaking in water.

INTERPRETATION OF ELECTROPHORETIC PATTERNS

Hoxter, G., Protides Biol. Fluids, Proc. 6th Colloq., 277–280 (1958).

By working under standardized conditions, reproducible patterns may be obtained both in free and in zone electrophoresis, and careful analysis of the geometrical characteristics of these patterns reveals three independent constants for every constituent: mobility, expressed by the mean path of the migration; concentration, expressed as a fraction of total dye density; and heterogeneity, expressed by the mobility distribution around the mean. The relative values of these constants may be transformed into absolute units using a mixture of known constitution as reference.

STUDIES OF THE SERUM PROTEINS: II. THE NITROGEN CONTENT OF PURIFIED SERUM PROTEINS SEPARATED BY CONTINUOUS FLOW ELECTROPHORESIS

Sunderman, F. W., Jr., F. W. Sunderman, E. A. Falvo, and C. J. Kallick, Am. J. Clin. Pathol., **30**, 112–119 (1958).

In measuring the nitrogen content of purified normal human serum proteins separated by continuous flow electrophoresis, it became apparent that the traditional 6.25 factor for the conversion of protein nitrogen to protein not only was inaccurate when applied to electrophoretic fractions, but introduced a significant error in estimating the total serum proteins as well. A base for the standardization of quantitative paper electrophoresis was provided by a presentation of the nitrogen factors for the serum protein moieties. The methods used in this study and the results are reported.

Cellulose-Acetate Electrophoresis

THYROGLOBULIN IN THE SERUM OF PARTURIENT WOMEN AND NEW-BORN INFANTS

Assem, E. S. K., Lancet **1**, 139–140 (1964).

Sera from 50 normal adults, 100 parturient women, and 101 newborn infants were examined for thyroglobulin by an improved version of the tanned-red-cell agglutination-inhibition technique of Hjort and Pedersen (1962) and by a method depending on the ability of unlabelled thyroglobulin to deviate ^{131}I-labelled thyroglobulin away from the γ-globulin zone on electrophoresis in the presence of a standard amount of thyroglobulin antibody. The results of the two techniques are reported. Electrophoresis on cellulose-acetate strips in a Veronal buffer (0.03M, pH 8.6) was performed for four hours on the labelled thyroglobulin added to the normal adult serum in the second method.

PROTEIN LOSS IN PERITONEAL DIALYSIS

Berlyne, G. M., V. Hewitt, J. H. Jones, and S. Nilwarangkur, Lancet **1**, No. 7339, 738–741 (1964).

In 12 patients studied, the amount of protein lost during peritoneal dialysis was found to vary from 10 to 207 g per dialysis, plasma-albumin and globulin concentrations were often reduced, and considerable hypoalbuminaemia, with generalized edema, was induced in three of the cases. Cellulose-acetate electrophoresis was carried out on the patients' serum, ascitic fluid before dialysis, and concentrated peritoneal dialysate. Various globulin fractions, as well as albumin, were shown by electrophoresis of the dialysate to be lost, and both α_2-glycoprotein (molecular weight 900,000) and β-lipoprotein (molecular weight $>$ 1,000,000) were identified immunologically. Significant quantities of γ-globulin, in particular, were lost.

RIBONUCLEASE OF BOVINE MILK: PURIFICATION AND PROPERTIES

Bingham, E. W., and C. A. Zittle, Arch. Biochem. Biophys. **106**, 235–239 (1964).

Ribonuclease A was isolated from bovine milk and found by chromatography, electrophoresis, specific activity, and amino acid composition to be identical to pancreatic ribonuclease A. This conclusion was supported by immunological studies. Column chromatography obtained a second ribonuclease as a distinct but minor component, and cellulose-acetate electrophoresis showed two components, one of which had the mobility of ribonuclease A.

INTERRELATIONS OF HUMAN SERUM PROTEIN FRACTIONS IN HEALTH AND DISEASE

Brackenridge, C. J., Nature **202**, No. 4933, 710–711 (1964).

This study was designed to determine the statistical associations between protein fractions of healthy subjects in the course of fluctuations within the normal range and to examine the possible effects of a variety of disease processes on these interrelations. Some results of this quantitative investigation of human serum proteins obtained from a series of patients with carcinoma, cirrhosis of the liver, myocardial infarction, pneumonia, and rheumatoid arthritis and compared with the results drawn from normal persons in good health and also in pregnancy are reported in this paper. Cellulose-acetate electrophoresis was employed in the investigation.

ELECTROPHORESIS OF SERUM PROTEIN WITH CELLULOSE ACETATE: A METHOD FOR QUANTITATION

Briere, R. O., and J. D. Mull, Am. J. Clin. Pathol. **42**, 547–551 (1964).

The authors describe a simple method of electrophoresis of serum proteins with cellulose acetate, and its quantitation with a densitometer in common use. Electrophoresis with cellulose acetate has advantages over the tedious and time-consuming technique of paper electrophoresis, because of its short time of migration, lack of trailing of albumin, and sharp separation. In the procedure, the Gelman electrophoresis chamber with a Spinco® Duostat power supply and a Beckman® model RB Analytrol adapted to scan the cellulose-acetate strips were used.

ROUTINE CLINICAL APPLICATION OF AN ELECTROPHORETIC METHOD FOR CHARACTERIZING THYROID FUNCTION

Burke, G., B. E. Metzger, and M. S. Goldstein, J. Lab. Clin. Med. **63**, No. 4, 708–714 (1964).

This paper describes a simple modification of a previously published technique for characterizing thyroid function by zone electrophoresis. The technique uses cellulose acetate rather than Whatman® paper as the supporting medium and measures radioactivity by the automatic gamma well counter instead of strip-scanning. The facility and speed of the test is increased while the consistency and diagnostic accuracy of the more elaborate procedure is reproduced. Plasmas of 66 euthyroid, 31 hyperthyroid, and 14 hypothyroid patients were enriched with radiothyroxine, and electrophoresis was performed on cellulose-acetate strips in both barbital and Tris maleate buffers, pH 8.6. Results of radiothyroxine partition among carrier plasma proteins were expressed as percent total radioactivity bound to thyroid-binding globulin (TBG). Electrophoretic characterization of thyroid function was best achieved by using data obtained from both buffers, since the separation of hyperthyroid from euthyroid sera was successful only in barbital buffer, while the delineation of hypothyroidism was more precise in Tris maleate buffer. The two-buffer system permits adequate delineation of the three levels of thyroid function and lends itself well to routine diagnostic studies.

EFFECTS OF α-METHYL-NORVALINE ON SYNTHESIS OF HAEMOGLOBIN IN THE AREA VASCULOSA OF THE CHICK EMBRYO

Deuchar, E., and A. M. L. Dryland, Nature **201**, No. 4921, 832–833 (1964).

Investigations on the effects of amino-acid analogues on explanted two-day chick embryos showed that α-methyl-norvaline, which has a structure closely similar to valine and leucine, caused a reduction of the quantity of hemoglobin formed in the area vasculosa. Methods, which included cellulose-acetate electrophoresis, and results of the study are described.

GROUPING OF BETA-HEMOLYTIC STREPTOCOCCI ON CELLULOSE ACE-TATE MEMBRANES

Goldin, M., and A. Glenn, J. Bacteriol. **87**, No. 1, 227–228 (1964).

The possibility of applying the microspot (Feinberg, Nature 195: 985, 1962) test for antigen-antibody reactions on cellulose membranes to the serological grouping of beta-hemolytic streptococci was investigated by the authors, and found to be successful. Their findings also indicate the method to be simple, sensitive, economical, specific, and readily adaptable to large-scale studies, and that the reactions were permanent and needed no further treatment. It appeared that the method also would be applicable to serological grouping of other microorganisms.

FOLIC-ACID DEFICIENCY IN RHEUMATOID ARTHRITIS

Gough, K. R., C. McCarthy, A. E. Read, D. L. Mollin, and A. H. Waters, Brit. Med. J. Part 1, 212–217 (1964).

Conventional voltage electrophoresis on cellulose-acetate strips was used to detect urinary FIGLU in six patients with rheumatoid arthritis and megaloblastic anemia who had been given an oral loading dose of 15 g of histidine. Investigations to determine the cause of megaloblastic anemia in these patients and results of a survey of the incidence of folic-acid deficiency in a randomly selected group of patients with rheumatoid arthritis are reported in this paper.

POSSIBLE SIGNIFICANCE OF THE ISOENZYMES OF LACTATE DEHYDROG-ENASE OF THE RETINA OF THE RAT

Graymore, C., Nature **201**, No. 4919, 615–616 (1964).

Cellulose-acetate electrophoresis was applied to the retina extracts of the rat and confirmed the existence of five isoenzymes, three of which were believed to represent hybrids of the major and discrete fractions designated as the M and H isoenzymes. Functions of the pure M and H isoenzymes are described, and results obtained by paper electrophoresis confirmed that LDH-5, the so-called M enzyme, predominates in muscle, but is relatively insignificant in heart tissue.

HEPARIN FRACTIONATION IN THE STUDY OF LYTIC ACTIVITY

Green, J., J. Clin. Pathol. **17**, 316–318 (1964).

In the observation of fibrinogenolysis and fibrinolysis in one of the heparin complexes prepared from fresh human plasma at low ionic strength, adapted for clinical study, lysis is more rapid than in any other system. Therefore, an isolating technique, felt to be of value in defining variable factors and in applying objective methods of measurements, was developed. A main fibrinogen band, some β globulin, and virtually no other fractions were revealed by electrophoresis of the re-precipitated fraction on cellulose acetate at pH 8.6. Other procedures used in the study and the findings are reported.

THE DETERMINATION OF URINARY 3-METHOXY 4-HYDROXYMANDELIC [VANILMANDELIC] ACID

Hermann, G. A., Am. J. Clin. Pathol. **41**, 373–376 (1964).

An electrophoretic method for the determination of urinary 3-methoxy-4-hydroxymandelic acid (VMA) is described in this paper. Cellulose-acetate membranes and conventional voltages were used. Fifty-seven patients without pheochromocytoma averaged 1.9 μg of VMA per mg of urinary creatinine. Age, sex, essential hypertension, or diet had no significant effect upon excretion.

QUANTITATION OF SOME ABNORMAL HEMOGLOBINS

Kelsey, J. R., and R. A. Kloss, Clin. Chem. **10**, No. 5, 424–428 (1964).

A method for the identification and quantitation of abnormal hemoglobins using cellulose acetate as the supporting medium was adopted for use in small clinical laboratories and by researchers unable to use elaborate and expensive equipment. Quantitation of the hemoglobins is followed by elution of each fraction into Drabkin's solution, and the resulting cyanmethemoglobin can be read in any standard photometer at 540 mμ. This technique is not useful for the more elaborate analytical approach to hemoglobin study, but has proved valuable in the routine identification and quantitation of hemoglobinopathies common in this country.

ELECTROPHORESIS ON CELLULOSE ACETATE

Kohn, J., and J. G. Feinberg, Das Ärztliche Laboratorium **10**, 233–248 and 269–278 (1964).

Cellulose-acetate membranes were introduced by J. Kohn as a support medium for zone electrophoresis. It was obvious from the start that cellulose-acetate membranes were not simply a substitute for filter paper in electrophoretic analyses, but had certain properties which made them superior to filter paper for such work. Kohn also showed that cellulose-acetate membranes were ideally suited for immunoelectrophoretic analysis, producing sharp, clear-cut, well-separated precipitation bands. In this paper, Kohn discusses properties and advantages of cellulose acetate membranes, techniques in using the membranes, and method of applying electrophoresis on cellulose-acetate membranes to serum glycoproteins, serum lipoproteins, hemoglobins, formimino-glutamil acid, haptoglobin determinations, isotope-labelled materials, nucleic acids, biologically active substances and enzymes.

NON-SPECIFIC HAEMOGLOBIN BINDING OF HUMAN SERUM GLOBULINS IN IMMUNOELECTROPHORESIS

Laurent, B., Nature **202**, No. 4937, 1121–1122 (1964).

When immunoelectrophoresis on cellulose-acetate paper, a useful tool for the detection of serum protein sub-fractions, was applied to the examination of human serum haptoglobins, hemoglobin binding to all other globulins always occurred. The precipitin arcs stained much less strongly, however, than those of the specifically hemoglobin-binding haptoglobins, a finding not reported for agar immunoelectrophoresis. The cellulose-acetate immunoelectrophoretic technique is described and results of the study are commented on.

LOCALIZATION OF LEUCINE AMINOPEPTIDASE ISOENZYMES

Meade, B. W., and S. B. Rosalki, J. Clin. Pathol. **17**, 61–63 (1964).

A simple and sensitive staining procedure for the localizing leucine aminopeptidase isoenzymes after electrophoretic separation on cellulose-acetate membranes, and its application to the study of certain aspects of hepatobiliary disease, are discussed. Cellulose-acetate electrophoresis of leucine aminopeptidase isoenzymes did not appear to be a rewarding diagnostic technique in the investigation of hepatobiliary or pancreatic disorders.

IMMUNOELECTROPHORETIC IDENTIFICATION OF HUMAN SERUM PROTEINS ON A CELLULOSE-ACETATE MEDIUM

Nelson, T. L., G. Stroup, and R. Weddell, Am. J. Clin. Pathol. **42**, 237–244 (1964).

Because immunoelectrophoretic identification of components of serum with agar gel is a relatively cumbersome procedure, it has been modified to permit its performance on a paper-like medium such as cellulose acetate which offers the advantages of convenience, uniformity, and the need for only small amounts of the serum specimen. In this study the authors use cellulose acetate to determine the electrophoretic patterns of human serums. The apparatus, materials, and methods used are described.

PROTEINS OF SERUM AND OEDEMA FLUID IN RHEUMATOID ARTHRITIS

Park, D. C., and K. Swinburne, Brit. Med. J. Part 1, 86–88 (1964).

In a study of the proteins of serum and edema fluid from patients with rheumatoid arthritis, it was noted that edema fluid had a low protein count in patients with uncomplicated rheumatoid arthritis. A relative increase in the concentration of smaller protein molecules and a relative decrease in the concentration of larger protein molecules was shown by cellulose-acetate electrophoresis of the serum and edema fluid proteins. These findings indicate that the capillary endothelium has a normal selective permeability to protein in rheumatoid arthritis, and that the edema, therefore, is not of inflammatory or allergic origin.

ELECTROPHORESIS OF SERUM PROTEINS ON CELLULOSE ACETATE

Ritts, R. E., and F. W. Ondrick, Am. J. Clin. Pathol. **41**, 321–331 (1964).

Despite the many advantages of cellulose acetate as a supporting medium for electrophoresis, one of the major difficulties inherent in this technique was the scarcity of data on the appropriate stains to dye the proteins on a photo-densitometer scanner-recorder that is easily secured. A study was made to find a reproducible and simple method of staining and scanning within the limits of optical density in accordance with the Beer-Lambert law. A technique for electrophoresis of human serum on cellulose acetate, using a Spinco-Analytrol for estimation of the fractions, is described. The method was used to analyze the serums of 48 normal men, and the results are summarized, evaluated, and compared with values observed in five other relevant studies.

ELECTROPHORETIC PATTERNS OF TUMOR TISSUE PROTEINS

Afonso, E., J. Clin. Pathol. **16**, 375–376 (1963).

A method of constructing pherograms is described in this paper. Protein extracts of human and animal tumor tissues were studied by electrophoretic techniques using cellulose-acetate membranes as the supporting medium. Preliminary results indicate that this technique, which was found to be simple, sensitive and quick, presents a new approach to the identification and classification of tumor tissues.

ELECTROPHORESIS AND IMMUNOELECTROPHORESIS OF NEONATAL TEARS

Allerhand, J., S. Karelitz, S. Penbharkkul, A. Ramos, and H. D. Isenberg, J. Pediat. **62**, 85–92 (1963).

A study was made on the tears of both full-term and premature babies to gain some insight into the lacrimal protein constituents of infants in the neonatal period. The identity and complexity of the various protein moieties in the infants' tears were ascertained by cellulose-acetate electrophoresis and immunoelectrophoresis and compared with those of older controls.

ELECTROPHORETIC STUDY OF SYNTHETIC FOOD DYES

Anwar, M. H., S. Norman, B. Anwar, and P. Laplaca, J. Chem. Educ. **40**, 537–538 (1963).

Past attempts to separate and study the impurities of the 11 synthetic dyes the U.S. government allows to be used in food products were done with various systems of paper chromatography. This paper reports an electrophoretic study of the effect of charge density and mass of dye molecules on the mobility of disperse dyes in an electric field, using cellulose acetate as the supporting medium. The apparatus used is described in detail and the results are discussed. Similar experiments using paper electrophoresis were unsuccessful because of the paper's high adsorptive power which caused tailing during separation. Cellulose-acetate membrane proved to be an excellent supporting medium because it has uniform propensity and minimal adsorptive power which eliminates tailing.

RAPID CELLULOSE ACETATE ELECTROPHORESIS: I. SERUM PROTEINS

Bartlett, R. C., Clin. Chem. **9**, No. 3, 317–324 (1963).

An electrophoretic technique using cellulose acetate as the supporting medium was developed to eliminate the problems of excessive time required, albumin trailing, and inconsistent reproducibility, and the requirement of additional apparatus for the determination of abnormal hemoglobin and quantitation of hemoglobin A2 encountered with filter paper electrophoresis. With this method, serum protein fractions can be quantitated by densitometry within two hours of the start of the procedure. Investigations using this technique are described in this paper, and showed that the densitometer had a linear response to Amido Black dye, that albumin and γ-globulin demonstrated equal dye-binding characteristics, and that excellent quantitations of serum protein fractions resulted. Further development of this cellulose-acetate electrophoretic method demonstrated that by increasing the buffer strength to 0.05M, comparable results could be obtained with the use of an inexpensive commercially available electrophoresis tank (Gelman Instrument Company, Ann Arbor, Michigan).

RAPID CELLULOSE ACETATE ELECTROPHORESIS: II. QUALITATIVE AND QUANTITATIVE HEMOGLOBIN FRACTIONATION

Bartlett, R. C., Clin. Chem. **9**, No. 3, 325–329 (1963).

A method of cellulose acetate electrophoresis is described which completes in two hours the qualitative and quantitative analysis of abnormal hemoglobins, including hemoglobin A_2, using the same apparatus used for the quantitative analysis of serum proteins. The quantitative values are comparable to those obtained by the starch-block technique. Further development of this method demonstrated comparable results with the use of an inexpensive commercially available electrophoretic tank (Gelman Instrument Company, Ann Arbor, Michigan), with optimal results obtained by increasing the buffer strength to 0.10M.

FOETAL MYOGLOBIN IN THE URINE OF AN ADULT

Benoit, F. L., G. B. Theil, and R. H. Watten, Nature **199**, No. 4891, 387 (1963).

The presence of fetal myoglobin, found in newborn animals and humans and showing a different electrophoretic mobility from myoglobin in adults, was discovered in the urine of a 28-year-old female with idiopathic myoglobinuria. Fetal myoglobin from the psoas muscle of stillborn infants and adult myoglobin from the psoas muscle of an adult who died of a non-myopathic disease were subjected to electrophoresis performed on cellulose acetate at 120 mV with barbital buffer (pH 8.6, ionic strength 0.1M) for 110 minutes. Samples of the patient's serum and hemoglobin were subjected to electrophoresis concurrently and the resulting separations were stained with benzidine for iron, as well as with nigrosin for protein. The results are described and evaluated.

HAPTOGLOBIN TYPE DETERMINATION BY CELLULOSE ACETATE ZONE ELECTROPHORESIS

Blackwell, R. Q., and C. S. Lin, Clin. Chim. Acta **8**, 868–871 (1963).

Cellulose-acetate zone electrophoresis, advantageous as a means of determining blood haptoglobin type because of the very small blood samples necessary, is described and evaluated. Only 10 μl of plasma or serum is required. This electrophoretic technique, useful for routine screening procedures, is employed with a slightly modified barbital-borate buffer system.

ELECTROPHORESIS ON CELLULOSE ACETATE OF INSULIN AND INSULIN DERIVATIVES: CORRELATION WITH BEHAVIOR ON COUNTERCURRENT DISTRIBUTION AND PARTITION-COLUMN CHROMATOGRAPHY

Carpenter, F. H., and S. L. Hayes, Biochemistry **2**, No. 6, 1272–1277 (1963).

The technique for paper electrophoresis of insulin derivatives in urea-containing buffers (Sundby, 1962) was modified by the substitution of cellulose acetate strips for paper and by the application of relatively high voltages (40 volts/cm) in an attempt to decrease the time of electrophoresis. Good separation of various charged forms of insulin was observed in three hours of electrophoresis at pH 6.5 in 0.065M phosphate buffer which is 7M in urea. The experimentation and results are discussed.

FOLIC-ACID DEFICIENCY IN COELIAC DISEASE

Dormandy, K., A. H. Waters, and D. L. Mollin, Lancet **1**, No. 7282, 632–635 (1963).

Nineteen coeliac children and 38 control children were investigated in order to ascertain the prevalence of folic acid deficiency in children with untreated coeliac disease by measuring the serum folate levels and the urinary excretion of FIGLU after histidine loading. The findings were related to the hematological features of the anemia which may develop in this condition. Conventional-voltage electrophoresis on cellulose-acetate strips was used to determine the urinary FIGLU. The main cause of folic-acid deficiency in the coeliac children over one year of age seemed to be intestinal malabsorption, but other factors may have contributed to the deficiency in those under one year because a subnormal serum folate level also appeared in one-third of the control children in this age group.

A COMPARISON OF THE PLASMA PROTEINS OF C₃H STRAIN MICE, FOUND BY CELLULOSE-ACETATE AND STARCH-GEL ELECTROPHORESIS

Duke, E. J., Nature **197**, No. 4864, 288–289 (1963).

Plasma from adult C_3H strain mice was used in an experiment comparing plasma protein patterns found with cellulose-acetate and with starch-gel electrophoresis. The electrophoretic techniques used are described and the findings reported.

PHARMACOLOGICALLY ACTIVE SUBSTANCES IN THE URINE OF BURNED PATIENTS

Goodwin, L. G., C. R. Jones, W. H. G. Richards, and J. Kohn, Brit. J. Exp. Pathol. **44**, 551–560 (1963).

Urine, serum, and blister fluid, collected in this study from patients with severe burns with precautions to minimize the production of artefacts, were found to contain substances which relaxed the isolated rat duodenum. The presence of ninhydrin-positive bands, which differed qualitatively and quantitatively from those given in normal urine, was revealed by cellulose-acetate electrophoresis of alcoholic extracts of burn urine. Column chromatography fractionation of alcoholic extracts of burn urine showed that it contained up to 500 times the histamine activity of normal urine, and a peptide fraction from burn urine had from 5 to 40 times the activity of a corresponding fraction from normal urine. Results of fractionation by paper chromatography are also given. Findings suggest that "toxaemia" and shock in burned patients may be attributed to histamine and pharmacologically active peptides released into the circulation as a result of the injury.

IMMUNOELECTROPHORESIS ON CELLULOSE ACETATE

Laurent, B., Scand. J. Clin. Lab. Invest. **15**, 98–101 (1963).

A detailed description of a procedure for immunoelectrophoresis on cellulose acetate, based on the method of Consden and Kohn (Nature *183*, 1512, 1959), is presented in this paper. The improvements in the procedure reported by the author consist mainly of the use of a detergent for the elution of excess protein and the staining of protein-bound carbohydrates after partial or complete removal of protein stain.

ULTRA-MICRO PRECIPITIN PROCEDURE ON CELLULOSE ACETATE

Markowitz, A. S., and J. Isenberg, Biochem. Biophys. Res. Commun. **12**, No. 1, 56–61 (1963).

The increased use of the immunochemical approach for the quantitation of substances present in biological fluids made apparent the limited practical applicability of the classical quantitative precipitin reactions due to the quantity of reactants consumed and the time and technical skill required. A different procedure was designed to resolve these difficulties. This report describes and evaluates the procedure which quantitates serum antibody as measured by the dye uptake of the immunologic precipitate on cellulose acetate.

IN-VITRO AND IN-VIVO STUDIES OF A PREPARATION OF UROKINASE

McNicol, G. P., S. B. Gale, and A. S. Douglas, Brit. Med. J. **1**, Part II, 909–915 (1963).

An account is given in this paper of *in vitro* and *in vivo* study of a commercial preparation of human urokinase, the physiological fibrinolytic substance present in normal urine available for clinical trial as a fibrinolytic agent in the treatment of intravascular thrombosis. At least four components of the preparation were indicated by electrophoresis on oxoid cellulose-acetate strips, and the fibrinolytic activity was seen to be largely located in one specific band. The value of the preparation is discussed.

CELLULOSE-ACETATE MEMBRANE AS A MEDIUM FOR DEMONSTRATING MILK PRECIPITINS: APPLICATION TO SELECTED PATIENTS

Nelson, T. L., C. T. Greene, and G. Stroup, Ann. Allergy **21**, 629–636 (1963).

A simplified method for studying milk precipitins by immunodiffusion using a cellulose-acetate medium, its application to representative clinical cases of milk precipitin disease and chronic gastrointestinal disorders, and use of the cellulose-acetate membrane for immunoelectrophoretic studies are described in this paper. Because the symptoms associated with the finding of milk precipitins often overlap those of classical atopic disease, patients suffering from milk precipitin disease are frequently referred to an allergist. The method described can readily be employed in the allergist's office laboratory for the study of patients suspected of this disease because it can be performed on simple equipment and can be learned easily by anyone with some basic knowledge of laboratory techniques.

DEHYDRATION OF CELLULOSE ACETATE STRIPS IN MICROELECTROPHO-RESIS

Padmore, G. R. A., and N. E. Emmett, Clin. Chim. Acta **8**, 325 (1963).

To eliminate the distortion sometimes caused by oven-drying of the strips after microelectrophoresis on cellulose acetate, a simple liquid dehydrating process was devised. The process is described in this report.

α-HYDROXYBUTYRATE DEHYDROGENASE

Rosalki, S. B., Lancet **1**, No. 7280, 554–555 (1963).

A simple enzyme staining technique is described for confirming the disproportionately high serum-α-hydroxybutyrate-dehydrogenase (SHBD) activity of the faster-moving serum-lactate-dehyrogenase (SLD) isoenzymes. The sera were separated by cellulose-acetate electrophoresis in a barbitone buffer, pH 8.6, using a constant current of 0.5 mA per cm for 1¼ hours.

RAPID MEASUREMENT OF HEMOGLOBIN A₂ BY MEANS OF CELLULOSE ACETATE MEMBRANE ELECTROPHORESIS

Rozman, R. S., R. P. Sacks, and R. Kates, J. Lab. Clin. Med. **62**, No. 4, 692–698 (1963).

The electrophoretic separation of hemoglobin A_2 from hemoglobin $A_1 + A_3$ on cellulose-acetate membranes is a simple and rapid procedure that can be used in smaller laboratories because less elaborate equipment is required. The hemoglobins were stained with Ponceau S stain, which is then eluted and measured by colorimetry. Normal human subjects and patients with hematologic disorders other than thalassemia minor showed values of 2.4 to 4.1 percent for hemoglobin A_2. Hemoglobin A_2 levels ranged from 5.4 to 7.8 percent in patients with thalassemia minor who exhibited an elevated A_2 level on starch block electrophoresis. Between these two groups no overlap of values was found and a statistically significant difference was demonstrated.

A METHOD OF QUANTITATIVE SERUM PROTEIN ELECTROPHORESIS

Sammons, H. G., and P. H. Whitehead, Clin. Chim. Acta **8**, 673–677 (1963).

A routine method of quantitatively estimating serum proteins by cellulose-acetate electrophoresis, which is suitable for most hospital laboratories, is described in this paper. The apparatus, reagents, and procedure employed and the results are reviewed.

PURIFICATION AND CHARACTERIZATION OF STAPHYLOCOAGULASE

Zolli, Z., Jr., and C. L. San Clemente, J. Bacteriol. **86**, No. 3, 527–535 (1963).

A study, described in this paper, was made to purify coagulase to a high degree and to characterize it chemically and serologically. Three cycles of dialysis in ethanol-water mixtures under controlled conditions, followed by molecular sieving through a column of Sephadex® G-200, were used in the separation and extreme purification of coagulase from *Staphylocoagulase aureus* strain 70. By manipulation of five variables (pH, ionic strength, temperature, protein and ethanol concentration), an approximate 3700-fold increase in activity per mg of protein was seen in the final preparation. Serological and chemical characterization was made of the successfully isolated coagulase, which contained 15.0 percent nitrogen. One zone of precipitation was obtained with the highly purified material through agar diffusion techniques, and purity was also confirmed by the appearance of a single peak with cellulose-acetate paper electrophoresis with a barbital buffer at pH 8.6. Through the four stages of purification, the progressive and eventual elimination of carbohydrate, deoxyribonuclease lipase, and phosphatase was observed. It was shown through temperature studies that the stability of each fraction was inversely related to its purity.

RAPID DETERMINATION OF A₂ HEMOGLOBIN BY REVERSE-FLOW ELECTROPHORESIS ON CELLULOSE ACETATE MEMBRANES

Afonso, E., Clin. Chim. Acta **7**, 545–549 (1962).

A simple, rapid method for determining Hb-A$_2$ by reverse-flow electrophoresis on cellulose-acetate strips that is ideal for routine use is described. The method also offers the advantage of a clear resolution of other minor components of the erythrocyte lysate. The results obtained on the determination of Hb-A$_2$ and nonheme proteins in adults and children, described in this paper, compare favorably with those of the starch-block method.

CHROMATOGRAPHY OF THE HUMAN TUBERCULIN DELAYED-TYPE HYPER-SENSITIVITY TRANSFER FACTOR

Baram, P., and M. M. Mosko, J. Allergy **33**, No. 6, 498–506 (1962).

Work to fractionate human white blood cells in order to isolate the factor responsible for passive transfer of tuberculin delayed hypersensitivity is discussed by the authors. Sonicated peripheral white blood cell supernatants from purified protein derivative (PPD)-hypersensitive and PPD-nonhypersensitive human beings were chromatographed on DEAE cellulose. Immune and cellulose-acetate electrophoresis revealed no difference in the number and position of the components from the cells of PPD-hypersensitive and PPD-nonhypersensitive donors. Using Fractions 0, 1, 2, and 3A, passive transfer was accomplished with varying frequency, with the greater number of transfers effected with Fractions 1 and 3A. Proteins, with mobilities similar to those of gamma globulin were determined by immune electrophoresis to be the only components common to all fractions.

INTERFIBRE FLUID FROM GUINEA-PIG MUSCLE

Creese, R., J. L. D'Silva, and D. M. Shaw, J. Physiol. (London) **162**, 44–53, (1962).

Small amounts of fluid were collected from the muscles of anesthetized guinea-pigs following intravenous injection with albumin labelled with ^{131}I. Separation by microelectrophoresis on cellulose-acetate strips revealed that all the major protein constituents of serum were present in the interfibre fluid.

'MICROSPOT' TEST APPLIED TO CELLULOSE ACETATE MEMBRANES

Feinberg, J. G., Nature **194**, No. 4825, 307–308 (1962).

The antigen-antibody microspot test on agar gel-coated cover-glasses was extended to permit use of cellulose-acetate membranes. The technique of the cellulose-acetate membrane microspot test and its advantages over the agar-gel microspot test are discussed in this report.

SERUM ELECTROPHORESIS CHANGES IN POLYMYOSITIS

Gavrilescu, K., and J. M. Small, Brit. Med. J. Part 2, 1720–1723 (1962).

Cellulose-acetate electrophoresis was applied in six cases of proven polymyositis to support the authors' belief that a study of the proteins is useful not only in diagnosis but also for a better understanding of the nature of the disease. In the six cases notable serum electrophoretic changes were a considerable increase in the γ-globulin fraction and a decrease in the albumin. Though this serum electrophoretic pattern may be found in other pathological conditions, its presence in a patient with a muscular disorder strongly suggests polymyositis.

IMMUNOELECTROPHORESIS WITH CELLULOSE ACETATE

Grunbaum, B. W., and L. Dong, Nature **194**, No. 4824, 185–186 (1962).

Immunoelectrophoresis of human serum and its equine antiserum has shown 25 serum constituents, but this technique is not widely used as an experimental tool mainly because of technical difficulties. An improved method, which is simple enough to produce outstanding results with a minimum of technical skill, and which has shown more than 35 precipitin lines, is reported in this paper.

DEMONSTRATION OF SEVERAL COMPONENTS IN HIGHLY PURIFIED HUMAN GROWTH HORMONE BY IMMUNODIFFUSION ON CELLULOSE ACETATE

Hayashida, T., and B. W. Grunbaum, Endocrinology **71**, 734–739 (1962).

The apparent homogeneity and the specificity of the reaction between highly purified human growth hormone (HGH) and its antiserum produced in rabbits had been demonstrated with various immunologic procedures including immuno-electrophoresis in agar. The authors employed immunodiffusion on cellulose acetate in their studies to determine the homogeneity of the growth hormone reaction with its antiserum. When the technique of immunoelectrophoresis in agar was used, employing antiserum to Raben or Li HGH, the presence of only one antigenic component upon reaction of the antiserum with the highly purified HGH of Raben or Li was demonstrated, while at least five to six components were shown in the same hormone preparations using immuno-diffusion on cellulose acetate. Immunoelectrophoresis on the cellulose medium showed no better separation of the components than was obtained with immunodiffusion alone. It appeared that the separation of the components was more dependent upon a phenomenon of differential diffusion than upon differences in net electrical charges. The results of their investigations are further discussed.

ELUTION OF LISSAMINE GREEN FROM CELLULOSE ACETATE ELECTRO-PHORESIS STRIPS

Laurent, B., Scand., J. Clin. Lab. Invest. **14**, 563–564 (1962).

In a device for scanning cellulose-acetate electrophoresis strips stained with Lissamine Green SF 150, it was found that considerable amounts of the Lissamine Green were still bound by the globulin fractions even after 15 hours' elution with phtalate buffer at pH 6.0 at room temperature. By adding absolute ethanol to the buffer in equal parts, complete removal of the dye was achieved in less than half an hour. In addition, alcohol has very good wetting properties which is important especially when the strips have been dried at high temperatures.

ACID MUCOPOLYSACCHARIDES OF THE HUMAN UTERUS

Loewi, G., and R. Consden, Nature **195**, No. 4837, 148–150 (1962).

The acid mucopolysaccharides of human uteri of several ages and conditions were analyzed in an effort to correlate any differences in such analyses with different conditions of the uterus. The methods used in the investigation, including the examination of the ethanol fractions (20-30 μgm each) for acid polysaccharides by cellulose-acetate electrophoresis (30 minutes at 15 volts/cm in M/40 phosphate buffer, pH 7, stained with Toluidine Blue), are described. The results of the study are also reported.

CELLULOSE ACETATE ELECTROPHORESIS OF MILK SERUM PROTEINS

Mhatre, N. S., J. G. Leeder, and G. N. Wogan, J. Dairy Sci. **45**, 717–723 (1962).

In the separation of milk serum proteins by cellulose-acetate electrophoresis in a Veronal buffer (pH 8.6, ionic strength 0.05) at 200 volts for two hours at 250, five distinct fractions were identified as blood serum albumin, beta-lactoglobulin, alpha-lactalbumin, pseudo-, and euglobulin in their decreasing rates of migration. A study was made on the choice of buffer, effect of ionic strength and pH of buffer, position of serum application, effect of voltage, effect of temperature, and concentration of the serum solution to determine the ideal conditions for the optimum resolution of the serum proteins. The method used was shown to be reliable by the standard deviations and coefficients of variation for the five fractions revealed, and seems to offer significant advantages for investigating milk serum proteins.

ELECTROPHORESIS OF SERUM AND SOLUBLE LIVER PROTEINS ON CELLULOSE ACETATE

Mullan, F. A., D. M. Hancock, and D. W. Neill, Nature **194**, No. 4824, 149–150 (1962).

Electrophoresis of soluble liver proteins to find a satisfactory method for the study of biopsies from patients with liver disease was carried out using rat liver. The duration of the electrophoresis, the site of application of the sample, and the buffer systems were varied during the course of the work. A better separation of the protein bands was obtained when the sample was applied 2.5 cm from the anode end of the strip ('anode application') than when it was applied the same distance from the cathode end ('cathode application'). This paper describes work done on anode application' of serum and soluble liver proteins to elucidate the patterns obtained by this procedure. 'Anode application' may be of value in studying electrophoretic separation of protein fractions from other tissues if satisfactory buffer systems and timing of the run can be arranged to suit this method.

CELLULOSE ACETATE MEMBRANES FOR THE ELECTROPHORESIS DEMONSTRATION OF HEMOGLOBIN A₂

Petrakis, N. L., M. A. Doherty, B. W. Grumbaum, and W. A. Atchley, Acta Haematol. **27**, 96–103 (1962).

This report describes a simple method for the rapid qualitative and quantitative electrophoresis of normal major and minor hemoglobin components using cellulose-acetate membranes and standard clinical electrophoresis equipment. In addition to speed, the use of cellulose-acetate strips in hemoglobin electrophoresis is advantageous because it can be done without the elaborate preparations required for starch-block and agar-gel techniques, it can be performed with standard laboratory equipment, and the patterns obtained are permanent and can be analyzed quantitatively, employing standard densitometric apparatus. This technique also offers a clinically useful method for the detection of thalassemia heterozygotes characterized by elevations in the A₂ fraction of hemoglobin.

A SIMPLE MODIFICATION OF A HORIZONTAL ELECTROPHORESIS TANK FOR USE WITH CELLULOSE ACETATE STRIPS

Ridler, M. A. C., J. Med. Lab. Technol. **19**, 174–176 (1962).

Due to inadequate sealing and a comparatively large air space in the electrophoresis chamber, some large horizontal electrophoresis tanks that were found to be suitable for use with filter paper, were quite unsuitable for cellulose-acetate strips because of water vapor leakage and varying saturation and temperature effects along the strips resulted in distortion of the protein bands. A specially designed, small, horizontal tank, commercially available from the Shandon Scientific Company, and an adaptation of the vertical type tank for the horizontal cellulose-acetate technique, were constructed to remedy these deficiencies. This paper describes a simple method of eliminating leaks in a Perspex® horizontal electrophoresis tank and the construction of an accessory tray for cellulose-acetate strips.

A DEVICE FOR THE APPLICATION OF SAMPLES IN CELLULOSE ACETATE ELECTROPHORESIS

Savaris, C. A., J. Electroanal. Chem. **3**, 350–351 (1962).

Zone electrophoresis with cellulose acetate as the supporting medium was a success in the application of samples in a straight line and exactly parallel to the path of migration, in making reproducibility far easier, and in allowing sharper, faster applications. A new technique of applying the sample, in which a movable Perspex® bridge is incorporated into the electrophoretic apparatus, is described in this paper.

USE OF CELLULOSE ACETATE STRIPS FOR ELECTROPHORESIS OF AMINO ACIDS

Scherr, G. H., Anal. Chem. **34**, No. 7, 777 (1962).

The entire operation of amino-acid electrophoresis on cellulose-acetate strips, using 300 volts and approximately 1.5 mA per strip, can be performed in 45 minutes including electrophoresis, drying and staining.

ON THE ELECTROPHORESIS OF PROTEINS ON CELLULOSE ACETATE MEMBRANES

Afonso, E. Clin. Chim. Acta **6**, 883–885 (1961).

In experiments employing cellulose-acetate electrophoresis, the use of a suitable buffer and an appropriate electrophoresis chamber allowed the application of much higher voltages than usual with the results being better and quicker than ordinarily obtained with this technique. Glycerol was added to the buffer in an amount adequate to cut down evaporation to a minimum. The method and apparatus used, which can separate protein fractions in migration times of 15-30 minutes at voltages up to 60 volts/cm measured between the electrodes, are described in this paper.

ELECTROPHORETIC AND ANALYTICAL ULTRACENTRIFUGE STUDIES IN SERA OF PSYCHOTIC PATIENTS: ELEVATION OF GAMMA GLOBULINS AND MACROGLOBULINS, AND SPLITTING OF ALPHA$_2$ GLOBULINS

Fessel, W. J., and B. W. Grunbaum, Ann. Internal Med. **54**, No. 6, 1134–1145 (1961).

The electrophoretic serum protein patterns of chronically psychotic patients, who had positive serologic tests for rheumatoid factor, and the patterns of certain control groups are reported. The results of the cellulose-acetate electrophoresis of sera from 162 patients with chronic psychosis, and the effects of chronic confinement, malnutrition, liver disease, ataractic drug therapy, and other factors coincidental to the psychosis in influencing serum protein changes are discussed. It was believed that some of these factors could be ruled out, and that there is a possibility that the macroglobulin disturbance plays an important role in the pathogenesis of some psychoses.

A STANDARDIZED PROCEDURE FOR SERUM PROTEIN ELECTROPHORESIS ON CELLULOSE ACETATE MEMBRANE STRIPS

Friedman, H. S., Clin. Chim. Acta **6**, 775–781 (1961).

A range of standard values for the serum protein fractions obtained by cellulose-acetate electrophoresis and a reproducible technique for the apparatus and materials at hand were established and are presented in this paper. Detailed directions are given for serum protein electrophoresis on cellulose-acetate strips using the EEL electrophoresis apparatus with adapter cells, for staining, and for quantitative determinations of the relative values of the fractions. Problems inherent in the method and the apparatus and a calculation of results are discussed. The values established are not to be construed as universally correct or useful in laboratories using different methods and equipment; they are clinically applicable only when the apparatus and methods described in this paper are used.

DENSITOMETRIC EVALUATION OF MICROELECTROPHORETIC SERUM: PROTEIN PATTERNS ON CELLULOSE ACETATE MEMBRANES

Grunbaum, B. W., W. J. Fessel, and C. F. Piel, Anal. Chem. **33**, 860–861 (1961).

A rapid, discriminative, and versatile technique of microelectrophoresis on cellulose acetate is described for the quantitation of serum protein fractions using a slight modification of available apparatus. Eight different protein fractions were revealed in normal sera, and results of quantification of these fractions in 50 sera are presented.

A LIPOPROTEIN STAINING METHOD FOR ZONE ELECTROPHORESIS

Kohn, J., Nature **189**, 312–313 (1961).

A new method, based on the use of Schiff's stain after preliminary ozonization, was developed in an attempt to overcome some of the disadvantages of the commonly used lipoprotein stains. It does not depend on the use of lipid stains, so washing the stain out of the background is not necessary, and it can be used equally well on cellulose acetate material and on filter paper. The separation of lipoprotein fractions in normal human serum was achieved by paper and cellulose-acetate electrophoresis, using a somewhat larger sample than that used for ordinary protein separation. Application of the new staining method is described.

IMMUNO-DIFFUSION TECHNIQUE ON CELLULOSE ACETATE

Kohn, J., Protides Biol. Fluid, Proc. 9th Colloq., 120–122 (1961).

A technique for the demonstration of precipitating antigen-antibody systems by immunodiffusion on cellulose-acetate paper is described. Cellulose acetate has advantages over agar because it is ready for immediate use, it requires smaller quantities of test material, and it is very easily stored and a permanent record kept. The main disadvantage of this technique is that the precipitation lines are visible only after staining, but this is of little importance since the optimum diffusion time for most reacting systems is between 24 and 48 hours. When using the cellulose-acetate strip for immunoelectrophoresis, it should be longer than the strips used for ordinary electrophoresis in order to provide enough clearance on both ends for the separation pattern.

CONVENTIONAL VOLTAGE ELECTROPHORESIS FOR FORMIMINOGLU-TAMIC-ACID DETERMINATION IN FOLIC ACID DEFICIENCY

Kohn, J., D. L. Mollin, and L. M. Rosenbach, J. Clin. Pathol. **14**, 345–350 (1961).

The application of conventional electrophoresis at 200 to 500 volts on cellulose-acetate strips on serum from 137 patients, resulting in 166 determinations, showed this method to be a simple, practical and apparently sensitive technique for determining urinary formiminoglutamic acid (FIGLU). The results of this investigation and those of the application of the measurement of urinary FIGLU with histidine loading as a test for folic acid deficiency are reported.

USE OF CELLULOSE ACETATE AND PONCEAU S FOR ELECTROPHORETIC SERUM PROTEIN ANALYSIS

Korotzer, J. L., L. M. Bergquist, and R. L. Searcy, Am. J. Med. Technol. **27**, 197–203 (1961).

The use of Ponceau S to stain serum separated electrophoretically on cellulose acetate was evaluated and compared to electrophoretic fractionations performed on filter paper and stained with Bromophenol Blue. The method was found to be efficient and reproducible for estimating proteins. The techniques and results are discussed.

CHANGES IN MICROSOMAL COMPONENTS ACCOMPANYING CELL DIF-FERENTIATION OF PEA-SEEDLING ROOTS

Loening, U. E., Biochem. J. **81**, 254–260 (1961).

Some changes found in the microsomes extracted by homogenization and centrifuging from small serial segments of seedling pea-root tips during growth, are described. Two fractions, a 'light' microme pellet consisting of a mixture of vesicles and ribosomes and a pellet of 'heavy' microsomes consisting largely of ribosomes, were found and are discussed in this paper. More nucleic acid fragments and proteins were given in the 'light' microsomes than in the 'heavy' by cellulose-acetate electrophoresis of the two fractions after disruption with ribonuclease. Also presented are electron micrographs of the two fractions.

RAPID METHOD FOR ZONE ELECTROPHORESIS OF SERUM AND PLASMA PROTEINS: THE USE OF CELLULOSE ACETATE STRIPS

Ohlesen, J. E., and G. P. Charlton, Nebraska State Med. J. **46**, 331–334 (1961).

The use of cellulose-acetate strips and simple equipment for the quantitative analysis of serum and plasma protein fractions offers a simple, economical technique with results similar to those using filter paper, but the bands are sharper, the albumin is more quantitatively recovered and the procedure is much less time-consuming. An investigation using this method is described and discussed.

OPTIMAL STAINING CONDITIONS FOR THE QUANTITATIVE ANALYSIS OF HUMAN SERUM PROTEIN FRACTIONS BY CELLULOSE ACETATE ELECTROPHORESIS

Brackenridge, C. J., Anal. Chem. **32**, No. 10, 1353–1356 (1960).

This report covers a fundamental study of dye-protein interactions on cellulose acetate during electrophoresis. The optimal conditions for staining with Lissamine Green on cellulose acetate have been clarified as a basis for the quantitative electrophoretic analysis of five major human serum protein fractions. The effects of pH on dye elution, and of pH, ionic strength, and staining time and temperature on dye uptake, were studied and applied to the measurements of the dye-binding capacities of five plasma proteins, one mucoprotein, and two lipoprotein preparations.

OPTIMAL FRACTIONATION CONDITIONS FOR THE QUANTITATIVE ANALYSIS OF HUMAN SERUM PROTEIN

Brackenridge, C. J., Anal. Chem. **32**, No. 10, 1357–1360 (1960).

In a study of several variables affecting the electrophoretic separation of human serum protein fractions on cellulose acetate to determine the optimal conditions, it was found that the choice of buffer, and the effects of ionic strength, pH and volume of buffer, length of membrane, position of serum application, voltage and running time influenced the over-all migration of the pattern and separation of protein bands. An accurate and complete procedure, based on the best conditions for staining and fractionation, was devised for the quantitative analysis of five fractions to yield results in 5 hours and 30 minutes. A determination was made of the normal protein ranges, based on a sampling of the general population.

VALUABLE DYE UPTAKE IN THE QUANTITATIVE ANALYSIS OF ABNORMAL GLOBULINS BY CELLULOSE ACETATE ELECTROPHORESIS

Brackenridge, C. J., Anal. Chem. **32**, No. 10, 1359–1360 (1960).

Twenty-eight serum samples from patients suffering from essential cryoglobulinemia, macroglobulinemia, and myeloma were subjected to quantitative electrophoresis on cellulose acetate to determine if the dye uptake of the serum paraproteins was regular or variable, and if regular, to which fraction they bore the most similarity. The investigation, reported in this paper, showed the paraprotein uptake to be variable owing to instability. The total protein content should be determined by an independent method to arrive at the concentration of the abnormal component because prediction of dye-binding behavior is extremely difficult.

INTERRELATIONS OF SERUM PROTEIN FRACTIONS IN NORMAL HUMANS

Brackenridge, C. J., Nature **188**, 155 (1960).

Interrelations between protein fractions in the course of variations within the normal human range were examined by cellulose-acetate electrophoresis. Blood for the study was obtained from 100 healthy men and women between the ages of 16 and 72. The results are reported in this paper. Little was known about the interdependence of serum proteins in the healthy state, although they frequently undergo significant alterations with the onset of disease and yield characteristic patterns on electrophoretic separation.

ELECTROPHORETIC PATTERN DURING AMPHIBIAN DEVELOPMENT

Denis, H., Nature **187**, No. 4731, 62 (1960).

Individual embryos of *Triturus alpestris* of every state from the uncleaved egg to stage Glasner 26, were sucked up into a capillary tube, frozen, thawed, and centrifuged at 7,000 g for 20 minutes, and the clear supernatant minus the fat layer was separated by cellulose-acetate electrophoresis. The investigation is described in this paper.

QUANTITATION OF PLASMA PROTEINS ON CELLULOSE ACETATE STRIPS

Albert-Recht, F., Clin. Chim. Acta **4**, 627–638 (1959).

In a search for an improved method for plasma or serum protein fractionation in a routine clinical laboratory, cellulose-acetate electrophoresis was evaluated. The characteristics of the migration patterns achieved indicated that this is a reliable routine method exhibiting accuracy, reproducibility, a well-defined normal range value, ease, and speed. The problem of the relatively high cost of the cellulose-acetate strip was partially solved by modifying the cellulose-acetate method of Kohn (Clin. Chim. Acta *3*, 450, 1958) which used small

(2.5 x 10 cm) strips. A means for assessing quantitative results for these small strips was worked out and tested to overcome the lack of information on the quantitative evaluation of results obtained by methods using cellulose acetate.

THE HAPTOGLOBIN CONTENT OF SERUM IN HAEMOLYTIC ANAEMIA

Brus, I., and S. M. Lewis, Brit. J. Haematol. **5**, 348–355 (1959).

Sera from 25 normal subjects and from 77 patients with various types of hemolytic anemia or other disorders were subjected to cellulose-acetate electrophoresis (modified method of Laurell and Nyman, 1957) to evaluate the importance of the estimation of the haptoglobin of sera as an indicator of an increased rate of hemolysis. This method was found to be a sensitive technique for detecting hemolysis. Haptoglobins were usually absent when hemolysis, as measured by the haptoglobin turnover per day, exceeded twice the normal rate. Malignancy or steroid therapy haptoglobins in the presence of infection, however, may be demonstrable even if hemolysis is taking place considerably faster than twice the normal rate. The study and the findings are reported.

CELLULOSE ACETATE AS A MEDIUM FOR IMMUNO-DIFFUSION

Consden, R. and J. Kohn, Nature **183**, 1512–1513 (1959).

The use of cellulose-acetate membrane instead of agar gel in the demonstration of antigen-antibody reactions had many advantages. Cellulose-acetate membranes are relatively simple to use, satisfactory results are obtained with about 1/100 of the amounts usually used for gel diffusion, and there is no difficulty in storing membranes. Electrophoretic and immunodiffusion experiments using this material are described and its performance is evaluated.

REGENERATION ON CELLULOSE ACETATE MEMBRANES USED FOR ZONE ELECTROPHORESIS

Jacobs, S., Nature **183**, 1326 (1959).

The analysis of serum proteins by zone electrophoresis or micro-electrophoresis on strips of cellulose-acetate membrane has proven this material to be important as a stabilizing medium. Workers may be deterred from using it, however, due to the relatively high expense of sheets of cellulose acetate as compared to filter paper. This report describes a suitable technique for treating the cellulose-acetate membrane strip so that it can be used several times.

SMALL-SCALE MEMBRANE FILTER ELECTROPHORESIS AND IMMUNO-ELECTROPHORESIS

Kohn, J., Clin. Chim. Acta **3**, 450–454 (1958).

A simple, rapid, sensitive and economical electrophoretic method for the separation of small quantities of protein is described. Cellulose-acetate filter membrane strips are used for the supporting medium.

A MICRO-ELECTROPHORETIC METHOD

Kohn, J., Nature **181**, 839-840 (1958).

A microelectrophoretic technique is described which is a combination of membrane filter electrophoresis and of the small scale technique with the use of a very sensitive stain, Nigrosin. It is sensitive, simple, rapid, does not require any elaborate apparatus, and provides very clear results. The supporting medium is cellulose-acetate membrane filter strips, 10-12 cm x 2.5 cm, and wider and longer strips can be used. Patterns obtained with this method performed on human serum are reported. This microtechnique can also be applied to immunoelectrophoresis with agar as the supporting medium.

SMALL-SCALE AND MICRO-MEMBRANE FILTER ELECTROPHORESIS AND IMMUNOELECTROPHORESIS

Kohn, J., Protides Biol. Fluids, Proc. 6th Colloq., 74–78 (1958).

A small-scale and micromembrane filter electrophoresis method was described, which is essentially the same as for full-scale membrane filter electrophoresis except that 12 x 2.5 cm strips and a special tank were used. When only very small quantities of protein were applied in the micromethod, a very sensitive stain, Nigrosin, was used. This method, when combined with the agar diffusion technique, can be used for small-scale immunoelectrophoresis with a great saving of antiserum.

MEMBRANE FILTER ELECTROPHORESIS

Kohn, J., Protides Biol. Fluids, Proc. 5th Colloq., 120–125 (1957).

Zone electrophoresis of serum proteins and hemoglobins is described using as the supporting medium cellulose-acetate membrane filter strips which were approximately 13 μ thick with pore sizes varying between 0.5 and 3.0 μ. The main advantages of this supporting medium are that it can be rendered completely transparent by immersion in a suitable clearing fluid; there is very little absorption and therefore practically no tailing; the stain can be washed out from the background without leaving any residual color; separation is very neat and distinct, the bands are narrower and of much finer grain than on filter paper; it can be dissolved in a suitable organic solvent; and a tougher and thicker material can be produced, rendering it less brittle and less sensitive to evaporation.

Immunoelectrophoresis

QUANTITATIVE IMMUNOELECTROPHORESIS OF SERUM PROTEINS

Afonso, E., Clin. Chim. Acta **10**, 114–122 (1964).

The principles and technique for quantitative immunoelectrophoresis are described in this paper. Immunoelectrophoretic fractions of serum proteins are determined by single gradient diffusion, and antiserum diffusion gradient is eliminated by either incorporating antiserum in the agar gel before electrophoresis, or by applying it in a uniform layer on the agar gel surface after electrophoresis. A calibration curve, from which the concentration of the unknown sample is read, is obtained from the plotting of transverse diameters of the precipitation patterns of standard protein solutions of different concentration.

ÉTUDE D'UNE β_{2A}-GLOBULINE CRYO-PRÉCIPITABLE

Auscher, C., and S. Guinand, Clin. Chim. Acta **9**, 40–48 (1964) **French**.

A cryo-precipitable globulin, with the electrophoretic properties of β_{2A}-globulin and which, in the ultracentrifuge, shows four components of S^{0}_{20}, w of 6.75, 9.4, 11.2, and 13.4 respectively, was isolated from pathologic serum. Two distinct groups of differing molecular weights, each retaining the antigenic specificity of β_{2A}-globulin, were isolated from the cryo-precipitable globulin by preparative ultracentrifugation. One group was a homogeneous component of S^{0}_{20}, w 6.75 with molecular weight 180,000 and rapid electrophoretic mobility; the other group was heterogeneous containing heavy components with electrophoretic mobility. When treated with cysteine, the two fractions lose their property of cryo-precipitation, and the S^{0}_{20}, w 11.7 and 13.4 fractions are shown to be polymeric forms of the other components. Normal β_{2A}-globulin, defined immunoelectrophoretically, is therefore concluded to be heterogeneous with components of differing molecular weights.

THE PROTEINS OF PSORIATIC SCALES: NONEXISTENCE OF A METACHRO-MATIC GLYCOPROTEIN REPORTED IN EXTRACTS OF PSORIATIC SCALES

Berrens, L., Clin. Chim. Acta **10**, 453–459 (1964).

Attempts to isolate a metachromatic glycoprotein from alkaline extracts of psoriatic scales are described in this paper. A number of complex mixtures which are contaminated with serum proteins are produced by the salt fractionation and isoelectric precipitation of the proteins in these scales. Results of electrophoretic and immunological studies, conducted on the fraction reported to contain the specific glycoprotein, indicate that no such compound exists and that the metachromasia in this special protein fraction, if detectable, must be attributed to the chondroitin sulfate coprecipitating with human serum albumin.

STUDIES ON γ-CRYSTALLIN FROM CALF LENS: II. PURIFICATION AND SOME PROPERTIES OF THE MAIN PROTEIN COMPONENTS

Björk, I., Exp. Eye Res. **3**, 254–261 (1964).

Four proteins of the γ-crystallin group, purified by chromatography on sulphoethyl-Sephadex® and phosphate-cellulose columns, were homogenous in gel and immunoelectrophoresis experiments and could be crystallized. Although their molecular weights, N-terminal amino acid sequences, and antigenic structures were all similar, certain dissimilarities were evident in their amino-acid compositions and in the sulphydryl groups contained. The experimental methods and materials and the results of the study are reported.

ANTIBODIES IN THE PIG AGAINST PIG INSULIN

Brunfeldt, K., and T. Deckert, Acta Endocrinol. **47**, No. 3, 367–370 (1964).

To determine the possibility of immunizing pigs by means of a pig insulin preparation, sera from three pigs injected subcutaneously for 87 days with pig, ox and horse insulin respectively were incubated with ^{131}I-pig insulin and investigated for the presence of antibodies by means of immunoelectrophoresis and autoradiography. There was a binding of ^{131}I-pig insulin to the gamma globulin fraction in all animals after the start of the injections, indicating a cross reactivity between pig and ox insulin and pig and horse insulin. A formation of antibodies against pig insulin in the pig was shown in the presence of a ^{131}I-pig insulin binding gamma fraction in the serum from the animal injected with pig insulin.

ÉTUDE IMMUNOLOGIQUE DES PROTÉINES DE LA SALIVI NORMALE ET DE MUCOVISCIDOSE

Burger-Girard, N., Schweiz. Med. Wochschr. **94**, 23–26 (1964) **French.**

The serum proteins albumin, α_1-antitrypsin, α_2-macroglobulin, siderophilin, γ1A-, γ1M- and γ-globulin were detected in human saliva by immunoelectrophoresis and Ouchterlony's double diffusion technique. Only four saliva-specific proteins were discernable with immunoelectrophoresis alone, whereas as many as nine could be identified with Ouchterlony's technique. When employed in an investigation of the saliva of patients with mucoviscidosis, using an immuno-serum prepared by immunizing rabbits with the patients' saliva, neither technique provided evidence to support the assumption that the antigenic pattern in the saliva of these patients differed from that of normal individuals.

QUANTITATIVE MICROFRACTIONATION OF SULPHOSALICYLIC ACID-SOLUBLE SUBSTANCES FROM HUMAN SERUM

Dolezalová, V., and Z. Broda, Clin. Chim. Acta **10**, 34–38 (1964).

Five components, orosomucoid, α-acid glycoprotein, haptoglobin, β_1-seromucoid and β_2-globulin, identified by immunoelectrophoretic analysis as occurring in the regions of α_1-, α_2- and β-globulins, were separated and determined from 0.5 ml of serum. The method for this quantitative microfractionation, which is based on the isolation of sulphosalicylic acid-soluble substances, removal of sulphosalicylic acid by gel filtration of Sephadex®, and fractionation of the isolated seromucoids on DEAE-cellulose, is described in this paper.

SERUM LEVELS OF BETA-1C GLOBULIN IN CHILDREN WITH ACUTE GLOME-RULONEPHRITIS AND IDIOPATHIC NEPHROTIC SYNDROME AND COR-RELATION WITH ELECTRON MICROSCOPIC AND IMMUNOELECTRON MI-CROSCOPIC STUDIES OF RENAL BIOPSIES

Hyman, L., C. G. Biava, and A. Grassman, J. Lab. Clin. Med. **64**, No. 5, 871–872 (1964).

Quantities of beta-1C globulin, a component of serum complement determined by immunoelectrophoresis, were found to be depressed in acute glomerulone-phritis and normal in idiopathic nephrotic syndrome (lipoid nephrosis). Electro-phoretic studies were made on equal quantities of pooled sera from normal children and from children with acute glomerulonephritis and idiopathic nephrotic syndrome. An inverse relation between deposits of gamma globulin-containing material and serum levels of beta-1c globulin was shown in pre-liminary studies by electron microscopy and immunoelectron microscopy on renal biopsies of representative patients from each group. It is suggested from these data that acute glomerulonephritis is an immune disease, but that idio-pathic nephrotic syndrome either is not an immune disease, or, if it is, it is one that employs an entirely different immune mechanism.

SERUM AND CEREBROSPINAL FLUID PROTEINS IN SCHIZOPHRENIA

Jensen, K., Acta Psychiat. Neurol. Scand., Suppl. **180**, 457 (1964).

Only a few unspecific changes in serum and CSF from schizophrenics were found in immunoelectrophoresis. No difference in quantity of serum cerulo-plasmin was found between schizophrenic patients and the normal controls.

APPARENT VARIATION IN THE CHARACTERISTICS OF HUMAN LOW-DENSITY LIPOPROTEINS DURING IMMUNOELECTROPHORESIS IN AGAR

Walton, K. W., Int. J. Immunochem. **1**, 279–288 (1964).

Low-density lipoproteins prepared by ultracentrifugal techniques were subjected to immunoelectrophoresis on agar, gelatine, and agarose. A pH 8.6 barbitone buffer was used in all cases. The lipoproteins demonstrated a variation in mobility which was dependent upon the concentration of lipoproteins and the agar.

ELECTROPHORETIC AND IMMUNOELECTROPHORETIC PATTERNS IN RHEUMATIC DISEASES

Allerhand, J., and S. H. Bernstein, Bull. N.Y. Acad. Med. **39**, No. 12, 802 (1963).

Electrophoresis of the sera of patients with acute rheumatic fever, streptococcal pharyngitis, and either rheumatoid arthritis or systemic lupus erythematosus made it possible to differentiate these diseases by their different protein patterns. Immunoelectrophoresis, using rabbit antirheumatic antiserum, was carried out against rheumatic fever sera to study the protein fractions immunochemically and to determine if the quantitative differences observed were due to increments in normal protein fractions or to new specific components. The preliminary results are given.

ÉTUDE DES α-LIPOPROTÉINES SÉRIQUES HUMAINES PAR IMMUNOÉLECTROPHORÈSE

Ayrault-Jarrier, M., G. Lévy, and J. Polonovski, Bul. Soc. Chim. Biol. **45**, 703–713 (1963) **French.**

The immunoelectrophoretic behavior of α-lipoproteins was studied on fractions purified by fractional ultracentrifugation, and the methods and materials used and the results obtained are presented in this paper. The existence of several distinct antigenic patterns was demonstrated by the presence of two or three lines colorable with Sudan Black or Amido Black.

ÉTUDE IMMUNO-ÉLECTROPHORÉTIQUE DE QUELQUES HÉMOGLOBINES HUMAINES

Boivin, P., and L. Hartmann, Ann. Soc. Belge Méd. Trop. **4**, 645–664 (1963) **French.**

The microelectrophoresis method of Scheidegger was adapted by the authors for the study of lysates of normal and pathological human erythrocytes. Immune sera were prepared by using Freund adjuvants from the erythrocytes hemolysates A, S, C, AS, AC, AD of umbilical cord blood and from crystalline hemoglobins A and F. Results of the investigation were presented. Adult anti-hemoglobin immune sera in the hemolysates containing a mixture of hemo-globins A and F (artificial mixtures, blood from the umbilical cord, thalassemia major) showed several important constituents of hemoglobin.

ÉTUDE ÉLECTROPHORÉTIQUE ET IMMUNOELECTROPHORÉTIQUE DES PRO-TÉINES DU RAT ATTEINT DE LEUCOSARCOME

Dickers, C., Protides Biol. Fluids, Proc. 11th Colloq., 105–108 (1963) **French.**

Four groups of reactions in rat with leucosarcoma were found by an electro-phoretic and immunoelectrophoretic analysis of the proteins: a group with para-protein of the γss-globulin type, a group with the γ_{1A}-globulin type, a group with the γ_{1M}-globulin type and a fourth group of animals with no typical protein pattern. The occurrence of these several paraprotein types is discussed.

MIT STÄRKE REAGIERENDE FRAKTIONEN IM MENSCHLICHEN SERUM: UNTERSUCHUNGEN IN EINER KOMBINATIONSELEKTROPHORESE

Gillert, K.-E., Protides Biol. Fluids, Proc. 11th Colloq., 388–392 (1963) **German.**

An extension of the principle of the immunodiffusion technique of Grabar and Williams by the use of substances other than immune sera as the diffusing reagent is called "combination-electrophoresis." Precipitation lines in the α_1- and α_2-region are yielded by the use of starch. The reacting proteins, which were chemically positive for starch, showed no lipid content.

RADIOIMMUNOELECTROPHORETIC CHARACTERIZATION OF ANTIBODY GLOBULINS IN HUMAN AND RABBIT SERA

Goodman, H. C., J. Robbins, and D. Exum, Arthritis Rheumat. **6**, 273–274 (1963).

Antithyroglobulin antibodies in guinea pig sera were characterized and gamma and macroglobulin (β_2M) anti-albumin antibody globulins in rabbit sera were identified by radioimmunoelectrophoresis. This technique permits the character-ization of serum antibody globulins by demonstrating the association of radio-active antigen with the globulins after the globulins are precipitated in gel during immunoelectrophoresis. Anti-thyroglobulin activity was identified in γ, β_2M (γ_1M) and $\cdot\beta_2$A (γ_1A) globulins in human sera from patients with chronic thyroditis. Radioimmunoelectrophoresis not only provides a means for distinguish-ing between these three types of immune globulins, but also will be useful for studying the increasingly evident heterogeneity within each group.

ANALYSE IMMUNOÉLECTROPHORÉTIQUE DES LIQUIDES GASTRIQUES NORMAUX ET PATHOLOGIQUES

Hirsch-Marie, H., and P. Burtin, Protides Biol. Fluids, Proc. 11th Colloq., 256–260 (1963) **French.**

An examination was made on 143 normal and pathological gastric juices in the fasting state and after histimine. Immunoelectrophoresis of the concentrated protein, performed in the presence of anti-human serum, anti-gastric juice serum, and anti-gastric mucosa serum, always revealed three serum proteins (albumin, β_{2A}- and γ-globulin) and sometimes showed 11 other lines of which five are similar to normal saliva. Clinical variations are briefly discussed.

IMMUNOELEKTROPHORETISCHE UNTERSUCHUNGEN VON UROPROTEIN-EN BEI PLASMOCYTOMEN UND NEPHROSEN

Huhnstock, K., and H. Weicker, Protides Biol. Fluids, Proc. 11th Colloq., 93–97 (1963) **German.**

The presence of a low-weight glycoprotein which migrated electrophoretically in the α_1-globulin region was demonstrated by immunoelectrophoretic analysis of stepwise prepared uroproteins in cases of plasmocytoma. A chemical and immunochemical difference was observed in a comparative study between the glycoproteins observed in plasmocytoma and in nephrosis, leading perhaps to new differential diagnostic possibilities.

ANTIGLOBULIN SERA: COMPARISON BETWEEN SEROLOGICAL TESTS AND IMMUNOELECTROPHORESIS

Jenkins, G. C., J. Clin. Pathol. **16**, 25–31 (1963).

In this paper the reactions of selected antiglobulin sera with sensitized cells and their relation to their precipitation patterns with normal serum on immuno-electrophoresis are discussed. The components which react with the absorbed complement are anti-β_1, further evidenced by the effect of the addition of human serum components on the antiglobulin reactions with 'complement-coated' cells. The findings of this study are commented on in the light of recent works by various authors.

INDIVIDUALITY OF HUMAN WHOLE DRY BLOOD: BY IMMUNOELECTRO-PHORESIS ON CELLULOSE ACETATE

Laudel, A. F., B. W. Grunbaum, and P. L. Kirk, J. Forensic Med. **10**, 57–64 (1963).

A search was made for laboratory methods applicable to problems of identification and individualization of sera, including whole dry blood, which would give precise and reliable results without the need of excessive expenditures of time or of costly and elaborate equipment. Immunoelectrophoresis on cellulose acetate was found to be a workable method within rather wide variations of experimental conditions. Practical application of this technique in the study reported in this paper was based on the relative stability and immunological integrity of the serum proteins, combined with their easy separation by electrophoresis. Although variations, believed to be technical in nature, occur from one run to the next to a degree that make it impossible to compare successive strips and makes it necessary to compare only the patterns formed on a single strip, forensic application of the technique should be readily possible.

PHYSICO-CHEMICAL AND IMMUNOLOGICAL STUDIES OF PATHOLOGICAL SERUM MACROGLOBULINS

Ratcliff, P., J. F. Soothill, and D. R. Stanworth, Clin. Chim. Acta **8**, 91–108 (1963).

Fourteen sera containing pathological macroglobulins were subjected to physico-chemical and quantitative gel-diffusion precipitin studies. The gel-diffusion precipitin technique, using a specific antiserum to the 19S component of γ-globulin, was shown to be a satisfactory substitute for ultracentrifugation in the identification and estimation of pathological macroglobulins, while other simple diagnostic tests studied were found to be of limited value. Findings suggested that above certain critical serum concentrations, the aggregation of pathological macroglobins to form even larger components occurs *in vivo*. Although acidification of four of the pathological sera (to pH 4.2) failed to dissociate their macroglobulin constituents, this was readily achieved by mercaptoethanol treatment.

THE EFFECT OF EXTRACTION PROCEDURES ON THE ELECTROPHORETIC AND IMMUNOCHEMICAL PROPERTIES OF THE SOLUBLE PROTEINS OF RAT LIVER MITOCHONDRIA

Sonnino, F. R., and P. P. Gazzaniga, Protides Biol. Fluids, Proc. 11th Colloq., 439–442 (1963).

Results of immunoelectrophoretic studies on the soluble proteins isolated from rat-liver mitochondria by use of different techniques are reported in this paper. The different methods of extraction employed—freezing and thawing, ultrasonic treatment, sodium deoxicholate, some ionic and non-ionic detergents, and barbiturate buffer—were believed to cause the discrepancies existing in the data given on investigations of these soluble proteins. Moving boundary electrophoresis runs were carried out at pH 7.5, ionic strength 0.2.

DIE IMMUNELEKTROPHORETISCHE TECHNIK AUF DER ZELLULOSE-ACETAT FOLIE

Südhof, H., and B. Walter, Clin. Chim. Acta **8**, 434–439 (1963) **German.**

A technique of immunoelectrophoresis on cellulose-acetate foil, described in this paper, shows in a reliable manner the precipitation lines of 14 different serum proteins, of which five had been identified. In contrast to agar-gel electrophoresis, the precipitation lines shown immunoelectrophoretically on the membrane foil concern mainly the proteins of the α_2, β, and γ-range, whereas the fast migrating proteins can be shown only incompletely. A horse anti-human serum should be used for the immunodiffusion in this technique.

THEORETICAL ANALYSIS OF IMMUNOELECTROPHORESIS

Aladjem, F., H. Klostergaard, and R. W. Taylor, J. Theoret. Biol. **3**, 134–145 (1962).

The processes that lead to the formation of the initial zone of precipitation between antigen and antibody during immunoelectrophoresis were analyzed. Equations were developed, with the aid of certain simplifying assumptions, which describe the concentration distribution of electrophoretically heterogeneous antigen after electrophoresis; the concentration of antigen and antibody at any point, at any time in the immunoelectrophoresis plate; and the amounts of antigen and antibody which accumulate in the region where the zone of precipitation forms.

ISOLATION AND PROPERTIES OF AN ALLERGEN FROM DWARF RAGWEED POLLEN

Callaghan, O. H., and A. R. Goldfarb, J. of Immunol. **89**, No. 5, 612–622 (1962).

A simplified method for preparing the partially purified material from dwarf pollen, referred to as Pool C, is described. Pool C is an ammonium sulfate fraction of a chromatographic cut of a water-soluble fraction derived by repeated extraction and precipitation in 90 percent methanol from a lyophilized aqueous extract of defatted dwarf pollen which had been previously methanol-leached. Pool C was subjected to ultracentrifugation and electrophoresis, and the effects of oxidizing and reducing agents on its chemical and biologic properties were studied. It was further fractionated to yield two allergenically active subfractions, of which Pool Cc (chromatographic fraction derived from Pool C) was by far the more copious. Experimentation using the authors' preparations of Pool C are reported. Immunoelectrophoretic patterns given by two samples of Pool C appeared to be very much alike after electrophoresis run in 0.01M borate buffer, pH 8.4, at 275 volts for 75 minutes. The concentration of Pool C was 10 mg/ml and the immunoelectrophoretic pattern for Pool Cc is virtually identical with that of Pool C.

ELECTROPHORESIS AND IMMUNO-ELECTROPHORESIS OF EXTRACTABLE PROTEINS IN BRAIN TISSUE

Gerhardt-Hansen, W., and J. Clausen, Danish Med. Bull. **9**, No. 1, 9–13 (1962).

Thirteen to fifteen protein fractions, which can be subdivided according to their electrophoretic mobilities into two to five pre-albumins, albumin, two to three alpha, three beta and three to five gamma globulins, were revealed by agar-gel microelectrophoresis performed on extracts of normal human brain tissue. After treatment of the specimens with methylene chloride, distinct fractionation of the anodic region was achieved. Precipitation curves were obtained in the tissue extracts corresponding to pre-albumin, albumin, α-1-glycoprotein, β-1-transferrin, and the intermediate part of the precipitation curve of gamma globulin in immunoelectrophoresis developed with horse and rabbit antiserum against whole normal human serum and with specific rabbit antisera.

ÉTUDE IMMUNOÉLECTROPHORETIQUE DE L'ACTIVITÉ ESTÉRASIQUE DES PROTÉINES SÉRIQUES DU RAT

Hermann, G., N. Talal, Ch. DeVaux St. Cyr, and J. Escribano, Protides Biol. Fluids, Proc. 10th Colloq., 186–189 (1962) **French.**

Immunoelectrophoretic analysis of normal rat serum revealed four precipitin arcs having esterase activity (slow and fast lipoprotein and two α_1-globulins). On simple electrophoresis, these esterases form three zones called ϵ-rapide, ϵ-lente and A-α, after their electrophoretic mobilities. Investigation of the action of certain inhibitors (DFP, eserine, copper sulfate and heat) on these esterases is also reported.

SIGNIFICANCE OF PRECIPITATING ANTIBODIES TO MILK PROTEINS IN THE SERUM OF INFANTS AND CHILDREN

Holland, N. H., R. Hong, N. C. Davis, and C. C. West, J. Pediat. **61**, No. 2, 181–195 (1962).

Precipitating antibodies to milk were found in the serum from 87 of 1,618 infants and children, with precipitins to bovine serum albumin and gamma globulin found more often than precipitins to other milk proteins. These 87 children, as compared to 87 hospital control patients, had increased incidence of recurrent respiratory tract disease, anemia, failure to thrive, and hepatosplenomegaly. Of 24 patients with precipitins, all but two showed marked improvement when placed on diets free of pasteurized milk and observed for a minimum of four months. Immunoelectrophoretic analysis was used to identify the antigens of milk, and disc electrophoresis, with acrylamide gel as the medium, was employed to resolve the protein components of the milk.

INDIVIDUALITY OF HUMAN SERUM BY IMMUNOELECTROPHORESIS

Laudel, A. F., B. W. Grumbaum, and P. L. Kirk, Science **137**, 862–864 (1962).

The direct comparison of serum samples from different individuals by simultaneous electrophoresis followed by simultaneous diffusion and precipitation by anti-human horse serum is described. An arbitrary reference grid was used to evaluate the results by comparing the numbers of precipitin bands present in paired immunoelectrophoretograms. Serum samples of like and unlike origin are distinguished by graphic representation of these differences.

ÜBER DIE LÖSLICHEN PROTEINE DER GEFÄSSENTIMA

Lohman, D., Protides Biol. Fluids, Proc. 10th Colloq., 61–65 (1962) **German.**

Several soluble and specific protein fractions are contained in the intima of the arterial wall, and some of the plasma proteins, namely α_1- and α_2-albumin, were identified by immunoelectrophoresis. The significance of the soluble intima proteins is unknown and it is unclear whether the increase of permeability of serum proteins is a primary or secondary phenomenon during intimal disease. Serum proteins in the intima in pathological cases and after death are discussed.

THE QUANTITATIVE DETERMINATION OF HUMAN SEROMUCOID FRACTIONS IN PAPER ELECTROPHORESIS

Morawiecka, B., and W. Mejbaum-Katzenellenbogen, Clin. Chim. Acta **7**, 722–728 (1962).

Seromucoids from sulphosalicylic acid filtrates of normal and pathological sera, concentrated by means of tannin and caffeine, were separated by paper electrophoresis into five fractions with mobilities corresponding to those of the total serum. The tannin micromethod, described in this paper, was used to determine the level of seromucoid in normal and pathological sera and the levels of the seromucoid electrophoretic fractions. Limits of normal ranges of individual seromucoid fractions and their content in serum proteins were established and their level in pathologic sera was found to be highly variable.

IMMUNOGLOBULIN LEVELS FROM THE NEWBORN PERIOD TO ADULT-HOOD AND IN IMMUNOGLOBIN DEFICIENCY STATES

West, C. D., R. Hong, and N. H. Holland, J. Clin. Invest. **41**, No. 11, 2054–2064 (1962).

The levels of the immunoglobulins in cord serum and in the serum of infants, children, and adults as determined by a method based on a combination of immunoelectrophoretic analysis and the quantitative precipitin reaction are described in this paper. The serum levels of the immunoglobulins in various immunoglobulin and antibody deficiency states, such as agammaglobulinemia, hypogammaglobulinemia, agammaglobulinemia with β_2 macroglobulinemia, isolated β_{2A} deficiency, various transient hypoimmunoglobulinemias, and the Aldrich syndrome, are also reported.

IMMUNOELECTROPHORETIC IDENTIFICATION OF GUINEA PIG ANTI-INSULIN ANTIBODIES

Yagi, Y., P. Maier, and D. Pressman, J. Immunol. **89**, No.5, 736–744 (1962).

Two anti-insulin antibodies from guinea pig sera, isolated by chromatography on diethylaminoethyl cellulose, were examined by immunoelectrophoresis and by radioautography with the aid of I^{131}-insulin and the rabbit antiserum against guinea pig serum components. This method for detecting antibodies is extremely sensitive and only a few microliters of the sample are necessary for the two antibodies to be clearly distinguishable.

AN IMMUNOELECTROPHORETIC ANALYSIS OF ANTIGENS REACTING WITH LUPUS SERUM

Atchley, W. A., Arthritis Rheumat. **4**, 471–479 (1961).

Serum samples from patients with systemic lupus erythematosus were studied by immunoelectrophoresis on cellulose-acetate membranes in order to analyze the antigen-antibody reactions in the serum. The antigens were obtained by extracting human leukocytes with 0.1 M glycine, which furnishes a preparation rich in deoxyribonucleic acid and associated proteins. Results of the study, described in this paper, showed that abnormal sera may contain at least two different precipitating antibodies, one directed against a deoxyribonucleohistone complex, and the other directed against an antigen that may be histone not attached to DNA.

IMMUNOLOGICAL PRECIPITATES IN AGAROSE GELS

Brishammar, S., S. Hjertén, and B. V. Hofsten, Biochim. Biophys. Acta **53**, 518–521 (1961).

Comparative studies of immunological precipitation reactions in ordinary agar and agarose gels are described in this report. When agarose was used, the strong electroendosmosis obtained in agar during immunoelectrophoresis is greatly reduced in the agarose gel, more distinct lines of precipitation of gel diffusion analysis of basic antigens were found, and no halo effects or false precipitation lines were observed.

IMMUNOELECTROPHORETIC ANALYSIS OF RAGWEED POLLEN EXTRACTS

Friedman, H., J. Spiegelman, M. A. Gershenfeld, and G. I. Blumenstein, J. Albert Einstein Med. Center **9**, 217–223 (1961).

The feasibility of using immunoelectrophoresis for analysis of the complex antigenic and allergenic constituents of ragweed pollen where other methods may not be as applicable is presented in this paper. Immunoelectrophoresis on cellulose-acetate membranes performed with whole ragweed pollen extract and rabbit anti-ragweed pollen sera revealed a pattern of at least 10 precipitin lines, indicating at least 10 antibody-antigen reactions across the electrophoretic membrane. Results of the study are evaluated.

COMBINAISON DE L'ÉLECTROPHORÈSE DE ZONE SUR AMIDON ET DE L'IMMUNOÉLECTROPHORÈSE DANS L'ÉTUDE DES PROTÉINES SERIQUES

Got, R., G. Levy, and R. Bourrillon, Clin. Chim. Acta **6**, 407–412 (1961) **French.**

By combining zone electrophoresis in starch for the separation of human and horse serum protein fractions and immunoelectrophoresis for their analysis, it is possible to determine the exact position of each serum protein on the starch blocks. In both types of electrophoresis, the γ-globulin and the serum albumin were obtained in the pure state, other proteins being in a heterogeneous condition. The experimental technique and the results are discussed.

IMMUNITY MECHANISMS IN NEUROLOGICAL DISEASE

McMenemey, W. H., Proc. Roy. Soc. Med. **54**, 127–136 (1961).

This paper is concerned with the globulins (the key to the clinical study of immunity), their significance, and the great importance of electrophoresis to neurologists. Specific topic areas include the protein pattern in cerebrospinal fluid as revealed by electrophoresis and the role of the choroid plexus in the formation of the fluid, transferred patterns in the cerebrospinal fluid in somatic disease and the value of serum patterns in the investigation of neurological disease, autochthonous patterns in the cerebrospinal fluid, multiple sclerosis, encephalitis, the possibility of 'overflow' patterns in the serum from brain disease, the significance and constancy of individual electrophoretic peaks in the serum and cerebrospinal fluid in terms of specific molecular weight with special reference to the γ-fraction, the roles of the plasma cell and the lymphocytic and the source of the antibody globulins in the cerebrospinal fluid, the β-globulins and the lipoproteins, brain proteins in the cerebrospinal fluid, and immunity mechanisms concerned in pathogenesis. Case studies are given in several of the areas under discussion and the role of electrophoresis in the investigations presented is reported.

ABNORMAL COMPONENTS OF PLASMA IN DISEASE: II. IMMUNOELECTROPHORETIC CHARACTERIZATION OF THE SERUM AND URINARY PROTEINS IN PLASMA CELL MYELOMA AND WALDENSTROM'S MACROGLOBULINEMIA

Osserman, E. F., and D. Lawlor, Ann. N.Y. Acad. Sci. **94**, 93–109 (1961).

This paper, reviewing immunoelectrophoretic analysis of the serum and urinary proteins of patients with plasma cell myeloma and primary Waldenstrom's macroglobulinemia, was presented at a conference on "Plasma Proteins in Health and Disease" held by the New York Academy of Sciences in 1961. Immunoelectrophoretic analysis is an extremely accurate and sensitive technique for the detection and characterization of these constituents and permits the classification of the myeloma serum globulins into two major immunological groups—those related to the major gamma fraction and those related to beta 2A. The demonstration of minute quantities of Bence-Jones protein in the serum of certain myeloma patients is made possible by the sensitivity of this technique. A three-reactant modification of it has permitted the precise identification of the several urinary proteins in cases of plasma cell myeloma with complex proteinuria, consisting of several different serum protein constituents. Serum immunoelectrophoresis was established as a valid substitute for ultracentrifugal analysis in the diagnosis of Waldenstrom's macroglobulinemia, and the serum immunoelectrophoretic pattern in primary macroglobulinemia was demonstrated to be specific and diagnostic.

RECURRENT BACTERIAL INFECTIONS AND DYSGAMMAGLOBULINEMIA: DEFICIENCY OF 7S GAMMA-GLOBULINS IN THE PRESENCE OF ELEVATED 19S GAMMA-GLOBULINS: REPORT OF TWO CASES

Rosen, F. S., S. V. Kevy, E. Merler, C. A. Janeway, and D. Gitlin, Pediatrics **28**, 182–195 (1961).

A report is given of two patients with recurrent bacterial infections associated with dysgammaglobulinemia, in whom there was a marked deficiency of 7S gamma-globulins in the presence of elevated concentrations of 19S gamma-globulins. The patients' sera were studied by several methods, including anti-typhoid O and H antibody titrations using a rapid slide test, isohemagglutinin titers, estimation of gamma-globulins, immunoelectrophoresis using 2.5 ml of agar in 0.1 molar borate buffer of pH 8.6, estimation of Forssman antibody, sucrose gradient ultracentrifugation for separation of 7S and 19S gamma-globulin, and analytical ultracentrifugation. These methods are described and the results of kidney, spleen, liver, lymph node and bone marrow biopsies, antibody studies, ultracentrifugal studies, and immunoelectrophoresis are reported.

IMMUNOELEKTROPHORETISCHE STUDIEN ZUR FRAGE DER IDENTITÄT MENSCHLICHER ORGAN-UND SERUMPROTEINE

Scheiffarth, F., H. Gotz, and G. Schernthaner, Clin. Chim. Acta **6**, 481–492 (1961) **German.**

Extracts of various human organs (liver, kidney, spleen, heart muscle, skeletal muscle and mucous membrane of the stomach) were compared with normal serum by means of immunoelectrophoresis in order to find out whether tissue proteins and serum proteins are identical, using the reaction to anti-human serum as the standard of comparison. The results, which indicated that each tissue extract examined possesses a specific protein spectrum with the protein components differing in number, amount and mobility, imply that in their behavior the tissue proteins differ not only from each other but also from those of normal serum. The immunological reactions were shown to be comparable to or identical with those of serum proteins in at least some of the proteins present in the extracts.

IMMUNOELECTROPHORESIS: METHODS, INTERPRETATION, RESULTS

Wunderly, C., Advan. Clin. Chem. **4**, 207–273 (1961).

The principle of immunoelectrophoresis; immunochemical reactions in agar gel, colloid chemistry of agar gel, and electrophoretic separation in agar gel; the technique; apparatus for macromethods, micro- and ultramicro- methods, preparation of gel and buffer, controls, run, diffusion of antiserum, and drying, staining of proteins, lipids and lipoproteins, photography, and interpretations; results of physiological applications; and biochemical applications are discussed in this paper.

ANALYSIS OF HUMAN LEUCOCYTE DEOXYRIBONUCLEOHISTONE BY IMMUNOELECTROPHORESIS

Atchely, W. A., Nature **188**, 579–581 (1960).

Immunoelectrophoresis on cellulose-acetate membranes, a laboratory technique found to be easily reproducible, was used to study the human leukemic leucocyte, with particular reference to the nucleus. The investigation, described in this report, constitutes a new approach to the problem made possible by a combination of histochemistry and cellulose-acetate electrophoresis. The findings are given.

THE THEORETICAL BASIS AND PRACTICAL EMPLOYMENT OF IMMUNO-ELECTROPHORESIS WITH SPECIAL REGARD TO SERUM PROTEINS

Clausen, J., Danish Med. Bull. **7**, No. 4, 88 (1960).

The theory and practical employment of immunoelectrophoresis as a diagnostic tool and as an instrument for continued exploration of proteins and their pathology is discussed. The protein pathology of man and mouse, concerning paraproteinemia, chronic stimulation (chronic infections and autoimmunisation), nephrosis, amyloidosis and hapatosis in mice, is described on the basis of the normal immunoelectrophoretical tracings.

IMMUNO-ELECTROPHORESIS AND AUTORADIOGRAPHY

Clausen, J., and T. Munkner, Protides Biol. Fluids, Proc. 8th Colloq., 147–151 (1960).

A combination of immunoelectrophoresis and autoradiography proved to be a convenient method for more extensive studies of the hitherto comparatively unknown area of the transport function of proteins in the biological fluids. The transport function of serum and cerebrospinal fluid of man and the serum of mouse, especially in regard to iron and thyroxine, was studied by these techniques and the results are reported in this paper.

ANTIGENIC RELATIONSHIPS BETWEEN THE GLOBULINS OF THE γ-SYSTEM

Heremans, J. F., Protides Biol. Fluids, Proc. 8th Colloq., 127–131 (1960).

Some of the common structural features of γ-globulin, β_{2A}-globulin, and β_2-macroglobulin, the proteins designated as "components of the γ-system," were deduced from double diffusion studies in Ouchterlony plates and by immuno-electrophoretic analyses.

IMMUNO-ELECTROPHORETIC DIFFERENTIATION OF HAPTOGLOBINS FROM ANOTHER GROUP-SPECIFIC INHERITABLE SYSTEM IN NORMAL HUMAN SERA

Hirschfeld, J., Nature **187**, No. 4732, 126–128 (1960).

A slightly modified technique of immunoelectrophoresis, which can characterize at least 25 precipitating components in a single human serum and seems to give better resolution of the components, is presented in this paper. Simple absorption techniques enable demonstrating the group-specific components and provide that an immune serum is available which contains precipitating anti-bodies against these components.

IMMUNOELECTROPHORETIC ANALYSIS OF ANTI-COMPLEMENT SPECIFIC-ITY IN IMMUNE ANTISERUM

Peetoom, F., and K. W. Pondman, Protides Biol. Fluids, Proc. 8th Colloq., 298–301 (1960).

Immune aggregates with bound complement evoke antibodies against the complement in rabbits, and in experiments to demonstrate the specificity of the antiserum, evidence indicated that the protein involved was a specific antibody against C_4. Immunoelectrophoretic experiments on fresh and old human serum and antiserum showed that one precipitation curve developed with the first serum and a double curved line with the second, and seemed to indicate that the precipitation reaction is dependent on antibodies against the complement. Results of the immunoelectrophoretic studies are discussed.

RADIOIMMUNOELECTROPHORESIS

Rejnek, J., and T. Bednarík, Clin. Chim. Acta **5**, 250–258 (1960).

By combining the methods of immunoelectrophoresis and isotopic labelling of compounds, a simple and sensitive technique for the investigation of the fate of proteins in complicated protein systems, was developed. This highly sensitive method can identify a number of proteins and differentiate proteins of varied origin. Hence, it can contribute to the solution of problems concerned with protein metabolism. The method and its application are described in this report.

IMMUNOELECTROPHORESIS

Williams, C. A., Jr., Sci. Am. **202**, 130–140 (1960).

The technique of immunoelectrophoresis, a great analytical tool which dis-tinguishes proteins from each other by first segregating them according to their electrophoretic mobility and then identifying them by the highly specific immune reaction, is evaluated in this paper. Other methods are reviewed, including moving-boundary electrophoresis developed by Tiselius and the modifications of zone electrophoresis devised by Oudin and Ouchterlony, which contributed to the development of immunoelectrophoresis.

THE IMMUNOGLOBULIN CONCEPT

Heremans, J. F., Clin. Chim. Acta **4**, 639–646 (1959).

The investigation reported in this paper is concerned mainly with the proteins constituting the classical γ peak of conventional electrophoresis, and with those which can be identified in the depression lying between this peak and the

β area. A review of the components of the β_2-γ area, including immunoelectrophoretic data and the pathology of the immunoglobulins is given. Three serum proteins, γ-globulin, β_{2A}-globulin and β_2-macroglobulin, selected for this study because of their close chemical, immunological and functional relationships, appear to be carriers of antibody activity and are termed immunoglobins. All three immunoglobins increase markedly in serum in chronic infections, collagen diseases and cirrhosis and hepatitis, which may be correlated to antibody synthesis. A fall in at least one of the three immunoglobins characterizes diseases associated with immune paralysis (primary and secondary agammaglobulinemia and biologically related syndromes).

ÉTUDES SUR LES PROTÉINES DU LIQUIDE CÉPHALO-RACHIDIEN

Burtin, P., Protides Biol. Fluids, Proc. 6th Colloq., 212-218 (1958) **French.**

The determination of total proteins and globulins in the cerebrospinal fluid by the biuret method and by an immunochemical method respectively, and immunoelectrophoretic studies of the proteins are described. It was possible to compare various methods of concentrating cerebrospinal fluid, as well as distinguish and sometimes identify the constituents, by these studies.

AN IMMUNO-ELECTROPHORETIC TECHNIQUE

Kohn, J., Nature **180**, 986–987 (1957).

An immunoelectrophoretic technique is described, which is comparatively simple, reliable, and very economical in the use of the antiserum, and has the added advantage of being able to separate very distinctly the alpha-1 fraction. By adjusting the current time, buffer concentration and pH and fluid levels in the buffer compartments, the electrophoretic pattern can be controlled according to requirements. The technique is a combination of cellulose-acetate filter membrane electrophoresis and the agar diffusion method. Application of this procedure to equine anti-human serum and the results are reported.

MÉTHODE IMMUNO-ÉLECTROPHORÉTIQUE D'ANALYSE DE MÉLANGES DE SUBSTANCES ANTIGÉNIQUES

Grabar, P., and C. A. Williams, Jr., Biochim. Biophys. Acta **17**, 67–74 (1955) **French.**

The technical details are described and the possibilities discussed of an electrophoretic method in which the substance to be studied is carried out in a 1.5 to 2 percent agar gel in a Veronal buffer solution of $\Gamma/2=0.05$ and with a drop in potential of 3 to 4 volts/cm in the gel. The precipitating immune serum is diffused perpendicularly to the electrophoretic migration axis. An independent specific precipitation band, which can be distinguished by its relative electrophoretic mobility, is given by each constituent of the mixture studied.

IMMUNOELECTROPHORETIC STUDIES ON SERUM PROTEINS: I. THE ANTI-GENS OF HUMAN SERUM

Williams, C. A., Jr., and P. Grabar, J. Immunol. **74**, 158–168 (1955).

An analytic method combining zone electrophoresis in gelified media with immunochemical analysis in gels is described and the results obtained by its use on normal human and horse serum are discussed. This immunoelectrophoretic method is of value in the study of normal antigen mixtures and of pathological and other modifications of these antigens, and is of interest for the control of fractionation of such mixtures.

IMMUNOELECTROPHORETIC STUDIES ON SERUM PROTEINS: II. IMMUNE SERA: ANTIBODY DISTRIBUTION

Williams, C. A., Jr., and P. Grabar, J. Immunol. **74**, 397–403 (1955).

The application of an immunoelectrophoretic method to the rapid characterization of antiserum types and the demonstration of the electrophoretic distribution of antibodies in such sera are reported in this paper. Immunoelectrophoresis can be used to determine the mobility of specific precipitating antibodies and to relate these to normal components of the immune serum, and is of value for the control of the separation and purification of antibody rich serum fractions. In this study, rabbit antisera against normal and immune horse serum, normal human serum, and diphtheria toxoid were used and the results are described.

IMMUNOELECTROPHORETIC STUDIES ON SERUM PROTEINS: III. HUMAN GAMMA GLOBULIN

Williams, C. A., Jr., and P. Grabar, J. Immunol. **74**, 404–410 (1955).

Immunoelectrophoretic analysis was applied to the study of human serum antigens in general and to the characterization of antisera to help clarify the nature of human γ-globulin. Antigenic homogeneity, or at least a high degree of antigenic correspondence over a broad range of electrophoretic mobilities, was suggested by immunoelectrophoretic analysis by horse or rabbit antibodies of human γ-globulin. The materials, methods, and results of the study are described and the significance of serum complexes in relation to the nature and function of γ-globulin is discussed.

Moving Boundary — Free Flow
Electrophoresis

APPLICATION OF PAPER IONOPHORESIS ELECTROCHROMATOGRAPHY TO THE STUDY OF METAL COMPLEXES IN SOLUTION

Blasius, Von E., and W. Preetz, Chromatog. Rev. **6**, 191–213 (1964).

High-voltage paper ionophoresis and electrochromatography were believed to be important methods for studying complex equilibria, particularly when a determination of the composition and structure of the ions becomes desirable in connection with their isolation. The general principles of high-voltage paper ionophoresis, the apparatus used, and its application to the separation and identification of the hydrolysis products of chloro-complexes and to the separation of mixed halogen complexes are presented in this paper.

PROTEIN COMPOSITION OF EARLY AMNIOTIC FLUID AND FETAL SERUM WITH A CASE OF BIS-ALBUMINEMIA

Brzezinski, A., E. Sadovsky, and E. Shafrir, Am. J. Obstet. and Gynecol. **89**, No. 4, 488–494 (1964).

In a continuing effort to determine the origin of human amniotic fluid proteins, usually done by comparing the electrophoretic distribution of the protein components at term, this study relates the protein distribution in maternal serum to that of fetal serum and amniotic fluid in early gestation. Maternal and fetal blood samples were obtained from 14 women whose pregnancy had been interrupted by abortion, hysterotomy or premature labor. Moving boundary electrophoresis apparatus was used to determine the protein composition of early amniotic fluid and the corresponding fetal and maternal serum. The results seemed to affirm the theory that fetal circulation is the main source of amniotic fluid proteins.

PRECAUTIONS REQUIRED IN INTERPRETATION OF MOVING BOUNDARY AND ZONE ELECTROPHORETIC PATTERNS

Cann, J. R., and W. B. Goad, Arch. Biochem. Biophys. **108**, No. 1, 171–172 (1964).

In discussing the electrophoretic patterns of purified proteins, the author stresses that in the study of the different molecular forms of enzymes and other biologically important macromolecules and their subunits, cognizance must be taken of the fact that multiple peaks or zones do not necessarily indicate heterogeneity and care must be taken to rule out interactions. He does feel, however, that a fundamental understanding of the electrophoretic transport of interacting systems, discussed in the article, makes electrophoresis a powerful method for investigating biological systems.

DÜNNSCHICHTIONOPHORESE ANOIGANISCHER STOFFE BEI NIEDERUND HOCHSPANNUNG UNTER VERWENDUNG VON RADIONUKLIDEN

Moghissi, A., Anal. Chim. Acta **30**, 91–95 (1964) **German.**

Thin-layer ionophoresis, used for separating inorganic substances, was carried out at 13 volts/cm or 45 volts/cm using radionuclides, and high-voltage ionophoresis was conducted either under cooling or in an atmosphere of water vapor.

THE AMINO ACID SEQUENCE AROUND THE REACTIVE SERINE RESIDUE OF SOME PROTEOLYTIC ENZYMES

Naughton, M. A., F. Sariger, B. S. Hartley, and D. C. Shaw, Biochem. J. **77**, 149–163 (1960).

The products of the partial acid hydrolysis of the diisopropoxy [^{32}P] phosphinyl derivatives of chymotrypsin, trypsin, and elastase were investigated after ionophoretic fractionation. A comparison of the radioactive peptides in the hydrolysates of the diisopropoxy [^{32}P] phosphinyl derivatives of the three enzymes revealed that elastase, like trypsin and chymotryspin, contains the sequence Asp-Ser-Gly around its reactive serine residue. The procedures used in the experimentation and the results obtained are described.

EFFECT OF BORATE BUFFER ON THE ELECTROPHORESIS OF SERUM

Cooper, D. R., Nature **181**, 713–714 (1958).

Some investigations are reported in which the protein components present in neutral or slightly alkaline extracts of bovine hide were compared with those of bovine serum by paper electrophoresis. A borate buffer (pH=8.6, I=0.05) was used and the papers were treated with Bromophenol Blue to detect the protein bands. Several runs revealed that the intense band corresponding to albumin was preceded by a less intense band that migrated just faster than the albumin, and a similar effect was also obtained with two samples of normal rat serum. No such effect was observed with the bovine serum when barbiturate buffer (pH=8.6, I=0.05) was used instead of the borate buffer, indicating that the borate buffer must increase the mobility of one or more of the components of the serum. Moving-boundary electrophoresis of bovine serum in barbiturate buffer further proved the effect of the borate ion on serum proteins.

AMYLASE IN ELECTROPHORETIC AND ULTRACENTRIFUGAL PATTERNS OF HUMAN PAROTID SALIVA

Patton, J. R., and W. Pigman, Science **125**, 1292–1293 (1957).

A study on the location of amylase in ultracentrifugal and electrophoretic patterns, which is a part of a more extensive investigation of the composition of human parotid and submaxillary gland secretions by electrophoretic and ultracentrifugal methods, is presented in this paper. α-Amylase, with an average electrophoretic mobility of -1.4 x 10^{-5}cm^2/volt-sec. and a sedimentation rate ($S_{20, w}$) of 4.1 Svedberg units in the Miller-Golder buffer of 0.1 ionic strength and pH 8.5, was located in both the electrophoretic and ultracentrifugal patterns of human parotid gland secretion.

A NEW HEMOGLOBIN VARIANT EXHIBITING ANOMALOUS ELECTROPHORETIC BEHAVIOR

Schneider, R. G., and M. E. Haggard, Nature **180**, 1486–1487 (1957).

The relative mobilities of the better-known hemoglobin variants in moving-boundary electrophoresis, cacodylate buffer, pH 6.5, μ0.1, are the same or only slightly different from those in paper electrophoresis, Veronal buffer, pH 8.6 μ0.05. Hemoglobin F, however, exhibits some discrepancy in behavior in free electrophoresis, as compared to paper electrophoresis. This paper deals with a hitherto undescribed type of hemoglobin, obtained from a Negro obstetric patient and one of her children, which has a completely different resolution in moving-boundary electrophoresis at pH 6.5 (cacodylate buffer) than it had in paper electrophoresis at pH 8.6 (Veronal buffer). This hemoglobin variant was tentatively designated 'Galveston type'.

Thin-Layer Electrophoresis

THIN-FILM ELECTROPHORESIS: I. THE ELECTROPHORETIC BEHAVIOR OF COAL TAR FOOD COLORS ON PAPER AND THIN-FILMS

Criddle, W. J., G. J. Moody, and J. D. R. Thomas, J. Chromatog. **16**, 305–359 (1964).

The common coal tar food colors permitted in the United Kingdom were separated by thin-layer electrophoresis on Whatman® No. 1 paper, Kieselguhr, alumina, and silica gel. The electrophoretic separation was achieved at 200 volts in 1N, 4.0N, 6N and 9.2N acetic acid as well as 0.1N ammonia. A high degree of resolution was possible on each medium.

USE OF THIN FILMS FOR ELECTROPHORESIS OF COAL-TAR FOOD COLOURS

Criddle, W. J., G. J. Moody, and J. D. R. Thomas, Nature **202**, No. 4939, 1327 (1964).

The authors describe a technique of electrophoresis with thin films of alumina, Kieselguhr, and silica gel as supporting absorbents used in the study of electrophoretic behaviour of coal-tar food colors.

NUCLEOTIDES: SEPARATION FROM AN ALKALINE HYDROLYSATE OF RNA BY THIN-LAYER ELECTROPHORESIS

De Filippes, F. M., Science **144**, 1350–1351 (1964).

Alkaline hydrolysis of RNA gives mixtures of 2′, 3′ nucleoside monophosphates, which can be separated from RNA mononucleotides by electrophoresis on thin layers of cellulose. Forty grams of MN cellulose powder were washed in 1N NaOH and 0.1N HCl alternately with water and brought to a final volume of 300 ml. Plates were made on a Desaga mounting board and dried at 105°C for 15 minutes. Ten μl of the sample was streaked onto the plate. A pH 3.4 sodium formate buffer, ionic strength 0.1, was used as the electrolyte for electrophoresis (450 volts, 75 minutes) and for soaking the paper wicks. The nucleosides were identified by μV light and eluted from the cellulose by 0.05M Tris–HC1 buffer, pH 8.2, at 50°C for ten minutes.

THIN-LAYER TECHNIQUES FOR MAKING PEPTIDE MAPS

Ritschard, W. J., J. Chromatog. **16**, 327–333 (1964).

A method is described for the preparation of peptide maps which is a combination of chromatography and electrophoresis for the two-dimensional separation of peptides using thin layer techniques instead of employing paper sheets. In short preliminary runs, the optimal experimental conditions (solvent systems, buffers, time and sample concentrations) were explored and chosen for the preparation of the peptide maps. Using this method, eight peptide maps are easily prepared per day.

FAST FINGERPRINTING ON THIN LAYERS WITH POLYACRYLAMIDE AS DIAPHRAGM

Stegemann, H., and B. Lerch, Anal. Biochem. **9**, 417–422 (1964).

An apparatus was constructed for a fast electrophoretic technique of fingerprinting. Paper or thin layers could be used as electrophoretic support media. Cellulose powder MN 300 containing silica gel H was the material used to prepare thin layers. Polyacrylamide gel poured into the cathode chamber was used as diaphragm and connected directly to the thin-layer plate. This technique was applied to the separation of tryptic digests of human and horse hemoglobin. A pH 5.2 buffer composed of pyridine–acetic acid–water (20:9.5:970) was used to effect separation. By applying 60 volts/cm for three to four hours, a completed fingerprint was possible.

THE QUANTITATIVE SEPARATION OF PERIODATE AND IODATE BY THIN-LAYER ELECTROPHORESIS ON STRIPS OF PLASTER OF PARIS

Dobici, F., and G. Grassini, J. Chromatog. **10**, 98–103 (1963).

The quantitative electrophoretic separations of milligram amounts and trace amounts of periodate and iodate are reported. The procedure was carried out on thin layers of plaster of paris with 0.05M ammonium carbonate as electrolyte with 300 to 400 volts for one and one-half to two hours.

Miscellaneous Methods

PROTEIN FRACTIONATION BY ZONE ELECTROPHORESIS ON PEVIKON® C-870: II. SEPARATION OF HUMAN SERUM PROTEINS

Bocci, V., Arch. Biochem. Biophys. **104**, 514–523 (1964).

Following fractionation of human serum proteins by electrophoresis on a Pevikon® C-870 (PVK) block, the protein fractions were recovered almost quantitatively from PVK segments and were measured for their radioactive, proteic, I^{131}-TCA- and I^{131}-TPA-soluble radioactivity contents. The electrophoretic mobilities and homogeneity of the protein fractions were characterized on agar-gel and starch-gel electrophoresis and double diffusion in agar gel. Human γ-globulin, albumin, prealbumin, and orosomucoid were isolated and, according to the tests in the study here reported, appeared homogeneous.

ALBUMIN-GEL AS A SUPPORTING MEDIUM IN ZONE ELECTROPHORESIS

Bocci, V., Experientia **20**, 234–235 (1964).

Colored proteins could be followed by electromigration of proteins in an albumin-gel, but detection of the colorless proteins was impossible with the usual staining method because of the supporting medium. The use of [131]-labelled proteins and detection of them by autoradiography resolved the difficulty. Investigation was then made on the separatory power for serum proteins of albumin-gel electrophoresis. Either albumin-gel electrophoresis or starch-gel electrophoresis was used to separate samples of guinea pig, rabbit and human serum and of human γ- and β-globulins labelled with [131]I. The methods and results are described.

SOLUBLE LIVER PROTEINS FRACTIONATION BY ZONE ELECTROPHORESIS ON PEVIKON® C-870

Bocci, V., Nature **203**, No. 4946, 775–777 (1964).

The use of Pevikon® C-870 (PVK) as a supporting medium for electrophoresis has numerous advantages. The results of an analysis of soluble proteins in rabbit liver, separated by PVK-block electrophoresis, are reported in this paper. Starch-gel electrophoresis was used to test the protein fractions after concentration and clarification. Findings showed that the resolution of the liver proteins in such a high number of fractions was due, at least in part, to the possibility of large loading of protein in the PVK-block.

ZONE ELECTROPHORESIS IN POLYVINYLCHLORIDE METHODOLOGICAL STUDIES

Böttiger, L. E., and R. Norberg, Clin. Chim. Acta **9**, 82–86 (1964).

The determination of electrophoretic separation in polyvinylchloride of proteins and protein-bound carbohydrates (hexoses, hexosamines, and sialic acids) and the evaluation of the components compared between zone and moving boundary electrophoresis (Tiselius method) are described. The reproducibility was very good for zone electrophoresis in polyvinylchloride, and the mean analytical error, calculated on the basis of duplicate analyses of five normal and five pathological sera, was six percent for serum proteins and eight percent for protein-bound carbohydrates. The results are very similar in the quantitative evaluation of the components for free electrophoresis and for zone electrophoresis.

POLYMORPHISM IN THE SERUM PROTEINS OF THE REINDEER

Braend, M., Nature **203**, No. 4945, 674 (1964).

Results from investigations of serum samples from 132 reindeer, using an electrophoretic method developed for cattle transferrins, are reported in this paper. It was concluded that the proteins dealt with in the study were transferrins and a genetic theory of six transferrin alleles, Tf^C, Tf^E, Tf^G, Tf^I, Tf^K, and Tf^M, was proposed.

PARTIAL PURIFICATION OF NUCLEASES FROM GERMINATING GARLIC

Carlsson, K., and G. Frick, Biochim. Biophys. Acta **81**, 301–310 (1964).

The authors describe the preparation and purification of a raw extract, with DNAase and RNAase activities, obtained from garlic bulbs. Subsequent fractionation, showing the complexity of the extract, was obtained through gel filtration and electrophoresis after the removal of the carbohydrate present in the raw extract. Some characteristics of the different fractions are given.

IMMUNOCHEMICAL STUDIES ON THE ANTIGENS OF BORDETELLA PERTUSSIS

Griffiths, B. W., and M. A. Mason, Can. J. Microbiol. **10**, 123–138 (1964).

A detailed description is given of chromatographic fractionation of *Bordetella pertussis* extracts on DEAE-cellulose and the analysis of mouse protective antigen (MPA), histamine-sensitizing factor (HSF), and precipitinogens (P) associated with the various fractions. A dissimilarity in the chemical natures of MPA and HSF is suggested by certain discrepancies between them based on the lack of proportionality of the activities from fraction to fraction.

CHARACTERISTICS OF SOLUTION FLOW IN SUPPORTING MEDIA IN THIN LAYER ELECTROPHORESIS IN MOIST CHAMBERS

Kowalczyk, J., J. Chromatog. **14**, 411–419 (1964).

Solution flow for different support media was characterized for thin-layer electrophoresis performed in moist chambers. Two systems were defined for solution support medium interactions. They were systems with hydrostatic equilibrium in the medium and systems having no hydrostatic zones. Systems which have no hydrostatic zones were of greater advantage.

PROCESSES ACCOMPANYING THE ELECTROPHORESIS IN CARRIER LAYERS: VI. THE INFLUENCE OF PROCESSES OCCURRING IN CARRIERS ON THE PARTICLES MOVEMENT IN THE ELECTRIC FIELD

Kowalczyk, J., Chem. Anal. **9**, 213–222 (1964).

Three types of carrier-solution systems in electrophoresis were distinguished: 1) systems in liquid and plate chambers; 2) moist chamber systems without π regions of hydrostatic equilibrium in the carrier layers; and 3) moist systems with the π region. The third system is difficult to control but is rarely met in practice.

MEASUREMENT OF FREE AMINO ACIDS IN PLASMA AND SERUM BY MEANS OF HIGH VOLTAGE PAPER ELECTROPHORESIS

Mabry, C. C., and E. A. Karam, Am. J. Clin. Pathol. **42**, 421–430 (1964).

A rapid method for separation of amino acids in serum is described. High-voltage paper electrophoresis was performed in a low ionic strength buffer made of 90% formic acid–glacial acetic acid–distilled water (6:24:170, pH 2.0). Electrophoresis was performed at 1800 (30 volts/cm; 6 mA/cm was used to effect separation. The electrophoretic run was for a period of 120 minutes at a temperature of 8–10°C. Quantitation and detection of the amino acids was possible by the use of selective staining reagents. Ten to 200 mμ moles of amino acids were detectable using this method.

THE PROTEOLYTIC ENZYMES OF ASPERGILLUS ORYZAE: II. PROPERTIES OF THE PROTEOLYTIC ENZYMES

Bergkvist, R., Acta Chem. Scand. **17**, No. 6, 1541–1551 (1963).

This is a presentation of the properties of three proteolytic enzymes which have been confirmed by electrophoresis, pH optimum, temperature optimum, substrate specificity, stability under different conditions, and effect of metal ions and other potential inhibitors. Though small amounts of the different proteases could be separated by continuous electrophoresis, the best results were obtained by the use of ion exchange cellulose.

ELECTROPHORETIC BEHAVIOUR OF RABBIT SERUM AMYLASE

Berk, J. E., M. Kawaguchi, R. Zeineh, I. Ujihira, and R. Searcy, Nature **200**, No. 4906, 372–373 (1963).

Electrophoretic patterns were established for normal rabbits for the purpose of elucidating the amylase distribution in the serum, a study previously carried out on normal serum of humans, dogs, mice and rats. Studies were also made on recrystalized hog pancreatic amylase diluted to the same saccharogenic activity of rabbit serum, and on diluted hog pancreatic amylase-normal rabbit serum mixtures. The techniques used and the findings are reported in this paper.

STARCH-PEVIKON® C-870 GEL AS A SUPPORTING MEDIUM IN ZONE ELECTROPHORESIS

Bocci, V., J. Chromatog. **11**, 515–523 (1963).

Pevikon® C-870 (PVK), a gel composed of hydrolysed starch and powdered co-polymer of polyvinyl chloride and polyvinyl acetate, is proposed as a supporting medium for the electrophoretic separation of serum proteins and soluble tissue proteins. The procedure for eluting proteins from starch-PVK gel is very simple and recoveries are much higher than from starch gel, and when proteins are recovered from starch PVK gel the ^{131}I-trichloroacetic acid soluble radio-activity is consistently lower. The estimation of the specific activity of ^{131}I-labelled proteins is not affected by the presence of soluble starch. The experiments, in which rabbit and human serum proteins were used, are discussed.

AEROBIC SPORULATING BACTERIA: I. GLUCOSE DEHYDROGENASE OF BACILLUS CEREUS

Bach, J. A., and H. L. Sadoff, J. Bacteriol. **83**, No. 4, 699–707 (1962).

Extracts of cells or spores of *Bacillus Cereus,* which had been grown at 30°C in G medium, were used in experiments to identify and purify an enzyme system which could be used as a model in the study of spore heat resistance. A heat resistant glucose hydrogenase occurs in cultures of *B. cereus* in the initial stages of sporulation, but is absent from logphase vegetative cells. In order to determine which properties the proteins had in common, an attempt was made to compare the vegetative and spore dehydrogenases by a variety of readily measured parameters. Identity lines were produced by both enzymes in two-dimensional immunodiffusion experiments, and their behavior in chromatographic and electrophoretic studies was also identical. They differed, however, in heat stability, with the vegetative hydrogenase possessing an inactivation constant only one-fifteenth that of the spore preparation. It is probable that the vegetative glucose hydrogenase is truly spore protein, but this cannot be stated conclusively. Because this work was of a preliminary nature, not much importance should be given to the differences between the stable and labile dehydrogenases.

ELECTROPHORETIC METHOD FOR CHARACTERIZING THYROID FUNCTION

Berger, S., M. S. Goldstein, and B. E. Metzger, New Engl. J. Med. **267**, No. 16, 801–805 (1962).

Plasmas of hyperthyroid, hypothyroid, and euthyroid subjects, enriched with radiothyroxine to yield three different levels of total thyroxine, were electrophoresed in barbital buffer, at pH 8.6. The procedure and results are described. The various disorders of thyroid function can be distinguished by this *in vitro* technique with a consistency necessary for a routine diagnostic test. This technique is a particularly suitable complementary parameter to the plasma protein-bound-iodine level, is reliable under circumstances that becloud other parameters, and provides clues to disordered plasma thyroxine-binding capacities as well as to abnormal thyroid function.

CRITERIA OF THE PURITY OF PROTEINS

Lontie, R., Protides Biol. Fluids, Proc. 9th Colloq., 1–10 (1961).

Zone electrophoresis and cellulose ion-exchange chromatography appeared to be the most powerful tools to reveal small differences in the purity of proteins, and immunological methods are indicated for the detection of trace contaminants.

ULTRAMICRO METHODS: ZONE ELECTROPHORESIS

van Haga, P. R., and J. de Wael, Advan. Clin. Chem. **4**, 339–341 (1961).

Included in this paper on ultramicromethods for clinical chemical analysis is a section on zone electrophoresis. Paper electrophoresis using filter paper and cellulose acetate as a medium, and the use of agar gel as a supporting medium in electrophoresis, are discussed and evaluated.

AMINOACIDURIA

Woolf, L. I., Brit. Med. Bull. **17**, No. 3, 224–229 (1961).

High- and low-voltage electrophoresis have been proven useful in the analysis of amino acid, as have paper chromatography, column chromatography on ion-exchange resins, and microbiological assay of single amino acids. Discussed in this paper are concentrations of amino acids in blood and urine and the genetically determined aminoacidurias which include phenylketo-NURIA, maple-syrup urine disease, hypophosphatasia, galactosemia, Wilson's disease, cystinosis, β-aminoisobutyric aciduria, cystathioninuria, argininosuccinic aciduria, Hartnup syndrome, benign familial aminoaciduria, glycinuria, Lowe-Terrey-MacLachlam syndrome, congenital hepatic and renal dysfunction, De Toni-Debré-Fanconi syndromes, and osteomalacia with amino-aciduria.

ZONE ELECTROPHORESIS OF ANTHOCYANINS

Markakis, P., Nature **187**, No. 4743, 1092–1093 (1960).

The typical anthocyanins of sour cherries, strawberries, plums and roses were found to move to the anode or the cathode, depending on the pH of the electrolyte solution, when placed within an electric field applied across filter paper or cellulose powder.

LE COMPORTEMENT ÉLECTROPHORÉTIQUE DES PHTALÉINES ISOLÉES ET EN PRÉSENCE DE PROTIDES

Delcourt, R., Protides Biol. Fluids, Proc. 6th Colloq., 121–126 (1958) **French.**

The factors involved in the formation of protein-phthalein complexes is described. The presence of a halogen atom in the phthalein is necessary for the formation of these complexes, and the affinity of the halogenated phthaleins for the protein depends on the nature of the halogen (Cl $<$ Br $<$ I). Iodophthalein, therefore, is more strongly attached to the protein and is able to displace bromo- and chlorophthaleins from their protein complexes.

NEW TRENDS IN TWO-DIMENSIONAL ELECTROPHORESIS

Peeters, H., and P. Vuylsteke, Protides Biol. Fluids, Proc. 6th Colloq., 53–58 (1958).

A new type of cellulose substrate was developed in which the buffer flow rate and the adsorption and diffusion during fractionation are determined by the quality of the supporting medium, in an attempt to improve the techniques of two-dimensional electrophoresis. Preparation of the new medium and results of its application on human serum are described.

ELECTROPHORETIC DEMONSTRATION OF A NON-HEMOGLOBIN PROTEIN (METHEMOGLOBIN REDUCTASE) IN HEMOLYSATES

Lonn, L., and A. G. Motulsky, Clin. Res. Proc. **5**, 157 (1957).

A previously unreported electrophoretic component was discovered in studies on the heterogeneity of normal human hemoglobin when hemolysates (about 10 percent hemoglobin concentration) were subjected to paper electrophoresis at acid pH and stained with Bromophenol Blue. Characteristics and the electrophoretic mobility of the component are discussed, and experiments which ruled out the possibility of the fraction being a hemoglobin are presented. Results of these studies indicated that electrophoretic techniques may be useful for demonstrating non-hemoglobin enzyme proteins of red cells, and that rigid criteria should be satisfied before labelling small electrophoretic components of hemolysates as heterogeneous hemoglobins.

ELEKTROPHORESE MIT ISOLIEITEN PLASMAPROTEINEN

Schultze, H. E., Protides Biol. Fluids, Proc. 5th Colloq., 24–33 (1957)
German.

The use of isolated plasma proteins of defined properties in electrophoretical studies proved suitable to make clear the influences on the mobility, depending on the different methods of electrophoresis (free and zone electrophoresis); the influence of microheterogeneities or of small admixtures of accompanying substances on the electrophoretic behavior of the plasma proteins; and the degree of mutual interaction of the plasma proteins in the various methods of electrophoresis.

NONDIALYZABLE MATERIAL IN NORMAL HUMAN URINE

Hamerman, D., F. T. Hatch, A. Reife, and K. W. Bartz, J. Lab. Clin. Med. **46**, 848–856 (1955).

In studies on the chemical nature of high molecular weight material in normal human urine, electrophoretic studies made on that portion of the total colloids soluble in saline demonstrated mobilities closely resembling those of albumin and several globulin components of plasma. This paper reports the results of analytic and chromatographic studies on normal human urine which had been dialyzed, concentrated at low temperature in vacuo, and lyophilized to obtain nondialyzable urine solids.

Books, Reviews and Symposia

HEMOGLOBINS OF PRIMATES

Buettner-Janusch, J., and V. Buettner-Janusch, Evolutionary and Genetic Biology of Primates, edited by Buettner-Janusch, John. Academic Press, New York (1964). Vol. II, Chapter 10, 75–90.

A study was made on the evolution and genetics of the hemoglobin molecule as one aspect of the biochemical genetics of the Primates and their closely allied taxonomic relatives, the Insectivora. Investigations were made into the variations in hemoglobin between genera and species by means of starch-gel electrophoresis. The materials and methods used, the differentiation of hemoglobin by electrophoresis, the alkali-resistant hemoglobins, and perspectives of the study are presented.

TECHNIQUES IN CHEMICAL PATHOLOGY

Cheyne, G. A., F. A. Davis Company, Philadelphia (1964).

This book, which is aimed at the trained technician in a medical laboratory, is intended to be a practical laboratory guide for the handling and analysis of biological samples. The contents of the book include a refresher course in basic inorganic chemistry; notes on chemical and physical separation; chemico-physical methods of analysis; instrumentation in the laboratory; measurement of weight, volume, temperature and concentration; collection and preservation of specimens; the spectroscopic, microscopic, and biochemical examination of urine; quantitative urine analysis; renal efficiency tests; estimation of enzymes in serum; enzyme tests in pancreatic disease; tests of liver function; estimation of vitamins; calcium in serum; nitrogen metabolism; metabolic balance experiments; basal metabolic rate; glucose metabolism; hemoglobin in blood; iron and iron-binding capacity of serum; hormones in urine; tests of gastric function; analysis of cerebrospinal fluid; acid-base metabolism; and special techniques.

IMMUNO-ELECTROPHORETIC ANALYSIS: APPLICATIONS TO HUMAN BIO-LOGICAL FLUIDS

Grabar, P., and P. Burtin, eds. Elsevier Publishing Company, Amsterdam (1964).

This monograph presents methods of immunoelectrophoresis, applications of immunoelectrophoresis to human plasma, and studies on other human media. This technique provides two means of investigation: the definition of a substance by its electrophoretic mobility based on electrochemical properties, and by its immunochemical specificity based on structural properties. A double or sometimes even triple definition of a substance in a single operation can be accomplished by immunoelectrophoretic analysis. Only a very small amount of the material is needed for the analysis and it may not be subjected to any preliminary treatment because of the danger of damaging alteration. The method has been used in a wide variety of studies, the greatest number being devoted to the analysis of human biological products.

THE APPLICATION OF ELECTROPHORESIS IN RADIOCHEMISTRY

Konrad-Jakovec, Z., Sb. Ref. Cetostatni Radiochem. Knof. **1964**, 74–103.

The applications of paper electrophoresis in radiochemistry are emphasized in this review. Uses of electrophoresis were discussed; literature and laboratory data were presented.

QUANTITATIVE IMMUNOPRECIPITATION FOR MEASUREMENT OF SINGLE PLASMA PROTEINS

Shwick, H. G., Colloq. Ges. Physiol. Chem. **15**, 55–72 (1964).

This is a discussion of the many trace proteins possessing specific biological activity which can be isolated by polyacrylamide, starch-gel, and paper electrophoresis. Accurate quantitative methods have been extensively developed for paper.

THE AGAR PRECIPITATION TECHNIQUE AND ITS APPLICATION AS A DIAGNOSTIC AND ANALYTICAL METHOD

Peetoom, F., Charles C. Thomas, Springfield, Ill. (1963).

This monograph is designed to acquaint the reader with immunoelectrophoretic analysis, now a routine technique in hospital laboratories; to give clinicians a detailed description of all the pathological abnormalities published to date, which may facilitate diagnosis in certain cases; and to offer information for the biologist and research worker which will be useful or which will stimulate further research. It opens with an introduction to precipitation reaction and agar precipitation techniques, followed by chapters on immunoelectrophoresis, immunoelectrophoretic pattern of plasma proteins, immunoelectrophoretic examination of various patients' sera, leukocyte antigens in human plasma, and other applications of the agar precipitation technique. A table giving a summary of serum protein abnormalities which have diagnostic significance is included.

CYANOGUM® GEL ELECTROPHORETIC STUDIES OF SERUM AND SYNOVIAL FLUID: PRELIMINARY REPORT

Hermans, Paul E., William P. Beetham, Jr., Warren F. McGuckin, Bernard F. McKenzie, Proc. Mayo Clinic Staff Meetings **37**, 311–312 (1962).

Electrophoresis in Cyanogum® gel, which permits the separation of the proteins into many components for specific analysis, was used in a preliminary investigation of the proteins present in the synovial fluid and serum of four normal persons and of eight patients with rheumatoid arthritis. Results of the study are reported.

IMMUNODIFFUSION

Crowle, A. J., Academic Press, Inc., New York (1961).

This monograph on the general subject of immunodiffusion contains ample references to immunoelectrophoresis with acrylamide gels and to the use of acrylamide gels in immunodiffusion analysis.

MULTIPLE MOLECULAR FORMS OF ENZYMES

Furness, F. N., ed., Ann. N.Y. Acad. Sci. **94**, 655–1030 (1961).

The 30 papers in this monograph resulted from the conference on Multiple Molecular Forms of Enzymes, held in 1961 by the New York Academy of Science. Enzymes from different species, long recognized as having the same substrate specificity and catalyzing the same reaction, may differ markedly in various other properties. Isoenzymes from diverse tissues of an organism also may differ. Enzyme heterogeneity is a common phenomenon, with more than 30 enzymes, existing in multiple forms within individual organisms having been distinguished on the basis of variety of characteristics including electrophoretic and chromatographic behavior, serological specificity, differential solubility, and differential response with coenzyme analogues.

ELECTROPHORETIC STUDIES OF SERUM PROTEINS IN CYANOGUM® GEL

Hermans, P. E., W. F. McGuckin, B. F. McKenzie, and E. D. Bayrd, Proc. Mayo Clinic Staff Meetings **35**, 792–808 (1960).

In the hope of overcoming some of the difficulties encountered with starch-gel electrophoresis, electrophoretic separation of serum proteins with cyanogum gel was studied. The trade name for the gel is Cyanogum®-41, a mixture of two acrylic-acid derivatives, acrylamide and N,N¹-methylenebisacrylamide. The gel is easier to handle, has high tensile strength, gives a final product having the transparency of glass, can be plasticized to a thin pliable strip after electrophoresis, staining and washing (which can be analyzed in the same way the paper electrophoretic patterns are analyzed); and can be stored at room temperature after staining and plasticizing. The preparation of the gel by means of polymerization of the monomers is simple and can be carried out at room temperature, and the gel strips then can be stored in the buffer solution. Electrophoresis of serum proteins using Cyanogum® gel as the supporting medium, with the aid of the correlation of paper and Cyanogum®-gel patterns, is presented in this paper, as well as the results of direct identification experiments with some of the protein fractions.

ELECTROPHORETIC RESEARCH IN MICROBIOLOGY AND IMMUNOBIOLOGY: A SURVEY OF THE LITERATURE

Ispolatovskaia, M. V., trans. by F. S. Freisinger, J. Microbiol., Epidemiol. Immunobiol. **31**, Part 1, 721–726 (1960).

This survey of the literature on electrophoresis shows that electrophoretic techniques can be successfully used in immunology and microbiology. These techniques also permit studying the degree of purity of several bacterial toxins and antigens, and localizing antibodies formed after various methods of immunization. It should be possible to select the most valuable methods of immunization with highly concentrated and rationally purified antigens using the techniques described.

ELECTROPHORESIS: THEORY, METHODS AND APPLICATIONS

Bier, M., ed., Academic Press Inc., New York (1959).

Contained in this book are treatises on the following subjects: electric potentials in colloidal systems; acid-base equilibrium of proteins; theory and practice of moving-boundary electrophoresis; paper electrophoresis; zone electrophoresis in various types of supporting media; preparative electrophoresis without supporting media; applications of moving boundary electrophoresis of viruses, bacteria, and cells, and the microscope method of electrophoresis; and applications of zone electrophoresis. Emphasis is placed on the fundamental principles of electrophoresis, its problems and the means of solving them, and detailed laboratory procedures illustrating these problems are included.

ZONE ELECTROPHORESIS IN STARCH GELS AND ITS APPLICATION TO STUDIES OF SERUM PROTEINS

Smithies, O., Advan. Protein Chem. **14**, 65–113 (1959).

This article describes the current technical problems associated with starch-gel electrophoresis, considers their possible solutions, summarizes the information already obtained by application of the method to studies of serum proteins and their inheritance, and puts this information into perspective in relation to future work. Following the introduction, Section II discusses the experimental procedures used for starch-gel electrophoresis with attention to technical details for the achievement of the best starch-gel separations. Section III is a summary of the data relevant to the specific identification of serum components demonstrated by starch-gel electrophoresis, and Section IV reports the results and significance of recent studies of serum proteins using this method. The nature of the physical factors governing the electrophoretic separations obtained in starch gels is discussed in Section V, and evidence to support the hypothesis that molecular size plays a particularly important role in the separation processes is presented.

STUDIES ON AGAR GEL ELECTROPHORESIS: TECHNIQUES—APPLICATIONS

Wieme, R. J., Arscia Uitgaven N.V., Brussels (1959).

This book on agar-gel electrophoresis is divided into two parts: development of the technique and its application to practical problems. Part A covers some physicochemical aspects of electrophoresis, a survey of electrophoresis methods used in protein chemistry, agar electrophoresis as a refined fractionating technique, agar-gel electrophoresis as an ultramicrotechnique (direct tissue protein electrophoresis), optical properties of an agar gel (enzymo-electrophoresis), and some special aspects of agar-gel electrophoresis. Part B presents an introduction to biochemical applications, electrophoretic fractionation of a mixture of adult and fetal hemoglobin, determination of the electrophoretic mobility of the complex hemoglobin/haptoglobin groups in man (electrophoretic study of haptoglobin groups in man), following an enzyme isolation from rabbit muscle, enzymo-electrophoretic study of lactate dehydrogenase and sorbitol dehydrogenase in plasma and tissues of mouse and rat, pherogram obtained on normal human serum by agar-gel electrophoresis (technique III), a few particular aspects intervening in the analysis of icteric sera, application of agar-gel electrophoresis to the study of plasmocytoma and primary macroglobulinemia, qualitative modifications at the level of γ-globulins, and clinical applications of LDH enzymo-electrophoresis.

A MANUAL OF PAPER CHROMATOGRAPHY AND PAPER ELECTROPHO-RESIS

Block, R. J., E. L. Durrum, and G. Zweig, Academic Press Inc., New York (1955).

Part I of this book concerns paper chromatography and includes the theory of paper chromatography; general and quantitative methods; amino acids, amines, and proteins; carbohydrates; aliphatic acids; steroids and bile acids; purines, pyrimidines and related substances; phenols, aromatic acids, and porphyrins; miscellaneous organic substances; antibiotics and vitamins; and inorganic separations. It is designed to be a practical manual for the average chemical laboratory in which tried and proven procedures, employing relatively simple equipment and available reagents, are summarized. Part II covers paper electrophoresis and presents the general theory; methods; two-dimensional technique; continuous electrophoresis; and some quantitative considerations. The emphasis of this section is on the basic principles and methodology and it deals mainly with the separation of protein mixtures, particularly blood serum, because in this area paper electrophoresis has not been challenged by paper chromatography.

ZONE ELECTROPHORESIS

Tiselius, A., and P. Flodin, Advan. Protein Chem. **8**, 461–486 (1953).

In this paper on zone electrophoresis, the types of apparatus, working conditions, sources of error, methods of zone localization, mobilities and isoelectric points, fields of application, preparative zone electrophoresis, electrochromatography, and zone electroultrafiltration are described, discussed and evaluated.

Spartan is the typeface principally used in this book. The typesetting was done by Pinckney Typesetting, Pinckney, Michigan. The printing and binding were done by LithoCrafters, Inc., Ann Arbor, Michigan.